SAC Auchincruive

036706

D1766502

Library
SAC Ayr
Riverside Campus, University Avenue
Ayr KA8 0SX
Tel: 01292 886413/4

withdrawn.

SCOTTISH AGRICULTURAL COLLEGE

AUCHINCRUIVE

LIBRARY

549.52 : 509.22
9b/g
Class

ORGANIZATION OF COMMUNITIES

PAST AND PRESENT

ORGANIZATION OF COMMUNITIES

PAST AND PRESENT

THE 27TH SYMPOSIUM OF
THE BRITISH ECOLOGICAL SOCIETY
ABERYSTWYTH 1986

EDITED BY

J.H.R. GEE

Department of Zoology
University College of Wales
Aberystwyth

P.S. GILLER

Department of Zoology
University College
Cork

LIBRARY
SCOTTISH AGRICULTURAL COLLEGE
AUCHINCRUIVE
AYR   KA6 5HW
TEL 01292 525209

BLACKWELL SCIENTIFIC PUBLICATIONS

OXFORD LONDON EDINBURGH

BOSTON PALO ALTO MELBOURNE

©1987 by
Blackwell Scientific Publications
Editorial offices:
Osney Mead, Oxford OX2 0EL
  (*Orders*: Tel. 0865 240201)
8 John Street, London WC1N 2ES
23 Ainslie Place, Edinburgh EH3 6AJ
52 Beacon Street, Boston
  Massachusetts 02108, USA
667 Lytton Avenue, Palo Alto
  California 94301, USA
107 Barry Street, Carlton
  Victoria 3053, Australia

All rights reserved. No part of this
publication may be reproduced, stored
in a retrieval system, or transmitted,
in any form or by any means,
electronic, mechanical, photocopying,
recording or otherwise without
the prior permission of the
copyright owner.

First published 1987

Photoset by Enset (Photosetting)
Midsomer Norton, Bath, Avon
Printed and bound
in Great Britain by
Richard Clay Ltd, Chichester

DISTRIBUTORS

USA and Canada
  Blackwell Scientific Publications Inc
  PO Box 50009, Palo Alto
  California 94303
  (*Orders*: Tel. (415) 965-4081)

Australia
  Blackwell Scientific Publications
  (Australia) Pty Ltd
  107 Barry Street
  Carlton, Victoria 3053
  (*Orders*: Tel. (03) 347-0300)

British Library
Cataloguing in Publication Data

British Ecological Society. *Symposium
(27th: 1986: Aberystwyth)*
Organization of communities past and
present: the 27th Symposium of the British
Ecological Society, Aberystwyth 1986. —
(Special publications series of the British
Ecological Society, ISSN 0262-7027; no. 27).
1. Biotic communities
I. Title II. Gee, John H.R. III. Giller,
Paul S. IV. Series
574.5′247   QH541

ISBN 0-632-01783-X hbk
    0-632-02143-8 pbk

Library of Congress
Cataloging-in-Publication Data

British Ecological Society. *Symposium
(27th: 1986: Aberystwyth, Dyfed)*
Organization of communities, past and
present.
(British Ecological Society Special
publication; no. 27)
Bibliography
1. Biotic communities — Congresses.
I. Giller, Paul S. II. Gee, John H.R.
III. Title. IV. Series: Special publication
of the British Ecological Society.
QH540.B75 1986   574.5′247
87-9344

ISBN 0-632-01783-X hbk
    0-632-02143-8 pbk

CONTENTS

PREFACE

The challenge to community ecologists is to identify the processes that determine the diversity of natural communities and the distribution and dynamics of their constituent species. An understanding of these processes should then form the basis for prediction of the characteristics and behaviour of communities. Reliable prediction is essential to the application of community ecology to practical problems such as crop or fishery management, conservation, and ecological impact assessment. As in forecasting the weather, prediction may be possible without a knowledge of the underlying processes, but the result is likely to be less accurate in the long run and certainly less satisfying intellectually.

There are two basic routes to understanding the peculiar properties of communities. The first, which might be termed the bottom-up route, originated with the explosion of interest in theoretical models of interactions between species. For a while, knowledge of the possible classes of behaviour of populations in theoretical communities outstripped the empirical data necessary for the validation of the models. This imbalance is being redressed by increasing adoption of the experimental approach in both the laboratory and the field. The second, top-down, route starts with the detection of consistent patterns in the structure of natural communities. Explanations for the patterns are then sought in the known characteristics of the organisms, the physico-chemical properties of the environment, or in possible interactions between species. A fruitful approach, one that has helped to generate the recent increase of interest in community ecology, is a 'melding' of these routes. Thus, the behaviour of community models can provide fresh conceptual perspectives for the seeking and interpretation of patterns in natural communities, and the detection of consistent patterns can feed back into the model building process.

Linking a perceived pattern with an ecological explanation of the pattern has always been contentious. Species abundances and distributions are dependent variables influenced by two sets of independent variables, the physical environment and other species (Diamond 1986). Herein lies the root of the current controversy concerning the existence and nature of community organization—which set of variables has the upper hand? Are communities structured by interactions between the species, or is the structure a consequence of the independent reactions of individual species to environmental factors? As May (1984) points out, this division in ecological thought can be traced back at least as far as the British Ecological Society Symposium of 1944. The division lives on, although there are signs that it is becoming more apparent than real and that each view may be correct in the appropriate circumstances (Schoener 1986).

One tangible outcome of the expansion of research in community ecology in the past dozen years has been the publication of several books, most of which are the edited proceedings of symposia. One of these (Salt 1984, although the papers originally appeared in *American Naturalist* for November 1983), concentrates on the philosophy and practicality of testing hypotheses in community ecology, principally those concerning the strength of interspecific competition. A 'new' ecology, proclaimed by Price, Slobodchikoff & Gaud (1984), also challenges the supposedly primal role of interspecific competition in community organization. In its place comes a gallery of alternative mechanisms, although much of the evidence in favour of these alternatives is drawn from studies of terrestrial insects, notably herbivores.

A similar theme emerges in the volume edited by Strong *et al.* (1984) in which there is a strong emphasis on the quality of the evidence adduced in favour of one community view or another. There is much discussion of the value of null or neutral models in 'subjective' tests for the existence of significant patterns in community data and for the influence of species interactions. As is often the case in contemporary community ecology there is a taxonomic bias towards terrestrial birds and insects, leavened by the occasional inclusion of other habitats and taxa.

Given the apparent richness of the literature, the reader might be forgiven for enquiring why the editors set about planning, in the spring of 1984, yet another symposium in community ecology. The rationale was that a number of factors had been firmly established as influential in shaping communities; these include stochastic environmental disturbances, past climatic conditions, availability of resources, competition, predation and, most recently, the transport processes controlling the supply of colonists. The next step would be to examine the distribution of organizational patterns and processes amongst a wide range of community types, in terms of both habitat and taxonomy. In addition, many of the patterns shown by communities are simply parts of the latest frame of a film recording community development and catastrophe over a span of time measured in millenia, yet ecologists have rarely given more than a passing nod to the importance of the past. Thus it was high time to abandon temporal myopia and bring together the palaeoecologists and the neoecologists. To this end we invited ecologists with research experience of communities in a wide range of environments, of organisms as different as protozoa and primates, and of patterns at diverse temporal and spatial scales, to meet in Aberystwyth in April 1986.

Early in the planning process we became aware that another meeting was due to take place in Los Angeles, the edited proceedings of which became available in Britain at about the same time as the Aberystwyth symposium was taking place. In the event the two publications have turned out to be complementary, despite having had roughly similar aims. Diamond & Case (1986) organized their material by pattern and process, whereas we have organized ours by habitat and taxon. In about two-thirds as many chapters, we have less emphasis on birds and terrestrial insects and cover their admitted taxonomic gaps (unicellular organisms and freshwater invertebrates). By another happy coincidence both texts include a palaeoecological perspective, but with little duplication of content. Our author list is truly international, but with a British flavour that befits the sponsoring society. For the record (and in the spirit of investigating community patterns), Sørenson's index of similarity of author lists is 0.25 for Strong *et al.* (1984) and Diamond & Case (1986), but 0.06 and 0.09 respectively between the two American volumes and the present British one!

The present book is divided into four sections. The first section discusses the concept and nature of the community (Chapter 1), explores the roles of spatial patchiness and seasonality in the organization of communities (Chapters 2 and 4) and examines the significance of species abundance distributions (Chapter 3). This sets the scene for the analysis of spatial and temporal organization of a wide range of contemporary communities in Section II. Together, the chapters on terrestrial (5–10), microbial and decomposer (11–13), and aquatic (14–17) assemblages present a broad sweep of habitats and taxa. In the third section (Chapters 18–20) attention switches to the longer view in time and establishes the debt that present day communities owe to their geologic past. The final section offers two new perspectives in the study of community ecology, the role of habitat selection (Chapter 21) and the influence of physical factors on the supply of recruits to open communities (Chapter 22). It finishes with a discussion of the equilibrium–non-equilibrium continuum and the problems of terminology and scale that beset community ecology. Each of these influences the variety and generality of identifiable patterns and processes in the organization of communities, past and present.

We are grateful for the generous support that we have had in the organization of the symposium and the publication of the proceedings. John Lawton, Peter Grubb and Tony Davy helped shape the symposium in the Steering Committee, and Edward Broadhead advised on matters concerning publication. Jonathan Roughgarden agreed to take on the difficult task of providing an overview at the meeting. The chapters of the proceedings benefited from the careful attention of numerous academic referees, and the authors themselves were sympathetic in their treatment of constructive editorial advice! Professors John Barrett at Aberystwyth and Maire Mulcahy at Cork made many facilities available to us in their respective departments. Finally, Anne Gee and Janet Giller kept the children at bay whilst their husbands wrestled with symposium business.

JOHN H. R. GEE
PAUL S. GILLER

REFERENCES

Diamond, J.M. (1986). Overview: laboratory experiments, field experiments and natural experiments. *Community Ecology* (Ed. by J.M. Diamond & T.J. Case), pp. 3–22. Harper & Row, New York.

Diamond, J.M. & Case, T.J. (Eds) (1986). *Community Ecology*. Harper & Row, New York.

May, R.M. (1984). An overview: real and apparent patterns in community structure. *Ecological Communities: Conceptual Issues and the Evidence* (Ed. by D.R. Strong, D. Simberloff, L.G. Abele & A.B. Thistle), pp. 3–16. Princeton University Press, Princeton, New Jersey.

Price, P.W., Slobodchikoff, C.N. & Gaud, W.S. (Eds) (1984). *A New Ecology: Novel Approaches to Interactive Systems*. Wiley, New York.

Salt, G.W. (Ed.) (1984). *Ecology and Evolutionary Biology: a Round Table on Research*. University of Chicago Press, Chicago.

Schoener, T.W. (1986). Overview: kinds of ecological communities—ecology becomes pluralistic. *Community Ecology* (Ed. by J.M. Diamond & T.J. Case), pp. 467–79. Harper & Row, New York.

Strong, D.R., Simberloff, D., Abele, L.G. & Thistle, A.B. (Eds) (1984). *Ecological Communities: Conceptual Issues and the Evidence*. Princeton University Press, Princeton, New Jersey.

I
THE ANALYSIS OF
COMMUNITY ORGANIZATION

1. THE CONCEPT AND NATURE OF THE COMMUNITY

T. R. E. SOUTHWOOD

Department of Zoology, University of Oxford, South Parks Road, Oxford OX1 3PS, UK

INTRODUCTION

A convenient division of biological sciences is based on the level of organization studied: ecology is concerned with the more complex levels—populations and communities. It is to the community, the most complex level, that this symposium is devoted. In this paper I will attempt to outline some of the questions and current controversies in the study of communities, more particularly in relation to the brief given to me by the organizers: the concept and nature of the community. The term 'concept' is widely used in ecology as a description of useful ideas: these need not be falsifiable theories (McIntosh 1980). In reviewing this area I have essentially sought to address the question 'What is a community?' Having recognized a community, then 'what is its nature?' What are the general properties that we can record, and 'what are the main determinants of the variation in these properties?'

THE CONCEPT OF THE ECOLOGICAL COMMUNITY

The science of ecology tends to use commonplace words for its concepts. This is an advantage in that it opens our subject to the layman, but it also provides a fertile field for misunderstanding and over-facile parallels between ecological and other organizations. In the first part of this century some ecologists, particularly those using German as their language, sought to bring precision by the development of a specialist terminology (Balogh 1958; Tischler 1975), but this work did not contribute significantly to the understanding of the functioning of ecological systems. The view may be taken that ecology cannot, by the very nature of its material, make all definitions precise any more than a cartographer can give the exact length of a coastline or the precise boundary between two vegetation types, without certain qualifications and assumptions. Indeed, even when modern ecology has coined jargon terms, like r and K-selection, their definition proves variable and elusive (Boyce 1984). Thus a consideration of the concept of the community in ecology can quite properly start with the *Oxford English*

Dictionary definition of 'community'. There are two main meanings: firstly a quality or state, such as being held by all in common, and secondly, a body of individuals. Ecologists have mostly used community in the second sense, in which communities may be, defined by their organization or by the occupancy of a location. It is about the relative emphasis on these two ideas (organizational and locational) that ecologists so often differ.

The idea of the community in ecology can be traced back, almost a century, to S.A. Forbes' (1887) paper 'The lake as a microcosm'. Around this time plant ecologists were much concerned with the description of vegetation and they soon recognized the role of competition in plant–plant interactions. F. E. Clements and his animal ecologist collaborator, V. Shelford, were impressed by this association of organisms, clearly closely interlinked and interlocked. This 'Clementsian view' was rejected by their contemporary H. A. Gleason who considered the association to be largely a random assemblage, i.e. the organisms in a location (McIntosh 1980).

In 1927 Charles Elton opened the main text of his seminal work *Animal Ecology* (p. 5): 'One of the first things with which an ecologist has to deal is the fact that each different kind of habitat contains a characteristic set of animals. We call these animal associations, or better, animal communities, for we shall see later on that they are not mere assemblages of species living together but that they form closely-knit communities or societies comparable to our own.' Thus he moved from the location definition to an emphasis on organization, and it was central to Elton's concept that in a community every animal is linked closely with other animals living around it (Elton 1927, p. 52). The extent and form of this linkage has been the subject of much debate, more especially in relation to interspecific competition. Strong *et al.* (1984b, p. vii) wrote: 'The contemporary questions in community ecology concern the existence, importance, looseness, transience and contingency of interactions'.

Linkage in communities

As pointed out in several chapters, the links may be separated into vertical or trophic links (predation, parasitism, herbivory and scavenging) and horizontal links between species at the same trophic level (competition) (see also May 1981 and Faeth 1987). In this framework the phenomenon of mutualism (Addicott 1986; Pierce & Young 1986) will often be seen to be providing a diagonal link. The now substantial corpus of knowledge on predation and parasitism (Hassell 1978; Anderson & May 1986; Strong 1986) indicates the variety and strength of such interactions. This is an area of ecology in which there has been a long-running controversy on the extent that such interactions are governed by density-dependent mechanisms, i.e. influenced by linkage with other animals (Andrewartha & Birch 1954, 1985; Bakker 1964; Huffaker & Messenger 1964; Richards & Southwood 1968; Dempster 1983; Hassell 1985). Is density dependence the rule or the exception?

Recently Strong (1986) has re-emphasized the impact of natural variance on population dynamics by coining a new term 'density vagueness'. As he points out, in nature there are many causes of such variation, so that it is unwise to argue that there is no density component in a particular relationship, because of the 'operational difficulties in demonstrating the absence of an effect'. This is well illustrated by a study on the viburnum whitefly (*Aleurotrachelus jelinikii*) on three bushes at Silwood Park, Ascot, Berkshire, that I commenced 24 years ago. Analysing the first 12 years by conventional key-factor analysis we were able to detect evidence of density dependent regulation in populations at the level of the bush (Southwood & Reader 1976; Southwood 1981). Early in the study it was apparent that it would be desirable to reduce the scale below that of a bush, to follow cohorts on individually identified leaves (Southwood & Reader 1978a, p. 5) and appropriate sampling was commenced in 1968–69. Professor M.P. Hassell is now participating in this work and we have been able to detect variable, and often weak, density-dependent influences for certain age-specific survivals at the scale of the leaf (Hassell, Southwood & Reader 1987). If a model is constructed based on these values for 1968–77, it gives a close fit to the data and to the current population size (Fig. 1.1).

There are, however, two periods when the population is not described by this model: 1963–67 and 1974–76. Both these periods followed catastrophic falls in population size during spells of unusual weather: a very cold spell in the winter of 1962–63 and, probably, an exceptionally heavy shower during a short, critical period for the establishment of the first instar larvae in July 1973. A number of general conclusions can be drawn, the most obvious but depressing being that the investigation of population processes may need a very long series of observations. However, for our central concern in this symposium, the lessons are that the links within a community (even between animals of the same sort) may be difficult to recognize, and furthermore that, even within the same community, they may not be consistent in their action, showing temporal and spatial variation. Lastly, the study shows the importance of the environment, the habitat; a theme to which I shall return.

Horizontal links between different species have generally been thought of in terms of interspecific competition. The evidence has often been derived from patterns of resource use (e.g. Cody & Diamond 1975; Mound & Waloff 1978). This approach has been criticized, principally by Simberloff

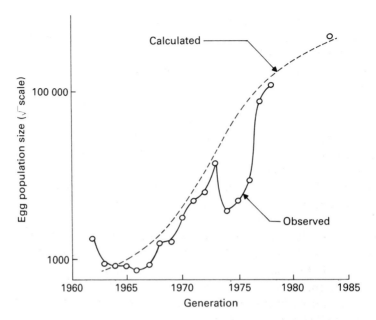

FIG. 1.1. Comparison of observed population growth with that calculated from a model
(incorporating density dependence), showing departures after climatic catastrophes in 1962
and 1973. Data for viburnum whitefly (*Aleurotrachelus jelinekii*). (After Hassell, Southwood &
Reader 1987.)

and his colleagues at Tallahassee, who have urged that however impressive
the pattern, one must test to see if it is significantly different from one
generated by random processes, i.e. the null hypothesis (Simberloff &
Connor 1979; Strong, Szyska & Simberloff 1979). These workers have
themselves emphasized that failure to falsify the null hypothesis does not
'prove' that biological processes have not influenced the pattern, but merely
that their existence cannot be assumed from the pattern. This point is
underlined by the computer models designed by Colwell & Winkler (1984);
one, termed GOD, generates a phylogenetic tree of species and the second,
WALLACE, assembles the species into a series of island communities. The
programme GOD has different speciation, extinction and character change
probabilities, whilst the subroutines of WALLACE allow for different levels of
biological interaction.

Colwell & Winkler sampled the resulting 'communities' and compared
the observed patterns with null hypotheses generated according to the
recipes of Strong, Szyska & Simberloff (1979) and Connor & Simberloff
(1978). They frequently found that the null hypothesis could not be rejected,
although they knew, from their programmes that had generated the data,

that there were many interactions between the 'organisms' in their 'communities'. The plea may be made that the models are not sufficiently realistic. Furthermore Colwell & Winkler (1984) found there were various conditions that would make it particularly difficult to detect competition, e.g. the weakness of the taxonomic constraints on sampling (the 'J. P. Morgan effect') or any correlation between morphology and vagility (the 'Icarus effect'). Following these ideas strictly, one might consider that studies should be limited to a particular taxon, to avoid the J. P. Morgan effect. But some of the most interesting interactions have been demonstrated right across taxonomic boundaries: a most striking example of this is the work by J. H. Brown, D. W. Davidson and their colleagues on seed-eating rodents and ants in the deserts of south-west USA (Brown & Davidson 1977; Davidson, Brown & Inouye 1980; Davidson, Sampson & Inouye 1985). It is my view that taxonomic myopia has been a major handicap in the study of ecological communities and to encourage it, thus to avoid the J. P. Morgan effect, would be a retrograde step. I will take up this theme later (p. 8), giving evidence of the insights to be gained by considering organisms from different groups and investigating their interactions experimentally (see West 1985a,b).

Many of the so-called neutral models (e.g. Lawlor 1980) fail to take account of the biological realities of the situation: for example, assumptions are made that all positions along a particular resource dimension are equally favourable. This is plainly nonsense; for instance, predator pressure is likely to be greatest at certain parts of the diel cycle and food in some areas will be better than in others. Some allowance may be made by constraining the model to non-zero resource states, but this allows only for the extreme of variability. It would be biologically more meaningful to take the curve of resource utilization in terms of total individuals (or biomass) and distribute the species randomly within this constraint. It is sometimes argued that to do this is to build too much of the biology into the neutral model, but we must always remember that the natural world is not a freely optimizing system. In spite of remarkable adaptations, species are constrained by their evolutionary history (a problem faced by Walt Disney's elephant 'Dumbo' that needed to fly)—a phenomenon often termed 'phylogenetic drag'.

However much neutral models are made biologically realistic they will seldom do more than caution us from drawing over-hasty conclusions from patterns of resource use (Harvey *et al.* 1983). The real search for competitive links in communities must also depend on experiments, especially field experiments, although these can pose problems of replication (Diamond 1986). Outstanding examples are provided by the work of J. H. Brown and D. W. Davidson, previously referred to in this chapter, and some of these will be described later (see Brown, Chapter 9). Further examples are given by Hildrew & Townsend (Chapter 16), Rosenzweig (Chapter 21), and Claridge (Chapter 7), among others in this volume.

A number of other recent studies, combining careful field observation with experiments, have also revealed subtle competitive interactions between members of insect communities. A group has been working at Oxford on the impact of defoliating caterpillars, particularly those of the winter moth (*Operophtera brumata*) and the oak tortrix (*Tortrix viridana*), on each other and on other phytophages. West (1985a,b) studied the distribution of *Phyllonorycter* leafminers and demonstrated that the greater the level of caterpillar damage to a leaf, the lower the survival of the leafminers to adulthood; this asymmetric competition appeared to be due to wound-induced changes in leaf chemistry. There is evidence that oak foliage quality for phytophages is highest in spring (Feeny 1970; Wint 1983), therefore why do the species of *Phyllonorycter* have life-cycles in which the larvae live in oak leaves only after mid-summer? One might postulate, to use the expression coined by Rosenzweig (1979) and Connell (1980), that this pattern was due to the 'ghost of competition past'. One of the experimental advantages of insects is that such 'ghosts' may be conjured up by experimental manipulation of the life-cycle. By warming *Phyllonorycter* pupae held in the laboratory, West was able to get adults to emerge in May and lay on young oak foliage. He found that in a caterpillar-free leaf, the larvae of this experimental generation survived well and produced heavier females (as would be predicted from the better leaf quality); however, the decline in survival of these larvae with caterpillar damage was steeper than that occurring later in the season. West calculated the relative fitnesses of the generations and showed that under normal conditions (i.e. with moth caterpillars) his experimental generation had a negative fitness—the 'ghost' was real enough! M. Hunter (personal communication) has found that not only does *O. brumata* affect *T. viridana* adversely, but the resulting combined defoliation reduces potential fecundity and survival of those other lepidopteran defoliators whose larvae feed later in the year. Also on the oaks are aphids (*Tuberculoides annulatus*), and I. Silva-Bohorquez (personal communication) has found that development of their populations is depressed on leaves previously damaged by defoliating caterpillars. These studies clearly show that even (or perhaps especially) when phytophagous insects are segregated in time there may be asymmetric competition between them, mediated through the host.

Another demonstration of host mediated competition (although it has not been demonstrated that this competition is responsible for the pattern of resource utilization) is provided by the studies of Waage (1985) and Waage

& Davies (1986) on the community of blood-sucking tabanids on horses in the Camargue, southern France.

The conclusion I draw from these recent studies is that subtle competitive linkages may be far more widespread in insect communities, even amongst phytophagous species, than is currently believed (Lawton & Strong 1981; Strong, Lawton & Southwood 1984a). Often we have sought synchronous and direct effects; perhaps they are more frequently asynchronous and host mediated (both in phytophages and haemotophages). It is likely, I believe, that we will find similarities between the detection of competitive interactions in communities and of density-dependence in population regulation. As in the study of the viburnum whitefly, we should not always expect the effect to reveal itself from a simple analysis of averaged population sizes, nor to be present in all populations all of the time. Chapters 2, 6, 12, 14 and 15 stress the importance of patchiness and spatial scale. By comparison with ourselves most organisms are small and therefore the appropriate scale for considering some aspects of spatial dynamics (Taylor 1986) may be small, e.g. at the level of a leaf.

The community as a super-organism

As mentioned above, the most neutral approach to the concept of the community is to define it simply as the organisms that occur together in a location: an assemblage (see O'Connor, Chapter 8). Although we must continue to be very cautious in attributing all patterns in communities to linkages I suggest that the evidence above and much that will be presented in subsequent chapters (e.g. those by Brown, Claridge, Doube, Fenchel, Hildrew & Townsend, Pearson & Rosenberg and Terbourgh & van Schaik) does indicate that linkages of the type envisaged by Elton (1927) are in fact a widespread feature of communities and therefore a valid component of the concept. From the time of Forbes (1887) some workers have gone further and argued that in this assemblage, organisms interact with each other in such a way that the community has properties of its own over and above the sum of its components and being itself exposed to selective forces (e.g. Dunbar 1960, Darnell 1970, Blandin 1980). The late R. H. Whittaker (1970) approached this position when he wrote that a natural community is ' a distinctive living system with its own composition, structure, environmental relations, development and function' (p. 1). Implicit in this concept is the action of selective forces on communities as a whole; there would be characteristics that have evolved, not because they maximize the fitness of the individuals who carry the responsible genes, but because in some way they benefit the community as a whole (Blandin 1980). Wilson (1986) also argues for 'adaptive indirect effects' occurring in communities. Whilst there is much evidence of organisms modifying their habitat (and hence the community) in ways that directly benefit them (Janzen 1967; Davidson, Sampson & Inouye 1985), there is little evidence for indirect effects, nor is it easy to envisage the mechanism. In saying this I am not arguing for extreme reductionism, the view that everything will be discovered by molecular biology. There are phenomena that will be discovered by those who work at higher levels; indeed such phenomena are the theme of this meeting and, as Maynard Smith (1986) has written, though 'one day (molecular biologists) will have to interpret (them) in their terms' (p. vii), they would never have discovered them.

I conclude that the concept of an ecological community is of a group of organisms (generally of wide taxonomic affinities) occurring together in a location; many of them will directly interact with each other within a framework of both horizontal and vertical linkages. If we could see the biosphere with these links traced out, it would look like a gigantic patchwork quilt of spiders' webs of different sizes and designs and with threads of different thicknesses. The webs are, of course, both the food webs (the trophic links), and the horizontal and diagonal links. In some places there would be very few strands, in others a tight concentration of links and junctions. Some strands will be broad (major links), some will be very fine (rare interactions). There is no one scale on which we should divide up this landscape into communities; both the smallest patches such as the acorn and the dung pat (see Doube, Chapter 12) and the largest such as the tropical rain forest (see Terborgh & van Schaik, Chapter 10) may be appropriate. The decision must depend on the organisms we are studying and the phenomenon being investigated (see Harris, Chapter 15). But if the phenomenon concerns the functioning of communities, it should encompass vertical and horizontal links.

THE NATURE OF COMMUNITIES

It is the features that constitute the nature of communities that provide the basis for their description. In early work they were often described in all their infinite variety, but some of the pioneers such as F. E. Clements, Charles Elton and G. E. Hutchinson sought generalities; they sought to recognize patterns, e.g. succession, the pyramid of numbers, size ratios. Insights into the functioning of biological systems depend not only on the recognition of such patterns, but also on the investigation of the factors that determine the variation in these patterns. The majority of the chapters in this symposium are concerned with the role of various factors in determining the form of community features.

The patterns

As mentioned above, an important component of ecological research is the recognition and description of patterns (Southwood 1980; May 1984; Wiens 1984, 1986) and three of the most important descriptions will be considered here.

Diversity

The most fundamental description of the nature of a community is provided by a measure of its diversity: the number of different species of organism and their abundance, generally in terms of individuals, but sometimes as genets or modules (Harper 1981) or biomass. As Gray shows in Chapter 3, the models of relative abundance are of two types: those concerned with the partitioning of resources and those, like the log series (Williams 1964), that are simply statistical descriptions. The relative ecological neutrality of the statistical model does mean that it provides a useful quantitative description for a variety of communities and the log series, originally devised by R. A. Fisher, is particularly useful (May 1975; Southwood 1978a; Taylor 1978). May (1975) has shown how it can be generated making certain assumptions concerning colonization and niche pre-emption, but the validity of its use is not dependent on this interpretation. Whittaker (1972) made a useful distinction between:

α-diversity: the diversity of species with a habitat or community;

β-diversity: a measure of the rate of change of species along a gradient from one habitat to another;

γ-diversity: the richness of species in a range of communities in a location or from one location to another on geographical scale (i.e. a combination of α and β).

The first two (α and β-diversities) have proved particularly useful; this was illustrated in a study of the vegetation along the gradient of a secondary succession in southern England (Southwood, Brown & Reader 1979). Whereas α-diversity rose and then fell with successional age, the magnitude of the β-diversity measure between communities was proportional to the successional age difference of those communities, age being on a logarithmic scale. This suggested that the turnover of plant species declined logarithmically with successional age.

The individual is generally the unit for diversity measurements, but it is obvious that with modular organisms the individual (the genet) is not a good unit if we are interested in diversity as an indication of resource use. Furthermore, with other types of organism, such as insects and birds, the total biomass may be partitioned differently between individuals in different habitats, i.e. the average individual size changes. Janzen & Pond (1975) pointed out various differences between the insect fauna of secondary vegetation in Britain, the USA and Costa Rica. Extensive studies of the fauna of various species of tree in Britain and South Africa (Southwood, Moran & Kennedy 1982) showed that the average individual biomass of the arthropods in South Africa was 4.5 times greater than in Britain (Table 1.1). As there were more individual insects per sample in Britain, the insect biomass per tree was far less variable than a consideration of either the number of individuals (population density) or weight alone would suggest. Another example is provided by a study of the use of different stages of a secondary succession by birds; it was found that larger birds were numerically dominant on open land and smaller ones in woodlands (Southwood *et al.* 1986) (Table 1.1).

TABLE 1.1. Some variations in the sizes of individuals in different communities

1   Arboreal arthropods* (average individual dry mass, mg)

	South Africa	UK
Phytophages (chewers)	20.82	3.43
Phytophages (suckers)	3.26	0.91
Insect predators	9.48	3.22
Other predators (mostly spiders)	6.43	0.86
Parasitoids	1.35	1.16

2   Birds† (average living weight (g), corrected to allow for time spent by bird)

	Young field	Old field	Woodland
All birds	457	135	19

*Data from Southwood, Moran & Kennedy 1982.
†Data from Southwood *et al.* 1986.

These differences in size must indicate differences in the pattern of resource utilization—a point emphasized by Harvey & Godfray (1987). They have demonstrated that species-abundance patterns may not reflect resource use because of two relationships: population density tends to decrease with increasing body size, but energy requirements (per unit of time) increase with body size. In particular a canonical lognormal species abundance distribution, with species of considerably varying size, will not reflect a canonical lognormal distribution of resource use.

Trophic structure

Food webs are individually interesting, but their graphic complexity is a powerful obstacle to generalizations. However, these are now emerging in an exciting manner. The earliest was the recognition of the relatively small number of vertical links in a food web (Hutchinson 1959). Pimm & Lawton (1977, 1980) argue that long chains would have excessive inherent instability, opposing the earlier view that they were limited by thermodynamic considerations (see also Kitching 1983; Pimm 1984). A theoretical analysis by Auerbach (1984) indicates that food webs generally have fewer interspecific interactions than would be due to chance alone.

Other approaches to the description of trophic structure involve the analysis of guild composition. Several workers have found some constancy in the proportion of total species in certain guilds within a community, more especially the ratio of predator species to prey species (Arnold 1972; Heatwole & Levins 1972; Cohen 1977; Cole 1980; Moran & Southwood 1982); see also Table 1.2. Moran & Southwood (1982) found that in the phytophage guild on broad-leaved trees the number of species of chewers and of sap-suckers appeared inversely related.

TABLE 1.2. The percentage of arthropod species in certain guilds or groups recorded from trees in South Africa (SA) and Britain (B) (from Moran & Southwood 1982)

	Total species	Phytophages	Predators	Others*
Betula (SA)	212	21.2	16.5	62.3
(B)	337	26.4	21.1	52.5
Buddleia (SA)	249	20.5	20.1	59.4
(B)	178	26.4	24.2	49.4
Erythrina (SA)	300	18.7	21.7	59.6
Quercus (SA)	149	24.8	18.8	56.4
(B)	465	22.6	20.9	56.5
Robinia (SA)	105	24.8	20.0	55.2
(B)	180	23.9	15.6	60.5
Salix cinerea (B)	322	26.7	21.7	51.6
†*Salix alba* (B)	176	35.8	22.2	42.0
†*Salix capensis* (SA)	122	34.4	16.4	49.2

*Others: parasitoids, epiphyte fauna, scavenging fauna, ants and tourists.
†Narrow leaved trees.

A constant predator–prey species ratio has now been shown to apply to a wide range of communities (Briand & Cohen 1984; Cohen & Briand 1984;

Jeffries & Lawton 1985) and certain quantitative empirical generalizations
have been recognized, e.g. the ratio of mean trophic links to species is about
1.86 and the mean proportions of basal, intermediate and top species are
approximately 0.19, 0.53 and 0.29 respectively (Cohen & Newman 1985). A
model, which assumes an ordering or cascade of species that constrains the
possible predators and prey of each species, but is otherwise stochastic, does
generate the observed proportions (Cohen & Newman 1985; Cohen,
Newman & Briand 1985). The assumed constraint is that a given species can
prey on only those species below it and can be preyed on only by those
species above it. However, developmental changes in size may reverse
predator–prey relationships (Southwood 1985; Roughgarden 1986); this
applies to most predatory organisms other than those where the adult feeds
the young until nearly full-grown. Notwithstanding this caveat, this
approach provides an important insight into trophic structure; Jeffries &
Lawton (1984) have suggested that the underlying biological explanation
may lie in competition for enemy-free space among victim species—another
boost for the potential importance of subtle forms of competition!

Stability and constancy

The variation in the composition of a community from time to time
(stability) or from place to place (constancy) is another descriptor of
interest. Either property may be viewed as evidence for a more deterministic
organization of communities, whilst variation implies a greater role for
stochastic processes. Taken over a relatively large area, community
composition often shows remarkable stability, especially the more abundant
species: this is shown by some data on heteropterous plant bugs caught in a
light trap at Rothamsted Experimental Station (Table 1.3). In the nine
seasons, over 23 years (1933–56) three species, out of a total of ninety-five,
were always amongst the six most abundant. Unfortunately we do not know
the spatial scale over which a light trap samples (Baker 1985), but there is no
doubt that if small patches of herbaceous vegetation had been examined
there would have been less stability. Variation in community composition is
greater the smaller the scale of the community studies (temporal and/or
spatial), the more frequently the habitat is disturbed, or the earlier it is in a
successional sequence (e.g. Meijer 1980). Hanski (1982) has termed widespread and abundant species, such as those in Table 1.3, core species. These
are always found in the relevant communities. Several chapters in this
volume present information on this topic (e.g. O'Connor, Chapter 8;
Brown, Chapter 9; Doube, Chapter 12; Reynolds, Chapter 14; Hildrew &
Townsend, Chapter 16 and Roughgarden, Gaines & Pacala, Chapter 22

TABLE 1.3 Stability in relative abundance over 23 years: the abundance rank of three most abundant of the ninety-five species of Heteroptera in the Rothamsted light trap (from Southwood 1960)

	1933	1934	1935	1936	1946	1947	1948	1949	1956
Lygus rugulipennis	1	2	2	5	2	1	1	3	1
Adelphocoris lineolatus	2	3	3	1	1	3	3	1	5
Megalocoleus molliculus	6	4	1	6	5	5	2	5	3

over ecological time and Brown, Chapter 9; Collinson & Scott, Chapter 18; Coope, Chapter 19 and Webb, Chapter 20 over evolutionary time).

Determinants of variation in the patterns

The organism and its evolutionary history

Every species is to some extent a prisoner of its evolutionary history (Southwood 1976). Just as contemporary communities reflect their recent geological history (see Coope, Chapter 19), so do species; their bionomic features will differ so that various aspects of the patterns of comunities will vary from taxa to taxa. Schoener (1986) has recently compared the assemblages of orb spiders, lizards and birds on various islands in the Bahamas and he has concluded that there are systematic differences; in particular, vertebrates have stronger species–area relationships and are altogether more predictable in occurrence. Spiders, with their high vagility in the early stages, are very dependent on colonization and recolonization, the extent of this being predictably related to distance over which immigration must occur. The evolution of vagility by ballooning in spiders must have had a dramatic influence on their community structure world-wide. Of course such an evolutionary pathway was not open to lizards! Roughgarden, Gaines & Pacala (Chapter 22) explore further the role physical transport processes can have in the organization of communities.

The role of evolutionary history can be explored through a consideration of past communities, particularly through the fossil record. We must note the changes in pattern associated with major evolutionary steps. Collinson & Scott (Chapter 18) will analyse the higher plants in detail and Webb (Chapter 20) the terrestrial vertebrates, so I will give an outline of the total picture, though not going back to the earliest origins in an anoxic world; Fenchel (Chapter 13) will provide some insights into that period. Raup (1972), Valentine (1973), Raup & Sepkoski (1984) and Sepkoski & Moller (1985) have estimated diversities of major taxa and extinction rates over the

Phanerozoic (the last 600 million years). There is a sharp rise from the early Cambrian, with marked extinctions at the end of the Permian and the end of the Cretaceous. As these workers and others (e.g. Pease 1985) have pointed out, there are many imperfections in the fossil record and one might expect the older fossils to be the most scarce because they have been exposed to more geological trauma that could obliterate them. Nevertheless, the general pattern of increase, with periodic extinctions, seems to be real. The main debate is between those who consider it is governed by various major steps in organic evolution and others who see it as a reflection of physical change. Gould (1977) has termed these latter explanations 'billiard ball models' for they treat organisms as inert substances, 'buffeted by an external environment and reacting immediately to physical stress without any counteracting resistance' (p. 21).

A major step in the development of the Metazoa was the evolution of biomineralization in the Cambrian (and incidentally this will have increased the probability of fossilization). What impact did this have on community structure? What were communities like prior to this development? In the last few decades numbers of Precambrian (Proterozoic) fossil assemblages 'Ediacaran faunas', have been recognized (Glaessner 1984): the organisms were all soft-bodied and often surprisingly large. With no skeleton, external or internal, for muscle attachment, movement would not have been fast; whilst prey did not have hard protective shells, neither did predators have jaws or other appendages to cut them up. Perhaps it was then quite possible for giant worm-type animals to survive as free-living organisms, and not as today when they are virtually restricted as parasites sheltered in vertebrate hosts (Kirchner, Anderson & Ingham 1980). (The only really large free-living worms found today are some Australasian earthworms; Lee 1985.) One can trace the evolution of swimming speed and other hunting and protective tactics through the two great groups of marine macro-predators, cephalopods and fish (Packard 1972). The modifications of some ammonite shells in the early Jurassic may have improved the efficiency of movement by reducing drag (Chamberlain & Westermann 1976), but in later forms the further elaboration of the shell would seem to have retarded movement. Perhaps selective pressures changed and shell-crushing predators were more important (Ward 1981), though other interpretations are possible (Kennedy & Wright 1985). The early Devonian fish emphasized armour, most modern species speed.

The colonization of land at the end of the Silurian and through the Devonian posed entirely different selective pressures. The terrestrial environment was then a two-dimensional world: organisms lived on or close to the ground in a humid environment. Not only do Amphibia recapitulate their evolutionary history by depending on humid conditions for early development, but many relatively primitive groups of insects lay their eggs in water or in the soil (e.g. Pyrrhocoridae). The evolution of the arborescent form, driven no doubt by competition for sunlight, transformed the terrestrial environment into one that was three-dimensional. An enormous amount of 'ecospace' became available for colonization by other organisms (Southwood 1978b, 1984) whilst at the same time posing hurdles: (i) attachment, remaining on the trees; (ii) avoidance of desiccation; (iii) movement other than by crawling would have major advantages in finding scattered host plants and moving from branch to branch (hence the evolution of flight); (iv) for those animals that became herbivores there were major biochemical hurdles, normally only overcome with the help of micro-organisms (Southwood 1973, 1984). Also in terms of temperature fluctuations, in both space and time, the terrestrial environment is much more patchy and variable than the seas: hence the evolution of diapause and other strategies that effectively enable a species to opt out of a community for a period. This may allow a species to complete its lifetime track by utilizing a series of fractional niches (niches that on their own would be inadequate for the life of an individual because of their seasonality and for limited nutritional composition; Southwood 1978b).

Thus at any time in the last 60 million years we would not have found the pattern of communities as we find them today, but we can suggest some organism-related hypotheses for the increasing species-richness indicated by the fossil record. There has been a positive feedback between organic evolution and the total number of niches. Biomineralization enabled organisms to develop their own refugia in space (shells, etc.) or time (speed), whilst corals and other reef-builders increased the ecospace in the marine environment, just as higher plants, especially those with an arborescent form, did on land. Therefore, given the biotic material and the abiotic conditions at the time, there is no reason to suppose that these fossil assemblages were any more or less at equilibrium than communities are today (MacArthur & Wilson 1967). Nor is it reasonable to conclude that today's arrangements represent a plateau; they are only a single frame in the infinite film of evolution. This is discussed further by Collinson & Scott, Chapter 18 and Webb, Chapter 20. Wilson (1969) has argued that there can be an evolutionary increase in species-richness beyond the equilibrium.

However, there has not been a continuous steady increase in the complexity of communities, but periods of extinction and relatively rapid increase. The most parsimonious explanation is surely that these have been governed by astronomical and/or geological events. Coope (Chapter 19) will show how we have evidence for quite rapid climatic change in the Quaternary. What happened to climate when the Tethys Ocean opened east to west, or when it closed? An asteroid impact, probable for several reasons, would have an even more dramatic effect (Allaby & Lovelock 1983; McLaren 1983). Such rapid changes would greatly increase the rate of selection and the pace of evolution. The phenomenon of punctuated equilibria in evolution is a reflection of the unevenness of change in the physical environment, but in saying this I am no more willing than Gould (1977) to accept the 'billiard ball model' of complete environmental dominance.

Founder populations and geographical history

If it is accepted that at least some of the extinctions have been due to catastrophes, then throughout geological history the whole earth has resembled one of Schoener's (1986) Bahamian islands after a hurricane. Only certain species survive and when more benign conditions return, there will be many empty niches. Thus chance and history, stressed by Hubbell & Foster (1986) as important mechanisms in structuring ecological communities, have played their part on a global as well as a local scale. Chapters 7, 9, 10, 19 and 22 all address aspects of the effect of these mechanisms on the nature of communities.

The number of species of insect associated with various trees in Britain is influenced by several factors (Southwood 1961; Strong 1974; Kennedy & Southwood 1984), but the history of the tree in Britain makes a significant contribution (Table 1.4). The accumulation of a fauna by a tree introduced into a new area provides a good model for the study of the roles of history, chance and pattern in community development. *Buddleia* has been in Britain for nearly 100 years; the most abundant species was introduced from China, but others came from India and southern Africa. About half the insect species now found on *Buddleia* in Britain are polyphagous. But the

TABLE 1.4. Determinants of insect species richness on different species of tree in Britain (from stepwise regression analysis, after Kennedy & Southwood 1984)

	Cumulative r^2 values
Log abundance ('area effect')	0.59
Time in Britain ('history')	0.67
Evergreenness	0.74
Taxonomic relatedness	0.78
Height	0.79
Leaf length	0.82

other fall into two categories: firstly those whose association may be thought to represent chance, their normal host plants neither taxonomically related to *Buddleia*, nor resembling it in leaf characters (e.g. pubescence); the 'normal hosts' of these species are often trees or shrubs found in gardens and areas where *Buddleia* grows. The second group came from plants that resemble *Buddleia*, especially *Verbascum* which may be related phylogenetically. What is perhaps even more interesting is that insects from this second group belong to the same genera or tribes as some of the most abundant insects living in endemic *Buddleia* species in South Africa (Southwood & Kennedy 1983).

The comparison of the fauna on birch (*Betula*) in Britain and South Africa, where it is introduced, also gave some indications of the same phenomenon. Chance has its role, but there are certain predilections, and these two factors came together in the historical framework to shape the community. In these insect communities on an introduced plant we do not know the effects of competition; do the polyphagous species just become proportionally rarer or are they actually rarer as the community develops with time? In Chapter 22 Roughgarden, Gaines & Pacala will consider all those aspects in communities where species interactions can be studied.

Habitat type

It has been suggested that the great variety of properties exhibited by communities or their components may be codified by reference to the features of the habitat (Grime 1977, 1979; Southwood 1977; Greenslade 1983; Sibly & Calow 1985). The habitat may be seen as a templet which produces a community with particular properties, but at the same time, like the original leather templet, may itself be slightly modified in the process. In assessing the properties of the habitat in time and space it is important to relate the habitat scales to those of the organisms (Southwood 1977; Grubb 1986; Kareiva 1986; Wiens 1986; Wiens *et al.* 1986). If this templet is restricted to two dimensions, like the Periodic Table of chemical elements, then the axes represent the two selective gradients, that of $r-K$ selection and that associated with adversity (Grime 1977; Southwood 1977; Greenslade 1983; Matthews & Kitching 1984; Lee 1985). There is still a variety of views about the exact definition of and appropriateness of $r-K$ selection (e.g. Barbault & Blandin 1980; Boyce 1984; Sibly & Calow 1985) and now is not the place for a full review; so I will merely set out my own views on a framework within which communities may be organized.

If habitats of equal productivity are arranged according to their durational stability, that is the period of time (in generation time) during which they remain suitable for the organisms that live in them, one gets a series that in general corresponds with succession. At one end are temporary, ephemeral habitats: rotting fruits, dung, patches of disturbed ground, temporary pools. Important features of such habitats are their patchy distribution (Swift will term them a mosaic of assemblages) and their low durational stability: they are very unapparent, and successful members of their communities will be adapted for dispersal and/or searching and colonization. Once the community is established it will be limited by time, for resources will become progressively depleted or unavailable. Three chapters of this symposium (Shorrocks & Rosewell, Chapter 2; Swift, Chapter 11 and Doube, Chapter 12), will describe and analyse communities from habitats of this type.

Some of these habitats arise from disturbance, as defined by Rykiel (1985), a physical force that reduces the biomass in the habitat. Grubb (Chapter 5) will emphasize the importance of the length of time without such disturbances in the formation of plant communities. Many attributes of communities and their constituent species change in a fairly predictable way as a habitat moves along the durational stability axes of the templet, i.e. as succession proceeds (Brown & Southwood 1987). S. Greenwood (personal communication) has studied features of the insect-plant interaction in a Far Eastern forest and shows how leaf toughness, water content and longevity change through succession and how these interact with herbivore feeding to give a falling rate of damage, but an increase in total damage with succession. An essentially similar pattern has been found in secondary succession in Britain (Southwood, Brown & Reader 1986).

The other axis on the templet is adversity, emphasized by Whittaker (1975). I used its converse 'habitat favourableness' (Southwood 1977), and Grime (1977, 1979) has referred to this axis as selection for the tolerance of stress. This concept has been challenged by those who point out that a penguin is not stressed in the Antarctic, nor a cactus in the desert, though these are adverse habitats, whilst Terborgh (1973) has argued that favourableness is really an expression of differences in speciation and extinction processes. I maintain that one is moving along this axis in the templet when primary productivity falls; in general this is a reflection of a combination of temperature and moisture or some other physical condition including a shortage of nutrients, that are sub-optimal for biological processes (i.e. for the homoeostasis of protoplasm, the functioning of enzyme systems and the integrity of living membranes). Communities in favourable habitats are predicted to be species-rich, diversity falling with increasing adversity. There are four chapters in this symposium (14–17) that consider acquatic communities, and we can see the extent to which these concepts, largely developed by terrestrial ecologists, are applicable in the aqueous environment.

A third variable that is important in the description of a community's habitat is the variation in the climatic conditions; the extent of seasonality and its predictability. Several studies (e.g. that by Walker & Ferreira 1985), but particularly the work of Wolda (1978, 1980 and Chapter 4) have shown that tropical environments are by no means always the stable, unvarying habitat that has sometimes been assumed, and tropical organisms are just as seasonal as those in temperate environments. This has considerable significance for community structure and pattern for Briand (1983), who shows that there is less trophic linkage in fluctuating than constant environments (see also Cohen, Newman & Briand 1985).

CONCLUSIONS

The current uncertainties and controversies in community ecology fall into two broad areas:

1   What is our concept? Are communities mere assemblages of organisms in a location or are they tightly linked and structured groups of interacting species?

2   To what extent is the nature of the community 'organism-driven' or 'environment-driven'? What are the roles of speciation and colonization (the biological history) and of habitat preference (see Chapters 8 and 21) on one hand, compared with seasonality, physical conditions and disturbance frequency on the other?

I forecast that we will find that within both questions neither alternative provides the whole answer, for ecology deals with a mixture of pattern and probabilism.

REFERENCES

Addicott, J.F. (1986). On the population consequences of mutualism. *Community Ecology* (Ed. by J.M. Diamond & T.J. Case), pp. 425–36. Harper & Row, New York.

Allaby, M. & Lovelock, J. (1983). *The Great Extinction*, 189 pp. Secker & Warburg, London.

Andrewartha, H.G. & Birch, L.C. (1954). *The Distribution and Abundance of Animals*. University of Chicago Press, Chicago.

Andrewartha, H.G. & Birch, L.C. (1985). *The Ecological Web*. University of Chicago Press, Chicago.

Anderson, R.M. & May, R.M. (1986). Vaccination and herd immunity to infectious diseases. *Nature (London)*, **318**, 323–9.

Arnold, S. (1972). Species densities of predators and their prey. *American Naturalist*, **106**, 220–36.

SCOTTISH AGRICULTURAL COLLEGE

AUCHINCRUIVE

LIBRARY

Auerbach, M.J. (1984). Stability, probability, and the topology of food webs. *Ecological Communities: Conceptual Issues and the Evidence* (Ed. by D.R. Strong, E. Simberloff, C.G. Abele & A.B. Thistle), pp. 413–38. Princeton University Press, Princeton, New Jersey.

Baker, R.R. (1985). Moths: population estimates, light-traps and migration. *Case Studies in Population Biology* (Ed. by L.M. Cook), pp. 188–211, Manchester University Press, Manchester.

Bakker, K. (1964). Backgrounds of controversies about population theories and their terminologies. *Zeitschrift für angewandte Entomologie*, **53**, 1187–208.

Balogh, J. (1958). *Lebensgemeinschaften der Landtiere*. Akademiai Kiado, Budapest.

Barbault, R. & Blandin, P. (1980). La notion de stratégie adaptative: sur quelques aspects énergétiques, démographiques et synécologiques. *Recherches d'Écologie Théorique Les Stratégies Adaptatives* (Ed. by R. Barbault, P. Blandin & J.A. Meyer), pp. 1–27. Maloine, Paris.

Blandin, P. (1980). *Evolution des écosystèmes et stratégies cénotiques*. *Recherches de'Écologie Théorique Les Stratégies Adaptatives* (Ed. by R. Barbault, P. Blandin & J.A. Meyer), pp. 221–35. Maloine, Paris.

Boyce, M.S. (1984). Restitution of r & K-selection as a model of density-dependent natural selection. *Annual Review of Ecology and Systematics*, **15**, 427–48.

Briand, F. (1983). Environmental centre of food web structure. *Ecology*, **64**, 253–63.

Briand, F. & Cohen, J.E. (1984). Community food webs have scale-invariant structure. *Nature (London)*, **307**, 264–6.

Brown, J.H. & Davidson, D.W. (1977). Competition between seed-eating rodents and ants in desert ecosystems. *Science*, **196**, 880–2.

Brown, V.K. & Southwood, T.R.E. (1987). Secondary succession: patterns and strategies. *Colonization, Succession and Stability*. Symposia of the British Ecological Society, 26 (Ed. by A.J. Gray, M.J. Crawley & P.J. Edwards). Blackwell Scientific Publications, Oxford.

Chamberlain, J.A. & Westermann, G.E.G. (1976). Hydrodynamic properties of cephalopod shell ornament. *Paleobiology*, **2**, 316–31.

Cody, M.L. & Diamond, J.M. (Eds) (1975). *Ecology and Evolution of Communities*, 545 pp. Harvard University Press, Cambridge, Massachusetts.

Cohen, J.E. (1977). Ratio of prey to predators in community food webs. *Nature (London)*, **270**, 165–6.

Cohen, J.E. & Briand, F. (1984). Trophic links of community food webs. *Proceedings of the National Academy of Sciences, USA*, **81**, 4105–9.

Cohen, J.E. & Newman, C.M. (1985). A stochastic theory of community food webs. I. Models and aggregated data. *Proceedings of the Royal Society of London. Series B*, **224**, 421–48.

Cohen, J.E., Newman, C.M. & Briand, F. (1985). A stochastic theory of community food webs. II. Individual webs. *Proceedings of the Royal Society of London. Series B*, **224**, 449–61.

Cole, B.J. (1980). Trophic structure of a grassland insect community. *Nature (London)*, **288**, 76–7.

Colewell, R.K. & Winkler, D.W. (1984). A null model for null models in biogeography. *Ecological Communities: Conceptual Issues and the Evidence* (Ed. by D.P. Strong, D. Simberloff, L.G. Abele & A.B. Thistle), pp. 344–59. Princeton University Press, Princeton, New Jersey.

Connell, J.H. (1980). Diversity and the coevolution of competitors, or the ghost of competition past. *Oikos*, **35**, 131–8.

Connor, E.F. & Simberloff, D. (1978). Species number and compositional similarity of the Galapagos flora and avifauna. *Ecological Monographs*, **48**, 219–48.

Darnell, R.M. (1970). Evolution and the ecosystem. *American Zoologist*, **10**, 9–15.

Davidson, D.W., Brown, J.H. & Inouye, R.S. (1980). Competition and the structure of granivore communities. *Bioscience*, **30**,(4), 233–8.

Davidson, D.W., Sampson, D.A. & Inouye, R.S. (1985). Granivory in the Chihuahuan desert: interactions within and between trophic levels. *Ecology*, **66**, 486–502.

Dempster, J.P. (1983). The natural control of populations of butterflies and moths. *Biological Reviews*, **58**, 461–81.

Diamond, J. (1986). Overview: laboratory experiments, field experiments and natural experiments. *Community Ecology* (Ed. by J.M. Diamond & T.J. Case), pp. 3–22. Harper & Row, New York.

Dunbar, M.J. (1960). The evolution of stability in marine environments: natural selection at the level of the ecosystem. *American Naturalist*, **94**, 129–36.

Elton, C.S. (1927). *Animal Ecology*, 209 pp. Sidgwick & Jackson, London.

Faeth, S.H. (1987). Community structure and folivorous insect outbreaks. *Insect Outbreaks* (Ed. by P. Barbosa & J.C. Schultz). Academic Press, New York, in press.

Feeny, P. (1970). Seasonal changes in oak leaf tannins and nutrients as a cause of spring feeding by winter moth caterpillars. *Ecology*, **51**, 565–81.

Forbes, S.A. (1887). The lake as a microcosm. *Bulletin of the Illinois Natural Survey*, **15**, Article IX, pp. 537–50.

Glaessner, M.F. (1984). *The Dawn of Animal Life. A Biohistorical Study*, 244 pp. Cambridge University Press, Cambridge.

Gould, S.J. (1977). Eternal metaphors of palaeontology. *Patterns of Evolution as Illustrated by the Fossil Record*, Developments in Palaeontology and Stratigraphy 5 (Ed. by A. Hallam), pp. 1–26. Elsevier, Amsterdam.

Greenslade, P.J.M. (1983). Adversity selection and the habitat templet. *American Naturalist*, **122**, 352–65.

Grime, J.P. (1977). Evidence for the existence of three primary strategies in plants and its relevance to ecological and evolutionary theory. *American Naturalist*, **11**, 1169–94.

Grime, H.P. (1979). *Plant Strategies and Vegetation Processes*, 222 pp. Wiley, New York.

Grubb, P.J. (1986). Problems posed by sparse and patchily distributed species in species-rich plant communities. *Community Ecology* (Ed. by J. Diamond & T.J. Case), pp. 207–25. Harper & Row, New York.

Hanski, I. (1982). Dynamics of regional distribution: the core and satellite species hypothesis. *Oikos*, **38**, 210–21.

Harper, H.L. (1981). The concept of population in modular organisms. *Theoretical Ecology*, 2nd edn. (Ed. by R.M. May), pp. 53–77. Blackwell Scientific Publications, Oxford.

Harvey, P., Colwell, R.K., Silvertown, J.W. & May, R.M. (1983). Null models in ecology. *Annual Review of Ecological Systematics*, **14**, 189–211.

Harvey, P. & Godfray, H.C.J. (1987). How species divide resources. *American Naturalist*. **129**, 318–20.

Hassell, M.P. (1978). *The Dynamics of Arthropod Predator–Prey Systems*, 237 pp. Princeton University Press, Princeton, New Jersey.

Hassell, M.P. (1985). Insect natural enemies as regulating factors. *Journal of Animal Ecology*, **54**, 323–34.

Hassell, M.P., Southwood, T.R.E. & Reader, P.M. (1987). The dynamics of the viburnum whitefly, *Aleurotrachelus jelinekii* (Fraunef.): a case study on population regulation. *Journal of Animal Ecology*, in press.

Heatwole, H. & Levins, R. (1972). Trophic structure stability and faunal change during recolonization. *Ecology*, **53**, 531–4.

Hubbell, S.P. & Foster, R.B. (1986). Biology, chance and history and the structure of tropical rain forest tree communities. *Community Ecology* (Ed. by J. Diamond & T.J. Case), pp. 314–29. Harper & Row, New York.

Huffaker, C.B. & Messenger, P.S. (1964). The concept of significance of natural control. *Biological Control of Insect Pests and Weeds* (Ed. by P. DeBach), pp. 47–117. Chapman & Hall, London.

Hutchinson, G.E. (1959). Homage to Santa Rosalia, or why are there so many kinds of animals? *American Naturalist,* **93,** 145–59.

Janzen, D.H. (1967). Interaction of the bull's horn acacia (*Acacia comigera L*) with an ant inhabitant (*Pseudomyrmex ferruginea* F. Smith) in eastern Mexico. *Kansas University Science Bulletin,* **47,** 315–558.

Janzen, D.H. & Pond, C.M. (1975). A comparison by sweep sampling, of the arthropod fauna of secondary vegetation in Michigan, England and Costa Rica. *Transactions of the Royal Entomological Society of London,* **127,** 33–50.

Jeffries, M.J. & Lawton, J.H. (1984). Enemy free space and the structure of ecological communities. *Biological Journal of the Linnean Society,* **23,** 269–86.

Jeffries, M.J. & Lawton, J.H. (1985). Predator–prey ratios in communities of freshwater invertebrates: the role of enemy free space. *Freshwater Biology,* **15,** 105–12.

Kareiva, P. (1986). Patchiness, dispersal, and species interactions: consequences for communities of herbivorous insects. *Community Ecology* (Ed. by J.M. Diamond & T.J. Case), pp. 192–206. Harper & Row, New York.

Kennedy, W.J. & Southwood, T.R.E. (1984). The number of species of insects associated with British trees: a re-analysis. *Journal of Animal Ecology,* **53,** 455–78.

Kennedy, W.J. & Wright, C.W. (1985). Evolutionary patterns in late Cretaceous ammonites. *Special Papers in Palaeontology,* **33,** 131–43. Palaeontological Association, London.

Kirchner, T.B., Anderson, R.V. & Ingham, R.E. (1980). Natural selection and the distribution of nematode sizes. *Ecology,* **61(2),** 232–7.

Kitching, R.L. (1983). Community structure in water filled tree holes in Europe and Australia—some comparisons and speculations. *Phytotelmata: Terrestrial Plants as Hosts of Aquatic Insect Communities* (Ed. by H. Frank & P. Loumobos), pp. 205–22. Plexus, New Jersey.

Lawlor, L.R. (1980). Structure and stability in natural and randomly constructed competitive communities. *American Naturalist,* **116,** 294–406.

Lawton, J.H. & Strong, D.R. (1981). Community patterns and competition in folivorous insects. *American Naturalist,* **118,** 317–38.

Lee, K.E. (1985). *Earthworms, Their Ecology and Relationship With Soils and Land Use,* 411 pp. Academic Press, Sydney.

Macarthur, R.H. & Wilson, E.O. (1967). *The Theory of Island Biogeography.* Princeton University Press, Princeton, New Jersey.

McIntosh, R.T. (1980). The background and some current problems of theoretical ecology. *Conceptual Issues in Ecology* (Ed. by E. Saarinen), pp. 1–61. Reidel, Dordrecht.

McLaren, D. (1983). Impacts that change the course of evolution. *New Scientist,* **100,** 588–92.

Matthews, E.G. & Kitching, R.L. (1984). *Insect Ecology,* 2nd edn., 211 pp. University of Queensland Press, Brisbane.

May, R.M. (1975). Patterns of species abundance and diversity. *Ecology and Evolution of Communities* (Ed. by M.L. Cody & J.M. Diamond), pp. 81–102. Harvard University Press, Cambridge, Massachusetts.

May, R.M. (1981). Patterns in multi-species communities. Theoretical Ecology, 2nd edn. (Ed. by R.M. May), pp. 197–227. Blackwell Scientific Publications, Oxford.

May, R.M. (1984). An overview: real and apparent patterns in community structure. *Ecological Communities: Conceptual Issues and the Evidence* (Ed. by D.R. Strong, D. Simberloff, L.G. Abele & A.B. Thistle), pp. 3–16. Princeton University Press, Princeton, New Jersey.

Maynard Smith, J. (1986). *The Problems of Biology,* 134 pp. Oxford University Press, Oxford.

Meijer, J. (1980). The development of some elements of the arthropod fauna. *Oecologia,* **45,** 220–35.

Moran, V.C. & Southwood, T.R.E. (1982). The guild composition of arthropod communities in trees. *Journal of Animal Ecology,* **51,** 289–306.

Mound, L.A. & Waloff, N. (Eds) **(1978).** *Diversity of Insect Faunas.* Symposia of the Royal Entomological Society of London, 9. Blackwell Scientific Publications, Oxford.

Packard, A. (1972). Cephalopods and fish: the limits of convergence. *Biological Reviews,* **47,** 241–307.

Pease, C.M. (1985). Biases in the durations and diversities of fossil taxa. *Palaeobiology,* **11,** 272–92.

Pierce, N.E. & Young, W.R. (1986). Lycaenid butterflies and ants: two species stable equilibria in mutualistic, commensal and parasitic interactions. *American Naturalist,* **128,** 216–27.

Pimm, S.L. (1984). Food chains and return times. *Ecological Communities: Conceptual Issues and the Evidence* (Ed. by D.R. Strong, D. Simberloff, C.G. Abele & A.B. Thistle), pp. 397–412. Princeton University Press, New Jersey.

Pimm, S.L. & Lawton, J.H. (1977). Number of trophic levels in ecological communities. *Nature (London),* **268,** 329–31.

Pimm, S.L. & Lawton, J.H. (1980). Are food webs divided into compartments? *Journal of Animal Ecology,* **49,** 879–98.

Raup, D.M. (1972). Taxonomic diversity during the Phanerozoic. *Science,* **177,** 1065–71.

Raup, D.M. & Spekoski, J.J. (1984). Periodicity of extinctions in the geological past. *Proceedings of the National Academy of Sciences of the USA,* **81,** 801–5.

Richards, O.W. & Southwood, T.R.E. (1968). The abundance of insects: introduction. *Insect Abundance.* (Ed. by T.R.E. Southwood), pp. 1–7. Symposia of the Royal Entomological Society of London, 4. Blackwell Scientific Publications, Oxford.

Rosenzweig, M.L. (1979). Optimal habitat selection in two-species competitive systems. *Fortschritte der Zoologie,* **25,** 283–93.

Roughgarden, J. (1986). A comparison of food-limited and space-limited animal competition communities. *Community Ecology* (Ed. by J. Diamond & T. Case), pp. 492–516. Harper & Row, New York.

Rykiel, E.J. (1985). Towards a definition of ecological disturbance. *Australian Journal of Ecology,* **10,** 361–5.

Schoener, T.W. (1986). Patterns in terrestrial vertebrate versus arthropod communities: do systematic differences in regularity exist? *Community Ecology* (Ed. by J.M. Diamond & T.J. Case), pp. 556–86. Harper & Row, New York.

Sepkoski, J.J. & Moller, A.I. (1985). Evolutionary faunas and the distribution of paleozoic marine communities in space and time. *Phanerozoic Diversity Patterns* (Ed. by J.W. Valentine), pp. 153–90. Princeton University Press, Princeton, New Jersey.

Sibly, R.M. & Calow, P. (1985). Classification of habitats by selection pressures: a synthesis of life cycle and r/K theory. *Behavioural Ecology: Ecological Consequences of Adaptive Behaviour* (Ed. by R.M. Sibly & R.H. Smith), pp. 75–90. Symposia of the British Ecological Society, 25. Blackwell Scientific Publications, Oxford.

Simberloff, D. & Connor, E.F. (1979). Q-mode and R-mode analyses of biogeographic distributions: null hypotheses based on random colonization. *Contemporary Quantitative Ecology and Related Ecometrics* (Ed. by G.P. Patil & M.L. Rosenzweig), pp. 128–38. International Publishing House, Fairland, Maryland.

Southwood, T.R.E. (1960). The flight activity of Heteroptera. *Transactions of the Royal Entomological Society of London,* **112,** 173–220.

Southwood, T.R.E. (1961). The number of species associated with various trees. *Journal of Animal Ecology,* **30,** 1–8.

Southwood, T.R.E. (1973). The insect/plant relationship—an evolutionary perspective. *Insect/ Plant Relationships* (Ed. by H.F. van Emden), pp. 3–30. Symposia of the Royal Entomological Society of London, 6. Blackwell Scientific Publications, Oxford.

Southwood, T.R.E. (1976). Bionomic strategies and population parameters. *Theoretical Ecology*, 1st edn. (Ed. by R.M. May), pp. 30–52. Blackwell Scientific Publications, Oxford.

Southwood, T.R.E. (1977). Habitat, the templet for ecological strategies? (Presidential address). *Journal of Animal Ecology*, **46**, 337–65.

Southwood, T.R.E. (1978a). *Ecological Methods*, 2nd edn. Chapman & Hall, London.

Southwood, T.R.E. (1978b). The components of diversity. *Diversity of Insect Faunas* (Ed. by L.A. Mound & N. Waloff), pp. 19–40. Symposia of the Royal Entomological Society of London, 9. Blackwell Scientific Publications, Oxford.

Southwood, T.R.E. (1980). Ecology—a mixture of pattern and probabilism. *Conceptual Issues in Ecology* (Ed. by E. Saarinen), pp. 230–4. Riedel, Dordrecht.

Southwood, T.R.E. (1981). Stability in field populations of insects. *The Mathematical Theory of the Dynamics of Biological Populations II* (Ed. by R.W. Hirons & D. Cooke), pp. 31–45. Academic Press, London.

Southwood, T.R.E. (1984). Insect-plant adaptations. *Origins and Development of Adaptation* (Ed. by B.C. Clarke), pp. 138–51. Ciba Foundation Symposium, 102. Pitman, London.

Southwood, T.R.E. (1985). Insect communities (Presidential address). *Antenna*, **9**, 108–16.

Southwood, T.R.E., Brown, V.K. & Reader, P.M. (1979). The relationship of plant and insect diversities in succession. *Biological Journal of the Linnean Society*, **12**, 327–48.

Southwood, T.R.E., Brown, V.K. & Reader, P.M. (1986). Leaf palatability, life expectancy and herbivore damage. *Oecologia*, **70**, 544–48.

Southwood, T.R.E., Brown, V.K., Reader, P.M. & Green, E.E. (1986). The use of different stages of a secondary succession by birds. *Bird Study*, **33**, 159–63.

Southwood, T.R.E. & Kennedy, C.E.J. (1983). Trees as islands. *Oikos*, **41**, 359–71.

Southwood, T.R.E., Moran, V.C. & Kennedy, C.E.J. (1982). The richness, abundance and biomass of the Arthropod communities on trees. *Journal of Animal Ecology*, **51**, 635–49.

Southwood, T.R.E. & Reader, P.M. (1976). Population census data and key factor analyses for the viburnum whitefly, *Aleurotrachelus jelinekii* (Frauef.) on three bushes. *Journal of Animal Ecology*, **45**, 313–25.

Strong, D.R. (1974). Nonasymptotic species richness models and the insects of British trees. *Proceedings of the National Academy of Sciences, USA*, **71**, 2766–9.

Strong, D.R. (1986). Density vagueness: the variance in the demography of real populations. *Community Ecology* (Ed. by J.M. Diamond & T.J. Case), pp. 257–68. Harper & Row, New York.

Strong, D.R., Lawton, J.H. & Southwood, T.R.E. (1984a). *Insects on Plants: Community Patterns and Mechanisms*, 313 pp. Blackwell Scientific Publications, Oxford.

Strong, D.R. Simberloff, D., Abele, L.G. & Thistle, A.B. (Eds) (1984b). *Ecological Communities: Conceptual Issues and the Evidence*. Princeton University Press, Princeton, New Jersey.

Strong, D.R., Szyska, L.A. & Simberloff, D. (1979). Tests of community-wide character displacement against null hypotheses. *Evolution*, **33**, 897–913.

Taylor, L.R. (1978). Bates, Williams, Hutchinson— a variety of diversities, *Diversity of Insect Faunas* (Ed. by A. Mound & N. Waloff), Symposia of the Royal Entomological Society, 9, pp. 1–18. Blackwell Scientific Publications, Oxford.

Taylor, L.R. (1986). Synoptic dynamics, migration and the Rothamsted Insect Survey (Presidential address). *Journal of Animal Ecology*, **55**, 1–38.

Terborgh, J.W. (1973). On the notion of favourableness in plant ecology. *American Naturalist*, **107**, 481–501.

Tischler, W. (1975). *Okologie*. (Wörterbücher der Biologie), 125 pp. Gustav Fischer, Jena.

Valentine, J.W. (1973). Phanerozoic taxonomic diversity: a test of alternate models. *Science*, **180**, 1078–9.

Waage, J.K. (1985). Report of RESL Meeting 3 April 1985. *Antenna,* **9(3),** 128–30.

Waage, J.K. & Davies, C.R. (1986). Host-mediated competition in a bloodsucking insect community. *Journal of Animal Ecology,* **55,** 171–80.

Walker, I. & Ferreira, M.J.N. (1985). On the population dynamics and ecology of the shrimp species (Crustacea, Decapoda, Natantia) in Central Amazonian river Tarama-Mirim. *Oecologia,* **66,** 264–70.

Ward, P. (1981). Shell sculpture as a defensive adaptation in ammonoids. *Paleobiology,* **7,** 96–100.

West, C. (1985a). *The effect on phytophagous insects of variations in defence mechanisms within a plant.* Unpublished DPhil. thesis, Oxford University.

West, C. (1985b). Factors underlying the late seasonal appearance of the lepidopterous leaf-mining guild on oak. *Ecological Entomology,* **10,** 111–20.

Whittaker, R.H. (1970). *Communities and Ecosystems,* 162 pp. Macmillan, London.

Whittaker, R.H. (1972). Evolution and measurement of species diversity. *Taxon,* **21,** 213–51.

Whittaker, R.H. (1975). The design and stability of some plant communities. *Unifying Concepts in Ecology* (Ed. by W.H. van Dobben & R.H. Lowe McConnell), p. 169. Junk, The Hague.

Wiens, J.A. (1984). On understanding a non-equilibrium world: myth and reality in community patterns and processes. *Ecological Communities: Conceptual Issues and the Evidence* (Ed. by D.R. Strong, D. Simberloff, L.G. Abele & A.B. Thistle), pp. 439–57. Princeton University Press, Princeton, New Jersey.

Wiens, J.A. (1986). Spatial scale and temporal variation in studies of Shrubsteppe birds. *Community Ecology* (Ed. by J.M. Diamond & T.J. Case), pp. 154–72. Harper & Row, New York.

Wiens, J.A., Addicott, J.F., Case, T.J. & Diamond, J.M. (1986). Overview: the importance of spatial and temporal scale in ecological investigations. *Community Ecology* (Ed. by J.M. Diamond & T.J. Case), pp. 145–53. Harper & Row, New York.

Williams, C.B. (1964). *Patterns in the Balance of Nature and Related Problems in Quantitative Ecology,* 324 pp. Academic Press, London.

Wilson, E.O. (1969). The species equilibrium. *Diversity and Stability in Ecological Systems,* pp. 38–47. Brookhaven National Laboratory, New York.

Wilson, D.S. (1986). Adaptive indirect effects. *Community Ecology* (Ed. by J.M. Diamond & T.J. Case), pp. 437–44. Harper & Row, New York.

Wint, G.R.W. (1983). The role of alternative host-plant species in the life of a polyphagous moth, *Operophtera brumata* (Lepidoptera: Geometridae). *Journal of Animal Ecology,* **52,** 439–50.

Wolda, H. (1978). Fluctuations and abundance of tropical insects. *American Naturalist,* **112,** 1017–43.

Wolda, H. (1980). Seasonality of tropical insects I. Leafhoppers (Homoptera) in Las Cumbres, Panama. *Journal of Animal Ecology,* **49,** 277–90.

2. SPATIAL PATCHINESS AND COMMUNITY STRUCTURE: COEXISTENCE AND GUILD SIZE OF DROSOPHILIDS ON EPHEMERAL RESOURCES

B. SHORROCKS AND J. ROSEWELL

*Department of Pure and Applied Zoology, The University,
Leeds LS2 9JT, UK*

. . . theories about population changes and interaction, which are nearly all based upon conceptions of mean density, must learn to take account of the fact that populations are split up into groups or centres of action. . . . I think it is even conceivable that the existence of certain scarcer species in such minor community units depends partly on stronger competitors being absent by chance. (Charles Elton 1949)

CENTRES OF ACTION

In 1949, Charles Elton gave his presidential address to the British Ecological Society in which he talked about the importance of 'minor community units', referring to them as 'centres of action'. Many animals, particularly arthropods, exploit such resources which are patchy, consisting of small, separate units, and are ephemeral in the sense that they support only one or two generations and are spatially unpredictable. These include dung and carrion (Doube, Chapter 12), fruit, seeds, fungi, sap-flows, decaying leaves, flowers, dead wood and small bodies of water held in the aerial parts of terrestrial plants (phytotelmata, Hildrew & Townsend, Chapter 16). The dominant arthropods utilizing such resources are usually Coleoptera and Diptera.

Flies of the genus *Drosophila* are primarily consumers of the yeasts and bacteria associated with the fermentation and decay of plant material, and the larvae typically live in ephemeral units such as fruit, fungi, flowers, sap and leaves. Over the past 16 years, extensive field and laboratory studies at Leeds on the ecology of *Drosophila* have led to conclusions about coexistence and guild size in these flies which this chapter will review and extend. We believe these conclusions have important implications for the study of all species utilizing such patchy and ephemeral resources.

THE TWO-SPECIES MODEL

In constructing the initial model (Shorrocks, Atkinson & Charlesworth 1979; Atkinson & Shorrocks 1981) we were especially interested in *Drosophila* species breeding in fruit or fungi (Atkinson & Shorrocks 1977; Shorrocks & Charlesworth 1980; Shorrocks 1982). Several workers have suggested that for drosophilid flies the major, if not the only, period of resource limitation occurs in the larval stage (Carson 1971; Shorrocks 1977; Atkinson 1979; Grimaldi & Jaenike 1984; Atkinson 1985) and in the model that follows we confine the competition to the larval stage within a patch. This competition is described by the competition equations of Hassell & Comins (1976):

$$N_i(t+1) = \lambda_i . N_i(t) . [1 + a_i . (N_i(t) + \propto_{ij} . N_j(t))]^{-b_i}$$

where the subscripts i and j refer to two species and $N(t)$ is the number of each species in a breeding site at time t, λ is the net reproductive rate, $\propto$ is a competition coefficient and a and b are constants. The parameter a is related to the population size per site at which density dependence starts to act and b describes the type of competition. A value of $b = 1$ simulates contest competition; as b increases, competition becomes less contest and more scramble (Nicholson 1954; Hassell 1975). When $b = 1$ these equations are a difference form of the Lotka–Volterra competition equations. With this system of equations, the carrying capacity per site (N^*) is equal to $(\lambda^{1/b} - 1)/a$ and only three of these four quantities can be varied independently. In this account, and in contrast to Atkinson & Shorrocks (1981), we do not vary a independently but treat it simply as a scaling factor.

In the model the eggs of both species are distributed over the available breeding sites according to a negative binomial distribution, which has an exponent, k, inversely related to the degree of aggregation. In the initial model this parameter was kept constant and the distributions were independent. For every combination of eggs of both species, the eggs produced by the survivors of competition are multiplied by the probability of occurrence predicted by the negative binomials. These figures are summed to give the mean number of eggs per site in the next generation. Notice that $N(t)$ here represents the density of eggs starting generation t, rather than the number of adults.

In the initial simulations (Atkinson & Shorrocks 1981) only a limited range of parameters was examined. For both species $\lambda = 5$, $b = 1$ (contest), and the competition coefficient of the superior species was equal to two and for the inferior species was equal to zero. Aggregation of eggs (k) and carrying capacity per site (N^*) were allowed to vary. Fig. 2.1 shows the effect

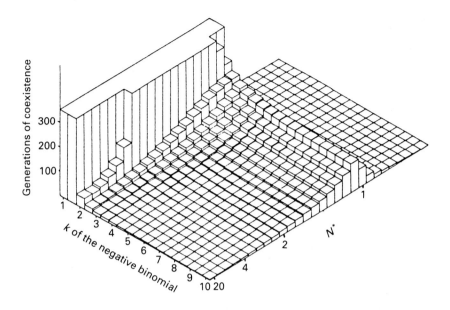

FIG. 2.1. Generations of coexistence predicted by the model of Atkinson & Shorrocks (1981) for values of k of the negative binomial and carrying capacity per breeding site N^*. Parameter values are given in the text. (After Atkinson & Shorrocks 1981.)

upon coexistence. Increasing aggregation (smaller k) of the superior species and decreasing N^* (to an optimum of near 1) resulted in prolonged coexistence. Sufficiently small values of k (less than about 2) resulted in coexistence for more than 350 generations.

If a stable equilibrium is the sole outcome of interest it is not necessary to simulate all the generations. By placing the superior species at equilibrium and seeing if the inferior species can invade, the parameter space producing coexistence can be mapped. Furthermore, since $\propto$ of the superior species is important in determining coexistence it is convenient to show the results of the model in terms of 'critical $\propto$' of the superior species. For values of $\propto$ below this critical point, the superior species will not exclude the inferior species and coexistence is assured. Figure 2.2 shows graphs of critical α against k of the negative binomial for a range of λ and N^* values.

The general conclusions that emerge from these simulations, confirmed by the analytical model of Ives & May (1985), are briefly as follows. On a divided and ephemeral resource, coexistence (invasion of the inferior species) is made easier with lower k of the negative binomial for the superior species. Under these conditions, there is more aggregation and therefore

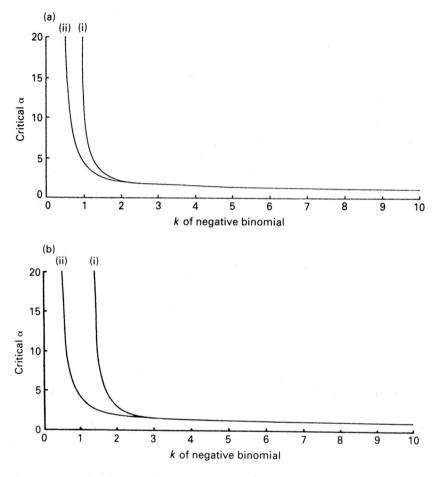

FIG. 2.2. Maximum value of the competition coefficient of the superior species that allows coexistence (critical α) as a function of aggregation (k of the negative binomial) of the superior species. Parameter values of both species are equal in each simulation. (a) (i) $\lambda = 50, N^* = 10, b = 1$; (ii) $\lambda = 5, N^* = 500, b = 1$. (b) (i) $\lambda = 50, N^* = 10, b = 5$; (ii) $\lambda = 5, N^* = 500, b = 5$.

more low density or empty sites in which the inferior species can increase. These we term probability refuges. The inferior species will also find it easier to invade when the superior species has a lower N^* relative to that of the inferior, when the superior species has a lower value of α, when the inferior species has a higher reproductive rate λ, and when values of b are higher. The influence of b acts through the number of refuges. With increased scramble competition (increase in b) and N^* kept constant, the point of onset of density dependence is delayed and the number of probability refuges is increased.

All the parameters have some effect, but usually only in one of the two species. However, k of the superior species is the best predictor of co-existence. Given sufficiently small values ($k < 1$), coexistence is virtually assured.

MEASURING THE PARAMETERS

In this section we summarize, for drosophilid flies, the available information on the parameter values used in the model.

Aggregation (k)

Figure 2.3 shows the distribution of k for twenty-five dipteran species under three conditions; field observations, field experiments and laboratory experiments. There is no significant difference between the values of k obtained from the three types of measurements ($x^2 = 4.13$, d.f. $= 4$).

Ideally, such measurements of k should be carried out on eggs laid on natural sites, in the field. However, it is difficult to count drosophilid eggs on field-collected breeding sites. The laboratory experiments counted eggs, but the field data came from emerging adults, that is, after pre-adult mortality may have changed the aggregation pattern. However, there are two things that suggest that the distributions of Fig. 2.3 are a realistic reflection of natural values. First, as already stated, there is no significant difference between the values of k obtained from the different types of experiment. Second, there are four measurements of eggs under natural conditions available for comparison. These are for *D. mimica* on *Sapindus* fruit with $k = 0.54$ (Kambysellis *et al.* 1980), *D. subobscura* on *Prunus* fruit with $k = 0.36$ (Atkinson & Shorrocks 1984), *D. phalerata* on *Phallus impudicus* with $k = 0.10$ (B. Shorrocks unpublished data), and *D. subobscura* on *Phallus impudicus* with $k = 0.25$ (B. Shorrocks unpublished data). It should also be noted that since larval mortality is likely to be more severe in crowded patches, such pre-adult reductions are likely to result in less aggregated distributions of adults compared to eggs.

Competition coefficient (α)

Figure 2.4 shows the distribution of fifty-two values of α for larval competition in seven drosophilid species (*D. busckii, D. melanogaster, D. nubulosa, D. pseudoobscura, D. serrata, D. simulans* and *D. willistoni*)

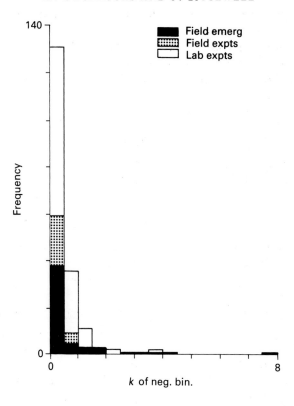

Fig. 2.3. Frequency histogram of values of *k* of the negative binomial for twenty-five dipteran species under three different conditions. Solid: field observations; stipple: field experiments; blank: laboratory experiments. Field observations come from eight species of Australian fruit (536 items), used by *D. buzzatii, D. fumida, D. immigrans, D. lativitatta, D. pseudotakahashii, D. simulans, D. specensis* and fourteen other dipterans, and from one species of British fungus (*Phallus impudicus*) used by *D. phalerata, D. cameraria* and *D. subobscura* (Atkinson & Shorrocks 1984; Shorrocks unpublished data). Field experiments are for *D. melanogaster, D. simulans, D. immigrans, D. subobscura, D. busckii* and *D. funebris*, with 4 cm² units of apple, lemon, onion and potato as oviposition sites (J. Rosewell & B. Shorrocks unpublished data). Laboratory experiments used *D. immigrans* and *D. melanogaster*, with 4 cm² and 9 cm² units of laboratory *Drosophila* medium as oviposition sites (Atkinson & Shorrocks 1984; K. Edwards & B. Shorrocks unpublished data).

maintained under laboratory conditions (Miller 1964; Ayala 1969, 1970, 1972; Ayala, Gilpin & Ehrenfeld 1973; DeBenedictis 1977. B. Shorrocks unpublished data). The confined nature of these experiments on artificial food is unlikely to invalidate their use in this context since the parameter we require is simply the species equivalent in a small patch. Although the nature of the food may well influence relative competitive ability in the case of any

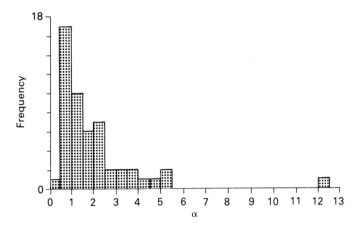

FIG. 2.4. Frequency histogram of values of α from laboratory experiments with *Drosophila* species (see text for details).

one pair of species, we feel it is unlikely to introduce a consistent bias into the range of values observed.

Type of competition (b)

There appears to be very little information about the form of density dependence (b) within natural drosophilid breeding sites. Values for laboratory experiments are *D. melanogaster* $b = 1.83$, 2.00 and 1.40; *D. simulans* $b = 3.30$ and 2.45; *D. immigrans* $b = 2.10$; *D. subobscura* $b = 2.13$; *D. busckii* $b = 1.02$; *D. funebris* $b = 2.10$ and *D. hydei* $b = 2.07$ (Shorrocks & Rosewell 1986; J. Rosewell unpublished). Only one value, $b = 0.9$, is for a natural site: *D. subobscura* on the berries of *Sorbus aucuparia* (Kearney 1979).

Reproductive output (λ)

The calculation of λ requires the day to day fecundity schedule and the female survivorship per day (ϕ). This latter value is ideally measured in the field using capture–recapture techniques. Table 2.1 shows values of ϕ measured on five species of drosophilid fly, all from studies in which suitable marking techniques (Crumpacker 1974; Shorrocks & Nigro 1981) and relatively closed populations were used. Seven sets of daily fecundity data (Shorrocks & Rosewell 1986), in conjunction with a mean ϕ of 0.63 from

Table 2.1. Data on daily adult survival rate (ϕ) in drosophilids taken from the literature. All *estimates (except D. limbata)* are for females only and Begon, Milburn & Fisher (which uses the negative method of Jackson 1937) come from the Fisher-Ford (1947) model for capture-recapture data

Author	Location	Species	Rate
Begon, Milburn & Turner (1975)	S. England, Rogate	*D. subobscura*	0.57
Begon (1978)	N. England, Leeds	*D. subobscura*	0.83 0.69
Hummel, van Delden & Drent (1979)	Netherlands	*D. limbata* (both sexes)	0.53 0.83
Begon, Krimbas & Loukas (1980)	Greece	*D. subobscura*	0.42 0.61
Rosewell & Shorrocks (unpublished data)	N. England, Leeds	*D. melanogaster/ D. simulans*	0.67 0.81 0.52 0.82
		D. immigrans	0.71 0.22

Table 2.1, give a range of values of λ: for the domestic fruit breeders *D. simulans* and *D. funebris* $\lambda = 35.4$ and 4.0 respectively; for the woodland fungal breeder *D. phalerata* $\lambda = 1.3$, and for four different populations of *D. subobscura*, a woodland fruit and fungal breeder, $\lambda = 4.8, 4.9, 3.8$ and 4.3.

Carrying capacity per breeding site (N^*)

Shorrocks & Rosewell (1986) give data on adult emergence per natural breeding site for a variety of classes of substrate. This data is reproduced in Fig. 2.5. Surprisingly, given the large amount of information on *Drosophila* breeding sites (Shorrocks 1982), this data is not extensive because most studies do not publish records for individual sites. The modal class is always less than fifteen, although species of fungi producing up to an average of forty emerging adults are quite common. It must be remembered, however, that N^* in the simulation model is actually numbers of eggs not adults. Taking a value of λ around 5 and assuming a steady population, this would suggest a maximum range of N^* between 10 and 250, but more usually between 10 and 100.

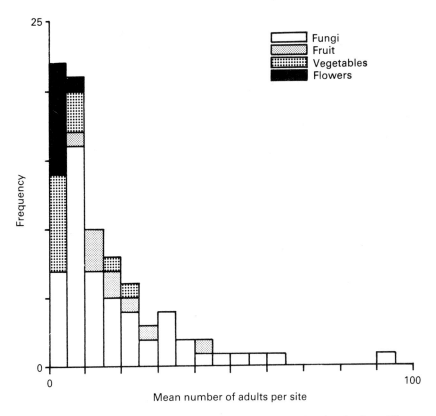

FIG. 2.5. Frequency histogram of mean numbers of drosophilid flies emerging from different species of breeding site. Blank: fungi; light stipple: fruit; heavy stipple: vegetables; solid: flowers.

The implications for drosophilid flies

The results presented in this section would suggest that the values of b, λ and N^* used to produce the graphs in Fig. 2.2 encompass all the realistic values for drosophilid flies. In fact the results for $\lambda = 5$, $N^* = 10$ and b between 1 and 5 are the most characteristic for these flies. Given these values, the initial model predicts that two competing species of drosophilid flies will coexist without any resource partitioning.

TESTING THE ASSUMPTIONS

In addition to using the parameters examined in the last section, the model of two-species competition in a patchy environment assumed a number of conditions as simple approximations to the real world. This section will examine these assumptions and where necessary modify the original model.

The negative binomial as a description of egg distribution

Atkinson & Shorrocks (1981) used the negative binomial as a convenient description of the aggregated distribution of competing stages starting any generation. It should be noted, however, that Ives & May (1985) in their analytical counterpart of the simulation model suggest that other aggregated distributions (variance $\gg$ mean) will have qualitatively similar results.

We have fitted the negative binomial to several field samples of drosophilids over resource patches using the χ^2 statistic. Only those examples with all expected class frequencies greater than 5 (Fisher 1970; Conover 1980) or greater than 1 (Cochran 1952, 1954; Nass 1959) were analysed. For those tests that were not significant we have carried out a power analysis (Cohen 1970) using a Poisson distribution as the alternative hypothesis. All the samples showed strongly aggregated distributions (variance: mean ratio test); most were not significantly different from the negative binomial and had sufficient power to reject the alternative hypothesis (Table 2.2).

TABLE 2.2. Results of χ^2 test for goodness of fit to a negative binomial distribution and subsequent power analysis. Figures in the table are the numbers of the data sets falling into the appropriate category. Figures in brackets are sets with low power ($P = 0.95$). Data set 1: *D. mimica* on *Sapindus* fruit, *D. subobscura* on *Prunus* fruit and *Phallus impudicus* fungal bodies and *D. phalerata* on *Phallus impudicus* fungal bodies. Data set 2: *D. melanogaster, D. simulans, D. immigrans, D. subobscura* and *D. funebris* on apple, lemon, onion and potato. Data set 3: *D. phalerata, D. cameraria* and *D. subobscura* on *Phallus impudicus*

Type of data	Min. exp of 5		Min. exp of 1	
	Sig.	Not sig.	Sig.	Not sig.
1 Eggs on field sites	1	2(0)	0	4(0)
2 Adults emerging from field experiments	2	13(1)	3	23(3)
3 Adults emerging from field sites	1	1(0)	0	10(0)

The fact that several mechanisms are each known to give rise to a negative binomial distribution has lead to criticism of its use on the grounds that it does not offer any information on the mechanisms producing it. However, this would be true of all other compound distributions (Pielou 1977), and in general, specific mechanisms can only be induced from specially designed experiments. Use of the negative binomial should be taken to be no more than a description rather than an explanation of observed events, although it may be profitable to speculate on mechanisms in order to suggest hypotheses (Atkinson & Shorrocks 1984).

Criticism of the negative binomial has also centred on its use under varying densities (Taylor, Woiwod & Perry 1979). Perry & Taylor (1985) have suggested that a new distribution, which they call Adès, provides a better description when a family of curves at different densities is involved. They use seven sets (families) of data to test for goodness of fit and conclude that both the Adès and negative binomial distributions fit equally well. They point out, however, that with n curves in a family, the Adès distribution uses only $n+2$ parameters while the negative binomial uses $2n$, a mean and exponent k for each distribution. The Adès is thus more parsimonious than the negative binomial. However, it is possible to fit Taylor's power law (log(variance) = log($\propto$)+β.log(mean); Taylor 1961) to a family of distributions and using the parameters $\propto$ and β, plus the mean from each distribution, predict the appropriate value of k for any density. This procedure will also only use $n+2$ parameters. We have compared the two methods using the data of McGuire, Brindley & Bancroft (1957) on the European corn borer, *Pyrausta nubilalis* which Perry & Taylor state are 'possibly the most extensive in the literature' and 'provide an invaluable test of the flexibility' of any family of distributions. Their six parameter Adès distribution gives a probability of 0.769, while the six parameter constrained negative binomials give a probability of 0.712. Both are more than adequate fits to the observed data.

We suggest that until such time as behavioural experiments indicate that another distribution is more appropriate there seems little point in abandoning the negative binomial.

Constant k of the negative binomial

There is plenty of field evidence that the degree of aggregation of animals changes with density (e.g. Taylor, Woiwod & Perry 1978). As mentioned in the previous section, Taylor (1961) has shown empirically that the variance (v) of the distribution of individuals over patches is related to the mean number of individuals per patch (m). We have fitted this relationship to the examples already detailed in Fig. 2.3 and obtained estimates of $\propto$ and β. Fig. 2.6 shows all the data combined for which $\propto = 7.44$ and $\beta = 1.57$ ($n = 134$, $P < 0.001$). From Taylor's power law, k of the negative binomial will respond to density according to the equation $1/k = \propto.m^{\beta-2}-1/m$ (Taylor, Woiwod & Perry 1979). This relationship predicts that, as mean numbers per patch increase, for $1 < \beta < 2$ the value of k initially decreases

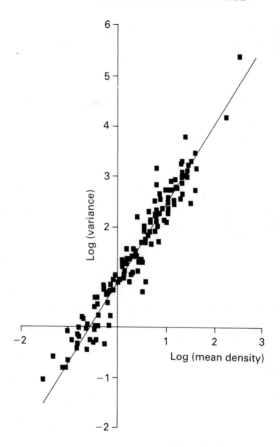

FIG. 2.6. Power law relationship of log(variance) against log(mean density) of 134 distributions of dipterans (117 of drosophilids) over breeding sites. ($v = 7.44 \times m^{1.57}$; $n = 134$, $P < 0.001$). Various subsets of the data are as follows. For field sites: Australian fruit $\propto = 4.91$, $\beta = 1.88$, $n = 34$, $P < 0.001$; British fungi $\propto = 7.30$, $\beta = 1.50$, $n = 18$, $P < 0.001$. For two field experiments: $\propto = 4.27$, $\beta = 1.21$, $n = 17$, $P < 0.001$; and $\propto = 6.04$, $\beta = 1.23$, $n = 14$, $P < 0.001$. For the laboratory studies: *D. melanogaster* $\propto = 12.04$, $\beta = 1.47$, $n = 37$, $P < 0.001$; *D. immigrans* $\propto = 6.67$, $\beta = 1.77$, $n = 18$, $P < 0.001$.

at low density but then increases at higher density. The turning point occurs at mean $= [1/(\propto(2-\beta))]^{1/(\beta-1)}$. For our drosophilid data as the mean numbers per patch increase beyond this point the value of k will rise implying that aggregation may be lost at high densities. However, we must remember that the model refers to patchy and ephemeral sites which have a limited size. For those drosophilids in Fig. 2.5 the mean density of eggs is usually much less than 100. With $\propto = 7.44$ and $\beta = 1.57$, a patch mean of 100 only gives $k = 0.98$ suggesting from Fig. 2.2 that coexistence is still the most likely outcome for these flies. This result is not surprising since natural values of k for drosophilids have already been measured (Shorrocks *et al.* 1984) and found to be mainly below one (Fig. 2.3).

Independent association between the species

The original model assumed that the two species were distributed over patches independently with respect to each other, that is they show neither positive association (e.g. attraction to the same site) nor negative association (e.g. resource partitioning). Positive association will make coexistence more difficult, tending to recover the predictions of the Hassell & Comins model when positive association is complete. In this section we will see if the original prediction of coexistence for drosophilids is seriously threatened.

For some of the examples in Fig. 2.3, where more than one species was present, we were able to measure k and association between pairs of species. We tested associations by presence and absence from patches using 2×2 contingency tables and the χ^2 statistic, excluding data sets where expected frequencies were below one. Some of the results (28%) were negative but none significant; the remaining positive results are largely not significant. None of the field experiments gave significant associations (30 data sets), five of the field observations gave a significant association (66 data sets) and two of the laboratory experiments gave a significant result (5 data sets). As a convenient index of association we have used the four point correlation coefficient (Conover 1980), $\phi = \sqrt{(\chi^2/N)}$ where N is the number of observations in the corresponding contingency table and χ^2 is the uncorrected statistic. This index has a range from $+1$ to -1.

An important point must be made here concerning the nature of the double absence cell in the contingency table. Numbers in this cell denote those sites which contain neither of the two species. These may represent either suitable patches that are not found by the two drosophilids by chance, or patches that are wholly unsuitable. The latter should ideally be recognized and excluded from the analysis and including them may well produce spurious positive associations. The field experiments and laboratory experiments used replicated standardized patches and so should avoid this problem. However, natural field sites show considerable variation, some of which excludes them as breeding sites. Where more than two species were present, we have calculated for the field data contingency χ^2 and ϕ between pairs of species using the full data, and again excluding those sites that are zero for *all* species. These latter sites are deemed to be unsuitable rather than not found. In all cases ($n = 55$) the second value of ϕ is less positive and in all but one case non-significant by χ^2, suggesting that this problem may frequently inflate associations measured on field sites.

We have run the two species simulation program using ϕ to adjust the joint matrix of probabilities for the two species. Figure 2.7 shows contours of critical $\propto$ for various combinations of ϕ and k and the empirical data excluding unsuitable sites as defined above. Three things may be concluded from this graph. First, the presence of positive association between the species does make coexistence more difficult. Second, the degree of

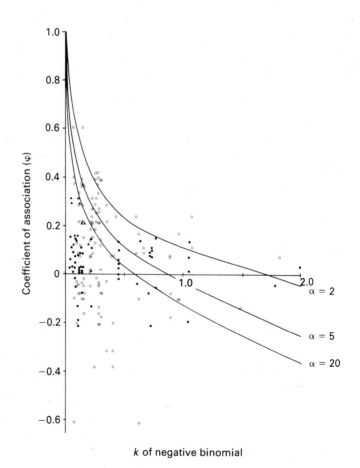

FIG. 2.7. Effect of aggregation and association between species on critical alpha. The curves are contours of equal critical alpha ($\propto$ = 2, 5, 20) as a function of aggregation (k) and association (ϕ). Shown also are observed values of k and ϕ from the datasets listed in Fig. 2.3. Filled circles: field experiments; open circles: field observations; filled triangles: laboratory experiments.

association measured for drosophilid flies, in conjunction with Figs 2.3 and 2.4, would still suggest coexistence as the most likely outcome. Third, aggregated distributions and probability refuges will also make coexistence easier when some degree of resource partitioning is present.

MORE THAN TWO SPECIES—GUILD SIZE IN DROSOPHILIDS

Unfortunately, organisms do not usually exist in groups of two species. For example, *Drosophila* tend to occur in the wild not in pairs of species, but in guilds of several species exploiting similar larval food resources (Shorrocks 1977, 1982). It is important therefore that the two-species model described earlier should be extended to a multi-species context. Shorrocks & Rosewell (1986) present a preliminary model which is described and modified below.

The multi-species model

In a multi-species system, due to the aggregation of competing species, small guilds are open to invasion but, as species are added, the joint occurrence of many species effectively reduces the number of probability refuges until invasion is no longer possible.

The model (Shorrocks & Rosewell 1986) effectively takes random groups of 2,3,4 . . . species, with parameters taken from our empirical distributions of k (Fig. 2.3) and $\propto$ (Fig. 2.4), and finds the probability of a stable guild of this size. Because full simulation of many species is impractical, multi-species systems are collapsed into a single super-species and a single invading species. The super-species is assumed conservatively to have the largest $\propto$ of the guild to represent the combined effect of all its species on the inferior, and to have a value of k which represents the combined distribution of its component species. It is assumed that potential competitors are independently distributed over resource units according to negative binomial distributions. The variance of the joint distribution is given by the sum of the variances of the component distributions and since the variance of a negative binomial is given by $m + (m^2/k)$, this implies that k of the joint distribution is given by

$$k_{1+2} = \frac{(1+m_1/m_2)^2 . k_1 . k_2}{(m_1/m_2)^2 . k_2 + k_1}$$

Notice that this method of calculating the combined value of k is more accurate than the simple sum used in Shorrocks & Rosewell (1986). In the simulation we have assumed that the means stay constant, implying that a site can support a certain number of drosophilids of any species, and the above expression reduces to $k_{1+2} = 4.k_1.k_2/(k_1+k_2)$.

The frequencies of guild size that result from running the model are set out in Fig. 2.8. Each curve on the graph is a frequency distribution of guild sizes, in cumulative form, for a particular combination of b, λ and N^*, obtained by using the appropriate critical $\propto$ graph of Fig. 2.2. Each curve is the result of 5000 simulations. The model class for number of drosophilids in a guild is moved to the right (more species in a guild) as b and λ increase and N^* decreases, as would be predicted from the effects upon two-species coexistence already discussed.

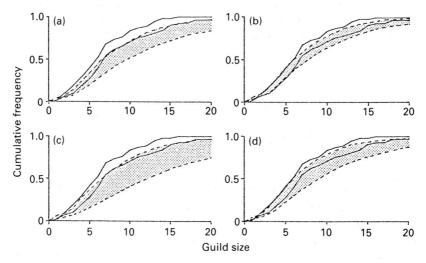

FIG. 2.8. Guild sizes predicted from the simulation model graphed as cumulative probability. Each graph is for a particular combination of parameters. (a) $b = 1, N^* = 10$. (b) $b = 1$, $N^* = 500$. (c) $b = 5, N^* = 10$. (d) $b = 5, N^* = 500$. The upper broken curve in each graph is for $\lambda = 5$ and the lower broken curve for $\lambda = 50$. Shown superimposed for comparison are the data of Fig. 2.9 for natural guild sizes, the top solid line representing the 'complex' and the lower solid line the 'simple' classification.

The empirical data

We have estimated guild size in wild drosophilids from the breeding records of flies in the literature (Shorrocks & Rosewell 1986, with additional data in Table 2.3). The larval habitat has been used to characterize the guild (Carson 1971; Shorrocks 1977, 1982). Traditionally the substrates utilized by drosophilids have been grouped into five main categories (Carson 1971; Kimura *et al.* 1977; Begon & Shorrocks 1978; Shorrocks 1982). These are

TABLE 2.3. Amendments and additions to the source material of Shorrocks & Rosewell (1986) for the data on guild size. Under guild size, 6 → (4) indicates a guild of 6 on the simple classification reducing to 4 on the complex classification. Asterisks mark alterations, other entries are additional to those given in Shorrocks & Rosewell (1986)

Guild size	Location	Habitat	Reference
3 → (3)	Portugal	Fungi	Rocha Pité & Brandão Ribeiro (1985)
4 → (3)	Japan, Kiyosumi	Sap	Toda (1984)
6 → (4)	Japan, Kiyosumi	Leaves	Toda (1984)
6 → (5)	Finland	Fungi	Hackman & Meinander (1979)*
8 → (3)	Japan, Kiyosumi	Fungi	Toda (1984)
9 → (9)	Japan, Morioka	Leaves	Toda (1984)
11 → (10)	Japan, Kiyosumi	Fruit	Toda (1984)
12 → (6)	Japan, Morioka	Sap	Toda (1984)
13 → (13)	Japan, Morioka	Fruit	Toda (1984)
15 → (8+3)	Japan, Hokkaido	Decaying plant material	Kimura *et al.* (1977), Toda, Kimura & Enomoto (1984)*
18 → (7)	Japan, Morioka	Fungi	Toda (1984)
26 → (11+10+3+2)	Panama, Cerro Campana	Flowers	Pipkin, Rodriguez & Leon (1966)*

fermenting fruits, sap fluxes, decaying plant material (such as leaves, stems and roots), fungi and flowers. In the main these have been used to obtain the guild sizes, unless authors specifically stated that some other classification should be used. However, we have estimated guild size twice. The 'simple' estimate was obtained by just counting the species recorded from a particular resource type. The 'complex' estimate was obtained by examining, where possible, the actual breeding records in detail and determining if the simple groups could be split or reduced because species had never actually been recorded sympatrically. Also, records for species represented by single individuals only were removed.

The frequency distributions of both the estimates of guild size in wild drosophilids are shown in Fig. 2.9. Both distributions are in fact rather similar, with modal guild size about seven in both cases. Examination of the guild size for the different substrates used by the larvae fails to suggest any difference in average size. Mean guild size ($\pm$SE) for the 'complex' data gives, flowers (5.6$\pm$0.9), fungi (6.3$\pm$0.7), sap (6.7$\pm$0.9), fruit (7.5$\pm$1.2), and rotten vegetation, mainly leaves (8.5$\pm$1.4). It is also not possible on the basis of this data to suggest any difference in guild size between temperate, tropical or Hawaiian drosophilids ('simple' guilds, $\chi^2 = 1.27$, $n = 4$, $P = 0.867$; 'complex' guilds, $\chi^2 = 6.22$, $n = 4$, $P = 0.182$).

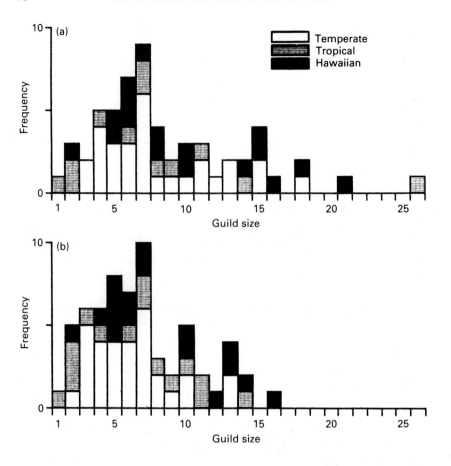

FIG. 2.9. Frequency histogram of natural guild sizes in drosophilids using (a) 'simple' and (b) 'complex' classifications (see text for explanation of these terms). Blank: temperate; stipple: tropical; solid: Hawaiian. (Shorrocks & Rosewell 1986, plus additional material.)

Probability refuges—a sufficient condition for guilds

The simulation model suggests that drosophilids should commonly coexist in guilds of between five and fifteen species. Natural guild sizes cover a similar but possibly smaller range. It should be noted that the simulation carried out with what we believe to be the most realistic parameters for drosophilids ($\lambda = 5$, $N^* = 10$ and b between 1 and 5) fits the field data extremely well. However, the precision of this pleasing result is less important in its details, which may alter with more data, than in its general implications. *The model* *predicts guilds of moderate size without any resource partitioning.* Traditional niche separation may be present but it is not a necessary condition for coexistence within these systems.

CONCLUSIONS

Although we have only talked about drosophilid flies, it would seem extremely likely that a similar mechanism might account for the rich diversity of species frequently associated with other 'centres of action'.

For example, the experiments of Kneidel (1985) demonstrate that patchiness and aggregation have similar effects in carrion systems. Using two carrion breeding Diptera, *Fannia howardi* and *Megaselia scalaris*, he was able to show that increased patchiness would prolong coexistence. Adult female flies were then allowed to distribute eggs over three 2 g sections of pork kidney placed in a large population cage (low patchiness) and over twelve 0.5 g sections (high patchiness). Interspecific competition, assessed in terms of the survival of the inferior species *F. howardi*, was shown to be an important mortality factor for *F. howardi* at low patchiness but not at high. Both species showed a strong tendency to aggregate their eggs and did so independently of one another. These results are consistent with the assumptions and predictions of the present model. Under high patchiness, where aggregation was high and the species were distributed independently, overlap was reduced, the effect of interspecific competition was reduced and the level of intraspecific competition was increased.

Hanski (1981, 1987) suggests similar effects on both carrion and dung-living arthropods, reporting that both dung beetles and carrion flies show independently clumped distributions with little positive association between emerging adults. Large samples of the local fly community in southern Finland (Hanski 1987) were enclosed in large outdoor cages and the larvae fed with 50 g of liver once or twice a week. The liver was cut into 1, 2, 4, 8 or 16 units depending upon the cage, and each treatment was replicated three times. Although under these confined conditions nine rare species were lost in all cages during the first two summers, two initially uncommon *Sarcophaga* species became quite abundant and coexisted with the dominant competitor *Lucilia illustris*. Significantly, the cages in which the *Sarcophaga* species were most successful were those in which food was divided into 8 and 16 units. *Sarcophaga* females lay first instar larvae instead of large clusters of eggs like *Lucilia*. They are thus well adapted to exploit a resource that varies spatially and do well on those pieces of liver that by chance receive the smallest number of *Lucilia* eggs which would otherwise outcompete them. In another experiment (Kouki & Hanski unpublished data) small plastic pots containing soil and 50 g of liver were left in the field for 3 days. After this oviposition period, the liver units were removed to closed containers which were kept in the field until adult flies emerged. The liver was either kept in single 50 g units (control) or grouped randomly into larger units of 100 g or 200 g. Removing patchiness (larger units) caused a marked decline in the number of emerging species per unit compared with the controls.

There are of course other mechanisms that may promote coexistence on patchy and ephemeral resources, in particular, resource partitioning (Atkinson & Shorrocks 1977; Shorrocks & Charlesworth 1980; Atkinson 1981). However, we believe that we have presented here a model, along with laboratory and field data, that provides a novel mechanism for coexistence in drosophilid flies, and probably in many other groups using patchy and ephemeral resources. As Charles Elton first intimated nearly 40 years ago, ideas about population interactions based upon conceptions of averages may be quite inappropriate (Lomnicki 1980; Begon 1982; Hassell & May 1985). 'Centres of action' are undoubtedly an important facet of many communities and we hope that probability refuges will receive due consideration in future studies of such systems.

ACKNOWLEDGMENTS

We would like to thank Will Atkinson, Mike Begon, Kathy Edwards, Illka Hanski, Mike Hassell, Tony Ives, Bob May, Joe Perry and Roy Taylor for their helpful discussions. This work was supported by Natural Environment Research Council grants GR3/4736 and GR3/5464 to B. Shorrocks and GT4/80/TLS/23 to J. Rosewell.

REFERENCES

Atkinson, W.D. (1979). A field investigation of larval competition in domestic *Drosophila*. Journal of Animal Ecology, **48**, 91–102.

Atkinson, W.D. (1981). An ecological interaction between citrus fruit, *Penicillium* molds and *Drosophila immigrans* Sturtevant (Diptera: Drosophilidae). *Ecological Entomology*, **6**, 339–44.

Atkinson, W.D. (1985). Coexistence of Australian rainforest Diptera breeding in fallen fruit. *Journal of Animal Ecology*, **54**, 507–18.

Atkinson, W.D. & Shorrocks, B. (1977). Breeding site specificity in the domestic species of *Drosophila*. Oecologia (Berlin), **29**, 223–32.

Atkinson, W.D. & Shorrocks, B. (1981). Competition on a divided and ephemeral resource: a simulation model. *Journal of Animal Ecology*, **50**, 461–71.

Atkinson, W.D. & Shorrocks, B. (1984). Aggregation of larval Diptera over discrete and ephemeral breeding sites: the implications for coexistence. *American Naturalist*, **124**, 336–51.

Ayala, F.J. (1969). Experimental invalidation of the principle of competitive exclusion. *Nature (London)*, **224**, 1076–9.

Ayala, F.J. (1970). Competition, coexistence and evolution. *Essays in Evolution and Genetics in Honour of Theodosius Dobzhansky* (Ed. by M.K. Hecht & W.C. Steere), pp. 121–58. Appleton-Century-Crofts, New York.

Ayala, F.J. (1972). Competition between species. *American Scientist,* **60,** 348–57.

Ayala, F.J., Gilpin, E.M. & Ehrenfeld, J.G. (1973). Competition between species: theoretical models and experimental tests. *Theoretical Population Biology,* **4,** 331–56.

Begon, M. (1978). Population densities in *Drosophila obscura* Fallen and *D. subobscura* Collin. *Ecological Entomology,* **3,** 1–12.

Begon, M. (1982). Density and individual fitness: asymmetric competition. *Evolutionary Ecology* (Ed. by B. Shorrocks), pp. 175–94. Blackwell Scientific Publications, Oxford.

Begon, M., Krimbas, C.B. & Loukas, M. (1980). The genetics of *Drosophila subobscura* populations. XV. Effective size of a natural population estimated by three independent methods. *Heredity,* **45,** 335–50.

Begon, M., Milburn, O. & Turner, D. (1975). Density estimates of *Drosophila* in southern England. *Journal of Natural History,* **9,** 315–20.

Begon, M. & Shorrocks, B. (1978). The feeding and breeding sites of *Drosophila obscura* Fallen and *D. subobscura* Collin. *Journal of Natural History,* **12,** 137–51.

Carson, H.L. (1971). The ecology of *Drosophila* breeding sites. *Arboretum lecture No. 2* (Ed. by H.L. Lyon), pp. 1–27. University of Hawaii, Honolulu.

Cochran, W.G. (1952). The χ^2 test of goodness of fit. *Annals of Mathematical Statistics,* **23,** 315–45.

Cochran, W.G. (1954). Some methods for strengthening the common χ^2 tests. *Biometrics,* **10,** 417–51.

Cohen, J. (1970). *Statistical Power Analysis for the Behavioural Sciences,* 2nd edn. Academic Press, New York.

Conover, W.J. (1980). *Practical Non-Parametric Statistics,* 2nd edn. Wiley, New York.

Crumpacker, D.W. (1974). The use of micronized fluorescent dusts to mark adult *Drosophila pseudoobscura. American Midland Naturalist,* **91,** 118–29.

DeBenedictis, P.A. (1977). The meaning and measurement of frequency dependent competition. *Ecology,* **58,** 158–66.

Elton, C. (1949). Population interspersion: an essay on animal community patterns. *Journal of Animal Ecology,* **37,** 1–23.

Fisher, R.A. (1970). *Statistical Methods for Research Workers,* 14th edn. Oliver & Boyd, Edinburgh.

Fisher, R.A. & Ford, E.B. (1947). The spread of a gene in natural conditions in a colony of the moth *Panaxia dominula (L.). Heredity,* **1,** 143–74.

Grimaldi, D. & Jaenike, J. (1984). Competition in natural populations of mycophagous *Drosophila. Ecology,* **65,** 1113–20.

Hackman, W. & Meinander, M. (1979). Diptera feeding as larvae on macrofungi in Finland. *Annales Zoologici Fennici,* **16,** 50–83.

Hanski, I. (1981). Coexistence of competitors in patchy environments with and without predation. *Oikos,* **38,** 210–21.

Hanski, I. (1987). Colonization of ephemeral habitats. *Colonization, Succession and Stability* (Ed. by A.J. Gray, M.J. Crawley & P.J. Edwards), pp. 155–86. Symposium of the British Ecological Society, 26. Blackwell Scientific Publications, Oxford.

Hassell, M.P. (1975). Density dependence in single-species populations. *Journal of Animal Ecology,* **44,** 283–95.

Hassell, M.P. & Comins, H.S. (1976). Discrete time models for two-species competition. *Theoretical Population Biology,* **9,** 202–21.

Hassell, M.P. & May, R.M. (1985). From individual behaviour to population dynamics. *Behavioural Ecology: Ecological Consequences of Adaptive Behaviour* (Ed. by R.M. Sibly

& R.H. Smith). Symposium of the British Ecological Society, **25**, 3–32. Blackwell Scientific Publications, Oxford.

Hummel, H.K., van Delden, W. & Drent, R.H. (1979). Estimation of some population parameters of *Drosophila limbata* V. Roser in a greenhouse. *Oecologia (Berlin)*, **41**, 135–43.

Ives, A.R. & May, R.M. (1985). Competition within and between species in a patchy environment: relations between microscopic and macroscopic models. *Journal of Theoretical Biology*, **115**, 65–92.

Jackson, C.H.N. (1937). Some new methods in the study of *Glossina morsitans*. *Proceedings of the Zoological Society of London*, **1936**, 811–96.

Kambysellis, M.P., Starmer, T., Smathers, G. & Heed, W.B. (1980). Studies of oogenesis in natural populations of Drosophilidae. II. Significance of microclimate changes on oogenesis of *Drosophila mimica*. *American Naturalist*, **115**, 67–91.

Kearney, J.N. (1979). *The breeding site ecology of three species of woodland* Drosophila. Unpublished Ph.D. thesis, University of Leeds.

Kimura, M.T., Toda, M.J., Beppu, K. & Watabe, H. (1977). Breeding sites of drosophilid flies in and near Sapporo, northern Japan, with supplementary notes on adult feeding habits. *Kontyu (Tokyo)*, **45**, 571–82.

Kneidel, K.A. (1985). Patchiness, aggregation, and the coexistence of competitors for ephemeral resources. *Ecological Entomology*, **10**, 441–8.

Lomnicki, A. (1980). Regulation of population density due to individual differences and patchy environment. *Oikos*, **35**, 185–93.

McGuire, J.U., Brindley, T.A. & Bancroft, T.A. (1957). The distribution of the European corn borer larvae *Pyrausta nubilalis* (Hbn.) in field corn. *Biometrics*, **13**, 65–78.

Miller, R.S. (1964). Larval competition in *Drosophila melanogaster* and *D. simulans*. *Ecology*, **45**, 132–48.

Nass, C.A.G. (1959). The χ^2 test for small expectations in contingency tables, with special reference to accidents and absenteeism. *Biometrika*, **46**, 365–85.

Nicholson, A.J. (1954). An outline of the population dynamics of animal populations. *Australian Journal of Zoology*, **2**, 9–65.

Perry, J.N. & Taylor, L.R. (1985). Adès: new ecological families of species-specific frequency distributions that describe repeated spatial samples with an intrinsic power-law variance-mean property. *Journal of Animal Ecology*, **54**, 931–53.

Pielou, E.C. (1977). *Mathematical Ecology.* Wiley, New York.

Pipkin, S.B., Rodriguez, R.L. & Leon, J. (1966). Plant host specificity among flower-feeding neotropical *Drosophila* (Diptera: Drosophilidae). *American Naturalist*, **100**, 135–56.

Rocha Pité, M.T. & Brandão Ribeiro, M.E. (1985). A preliminary note on Portugese fungal breeding Drosophilidae (Insecta, Diptera). *Boletim da Sociedade Portugesa de Entomologia (Supplemento 1, Actas do II Congresso Ibérico de Entomologia)*, **1**, 189–99.

Shorrocks, B. (1977). An ecological classification of European *Drosophila* species. *Oecologia (Berlin)*, **26**, 335–45.

Shorrocks, B. (1982). The breeding sites of temperate woodland *Drosophila*. *The Genetics and Biology of Drosophila, Vol. 3b*. (Ed. by M. Ashburner, H.L. Carson & J.N. Thompson Jr), pp. 385–428. Academic Press, London.

Shorrocks, B., Atkinson, W.D. & Charlesworth, P. (1979). Competition on a divided and ephemeral resource. *Journal of Animal Ecology*, **48**, 899–908.

Shorrocks, B. & Charlesworth, P. (1980). The distribution and abundance of the British fungal-breeding *Drosophila*. *Ecological Entomology*, **5**, 61–78.

Shorrocks, B. & Nigro, L. (1981). Microdistribution and habitat selection in *Drosophila subobscura* Collin. *Biological Journal of the Linnean Society*, **16**, 293–301.

Shorrocks, B. & Rosewell, J. (1986). Guild size in drosophilids: a simulation model. *Journal of Animal Ecology*, 55, 527–41.

Shorrocks, B., Rosewell, J., Edwards, K. & Atkinson, W.D. (1984). Interspecific competition is not a major organizing force in many insect communities. *Nature (London)*, 310, 310–12.

Taylor, L.R. (1961). Aggregation, variance and the mean. *Nature (London)*, 189, 732–5.

Taylor, L.R., Woiwod, I.P. & Perry, J.N. (1978). The density dependence of spatial behaviour and the rarity of randomness. *Journal of Animal Ecology*, 47, 383–406.

Taylor, L.R., Woiwod, I.P. & Perry, J.N. (1979). The negative binomial as a dynamic ecological model for aggregation, and the density dependence of k. *Journal of Animal Ecology*, 48, 289–304.

Toda, M.J. (1984). Guild structure and its comparison between two local drosophilid communities. *Physiology and Ecology (Japan)*, 21, 131–72.

Toda, M.J., Kimura, M.T. & Enomoto, O. (1984). Bionomics of Drosophilidae (Diptera) in Hokkaido. VI. Decayed herbage feeders, with special reference to their reproductive strategies. *Japanese Journal of Ecology*, 34, 253–70.

3. SPECIES–ABUNDANCE PATTERNS

JOHN S. GRAY

*Biology Institute, Section of Marine Biology and Marine Chemistry,
University of Oslo, PB 1064, 0316 Blindern, Oslo 3, Norway*

INTRODUCTION

Community ecologists have sought common patterns in their data sets for decades. One of the prevailing themes has been to examine for patterns in the distribution of individuals amongst species (i.e. species–abundance patterns) in the hope that exact quantitative models could be fitted which would be of general applicability and would help in our understanding of community organization. The models that have been applied have been reviewed by Pielou (1975), May (1975), Engen (1978) and Frontier (1985) and are listed in Table 3.1. The four commonest models are the geometric series of Motomura (1932), the broken-stick model of MacArthur (1957), the log series model of Fisher (in Fisher, Corbet & Williams 1943) and the lognormal model of Preston (1948). The Zipf model was not covered by Pielou or May but was treated by Engen (1978) and modified by Frontier (1985) to include arguments from Mandelbrot (1982) as the Zipf–Mandelbrot model. The generality of the gamma distribution and its relation to the negative binomial (which has also been used to describe species–abundance patterns) and log series models have been treated by Kempton & Taylor (1974) and Engen (1978).

TABLE 3.1. Recent reviews of models of the distribution of individuals among species (* = model treated in paper)

Model	Author			
	Pielou (1975)	May (1975)	Engen (1978)	Frontier (1985)
Geometric series	*	*	*	*
Broken-stick	*	*	*	*
Zipf–Mandelbrot			*	*
Negative binomial	*		*	*
Log series	*	*	*	*
Lognormal	*	*	*	*

The classification of models is not uniform from author to author. Engen (1978) regards the broken-stick as a specialized form of the negative binomial with k of the negative binomial equal to 1. The negative binomial is treated as the discrete analogue of the more general gamma distribution by both Pielou (1975) and Engen (1978). The log series is regarded as a special case of the negative binomial with k equal to 0 by Pielou (1975) and Elliott (1971), although Engen (1978) states that the lower bound for the log series is $k = -1$.

The ecological theories on which the models are built often have rather tenuous links to ecological reality and there are many mathematical and statistical problems associated with model fitting to ecological data. In this chapter the most often used models are examined, and where applicable, the ecological theories on which they are based are discussed and examples of the application of the model to ecological data are given.

MODELS OF THE DISTRIBUTION OF INDIVIDUALS AMONGST SPECIES

It is not my intention to review in detail the models that have been proposed nor to cover the mathematical bases for the models. These aspects are well covered in the reviews cited in Table 3.1.

Two models, the geometric series and the broken-stick, have their theoretical base in ecology, whereas the remaining models were originally proposed as purely statistical descriptors, although some attempts have been made to apply biological explanations to the distributions.

The geometric series (Motomura 1932)

In this model it is assumed that the first species colonizing an area appropriates a fraction of the resource and by competitive interaction preempts that fraction. The second colonizing species preempts a similar fraction of the remaining resource and so on with further colonists. Assuming the abundance of species is proportional to the respective fractions of resource preempted, a geometric series of frequencies of species in the community is obtained. The data are usually plotted as ranks of species on the abscissa and abundance on the ordinate. However, if logarithms of abundance are used a straight line is obtained. Fig. 3.1a shows a typical plot.

This model, when proposed, was thought to account for successional processes and to be an acceptable model of how a limiting resource (usually space) was proportioned among species. Fits to this model have been found for plants from a sub-alpine fir forest community (Whittaker 1975), and

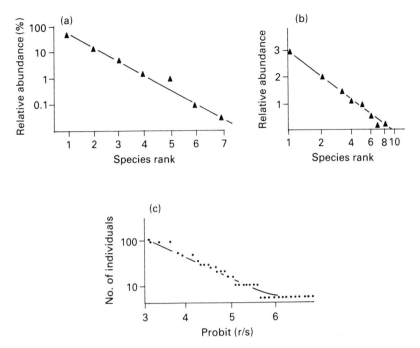

FIG. 3.1. Species–abundance models: (a) the geometric series, data on marine benthos from a heavily polluted part of Oslofjord (Gray 1981). (b) The broken-stick model, data on ophiuroids from Eniewetak atoll (King 1964). (c) The Zipf–Mandelbrot model; data on terrestrial insects from Brunel 1983 in Frontier (1985), (where (r/s) is the ranked cumulative frequency in probits).

Gray (1981) has shown a good fit to benthos of a polluted fjord (Fig. 3.1a). In general the model is only found to fit communities poor in species such as those of the earliest successional stages or under extreme pollution; in both cases dominance is extremely high. Furthermore, the ecological realism of the model is rather dubious, in that it assumes that each species appropriates a similar fraction of the resource available to it.

The broken-stick model (MacArthur 1957)

This model has a similar postulate to the geometric series. Resources (the stick) are divided at the same time and at random into segments over the whole stick length. The segments are ranked into decreasing length order and the theoretical distribution of lengths can be calculated. MacArthur argued that each segment represented an ecological niche where no two species occupy the same niche space, and that all the niches are occupied.

The abundance of each species is assumed to be proportional to the length of the stick segment representing its niche. Pielou (1975) has pointed out that 'a census of individuals in a single community can never provide evidence either for or against the broken-stick model' as the model predicts an average over several communities.

The model is also unclear on whether or not niche space is divided amongst species according to local competitive interactions or has arisen by evolutionary adaptation to particular tolerance ranges over time and is thus predetermined.

Fits to this model have been found with birds (MacArthur 1960), minnow species and ophiuroids (King 1964), littoral gastropods (Kohn 1969) and microcrustaceans deposited in lake-bed sediments (Goulden 1969; Deevey 1969 and Tsukada 1972). However, the model in general has been abandoned as being unrealistic, even by MacArthur (1966). Fig. 3.1b shows a typical fit to the model.

Recently De Vita (1979) has suggested that rather than the broken-stick model being proportional to relative abundance, segment lengths are proportional to niche dimensions. Empirical data from stem-boring insects, birds and gastropods were found to be in good agreement with the revised model. Thus there may be a case for re-examining the broken-stick model.

The Zipf–Mandelbrot model (Zipf 1949; Mandelbrot 1977)

Zipf's original model was applied to linguistic and socio-economic data (1949, 1965), but modified by Mandelbrot (1977, 1982) to relate information theory and ecology using 'fractal structure', which is currently being examined in a variety of ecological contexts (see Reichelt & Bradbury 1984).

Mandelbrot applied the model to the distribution of words within languages. The analogy of the distribution of individuals among species is clear. The proposed ecological explanation for the model relates to successional processes (see Frontier 1985). Initial colonizers are held to have few ecological requirements for them to be able to colonize and reproduce, and can therefore become common. Later colonists have more specific requirements; often the presence of previous colonists is essential, and such species are accordingly rarer than initial colonists. The model describes the expected frequencies of the species in terms of their overall 'costs'. It is assumed that the greater the number of preconditions that a species requires before it can colonize, the greater is the 'cost' to the species. Costs include the energetic costs of colonizing, time lags necessary before colonization and information content required on previous settlers which, although not quantitatively included in the model, are assumed in the model's structure (see Frontier 1985 for a more detailed explanation). Frontier (1985) gives many examples of the application of this model in ecology. Figure 3.1c shows a fit to data on terrestrial insects. The postulate that successional processes always follow repeatable patterns for given communities is perhaps unrealistic and in most cases the model gives poor fits to the rare species (see plots in Frontier 1985).

The remaining three models were originally suggested as statistical descriptors of the distribution of individuals amongst species and need not necessarily have any implied biological base.

The negative binomial (and gamma) distribution (Arrhenius 1922; Gleason 1922)

The negative binomial model is usually fitted to the spatial distribution of individuals in a population, but has been applied to the distribution of individuals amongst species by Arrhenius (1922), Gleason (1922) and Brian (1953). The model is closely related to the gamma distribution, and Pielou (1975) and Engen (1978) should be consulted for the mathematical background. Figure 3.2 shows the negative binomial fitted to data on bird species in the US (Brian 1953).

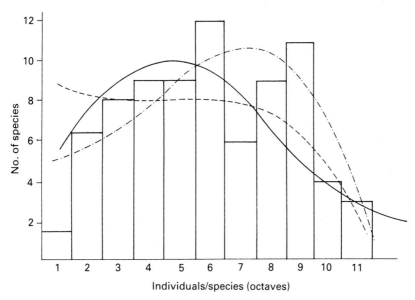

FIG. 3.2. Fits of species abundance models to data on nesting birds in Quaker Run Valley, USA; data from Brian 1953. (Solid line: lognormal; broken line: log series; broken line with dots: negative binomial; octaves $\log_2$ abundance range, e.g. 1 = 1 individual/spp., 2 = 2–3 individuals/spp., 3 = 4–7 individuals/spp., etc.)

This model is entirely statistical and is not based on ecological theory. The fitting of the model thus quantifies the pattern of individuals amongst species and can summarize the data in a few parameters. If the exercise proves repeatable, the parameters can be used to compare different communities. This is a property common to all statistical models.

The log series model (Fisher 1943)

R. A. Fisher (in Fisher, Corbett & Williams 1943) suggested that plots of the distribution of individuals against the number of species followed a decreasing logarithmic series. In this distribution the number of species represented by one individual is always the largest class. The distribution has been fitted to a wide range of ecological data (see Williams 1964 for a summary). Figure 3.2 shows the log series applied to bird data. This distribution is a statistical one and no theory has been proposed to explain the pattern observed. One of the parameters of the model is widely used as a diversity index, but the model is seldom fitted when applying the diversity index.

The lognormal model (Preston 1948)

Preston (1948) observed that in many data sets the number of species represented by one individual was not the most abundant frequency class. He proposed another model, the truncated lognormal, in which plots of individuals per species, on a geometric scale, against the number of species followed a normal distribution but were truncated to the left of the mode. The truncation was explained as being due to species that were present in the habitat but not present in the sample; if larger samples were to be taken, more species would be obtained and the mode would move to the right. He illustrated this point by plotting the distribution for cumulative samples taken at 3, 6, 9 and 12-month intervals. Preston (1962) provided a large number of data sets that fitted the lognormal model. Figure 3.2 shows a typical data example (from Brian 1953). The lognormal is probably the most widely applied of all species–abundance distributions.

A normal distribution can be transformed to a straight line by plotting cumulative percentage species against individuals per species in geometric classes on probability paper (or transformed to probits) (Williams 1964; Gray & Mirza 1979). A straight line indicates a fit to the lognormal model, but care must be taken, as the transformation assumes a complete normal distribution, whereas the data is always truncated to a greater or lesser degree. A more exact model can be fitted by using the Gaussfit programme

(Gauch & Chase 1975) or the maximum-likelihood method (Slocomb, Stauffer & Dickson 1977).

There are three main ecological arguments proposed to explain the fit to the lognormal model. The first is a hypothesis related to the broken-stick model, but the stick is broken sequentially rather than instantaneously (Pielou 1975). The stick is first broken at random into two parts, and then one part is chosen at random and broken again giving three parts; one of the three is chosen at random and broken again and so on. After a large number of breaks the product is a lognormal distribution of lengths. Sugihara (1980) has developed an ecological analogue, the minimum community structure model, where the lengths of the sequentially broken stick are regarded as niche size, which may be apportioned on several different sets of axes. The second hypothesis (see Pielou 1975; May 1975) is that populations of species grow logarithmically not arithmetically and the product of mixing many species is a lognormal distribution of individuals among species. The third hypothesis (see Pielou 1975; May 1975) is that individuals of any species are influenced by a wide range of randomly acting environmental factors. From the central limit theorem the product of the effects of a large number of random factors acting on the individuals and species gives a lognormal distribution. Few attempts have been made to distinguish between these ecological arguments. Rather, the lognormal has been used purely as a statistical descriptor.

Ugland & Gray (1982) give another hypothesis for the lognormal distribution. They hypothesize that the lognormal distribution results from the summation of three or more underlying symmetrical groups of species. The first group contains many species which are, over ecological time, consistently rare. The second group contains a smaller number of moderately common species and the third group an even smaller number of species with high abundances. Figure 3.3 illustrates the model showing how the three symmetrical (normal) distributions sum to give a truncated lognormal. This pattern, Ugland & Gray maintain, represents an undisturbed community at an equilibrium state. The equilibrium is, however, dynamic; increases in abundance of some species are compensated by decreases in abundance of others and the pattern maintained. In a marine benthic community over 2 years the pattern remained constant despite the fact that the species that dominated varied greatly (Gray, Valderhaug & Ugland 1985).

Data from undisturbed communities (Fig. 3.4) supports the suggestion that there may be three or more underlying groups of species. An analysis of the data in Fig. 3.4 using probability paper reveals that at least three underlying normal distributions can be discerned. A test that would permit rejection of the multiple group model would be to suddenly disturb the

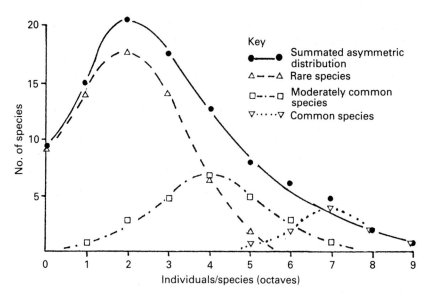

FIG. 3.3. Asymmetrical model of lognormal distribution of Ugland & Gray (1982) for undisturbed communities. The model comprises three symmetrical groups which summate to give the asymmetric model. Octaves as in Fig. 3.2.

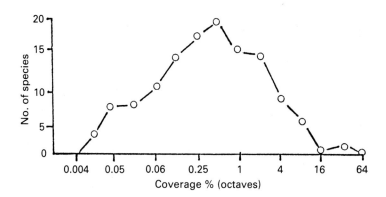

FIG. 3.4. Species–abundance pattern of Sonoran plant species (from Whittaker 1965) suggesting possible underlying symmetric models.

equilibrium state by adding resources (organic matter). The multiple group model predicts that a few opportunistic species will increase greatly in abundance, some additional species adapted to the new conditions will increase, whereas less well adapted species will decrease in abundance. The groups of species will move apart and become more apparent. Yet species constituting the new groups may be different to those in the original groups in the undisturbed assemblage. The hypothesis is simply that the response to disturbance varies among groups of species which may originally be in different abundance classes. If the single group model is correct then the response will not be by groups but by a continuum of species, and only a single group will be observed. Figure 3.5a,b shows the effect of disturbance on a freshwater community and Fig. 3.5c,d,e shows the response of a marine benthic community to progressive enrichment on moving from an undisturbed habitat to a habitat with organic enrichment. The multiple group model cannot be rejected. Pearson, Gray & Johannessen (1984) give many similar examples.

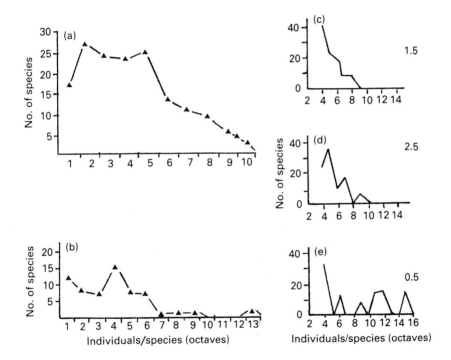

FIG. 3.5. Effect of disturbance on species–abundance patterns. (a) Diatoms on slides from an undisturbed riverine community, Ridley Creek (from Patrick 1954). (b) Diatoms on slides from a polluted river, Lititz Creek (from Patrick 1954). (c)–(e) Effect of gradient of organic enrichment on marine benthic communities (from Pearson, Gray & Johannesen 1984): (c) 6 km from pollutant source, (d) 2.5 km from source, (e) 0.5 km from source.

The fitting of normal distributions to the underlying groups is difficult as the number of data points and hence degrees of freedom for statistical goodness-of-fit tests is low. By simply plotting number of species against individuals per species the multiple group hypothesis appears to be preferable to the single group model.

FITTING THE MODELS TO FIELD DATA

Fitting any of the models described above to data is fraught with problems. Daget (1976) states that 'authors generally are satisfied with verifying the possible fit to their preferred model or, at most, trying to fit a few classical models in order to see which fits the best, but rarely using a statistical test'. Yet the use of a statistical test is itself problematic. Often it is not possible to distinguish between fits to two or more competing models within the limits of the chi-square goodness-of-fit test. Furthermore the chi-square test itself may not be the appropriate test to use in goodness-of-fit tests (see Engen 1978 for a fuller discussion of this point). Figure 3.2 shows that none of the models adequately fits the data. The general lack of statistical fit of models has been discussed by Gray (1979), Hughes (1984) and Lambshead & Platt (1985). Amanieu, Gonzalez & Guerlorget (1981) tried a novel approach to test the goodness-of-fit of the geometric series, lognormal and Zipf-Mandelbrot models to benthic marine data by computing the Hellinger distance (D) for ranks of samples where

$$D^2_{(i,j)} = \sum_{r=1}^{s} (\sqrt{p_{r_i}} - \sqrt{p_{r_j}})^2$$

and p_{r_i} and p_{r_j} are the proportions or the frequencies of the species of rank r in samples i and j. Amanieu, Gonzales & Guerlorget applied this statistic to a comparison of data from untransformed benthic samples with samples to which were fitted the three models mentioned above. They concluded from a principal components analysis of the Hellinger distance that for eleven separate benthic samples no model fitted the untransformed data well and that the analysis grouped each model tightly together with no overlap between models.

Further problems arise in interpreting the ecological aspects of data that do fit a given model. Much controversy has recently arisen over the interpretation of data that fit the log series and lognormal distributions. In his review, May (1975) suggested that 'equilibrium' communities usually fit the lognormal distribution and gave a number of examples of such fits. Gray (1979, 1980, 1981) has applied this idea to the marine benthos. It has been argued, however, that the log series fits data from undisturbed (i.e. equilibrium) communities whereas the lognormal fits data from disturbed communities (Kempton & Taylor 1974; Lambshead & Platt 1985; Hughes 1985). Yet in his review, May also stated that 'the lognormal abundance distributions in communities of opportunistic creatures reflect nothing about the structure of the community. Dominant species simply are those that recently enjoyed a large *r*'. Such patterns are found in diatom communities on slides (Patrick, Hohn & Wallace 1954; Patrick 1972). The lognormal is suggested, therefore, either to represent an assemblage of opportunists or an equilibrium community.

Similar problems have arisen in the interpretation of the biological meaning of fits to the log series distribution. Stenseth (1979) regards the log series as representing a perturbed community whereas Caswell (1976) suggests it represents a community in a 'neutral' state where biological interactions such as competition and predation are not important. These two arguments are not necessarily contradictory, but it remains unclear what ecological interpretation should be given to an acceptable fit to either the log series or lognormal distribution. In addition, the term equilibrium is often not defined (May 1975; Gray 1979, 1980, 1983; Ugland & Gray 1982) and disturbance not quantified (Kempton & Taylor 1974; Lambshead & Platt 1985; Hughes 1984, 1985). Thus, the ecological explanations erected to explain fits to the models are not testable experimentally because of lack of definition of terms employed. This criticism also applies to a recent model of species–abundance pattern proposed by Hughes (1984, 1985). This model simulates the distribution of individuals among species of marine benthic organisms based on assumptions concerning recruitment potential, number of established conspecifics, gregarious behaviour of larvae, and stochastic mortality. The model produces patterns which are similar to those of natural marine communities but it cannot be verified since few of its parameters have been or can be measured in the field.

MODELS OF SPECIES–ABUNDANCE PATTERNS AND SUCCESSION

Most evidence suggests that early successional stages fit the geometric series and later, as more species appear, the community develops through a log series or lognormal model and may then return to a geometric series at the climax (Whittaker 1975).

Trends similar to the successional sequence but in a reverse direction can be found along pollution gradients or following pollution-induced changes in assemblages over time. In such cases the assemblage changes from a complex, mature assemblage possibly following lognormal, log series or

Zipf–Mandelbrot distributions to a species-poor community fitting a geometric series under gross pollution (Gray 1978; Frontier 1985). Thus the fits to the models alone cannot be interpreted in a successional or retrogressional sequence; comparative data is needed over wide temporal or spatial scales on the changes in species–abundance patterns. Whittaker (1975) also gives a cautionary warning that samples from the same community can be obtained which appear to fit all models, simply by varying sample size.

SPECIES–ABUNDANCE PATTERNS WITHOUT MODELS

An alternative approach, which may have merit, is not to fit an exact mathematical model but simply to plot the number of species against the number of individuals among species on a geometric scale to see if there is consistency of pattern over time or along environmental gradients. Such an approach has been applied to the benthos of a subtidal area of Oslofjord. Over a period of 2 years (June 1981 to July 1983) the pattern of individuals among species remained constant despite the fact that the dominant species showed considerable variation (Gray, Valderhaug & Ugland 1985). An analysis of variability in the species groups showed that the rare species fluctuated more in abundance than the common and moderately common species, yet overall species number remained constant (95±4) over the 2 year period. The consistent pattern of the distribution of individuals among species suggests that increases in abundance of some species are compensated by decreases in others, and the pattern is maintained. This adjustment stability is in our opinion (Gray, Valderhaugh & Ugland 1985) the rule for marine soft-sediment benthos, and implies that competitive interactions are important in maintaining the overall pattern. The pattern remains constant; fits to an exact model would not further our ecological understanding.

Based on these patterns Gray & Pearson (1982) have proposed an objective method of identifying sensitive species, which can illustrate ecological gradients. In an undisturbed marine benthic community there are typically over 100 species. By selecting those species which occur in the moderately common group and plotting their occurrence over the gradient clear trends were observed over a variety of marine gradients (Pearson, Gray & Johannessen 1984). The moderately common group was defined from Preston's (1984) octaves as species having abundance of sixteen to sixty-three individuals per 0.1 m^2 (the sampling unit) in the undisturbed community.

Other methods which have been used are plots of dominance–diversity curves, e.g. the *k*-dominance plots used by Platt, Shaw & Lambshead (1984). Here the slope of the straight-line plots is used to distinguish differing community structure. The problem is that the abundance of the most dominant species largely dictates the slope of the line and thus the ecological information content relates almost exclusively to the most dominant species. Simple plots of number of species against geometric abundance classes may be preferable.

CONCLUSION

One can conclude that, in general, models of species–abundance patterns are poor fits to ecological data sets. There does not seem to be any single distribution that is universal. Frontier (1985) believes that this lack of success in fitting models suggests that the catalogue of models is as yet insufficient. I do not agree with this view and believe that the search for yet more models is unlikely to give any new insights into factors structuring biological assemblages.

Whittaker's (1975) summary of species–abundance models, 'the study of importance values has not produced the single mathematical form and choice . . . that the early work suggested might be possible' is one with which I agree. The consensus is probably that models of species–abundance distributions may be adequate as statistical descriptors. Engen (1978) suggests that model fitting is done 'to establish what the population structure is before putting too much effort into the more difficult task of explaining its evolution'. Certainly the change in species–abundance pattern from a geometric series to a lognormal or vice versa 'expresses something significant about the groups of species the curves represent' (Whittaker 1975, p. 94). If species–abundance patterns in undisturbed communities remain the same whilst dominance patterns change, as Ugland & Gray (1982) suggest, then it may be important to pay more attention to competitive and other interactions between the moderately common species rather than concentrating almost exclusively on the few dominant species that are the usual focus of attention of population dynamicists.

REFERENCES

Amanieu, M., Gonzalez, P.L. & Guerlorget, O. (1981). Criteres de choix d'un modèle de distribution d'abondances. Application à des communautés animales en écologie benthique. *Acta Oecologica–Oecologia Generalis*, 2, 265–86.

Arrhenius, O. (1922). Species and area. *Journal of Ecology*, 9, 95–9.

Brian, M.V. (1953). Species frequencies in random samples from animal populations. *Journal of Animal Ecology*, 22, 57–64.

Caswell, H. (1976). Community structure: a neutral model analysis. *Ecological Monographs,* **46,** 327–54.

Daget, J. (1976). *Les Modèles Mathématiques en Écologie.* Masson, Paris.

Deevey, E.S. Jr. (1969). Specific diversity in fossil assemblages. *Diversity and Stability in Ecological Systems* (Ed. by G.M. Woodwell & H.H. Smith), pp. 224–41. Brookhaven Symposium in Biology No. 22. US Department of Commerce, Springfield, Virginia.

De Vita, J. (1979). Niche separation and the Broken Stick Model. *American Naturalist,* **114,** 171–8.

Elliott, J.M. (1971). *Some Methods for the Statistical Analysis of Samples of Benthic Invertebrates.* Freshwater Biological Association. Scientific Publication No. 25.

Engen, S. (1978). *Stochastic Abundance Models with Emphasis on Biological Communities and Species Diversity.* Chapman & Hall, London.

Fisher, R.A., Corbet, A.S. & Williams, C.B. (1943). The relation between the number of species and the number of individuals in a random sample from an animal population. *Journal of Animal Ecology,* **12,** 42–58.

Frontier, S. (1985). Diversity and structure in aquatic ecosystems. *Oceanography and Marine Biology–an Annual Review,* 23 (Ed. by M. Barnes), pp. 253–312. Aberdeen University Press, Aberdeen.

Gauch, H.G. & Chase, G.B. (1975). Fitting the Gaussian curve in ecological applications. *Ecology,* **55,** 1377–81.

Goulden, C.E. (1969). Temporal changes in diversity. *Diversity and Stability in Ecological Systems* (Ed. by G.M. Woodwell & H.H. Smith), pp. 96–102. Brookhaven Symposium in Biology No. 22. US Department of Commerce, Springfield, Virginia.

Gleason, H. (1922). On the relation between species and area. *Ecology,* **3,** 156–62.

Gray, J.S. (1978). The structure of meiofauna communities. *Sarsia* **64,** 265–72.

Gray, J.S. (1979). Pollution-induced changes in populations. *Philosophical Transactions of the Royal Society, London, Series B,* **286,** 545–61.

Gray, J.S. (1980). The measurement of effects of pollutants on benthic communities. *Rapports Procès Verbaux des Réunions du Conseil International l'Exploration de la Mer* **179,** 188–93.

Gray, J.S. (1981). *The Ecology of Marine Sediments.* Cambridge University Press, Cambridge.

Gray, J.S. (1983). On the use and misuse of the lognormal plotting method for detection of effects of pollution. *Marine Ecology Progress Series,* **11,** 203–4.

Gray, J.S. & Mirza, F.B. (1979). A possible method for detecting pollution-induced disturbance on marine benthic communities. *Marine Pollution Bulletin,* **10,** 142–6.

Gray, J.S. & Pearson, T.H. (1982). Objective selection of sensitive species indicative of pollution-induced change in benthic communities. 1. Comparative methodology. *Marine Ecology Progress Series,* **9,** 111–19.

Gray, J.S., Valderhaug, V. & Ugland, K.I. (1985). The stability of a benthic community of soft sediment. *Proceedings of the 19th European Marine Biology Symposium, Plymouth, 1984* (Ed. by P.E. Gibbs), pp. 245–53. Cambridge University Press, Cambridge.

Hughes, R. (1984). A model of the structure and dynamics of benthic marine invertebrate communities. *Marine Ecology Progress Series,* **15,** 1–11.

Hughes, R. (1985). A hypothesis concerning the influence of competition and stress on the structure of benthic communities. *Proceedings of the 19th European Marine Biology Symposium, Plymouth, 1984* (Ed. by P.E. Gibbs), pp. 391–400. Cambridge University Press, Cambridge.

Kempton, R.A. & Taylor, L.R. (1974). Log-series and log-normal parameters as diversity discriminants for the Lepidoptera. *Journal of Animal Ecology,* **43,** 381–99.

King, C.E. (1964). Relative abundance of species and MacArthur's model. *Ecology,* **45,** 716–27.

Kohn, A.J. (1969). The ecology of the genus *Conus* in Hawaii. *Ecological Monographs*, **29**, 47–90.

Lambshead, J. & Platt, H.M. (1985). Structural patterns of marine benthic assemblages and their relationships with empirical statistical models. In *Proceedings of the 19th European Marine Biology Symposium, Plymouth, 1984* (Ed. by P.E. Gibbs), pp. 371–80. Cambridge University Press, Cambridge.

MacArthur, R.A. (1957). On the relative abundance of bird species. *Proceedings of the National Academy of Science, USA*, **43**, 293–5.

MacArthur, R.A. (1960). On the relative abundance of species. *American Naturalist*, **94**, 25–36.

MacArthur, R.A. (1966). A note on Mrs. Pielou's comments. *Ecology*, **47**, 1074.

Mandelbrot, B.B. (1977). *Fractals. Form, Chance and Dimension*. W.H. Freeman, San Francisco.

Mandelbrot, B.B. (1982). *The Fractal Geometry of Nature*. W.H. Freeman, San Francisco.

May, R.M. (1975). Patterns of species abundance and diversity. *Ecology and Evolution of Communities* (Ed. by M.L. Cody & J.M. Diamond), pp. 81–120. Belknap Press, Harvard.

Motomura, I. (1932). A statistical treatment of associations (in Japanese). *Zoological Magazine, Tokyo*, **44**, 379–83.

Patrick, R. (1972). Benthic communities in streams. *Transactions of the Connecticut Academy of Arts and Science*, **44**, 271–82.

Patrick, R., Hohn, M. & Wallace, J. (1954). A new method of determining the pattern of the diatom flora. *Notulae Natura Academy of Natural Sciences, Philadelphia*, **259**, 12 pp.

Pearson, T.H., Gray, J.S. & Johannessen, P. (1984). Objective selection of sensitive species indicative of pollution-induced changes in benthic communities. 2. Data analyses. *Marine Ecology Progress Series*, **12**, 237–55.

Pielou, E.C. (1975). *Ecological Diversity*. Wiley Interscience, New York.

Platt, H.M., Shaw, K.M. & Lambshead, P.J.D. (1984). Nematode species abundance patterns and their use in the detection of environmental perturbations. *Hydrobiologia*, **118**, 59–66.

Preston, F.W. (1948). The commonness and rarity of species. *Ecology*, **29**, 254–83.

Preston, F. (1962). The canonical distribution of commonness and rarity. *Ecology*, **43**, 410–32.

Reichelt, R.E. & Bradbury, R.H. (1984). Spatial patterns in coral reef benthos: multiscale analysis of sites from three oceans. *Marine Ecology Progress Series*, **17**, 251–7.

Slocomb, J., Stauffer, B. & Dickson, K.L. (1977). On fitting the truncated log-normal distribution to species–abundance data using maximum likelihood estimation. *Ecology*, **58**, 693–6.

Sugihara, G. (1980). Minimal community structure: an explanation of species abundance patterns. *American Naturalist*, **116**, 770–87.

Stenseth, N.C. (1979). Where have all the species gone? On the nature of extinction and the Red Queen hypothesis. *Oikos*, **33**, 196–227.

Tsukada, M. (1972). The history of Lake Nojori, Japan. *Transactions of the Connecticut Academy of Arts and Science*, **44**, 337–65.

Ugland, K.I. & Gray, J.S. (1982). Lognormal distributions and the concept of community equilibrium. *Oikos*, **39**, 171–8.

Whittaker, R.H. (1965). Dominance and diversity in land plant communities. *Science*, **147**, 250–60.

Whittaker, R.H. (1975). *Communities and Ecosystems*, 2nd edn. Macmillan, New York.

Williams, C.B. (1964). *Patterns in the Balance of Nature*. Academic Press, London.

Zipf, G.K. (1949). *Human Behaviour and the Principle of Least Effort*. Hafner, New York.

Zipf, G.K. (1965). *Human Behaviour and the Principle of Least Effort*. 2nd edn. Hafner, New York.

4. SEASONALITY AND THE COMMUNITY

HENK WOLDA

Smithsonian Tropical Research Institute, P.O. Box 2072, Balboa,
Republic of Panama

INTRODUCTION

Life cycles of plants and animals tend to be organized temporally. Between species, or between different populations of the same species, appreciable differences occur in the sequence of the different phases of the life cycle: moulting in birds can precede, coincide with, or follow nesting or breeding; leafing in plants may occur before, during or after flowering or fruiting. However, within a population the components of the life cycle follow each other in a certain sequence and each tends to occur at the same time of the year: they tend to be seasonal.

Descriptions of seasonality patterns are available for a large number of organisms in a great variety of habitats and climates. These patterns are set by the physiological and ecological constraints of the species concerned and by the interactions with the rest of the community and its environment.

In this chapter I will give a brief summary of observed seasonality patterns, and will discuss some of the consequences of seasonality for the species concerned and for the community as a whole. In order not to add to the existing confusion in ecological terminology, I will start with some definitions.

DEFINITIONS

A phenomenon is 'seasonal' if it, or its maximum expression, predictably occurs at roughly the same time of the year, each year it does occur. A phenomenon is 'aseasonal' or 'non-seasonal' if it is not seasonal. The 'phenology' of a phenomenon is the temporal distribution of that phenomenon over the year. The 'seasonality' of a phenomenon is its phenology and the degree to which it is seasonal. The term 'season' refers entirely to the phenomenon under discussion, as in 'mating season', 'breeding season' or 'growth season' and has no *a priori* connection with the conventional seasons such as 'spring' or 'rainy season'.

Some types of seasonality are exemplified in Fig. 4.1. Only types 3A, 3B and 3C are considered aseasonal. A phenomenon which occurs year-round, but has seasonal peaks, such as types 2A, 2B, and 2C, are, by the above

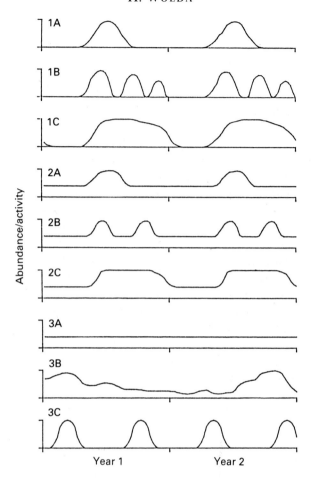

FIG. 4.1. A classification of types of seasonality patterns. 1A: Distinct, relatively short season. 1B: Multiple seasons or multiple generations per year. 1C: Distinct, broad, season. 2A: Year-round with a distinct seasonal peak. 2B: Year-round with multiple seasonal peaks. 2C: Year-round with a broad seasonal peak. 3A: Aseasonal, constant. 3B: Aseasonal, irregular. 3C: Aseasonal, synchronous subannual cycles; peaks occur at different times each year.

definition, seasonal, and so are phenomena which are not found every year, but which happen at the same time of year in the years in which they do occur. Examples of such patterns abound (e.g. Croat 1978; Wolda 1978; Wolda & Fisk 1981; Wolda & Galindo 1981).

The term 'community' is used here as 'the sum total of all living organisms in a given area', without reference to interactions, and without taxonomic restrictions.

THE PHYSICAL SETTING

One of the most salient seasonal features of the temperate zone is the changing temperature. Colder winters alternate with warmer summers, a cycle which is more pronounced in continental than in maritime climates, and whose amplitude increases with increasing latitude. Except for the high Arctic or the Antarctic, however, it is not temperature but rainfall which determines the physiognomy of the landscape. The combination of the total annual rainfall and its seasonality determines whether a given area is a desert, a savanna or a rainforest. Seasonality in rainfall is especially important if the dry season coincides with the warm season, such as in mediterranean climates. Temperature values below a given threshold prohibit most activities in organisms so that the 'growing season', that part of the year where the environment allows activity in the vast majority of organisms in a community, decreases in length with increasing latitude.

Towards lower latitudes there is less seasonal variation in temperature, and variations in rainfall become increasingly more important. In the tropics the seasonal variation in temperature is very small or absent and the year is usually characterized by an alternation of rainy and dry seasons, once or twice per year. In some areas there is little or no seasonality in either rainfall or in temperature, e.g. Fortuna in Panama (Wolda & Fisk 1981). However, even in Fortuna organisms do experience seasonal cues and many of them have highly seasonal activity patterns.

Seasonal variation in day-length, an important environmental factor for many organisms, decreases towards lower latitudes, yet some organisms are known to respond to slight changes in day-length (e.g. Denlinger 1986; Tanaka, Denlinger & Wolda 1987). To the best of my knowledge, areas that are completely devoid of seasonal cues, such as day-length, temperature or rainfall, do not exist.

SEASONAL PATTERNS

The temperate zone

Within the limits set by physical factors, a large variation in seasonal patterns can be found. Most species have short reproductive episodes, which tend to be staggered over the warm season (Malaisse 1974; Mooney, Parsons & Kummerow 1974; Taylor 1974; Wielgolaski 1974).

Relatively long reproductive seasons are not a regular feature of the temperate year. However, they do occur. The common crossbill *Loxia curvirostris* in northern Russia breeds from August through May (Newton

1973; Nethersole-Thompson 1975). Zebrafinches (*Poephila guttata*) have a very long breeding season in New South Wales (Kikkawa 1980). Some temperate rodents have very long reproductive seasons (Asdell 1964; Smith 1974). In Spain, male individuals of the parasitic shrub *Osyris quadripartita* flower and grow throughout much of the year (Herrera 1984). The bug *Oncopeltus fasciatus* in Florida (Miller & Dingle 1982), the collared peccari *Tayassu tajacu* in Arizona (Sowls 1984), the red kangaroo *Megaleia rufa*, and the grey kangaroo *Macropus giganteus* (Frith & Sharman 1964; Poole 1983) all breed around the year. In the Australian Mallee, most birds breed regularly and seasonally in spite of the dry conditions and the unpredictability of the rain, but some breed around the year (Schodde 1981). On the Bonin Islands (27°N) thirteen out of sixty-four plant species flush new leaves, and some species flower, around the year (Shimizu 1983).

Not all temperate organisms avoid activity in the cold season. Bears give birth in winter while hibernating (Ramsay & Dunbrack 1986). In the Netherlands the snowspringer *Boreus hyemalis* (Mecoptera) is active as an adult only, and the carabid *Bembidion nigricorne* mostly, during the winter (Den Boer 1967). Adults of the winter moth *Operophtera brumata* are active in late autumn and early winter (Holliday 1985). In parts of northern Europe the common crossbill *Loxia curvirostris* breeds in the winter (Newton 1973; Nethersole-Thompson 1975). Some plants start flowering while there is still snow on the ground. Cold weather makes life difficult for most species, but not all, as is dramatically demonstrated by a chironomid fly which lives in a glacier in the Himalayas (Kohshima 1984).

There may be more than one activity season each year. Some oak trees (*Quercus* sp.) usually produce new leaves twice per year. Birds tend to have one breeding season per year with one or more broods, but some have a second breeding season, separated from the first by a regression of the gonads. At temperate latitudes this 'breeding out of season' is rarely successful, but towards more subtropical latitudes it is (Thomson 1950; Payne 1969; Immelmann 1971; Ligon 1971; Murton & Westwood 1977).

Some vertebrates, in at least part of their breeding range, have two equally important breeding seasons, for instance the New Holland honeycreeper *Phylidonyris novaehollandiae* in some populations in Victoria, Australia (Paton 1985), the common crossbill *Loxia curvirostris* in the southern Ural (Newton 1973; Nethersole-Thompson 1975), the red-tailed black cockatoo *Calyptorhynchus magnificus* (Saunders 1977) and the tern *Sterna bergii* (Dunlop 1985) in south-western Australia and the lorisid primate *Galago senegalensis* in South Africa (Charles-Dominique 1977). Some species of *Viola* seem to have an occasional second season in autumn (Fernald 1950; Fournier 1961). Such multiple reproductive seasons are, however, extremely rare, and they tend to be concentrated at lower temperate latitudes. On the Bonin Islands (27°N) some plant species leaf or flower twice per year (Shimizu 1983).

Short-lived organisms, such as insects, often have more than one generation per year, but this is not always directly comparable to the multiple seasons just referred to. Real multiple seasons in insects are rare, but do occur: the carabid beetles *Nebria brevicollis* and *Patrobus atrorufus* in Europe have two activity periods, but only one reproductive period, per year (Thiele 1969).

The tropics

The tendency, within temperate areas, towards longer reproductive seasons and more frequent and successful multiple breeding seasons at lower latitudes, might lead one to expect that in the tropics breeding is year-round, often aseasonal, and that multiple breeding seasons are commonplace. Individuals might be expected to reproduce with a 'conventional periodicity' (Simmons 1967), that is, as often as physiologically possible, with no timing effect by the environment. This expectation is not fulfilled.

There is a vast literature on seasonal patterns in the tropics, but I will review only some of the literature on two groups of organisms: birds and plants.

Marine birds

Tropical oceanic islands seem to provide one of the most aseasonal environments imaginable for seabirds, and yet, at each of these sites, many species have well-defined breeding seasons, often long, sometimes rather short (Moreau 1950b; Dorward 1962; Stonehouse 1962; Ashmole 1965, 1971; Snow & Snow 1966; Nelson 1967; Harris 1969; Schreiber & Ashmole 1970; Diamond 1976; Murton & Westwood 1977). Some species do breed year-round (Stonehouse 1962; Nelson 1967; Harris 1969; Snow & Snow 1969). Individuals of such species may breed at intervals other than 12 months, varying, between species, from 8 to 13 months. Other species breed around the year, but have clear peaks of breeding activity at intervals less than one year (Snow 1965; Snow & Snow 1967; Harris 1969; Diamond 1976).

Most fascinating of all are the species which breed synchronously at subannual intervals. The classical example is the wideawake, or sooty tern, *Sterna fuscata*, on Ascension (Chapin & Wing 1959), but the phenomenon has since been found in other species and in other places (Dorward 1962;

Simmons 1967; Diamond 1976; Murton & Westwood 1977). The intervals between successive breeding episodes vary from 7 to 11 months. Some of these species show a pronounced geographic variation. For instance, *Sterna fuscata* has a subannual breeding cycle on Ascension (Chapin & Wing 1959), one breeding season per year in the Seychelles (Diamond 1976) and two breeding seasons per year on Christmas Island (Ashmole 1965; Schreiber & Ashmole 1970).

Landbirds

Virtually all tropical terrestrial birds have distinct breeding seasons, or well-defined seasonal breeding peaks. (Skutch 1950; Voous 1950; Davis 1953; Frith 1956; Snow 1964; Snow & Snow 1964; Lavery, Seton & Bravery 1968; Fogden 1972; Whitmore 1975; Britton 1978; Gaston, Mathew & Zacharias 1979; Langham 1980; Stiles 1980; Mader 1981; Serle 1981; Bell 1982a,b; Dyrcz 1983, 1984; Turner 1983; Hails 1984; Lenton 1984). There are some exceptions to this general picture: Moreau (1950a) in Zaïre, and Miller (1955) in Colombia found no distinct breeding seasons, Snow & Snow (1964) mention three species which breed and moult at any time of the year in Trinidad, and Fogden (1972) lists two species in Sarawak, the babbler *Stachyris erythroptera* and the spider hunter *Arachnothera longirostris*, which may have subannual cycles.

Multiple breeding seasons in the tropics are surprisingly rare. Medway (1962) found that the swiftlet *Collocalia esculenta* in Sarawak has three breeding seasons. The southern race of the housewren, *Troglodytes aedon*, in Panama has two breeding seasons (Freed, personal communication), and so have the chestnut-bellied starling *Spreo pulcher* in Nigeria (Wilkinson 1983), and the Andean sparrow *Zonotrichia capensis* in Colombia (Miller 1962) and in Costa Rica (Wolf 1969). Some babblers in India may also have two distinct breeding seasons (Gaston, Mathew & Zacharias 1979).

Plants

The seasonality of plants on Barro Colorado Island (BCI), Panama, is exceptionally well known (Croat 1969, 1975, 1978; Augspurger 1982, 1983, 1985; Foster 1982a,b, 1985). Most species have one leaf flushing, flowering and fruiting period per year, although the order in which these phenophases occur varies among species. Some species have a period of being leafless, others produce new leaves as soon as the old ones fall, and thus remain green. Some trees flower after new leaves are produced; some produce leaves after, or during, flowering. Most of the deciduous species drop their leaves in the dry season, but a few are deciduous in the rainy season.

A few species flush leaves or flower more than once a year; for example, *Quararibea asterolepis* flowers once, but flushes new leaves twice per year. A number of species flower twice a year and *Hirtella triandra* (Chrysobalan-aceae) has three flowering periods per year. A few trees flower synchronously once every other year (Foster 1982b), and one, *Tachigalia versicolor*, flowers only once, at the end of its lifetime (Foster 1977). Croat (1978) lists many species which flower or fruit year-round, and several of these, mostly herbs and lianas, have no seasonal peak at all. A frequency distribution of a number of seasonality patterns in these plants is given in Fig. 4.2.

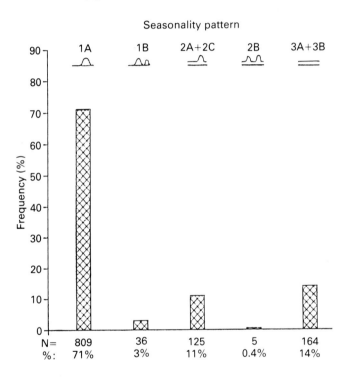

FIG. 4.2. Frequency distribution of seasonality patterns of flowering of plants on Barro Colorado Island, Panama, according to Croat (1978).

Throughout the tropics, a great diversity of seasonality patterns is found, with between site variation in the relative frequencies of the different seasonality type (e.g. Coster 1923, 1926; Davis 1945; Davis 1953; Snow

1964; Fogden 1972; Medway 1972; Frankie 1973; Frankie, Baker & Opler 1974; Whitmore 1975; Crome 1975; Hladik 1978; Jackson 1978; Stiles 1978; Struhsaker 1978; Cheke, Nanakorn & Yankoses 1979; Dressler 1981; Leighton & Leighton 1983; Savage & Ashton 1983; Wong 1983). All these studies demonstrate a clearly seasonal pattern in a large percentage of the species. The one exception to this general rule is a study by Putz (1979) who found that 66% of the species in a Malaysian forest 'flowered neither for long periods nor at regular intervals'. Only four of his species showed a clear seasonal pattern. Multiple flowering or leafing seasons are not uncommon (Coster 1923; Croat 1969, 1975, 1978; Whitmore 1975; Putz 1979; Borchert 1980, 1983). Related species with relatively short flowering seasons may be staggered over the year, good examples being *Miconia* in Trinidad (Snow 1964) and *Coelogyne* orchids in India (see Dressler 1981).

Fig trees (*Ficus* sp.) are especially fascinating. Fogden (1972) reports some fig trees in Sarawak to have several irregular fruiting episodes per year, and so do figs elsewhere (Coster 1923; McClure 1966; Crome 1975; Shimizu 1983). Milton *et al.* (1982) analyse seasonality data for a large number of fig trees on BCI, Panama, covering 7 years. Both *Ficus yoponensis* and *F. insipida* populations tend to have bimodal peaks in fruit production, but individuals have more or less regular subannual or supraannual fruiting cycles (Fig. 4.3). The average intervals between fruiting episodes vary from 19.5 to 93 weeks in *F. yoponensis* and from 28.7 to 107.3 weeks in *F. insipida* individuals. These cycles are reminiscent of the breeding cycles of the wideawake tern on Ascension. Subannual cycles, such as the ones found in figs, have also been found in some tree species in Malaysia (Putz 1979).

In summary

Reviews of other organisms, including fishes, frogs, reptiles, mammals, insects and other invertebrates, produce patterns similar to those described above. At any one site, however aseasonal the physical setting seems to be, some species may be aseasonal, or nearly so, others breed year-round but with a clear seasonal peak, and again others have one or more very distinct, sometimes surprisingly short, reproductive seasons. One cannot classify a given locality by choosing any one of the seasonality types shown in Fig. 4.1; rather, one needs to give a frequency distribution of all the types found (Fig. 4.2).

Within a species, or among related species, the timing and length of the breeding season often varies with latitude (Lack 1950; Ricklefs 1969; Murton & Westwood 1977) and may change in time as conditions change (van Noordwijk, van Balen & Scharloo 1981; Kikukawa & Chippendale

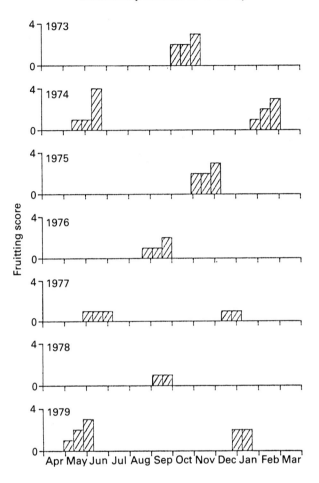

FIG. 4.3. Subannual cycles in a fig tree, *Ficus insipida* number 83 in Milton *et al.* (1982), on Barro Colorado Island. (Data provided by D.M. Windsor.)

1983; Beintema, Beintema-Hietbrink & Müskens 1985). In other cases, however, while geographic variation does occur, no such latitudinal correlation is obvious, as exemplified by the little penguin *Eudyphtula minor* in Australia and New Zealand (Gales 1985), and the common crossbill *Loxia curvirostris* in Europe and North America (Newton 1973; Nethersole-Thompson 1975).

Outside the tropics aseasonal patterns do seem to occur, but they are extremely rare and within tropical latitudes they are also surprisingly less common than one might expect. In only a few studies was aseasonality found to be the rule rather than the exception (Putz 1979, plants, Bukit Lanjan,

Malaysia; Moreau 1950a, landbirds, Zaïre; Inger & Greenberg 1966, lizards, Sarawak; Inger & Bacon 1968, frogs, Sarawak).

Subannual cycles in individuals are probably fairly common, because in many of the species which breed year-round, individuals may reproduce at subannual intervals. Even populations which have bimodal seasonal reproduction (type 2B) may consist of individuals which breed at subannual cycles, as was shown for Panamanian figs (Milton *et al.* 1982). Synchronous subannual cycles in populations, however, are extremely rare. Among seasonal species the frequency of long seasons is higher at low latitudes, but even there and even in areas with a largely aseasonal climate, some species have a very distinct and short season. Multiple reproductive seasons occur sporadically in some frogs, birds, mammals and, more commonly, in plants.

THE REASONS WHY

Most of the papers cited here discuss the causes of the seasonal patterns observed. (For summaries see Lack 1966; Harris 1969; Immelman 1971; Whitmore 1975; Giesel 1976; Harper 1977; Murton & Westwood 1977; Sinclair & Norton-Griffith 1979; Pierce 1984; Thresher 1984; Foster 1985; Grigg, Shine & Ehmann 1985.) In most cases the availability of resources, especially food, is shown to be, or at least is invoked as, the ultimate reason behind the seasonal patterns observed. Prevention of crossbreeding is sometimes seen as the reason why different closely related species reproduce at different times of the year (Snow 1964; Dressler 1981). Seasonal changes in predation pressure may also affect reproductive seasons: terns in Massachusetts breed relatively late because early breeders suffer heavy predation (Nisbet & Welton 1984) and the robin *Turdus grayi* in Panama breeds only in the dry season. Increased predation pressure during the early rainy season prevents breeding at that time although abundant food is available (Morton 1971; Dyrcz 1983).

Proximate cues such as day-length, temperature or rainfall are often used to ensure that reproductive activities occur at the appropriate time of the year. Sometimes the ultimate factors themselves, for instance food availability, are used as a proximate cue (Germain, Huignard & Monge 1985; H. Wolda & S. Tanaka unpublished). In many cases a combination of factors is employed, producing flexibility and a better adjustment of reproduction to a variable and unpredictable environment (e.g. Maclean 1976; Bolton, Newsome & Merchant 1982; Dingle & Baldwin 1983; Dingle 1984).

The seabirds on a tropical oceanic island illustrate the point. They all experience the same day-length and climate and feed in the same ocean.

They are, of course, not identical. They may vary in their hunting technique, their prey selection, etc. But does this explain why some species are highly seasonal and others are completely aseasonal? Or is it that one strategy is as good as another and that many of the differences we now observe are consequences of evolutionary history and historical 'accidents'?

THE COMMUNITY

What are the consequences of the degree of seasonality or of changes in intensity of seasonal factors on the community? Most events, but not all, that one reads at the community level, occur at the species level, but may spread as a ripple effect throughout the community. If a herbivorous insect reaches a low abundance in a given year, this directly affects the species diversity of the community. It also may release the host plant from herbivory pressure and species of predators and parasitoids may suffer decreased population sizes through lack of food and/or hosts. Alternatively, these natural enemies may turn to alternate prey or host species which may then, in turn, suffer increased mortality, etc. Much of the following review will, therefore, be devoted to individual species.

Species characteristics and population size

A number of species characteristics have been attributed to seasonality. For instance, Bergman's rule, the tendency among homoeotherms to increase in size with increasing latitude, has been attributed to 'environmental seasonality' (Lindstedt & Boyce 1985), but the arguments for that explanation are weak. At any one site some species are active during the entire growing season while others use only a small part of that season. As discussed above, there is not one 'environmental seasonality' in a given site, but a suite of environmental seasonalities, varying between species. To base predictions on an anthropomorphic 'environmental seasonality' is unconvincing unless one can demonstrate that, within each site, differences in seasonality in similar organisms also correlate with body size. To the best of my knowledge no such correlation exists.

The severity of the drought in the dry part of the year and/or the wetness of the rainy period, vary between years in many places and can have large effects. The lizard *Lampropholis quichenoti* near Sydney, Australia, produces two clutches of eggs per year in relatively wet years and only one clutch in dry years, probably because in wet years more food is available, as evidenced by the higher weight of the livers of the adults (Joss & Minard 1985). Drought in the interior of Australia strongly reduces the number of suitable rabbit warrens and the number of rabbits (Myers & Parker 1975). Similarly red kangaroos *Megaleia rufa* are greatly reduced in numbers by a long drought (Newsome, Stephens & Shipway 1967).

In a relatively dry year in England the size of butterfly populations may be severely reduced (Pollard 1982). The same is true in California as documented by Shapiro (1979): in a lowland site *Pieris napi microstriata* was observed in only small numbers after the 1976/77 winter, which was grossly deficient in rainfall. After heavy rains in the 1977/78 winter the species was back to pre-drought numbers. At a higher elevation the opposite occurred. In 1977 there still were a fairly large number of adults, but the entire reproductive output was lost due to cold rain and snow in May. This species' seasonality is a compromise between lateness of emergence to reduce the probability of a weather induced catastrophe, and earliness to match the seasonality of its hostplants, vernal crucifers.

Breeding success in birds may strongly depend on the intensity of the rainfall, as shown by the tern *Sterna hirundo* in Germany (Becker & Finck 1985). The same is true for birds in a tropical dry forest in Puerto Rico (Faaborg, Arendt & Kaiser 1984) but their main breeding period is April to July, when the rainfall is relatively unreliable. They do not delay their breeding until the more reliable rains of August–November, probably due to another seasonal factor, the presence of many temperate zone migrants.

Physical and biological factors may interact. Plants may be stressed by unusually dry or wet conditions, by lightning, or by other seasonal weather factors. Such stresses may result in reduced defences against herbivores and at the same time may make the plants more nutritious through an increase in the amount of available nitrogen, thus herbivore populations may flourish (White 1984; Rhoades 1985).

Most papers on seasonality discuss the possible causes of the observed patterns in some detail. Often these causes are found, or assumed, to be some form of maximizing resource availability or predator avoidance. In a community, however, this maximization is a two-way street. A predator may try to adjust its activity pattern to that of its preferred prey while, at the same time, it is to the advantage of the prey to avoid the predator. Biological interactions in a community are closely intertwined and causes and effects are inseparable. The seasonality of a given organism has evolved, or is phenotypically adjusted, to the totality of selective forces active on it in the community, usually with satisfactory results.

The tropical shrub *Hybanthus prunifolius* flowers in Panama 6–7 days after heavy rain subsequent to a period of drought. This sequence normally occurs at the beginning of the rainy season, but it may also happen in the middle of the dry season. The entire population is highly synchronous and the total flowering period is less than 1 week (Augspurger 1981). By experimentally watering a number of shrubs in the forest, selected individuals were triggered to flower asynchronously with the rest of the population. Those asynchronous individuals suffered much more seed predation by lepidopteran larvae and much less pollination success than individuals in the synchronous population, demonstrating that mass flowering attracts especially large numbers of pollinators and that the seed predators are normally swamped by the mass production of seeds (Augspurger 1981). Although rarely proven, such factors are often invoked as causes of an observed seasonal pattern.

Species which are not contemporary in a community can still affect each other. For instance, survival of some tadpoles can be diminished seriously by the past presence of other species of tadpoles in the same pool (Wilbur & Alford 1985).

Many organisms avoid an adverse season in their breeding grounds by migrating to more hospitable areas. Seasonal migrations of birds, butterflies and mammals are well known and can be observed anywhere in the world. In the wintering areas, these migrants play an integral part in the community. Songthrushes *Turdus philomelos* from western and northern Europe, winter in mediterranean areas, and are important dispersing agents for seeds of many plants there (Debussche & Isenmann 1985; Izhaki & Safriel 1985). These thrushes, wintering in Israel, brought a catastrophe to a landsnail in 1972. They ate large numbers of *Sphincterochila zonata* and *Trochoidea seetzenii* near Sede Boqer in the Negev desert, consuming 65 per cent of the individuals of *S. zonata* and 10 per cent of *T. seetzenii*. The population of the latter species recovered quickly but *S. zonata* had not yet recovered 7 years after the incident (Shachak, Safriel & Hunum 1981).

As shown, there is a great variety in seasonal patterns. Interestingly, different species with similar ecological requirements living in the same area may have different solutions to seasonality problems. Five species of rodents in the Californian desert are relatively similar in their ecological requirements, but they are very different in the seasonality of their reproduction (Kenagy & Bartholomew 1985). In spite of these differences all five species maintain stable population densities. There is obviously more than one solution to similar ecological problems.

Seasonal patterns often seem to be perfect adaptations: '. . . everything is for the best in the best of all possible worlds . . .' (Harper 1982), but one should be careful in *a priori* assuming that everything is well-adapted (Gould & Lewontin 1979). Males of the dioecious shrub *Osyris quadripartita* in Spain flower virtually around the year, while females flower only during a shorter period (Herrera 1984). Why 'waste energy' by producing male flowers when there is no chance of pollination? Such behaviour is not necessarily non-adaptive (Herrera 1984), but it might be. The saltwater crocodile *Crocodylus porosus* in northern Australia breeds at the onset of the wet season, thus maximizing mortality of its offspring due to flooding of the nest. 'The adaptive significance of this seasonal timing is problematical' (Shine 1985). It cannot simply be assumed that the seasonality of breeding in this crocodile is non-adaptive, after all, the species has been around for a long time. However, by the same token one would not assume without further evidence that any pattern observed is adaptive. It is conceivable that many organisms are a long way from a perfect adaptation to their present environment and that they do well 'in spite of' rather than 'because of' a given feature of their life history strategy.

Species-richness

Species which cannot tolerate conditions in a given habitat obviously cannot live there. In many cases these conditions are determined by seasonal factors. A species which is maladjusted seasonally to fluctuations in resource abundance or predation is unlikely to persist in the community concerned. Usual or occasional extreme intensities of seasonal physical factors may also prevent species from flourishing or even from becoming established permanently.

In the Gulf of Chiriquí (Panama) there are, or at least there were until recently (Glynn 1983, 1984), large coral reefs, while in the Bay of Panama there are only a few small ones (Glynn & Stewart 1973). The Bay of Panama undergoes a seasonal upwelling and the Gulf of Chiriquí does not. The colder seawater temperatures during the upwelling season prevent the development of such reefs. The existence of a seasonal change in one physical factor, here water temperature, has a tremendous effect on species diversity.

The endomychid Beetle *Stenotarsus rotundus* in Panama is programmed, mainly through its response to day-length, to terminate diapause in April, at the beginning of the rainy season (Wolda & Denlinger 1984; Tanaka, Denlinger & Wolda 1987). Without changes the species could not live in places where the timing of the response to day-length and the beginning of the rainy season did not coincide.

Species, of course, do change. Those that occur over a large geographic range and/or in a variety of habitats are either very flexible in their seasonality or genetically variable over their range. Genetic variation has been shown to exist. Van Noordwijk, van Balen & Scharloo (1981) demonstrate a strong heritability of the beginning of the egg-laying season in the great tit,

Parus major, and estimate that the population mean could shift as much as a week in five generations, over a decade. Beintema, Beintema-Hietbrink & Müskens (1985) demonstrate that such changes have occurred in the past decades in a number of meadow birds in the Netherlands. Between-population genetic differences in seasonality characters have been demonstrated in many organisms (e.g. Giesel 1976; Kikukawa & Chippendale 1983; Dingle 1984; Reinartz 1984; Denlinger 1986).

In some cases organisms respond directly to food and can adjust virtually instantaneously to different food conditions. The seed-eating lygaeid bug *Oxycarenus hyalinipennis* in Egypt reproduces if suitable seeds are available, whenever this happens during the year. In areas where cotton is their host, they feed from August to October after which they go into diapause until May. In other areas *Hibiscus mutabilis* is their host and their ripe seeds are not available until November, so that the breeding season occurs during the winter. The breeding season on hollyhocks is in early spring and diapause in July (Kirkpatrick 1957). The common crossbill *Loxia curvirostris* has a large breeding range and is extremely variable in its breeding seasonality, depending on the availability of conifer seeds (Newton 1973; Nethersole-Thompson 1975). At times the species undergoes extensive dispersal and it is highly likely that the same individuals breed in completely different seasons in different localities.

It has been suggested that the much larger species-richness of tropical communities is related to the longer growing season. However, Ricklefs (1966) found that 'the average length of the breeding season of individual species of birds occupies very nearly the same proportion of the total breeding season' in the tropics and in temperate areas, suggesting that the longer growing season is not a contributing factor in the greater species-richness of the tropics, at least not in birds.

In some cases events which occur above the species level determine the composition of the community. The seasonal patterns of a series of species together affect their own continued presence.

In a wet tropical forest in Costa Rica, plants that are pollinated by hummingbirds tend to have very short flowering seasons which are distributed over most of the year (Fig. 4.4), probably because of competition for pollinators (Stiles 1977, 1978). In this way the plant species together ensure that there are enough hummingbirds available to pollinate their flowers, and that the birds have a source of nectar for most of the year, with a period of possibly severe shortage in November–December (Stiles 1978). Plants that bridge such gaps in food availability are vitally important to the existence of a series of hummingbird species that depend on that food source. If the gap is too large, the hummingbirds cannot survive in the

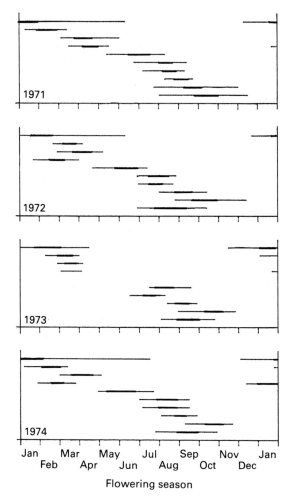

FIG 4.4. Flowering phenology of hermit pollinated plants in a Costa Rican forest. Plants in the same order each year, from top to bottom (*H.* = *Heliconius*): *H. pogonantha, Passiflora vitifolia, H. wagneriana, Jacobinia umbrosa, Costus ruber, H. irrasa, H. umbrophila, Aphelandra sinclairiana, Costus malortieanus,* and *H. mathiasi* (after Stiles 1978).

habitat and the plants have a pollinator problem. How fragile such a system can be is illustrated by the data for 1973 when 'evidently as a result of the extreme drought of that year' the regular sequence of flowering was disturbed and one species, *Costus ruber* failed to flower at all, producing a conspicuous gap in the availability of flowers over the year. Howe (1984) identifies two species of trees, *Casearia corymbosa* in Costa Rica and *Virola sebifera* in Panama, which bridge such a seasonal gap. They bear fruits during annual periods of fruit scarcity and thus enable a number of species of frugivorous mammals and birds to live in these forests. The animals are critical for the dispersal and recruitment of many plant species at other times of the year. Such 'pivotal' species, with their aberrant seasonal pattern of fruit production, are vitally important in the community to maintain diversity and proper recruitment in birds, mammals and plants.

In some rocky intertidal habitats species diversity is maintained by winterstorms that overturn the boulders on which many organisms grow (Sousa 1979). The effects depend on the successional stage of the communities on the boulders and on the season in which the overturning occurs (Osman 1978; Sousa 1980). A tropical forest is also a very dynamic system. Treefalls occur frequently and the gaps formed allow seeds to germinate and new seedlings and saplings to grow up. In Panama most treefalls tend to occur during the rainy season, and the exact timing of a particular treefall determines which seeds have the best chance to germinate and which saplings may grow up (Brokaw 1982).

The diversity of macroinvertebrates in the Dutch Wadden Sea is dependent on the water temperature during the winter. After a few years of relatively mild winters species-richness increases and the relative abundance of the species has changed dramatically (Beukema, de Bruin & Jansen 1978).

Stability

The stability of a population, in the present context, is the inverse of the variance of the number of individuals in that population (Wolda 1983). The more constant the number of individuals, the more stable the population. Stability of a community is the degree of maintenance of the species-richness and the abundances of the component species; a maintenance of the status quo. The part of a rocky shore community which lives on boulders, such as algae and invertebrates, lives in a highly unstable physical environment in which winterstorms constantly overturn boulders (Osman 1978; Sousa 1979, 1980). This community does not have the opportunity to go through many successional stages and is maintained at an early level of succession and as such can be highly stable. If one year the storms did not come, that calm would be a major disturbance causing extensive temporary changes in the community, a sign of instability.

Rodents in northern Europe demonstrate an increased variability (cyclicity) in their numbers with an increased snow cover in winter. Generalist predators, which keep the rodent populations stable in areas with zero to moderate snow cover, are no longer effective when the snow is deep, and they give way to specialist predators such as mustelids which allow the rodent populations to vary in abundance (Hansson & Henttonen 1985). Insects with shorter seasons tend to be more variable in abundance from year to year (Wolda 1978; 1980). A short reproductive season has a relatively large chance of failure if environmental factors happen not to be right at the time (Kenagy & Bartholomew 1985).

The 1970 dry season in Panama was extraordinarily short and wet, with especially heavy showers in January. As a result, many important fruit trees in the forest on Barro Colorado Island did not flower and, consequently, did not produce the fruits normally used by a wide variety of animals. The total weight of the fruit which fell between August 1970 and February 1971 was only one-third of that in the previous year (Foster 1982b). The result was a widespread famine among the forest animals, lasting from July 1970 to August 1971. Animals foraged longer and fed on novel foods, dead animals abounded and 'at times it was difficult to avoid the stench: neither the turkey vultures nor the black vultures seemed able to keep up with the abundance of carcasses'. Similar situations probably occurred in 1931 and 1958. 'The rhythm of a large part of the plant community seems to depend on dry seasons of the length and severity currently prevailing on the island. Excessive rain out of season is far more disruptive to the forest community than any other aberration of climate so far observed' (Foster 1982b). This is a good example of the 'ripple effect' alluded to earlier, how a small seasonal change in a weather factor can affect the stability of the entire community. The following example also illustrates this point on a larger scale.

Off the pacific coast of South Africa there is a seasonal upwelling. Seabirds as well as fishermen depend upon the cold-water season for their catches. Every now and then this pattern is disturbed when an 'El Niño' event occurs, with higher water temperatures and with sometimes disastrous results. The very severe 1982/83 Niño caused extensive damage to the fishing industry and to the seabirds and other organisms in that region (Barber & Chavez 1983; Feldman, Clark & Halpern 1984; Schreiber & Schreiber 1984; Glantz 1985; Jordan 1985), and beyond. Extensive coral mortality occurred all over the eastern Pacific(Glynn1983, 1984) and the dead reefs were soon covered by a dense mat of algae (Wellington & Victor 1985). Seabirds as far north as Oregon were affected (Hodder & Graybill 1985). It will take a long time for all this damage, especially that to the coral reefs, to be repaired.

Storms during the winter of 1982/83 destroyed most of the canopy of the giant kelp *Macrocystis pyrifera* off the coast of southern California; understory kelp species were not affected. During the subsequent summer, affected by the same Niño event just mentioned, a massive recruitment of

Macrocystis juveniles occurred, but the subsequent survival of these juveniles and that of the adults that had survived the storm was poor. This sequence of disturbances may have long-lasting consequences for the community, because the understory patches, once established, can resist invasion by *Macrocystis* (Dayton & Tegner 1984).

How pervasive the results of changes in seasonality may be, is illustrated in a very interesting way by a study on the effects of the distribution of landmasses over the earth on the seasonal cycle of temperatures on the past 100 million years. It was suggested that the difference between a permanent icecap and ice-free conditions may have been due to seasonality in temperature rather than year-round warmth. Palaeontological evidence provides some support for this hypothesis (Crowley *et al.* 1986).

DISCUSSION

The ultimate study on the effects of seasonality on the community would be a comparison between seasonal and completely non-seasonal areas. However, as said before, areas devoid of any seasonal physical cues (rainfall, temperature, wind, day-length, etc.) very probably do not exist. Completely aseasonal communities, communities where all members regard the environment as effectively aseasonal, are highly unlikely, because inter-specific interactions tend to keep reproductive seasons, and other phenophases, apart. Seasonality is thus an integral part of all communities, the above comparison cannot be made and a review of the effects of seasonality on communities has to be restricted to the effects of changes in seasonal patterns or of changes in intensity of seasonal factors.

Each link in the network of relationships in the community has its phenophases which are more or less seasonal. If these links are not mutually adjusted, seasonally or otherwise, or if one or more of the physical or biotic factors in the community reaches extreme values, the chances to survive and multiply of component species will be affected and the species-richness and the stability of the community as a whole will be altered. A seemingly small change in a weather factor, such as in the rainfall in the dry season in Panama or in the seawater temperature during the upwelling season in the eastern Pacific, may have a tremendous effect on the species diversity and stability of communities living in those areas. In addition a whole suite of species must have their seasonalities adjusted to each other if any of them is going to survive, as exemplified by hummingbirds and their hostplants in Costa Rica. Such cases may, in fact, be common.

The seasonal adjustment of populations to their environment is often a compromise between opposing forces, as was shown for the breeding seasons of birds on Puerto Rico (Faaborg *et al.* 1984) and of *Turdus grayi* in Panama (Dyrcz 1983). In some cases the word maladjustment seems to be more appropriate, as for the saltwater crocodile in Australia (Shine 1985). In these cases, however, the adjustment is still good enough for the species to survive and multiply and remain an integral part of the community.

ACKNOWLEDGMENTS

I am grateful to the British Ecological Society for inviting me to write this paper, and to Annette Aiello for carefully reading the manuscript. Don Windsor and the Environmental Sciences Program of the Smithsonian Institute kindly provided the data for Fig. 4.3.

REFERENCES

Asdell, S.A. (1964) *Patterns of Mammalian Reproduction.* Cornell University Press, Ithaca, New York.

Ashmole, N.P. (1965). Adaptive variation in the breeding regime of a tropical seabird. *Proceedings of the National Academy of Sciences of the USA,* **53,** 311–18.

Ashmole, N.P. (1971). Sea bird ecology and the marine environment. *Avian Biology, Vol. 1* (Ed. by D.S. Farner, J.R. King & K.C. Parkes), pp. 223–86. Academic Press, New York.

Augspurger, C.K. (1981). Reproductive synchrony of a tropical shrub: experimental studies on effects of pollinators and seed predators on *Hybanthus prunifolius* (Violaceae). *Ecology,* **62,** 775–88.

Augspurger, C.K. (1982). A cue for synchronous flowering. *The Ecology of a Tropical Forest. Seasonal Rhythms and Longterm Changes* (Ed. by E.G. Leigh, A.S. Rand & D.M. Windsor), pp. 133–50. Smithsonian Institution Press, Washington, DC.

Augspurger, C.K. (1983). Phenology, flowering synchrony, and fruitset of six neotropical shrubs. *Biotropica,* **15,** 257–67.

Augspurger, C.K. (1985). Flowering synchrony in neotropical plants. *The Botany and Natural History of Panama* (Ed. by W.G. D'Arcy & M.D. Correa), pp. 235–43. Missouri Botanical Gardens, Saint Louis, Missouri.

Barber, R.T. & Chavez, F.P. (1983). Biological consequences of El Niño. *Science,* **222,** 1203–10.

Becker, P.H. & Finck, P. (1985). Witterung und Ernährungssituation als entscheidene Faktoren des Bruterfolges der Flussseeschwalbe (Sterna hirundo). *Journal of Ornithology,* **126,** 393–404.

Beintema, A.J., Beintema-Hietbrink, R.J. & Müskens, G.J.D.M. (1985). A shift in the timing of breeding in meadow birds. *Ardea,* **73,** 83–9.

Bell, H.L. (1982a). A bird community of lowland rain forest in New Guinea. 2. Seasonality. *Emu,* **82,** 65–74.

Bell, H.L. (1982b). Abundance and seasonality of the savanna avifauna at Port Moresby, Papua New Guinea. *Ibis,* **124,** 252–74.

Beukema, J.J., Bruin, W. de & Jansen, J.J.M. (1978). Biomass and species richness of the macrobenthic animals living on the tidal flats of the Dutch Wadden Sea: long-term changes during a period with mild winters. *Netherlands Journal of Sea Research,* **12,** 58–77.

Bolton, B.L., Newsome, A.E., Merchant, J.C. (1982). Reproduction of the agile wallaby *Macropus agilis* (Gould) in the tropical lowlands of the Northern Territory: opportunism in a seasonal environment. *Australian Journal of Zoology*, **7**, 261–77.

Borchert, R. (1980). Phenology and ecophysiology of tropical trees: *Erythrina poeppigiana*. O.F. Cook. *Ecology*, **61**, 1065–74.

Borchert, R. (1983). Phenology and control of flowering in tropical trees. *Biotropica*, **15**, 81–9.

Britton, P.L. (1978). Seasonality, density and diversity of birds of a papyrus swamp in western Kenya. *Ibis*, **120**, 450–66.

Brokaw, N.V.L. (1982). Treefalls: frequency, timing and consequences. *Ecology of a Tropical Forest. Seasonal rhythms and long-term changes* (Ed. by E.G. Leigh, A.S. Rand & D.M. Windsor), pp. 101–8. Smithsonian Institution Press, Washington, DC.

Chapin, J.P. & Wing, L.W. (1959). The wideawake calendar, 1953 to 1958. *Auk*, **76**, 153–8.

Charles-Dominique, P. (1977). *Ecology and Behaviour of Nocturnal Primates. Prosimians of Equatorial West Africa* (Translated by R.D. Martin). Columbia University Press, New York.

Cheke, A.S., Nanakorn, W. & Yankoses, C. (1979). Dormancy and dispersal of seeds of secondary forest species under the canopy of a primary tropical rain forest in Northern Thailand. *Biotropica*, **11**, 88–95.

Coster, C. (1923). Lauberneuerung und andere periodische Lebensprozesse in dem trockenen Monsun-Gebiet Ost-Java's. *Annales du Jardin Botanique de Buitenzorg*, **33**, 117–89.

Coster, C. (1926). Periodische Blütenerscheinungen in den Tropen. *Annales du Jardin Botanique de Buitenzorg*, **35**, 125–62.

Croat, T.B. (1969). Seasonal flowering behaviour in Central Panama. *Annals Missouri Botanical Gardens*, **56**, 295–307.

Croat, T.B. (1975). Phenological behaviour of habit and habitat classes on Barro Colorado Island (Panama Canal Zone). *Biotropica*, **7**, 270–7.

Croat, T.B. (1978). *Flora of Barro Colorado Island*. Stanford University Press, Stanford, California.

Crome, F.H.J. (1975). The ecology of fruit pigeons in tropical northern Queensland. *Australian Wildlife Research*, **2**, 155–85.

Crowley, T.J., Short, D.A., Mengel, J.G. & North, G.R. (1986). Role of seasonality in the evolution of climate during the last 100 million years. *Science*, **231**, 579–84.

Davis, D.E. (1945). The annual cycle of plants, mosquitos, birds, and mammals in two Brazilian forests. *Ecological Monographs*, **15**, 243–95.

Davis, T.A.W. (1953). An outline of the ecology and breeding seasons of the lowland forest region of British Guiana. *Ibis*, **95**, 450–67.

Dayton, P.K. & Tegner, M.J. (1984). Catastrophic storms, El Niño, and patch stability in a southern Californian kelp community. *Science*, **224**, 283–5.

Debussche, M. & Isenmann, P. (1985). Le régime alimentaire de la grive musicienne (*Turdus philomelos*) en automne et en hiver dans les garrigues de Montpellier (France méditerranéenne) et ses relations avec l'ornithochorie. *Revue d'Ecologie (La Terre et la Vie)*, **40**, 379–88.

Den Boer, P.J. (1967). De relativiteit van zeldzaamheid. *Entomologische Berichten (Amsterdam)*, **27**, 52–60.

Denlinger, D.L. (1986). Dormancy in tropical insects. *Annual Review of Entomology*, **31**, 239–64.

Diamond, A.W. (1976). Subannual breeding and moult cycles in the bridled tern *Sterna anaethetus* in the Seychelles. *Ibis*, **118**, 414–19.

Dingle, H. (1984). Behavior, genes, and life histories: complex adaptations in uncertain environments. *A New Ecology: Novel Approaches To Interactive Systems* (Ed. by P.W. Price, C.N. Slobodchikoff & W.S. Gaud), pp. 169–94. Wiley, New York.

Dingle, H. & Baldwin, J.D. (1983). Geographic variation in life histories: a comparison of tropical and temperate milkweed bugs (*Oncopeltus*). *Diapause and Life Cycle Strategies in Insects* (Ed. by V.K. Brown & I. Hodek), pp. 143–65. Junk, The Hague.

Dorward, D.F. (1962). Comparative biology of the white booby and the brown booby *Sula* spp. at Ascension. *Ibis*, **103b**, 174–220.

Dressler, R.L. (1981). *The Orchids. Natural History and Classification.* Harvard University Press, Cambridge, Massachusetts.

Dunlop, J.N. (1985). Reproductive periodicity in a population of crested terns, *Sterna bergii* Lichtenstein, in South-Western Australia. *Australian Wildlife Research*, **12**, 95–102.

Dyrcz, A. (1983). Breeding ecology of the clay-coloured robin *Turdus grayi* in lowland Panama. *Ibis*, **125**, 287–304.

Dyrcz, A. (1984). Breeding biology of the mangrove swallow *Tachycineta albilineata* and the greybreasted martin *Progne chalybea* at Barro Colorado Island, Panama. *Ibis*, **126**, 59–66.

Faaborg, J., Arendt, W.J. & Kaiser, M.S. (1984). Rainfall correlates of bird population fluctuations in a Puerto Rican dry forest: a nine year study. *Wilson Bulletin*, **96**, 575–93.

Feldman, G., Clark, D. & Halpern, D. (1984). Satellite color observations of phytoplankton distribution in the eastern equatorial Pacific during the 1982–1983 El Niño. *Science*, **226**, 1069–71.

Fernald, M.L. (1950). *Gray's Manual of Botany*, 8th edn. American Book Company, New York.

Fogden, M.P.L. (1972). The seasonality and population dynamics of equatorial forest birds in Sarawak. *Ibis*, **114**, 307–43.

Foster, R.B. (1977). *Tachigalia versicolor*, a suicidal neotropical tree. *Nature*, **268**, 624–6.

Foster, R.B. (1982a). The seasonal rhythm of fruitfall on Barro Colorado Island. *The Ecology of a Tropical Rainforest. Seasonal Rhythms and Long-term Changes* (Ed. by E.G. Leigh, A.S. Rand & D.M. Windsor), pp. 151–72. Smithsonian Institution Press, Washington, DC.

Foster, R.B. (1982b). Famine on Barro Colorado Island. *The Ecology of a Tropical Rainforest. Seasonal Rhythms and Long-term Changes* (Ed. by E.G. Leigh, A.S. Rand & D.M. Windsor), pp. 201–12. Smithsonian Institution Press, Washington, DC.

Foster, R.B. (1985). Plant seasonality in the forests of Panama. *The Botany and Natural History of Panama* (Ed. by W.G. D'Arcy & M.D. Correa), pp. 255–62. Missouri Botanical Gardens, Saint Louis, Missouri.

Fournier, P. (1961). *Les quatre flores de France.* Paul Lechevalier, Paris.

Frankie, G.W. (1973). Tropical forest phenology and pollinator plant coevolution. *Coevolution of Animals and Plants.* Symposium V, First International Congress of Systematic and Evolutionary Biology (Ed. by L.E. Gilbert & P.H. Raven), pp. 192–209. University of Texas Press, Austin.

Frankie, G.W., Baker, H.G., Opler, P.A. (1974). Comparative phenological studies of trees in tropical wet and dry forests in the lowlands of Costa Rica. *Journal of Ecology*, **62**, 881–913.

Frith, H.J. (1956). Breeding habits in the family Megapodidae. *Ibis*, **98**, 620–40.

Frith, H.J. & Sharman, B.G. (1964). Breeding in wild populations of the red kangaroo *Megaleia rufa*. *CSIRO Wildlife Research*, **9**, 86–114.

Gales, R. (1985). Breeding seasons and double brooding of the little penguin *Eudyptula minor* in New Zealand. *Emu*, **85**, 127–30.

Gaston, A.J., Mathew, D.N. & Zacharias, V.J. (1979). Regional variation in the breeding season of babblers (*Turdoides* spp.) in India. *Ibis*, **121**, 512–16.

Germain, J.F., Huignard, J. & Monge, J.P. (1985). Influence des inflorescences de la plante hôte (*Vigna unguicula*) sur la levée de la diapause réproductrice de *Bruchidius atrolineatus*. *Entomologia experimentalis et applicata*, **39**, 35–42.

Giesel, J.T. (1976). Reproductive strategies as adaptations to life in temporally heterogeneous environments. *Annual Review of Ecology and Systematics*, **7**, 57–79.

Glantz, M.H. (1985). Climate and fisheries: a Peruvian case study. *ERFEN Boletin,* **15,** 13–31.

Glynn, P.W. (1983). Extensive 'bleaching' and death of reef corals on the Pacific coast of Panama. *Environmental Conservation,* **10,** 149–54.

Glynn, P.W. (1984). Widespread coral mortality and the 1982/83 El Niño warming event. *Environmental Conservation,* **11,** 133–46.

Glynn, P.W. & Stewart, R.H. (1973). Distribution of coral reefs in the Pearl islands (Gulf of Panama) in relation to thermal conditions. *Limnology and Oceanography,* **18,** 367–79.

Gould, S.J. & Lewontin, R.C. (1979). The spandrels of San Marco and the Panglossian paradigm: a critique of the adaptationist programme. *Proceedings of the Royal Society of London. Series B,* **205,** 581–98.

Grigg, G., Shine, R. & Ehmann, H. (Eds) (1985). *Biology of Australasian frogs and reptiles.* Surrey Beatty & Sons and Royal Zoological Society of New South Wales, Chipping Norton, New South Wales.

Hails, C.J. (1984). The breeding biology of the Pacific swallow *Hirundo tahitica* in Malaysia. *Ibis,* **126,** 198–211.

Hansson, L. & Henttonen, H. (1985). Gradients in density variations of small rodents: the importance of latitude and snow cover. *Oecologia (Berlin),* **67,** 394–402.

Harper, J.L. (1977). *Population Biology of Plants.* Academic Press, London.

Harper, J.L. (1982). Panglossian botany. *Nature (London),* **295,** 470.

Harris, M.P. (1969). Breeding seasons of sea-birds in the Galapagos Islands. *Journal of Zoology, London,* **159,** 145–65.

Herrera, C.M. (1984). The annual cycle of *Osyris quadripartita,* a hemiparasitic dioecious shrub of Mediterranean scrublands. *Journal of Ecology,* **72,** 1065–78.

Hladik, A. (1978). Phenology of leaf production in rain forest of Gabon: distribution and composition of food for folivores. *The Ecology of Arboreal Folivores* (Ed. by G.G. Montgomery), pp. 51–71. Smithsonian Institution Press. Washington, DC.

Hodder, J. & Graybill, M.R. (1985). Reproduction and survival of sea birds in Oregon during the 1982–1983 El Niño. *Condor,* **87,** 535–41.

Holliday, N.J. (1985). Maintenance of the phenology of the winter moth (Lepidoptera: Geometridae). *Biological Journal of the Linnean Society,* **25,** 221–34.

Howe, H.F. (1984). Implications of seed dispersal by animals for tropical reserve management. *Biological Conservation,* **30,** 261–81.

Immelmann, K. (1971). Ecological aspects of periodic reproduction. *Avian Biology. Vol. 1* (Ed. by D.S. Farner, J.R. King & K.C. Parkes), pp. 341–89. Academic Press, New York.

Inger, R.F. & Bacon Jr, J.P. (1968). Annual reproduction and clutch size in rain forest frogs from Sarawak. *Copeia,* **68,** 602–6.

Inger, R.F. & Greenberg, B. (1966). Annual reproductive patterns of lizards from a Bornean rain forest. *Ecology,* **47,** 1007–21.

Izhaki, I. & Safriel, U.N. (1985). Why do fleshy-fruit plants of the mediterranean shrubs intercept fall-, but not spring passage of seed-dispersing migrating birds? *Oecologia (Berlin),* **67,** 40–3.

Jackson, J.F. (1978). Seasonality of flowering and leaf-fall in a Brazilian subtropical lower montane moist forest. *Biotropica,* **10,** 38–42.

Jordan, R. (1985). Ecological changes and economic consequences of the El Niño phenomenon in the south-eastern Pacific. *ERFEN Boletin,* **12,** 2–4.

Joss, J.M.P. & Minard, J.A. (1985). On the reproductive cycles of *Lampropholis guichenoti* and *L. delicata* (Squamata: Scincidae) in the Sydney region. *Australian Journal of Zoology,* **33,** 699–704.

Kenagy, G.J. & Bartholomew, G.A. (1985). Seasonal reproductive patterns in five coexisting California desert rodent species. *Ecological Monographs,* **55,** 371–97.

Kikkawa, J. (1980). Seasonality of nesting by zebra finches at Armidale, NSW. *Emu,* **80,** 13–20.

Kikukawa, S. & Chippendale, G.M. (1983). Seasonal adaptations of populations of the southwestern corn borer, *Diatraea grandiosella*, from tropical and temperate regions. *Journal of Insect Physiology*, **29**, 561–7.

Kirkpatrick, T.W. (1957). *Insect Life in the Tropics.* Longmans, London.

Kohshima, S. (1984). A novel cold-tolerant insect found in a Himalyan glacier. *Nature*, **310**, 225–7.

Lack, D. (1950). The breeding seasons of European birds. *Ibis*, **92**, 288–9.

Lack, D. (1966). *Population Studies of Birds.* Clarendon Press, Oxford.

Langham, N. (1980). Breeding biology of the edible-nest swiftlet *Aerodramus fuciphagus. Ibis*, **122**, 447–61.

Lavery, H.J., Seton, D. & Bravery, J.A. (1968). Breeding seasons of birds in north-eastern Australia. *Emu*, **68**, 133–47.

Leighton, M. & Leighton, D.R. (1983). Vertebrate responses to fruiting seasonality within a Bornean rain forest. *Tropical Rainforest: Ecology and Management* (Ed. by S.L. Sutton, T.C. Whitmore & A.C. Chadwick), pp. 181–96. Special Publications of the British Ecological Society, 2. Blackwell Scientific Publications, Oxford.

Lenton, G.M. (1984). The feeding and breeding ecology of barn owls *Tyto alba* in Peninsular Malaysia. *Ibis*, **126**, 551–75.

Ligon, J.D. (1971). Late summer–autumnal breeding of the Piñon Jay in New Mexico. *Condor*, **73**, 147–53.

Lindstedt, S.L. & Boyce, M.S. (1985). Seasonality, fasting endurance, and body size in mammals. *American Naturalist*, **125**, 873–8.

McClure, H.E. (1966). Flowering, fruiting and animals in the canopy of a tropical rainforest. *Malaysian Forester*, **29**, 192–203.

Maclean, G.L. (1976). Rainfall and avian breeding seasons in north-western New South Wales in spring and summer 1974–75. *Emu*, **76**, 139–42.

Mader, W.J. (1981). Notes on nesting raptors in the llanos of Venezuela. *Condor*, **83**, 48–51.

Malaisse, F.P. (1974). Phenology of the Zambezian woodland area with emphasis on the Miombo ecosystem. *Phenology and Seasonality Modelling* (Ed. by H. Lieth), pp. 269–86. Springer-Verlag, New York.

Medway, Lord (1962). The swiftlets (*Collocalia*) of Niah cave, Sarawak. *Ibis*, **104**, 45–66.

Medway, Lord (1972). Phenology of a tropical rain forest in Malaya. *Biological Journal of the Linnean Society*, **4**, 117–46.

Miller, A.H. (1955). Breeding cycles in a constant equatorial environment in Colombia, South America. *Proceedings of the XI Ornithological Congress Basel*, 495–503.

Miller, A.H. (1962). Bimodal occurrence of breeding in an equatorial sparrow. *Proceedings of the National Academy of Sciences of the United States of America*, **48**, 396–400.

Miller, E.R. & Dingle, H. (1982). The effect of the host plant phenology of reproduction of the milkweed bug, *Oncopeltus fasciatus*, in tropical Florida. *Oecologia (Berlin)*, **52**, 97–103.

Milton, K., Windsor, D.M., Morrison, D.W., Estribí, M.A. (1982). Fruiting phenologies of two neotropical *Ficus* species. *Ecology*, **63**, 752–62.

Mooney, H.A., Parsons, D.J. & Kummerow, J. (1974). Plant development in Mediterranean climates. *Phenology and Seasonality Modelling* (Ed. by H. Lieth), pp. 255–67. Springer-Verlag, New York.

Moreau, R.E. (1950a). The breeding seasons of African Birds. 1. Land birds. *Ibis*, **92**, 223–67.

Moreau, R.E. (1950b). The breeding seasons of African birds. 2. Sea birds. *Ibis*, **92**, 419–33.

Morton, E.S. (1971). Nest predation affecting the breeding season of the Clay-coloured Robin, a tropical song bird. *Science*, **171**, 920–1.

Murton, R.K. & Westwood, N.J. (1977). *Avian Breeding Cycles.* Clarendon Press, Oxford.

Myers, K. & Parker, B.S. (1975). Effects of severe drought on rabbit numbers and distribution in a refuge area in semiarid north-western New South Wales. *Australian Wildlife Research*, **2**, 103–20.

Nelson, J.B. (1967). Etho-ecological adaptations in the Great Frigate-bird. *Nature (London),* **214,** 318.

Nethersole-Thompson, D. (1975). *Pine Crossbills. A Scottish Contribution.* Poyser, Berkhamsted, UK.

Newsome, A.E., Stephens, D.R. & Shipway, A.K. (1967). Effects of a long drought on the abundance of red kangaroos in central Australia. *CSIRO Wildlife Research,* **12,** 1–8.

Newton, I. (1973). *Finches.* Taplinger, New York.

Nisbet, I.C.T. & Welton, M.J. (1984). Seasonal variations in breeding success of common terns: consequences of predation. *Condor,* **86,** 53–60.

Noordwijk, A.J. van, Balen, J.H. van & Scharloo, W. (1981). Genetic variation in the timing of reproduction in the Great Tit. *Oecologia (Berlin),* **49,** 158–66.

Osman, R.W. (1978). The influence of the seasonality and stability on the species equilibrium. *Ecology,* **59,** 383–99.

Paton, D.C. (1985). Do New Holland honeyeaters *Phylidonyris novaehollandiae* breed regularly in spring and autumn? *Emu,* **85,** 130–3.

Payne, R.B. (1969). *Breeding Seasons and Reproductive Physiology of Tricolored Blackbirds and Redwinged Blackbirds.* University of California Press, Berkeley & Los Angeles.

Pierce, S.M. (1984). A synthesis of plant phenology in the Fynbos biome. *South African National Scientific Programs Report,* **88,** 1–57.

Pollard, E. (1982). Monitoring butterfly abundance in relation to the management of a nature reserve. *Biological Conservation,* **24,** 317–28.

Poole, W.E. (1983). Breeding in the grey kangaroo, *Macropus giganteus,* from widespread locations in eastern Australia. *Australian Wildlife Research,* **10,** 453–66.

Putz, F.E. (1979). Aseasonality in Malaysian tree phenology. *The Malaysian Forester,* **42,** 1–24.

Ramsay, M.A. & Dunbrack, R.L. (1986). Physiological constraints on life history phenomena: The example of small bear cubs at birth. *American Naturalist,* **127,** 735–43.

Reinartz, J.A. (1984). Life history variation of common mullein (*Verbascum thapsus*). I. Latitudinal differences in population dynamics and timing of reproduction. *Journal of Ecology,* **72,** 897–912.

Rhoades, D.F. (1985). Offensive–defensive interactions between herbivores and plants: their relevance in herbivore population dynamics and ecology theory. *American Naturalist,* **125,** 205–38.

Ricklefs, R.E. (1966). The temporal component of diversity among species of birds. *Evolution,* **20,** 235–42.

Ricklefs, R.E. (1969). The nesting cycle of songbirds in tropical and temperate regions. *The Living Bird,* **8,** 165–75.

Saunders, D.A. (1977). Red-tailed Black Cockatoo breeding twice a year in the south-west of Western Australia. *Emu,* **77,** 107–10.

Savage, A.J.P. & Ashton, P.S. (1983). The population structure of the double coconut and some other Seychelles palms. *Biotropica,* **15,** 15–25.

Schodde, R. (1981). Bird communities in the Australian Mallee: composition, derivation, distribution, structure and seasonal cycles. *Mediterranean-type Shrublands* (Ed. by F. di Castri, D.W. Goodall & R.L. Specht), pp. 387–415. Elsevier, Amsterdam.

Schreiber, R.W. & Ashmole, N.P. (1970). Sea-bird breeding seasons on Christmas Island, Pacific Ocean. *Ibis,* **112,** 363–94.

Schreiber, R.W. & Schreiber, E.A. (1984). Central Pacific seabirds and the El Niño oscillation: 1982 to 1983. *Science,* **225,** 713–16.

Serle, W. (1981). The breeding season of birds in the lowland rainforest and in the montane forest of West Cameroon. *Ibis,* **123,** 62–74.

Shachak, M.B., Safriel, U.N. & Hunum, R. (1981). An exceptional event of predation on desert snails by migratory thrushes in the Negev desert, Israel. *Ecology,* **62,** 1441–9.

Shapiro, A.M. (1979). The phenology of *Pieris napi microstriata* (Lepidoptera). *Psyche,* **86,** 1–10.

Shimizu, Y. (1983). Phenological studies of the subtropical broad-leaved evergreen forests at Chichijima island in the Bonin (Ogasawara) Islands. *Japanese Journal of Ecology,* **33,** 135–47.

Shine, R. (1985). The reproductive biology of Australian reptiles: a search for general patterns. *Biology of Australasian frogs and reptiles* (Ed. by G. Grigg, R. Shine & H. Ehmann), pp. 297–303. Surrey Beatty & Sons and Royal Zoological Society of New South Wales, Chipping Norton, New South Wales.

Simmons, K.E.L. (1967). Ecological adaptations in the life history of the brown booby at Ascension island. *The Living Bird,* **6,** 187–212.

Sinclair, A.R.E. & Norton-Griffith, M. (1979). *Serengeti. Dynamics of an Ecosystem.* University of Chicago Press, Chicago.

Skutch, A.F. (1950). The nesting seasons of Central American birds in relation to climate and food supply. *Ibis,* **92,** 185–222.

Smith, M.H. (1974). Seasonality in mammals. *Phenology and Seasonality Modelling* (Ed. by H. Lieth), pp. 149–62. Springer-Verlag, New York.

Snow, B.K. & Snow, D.W. (1969). Observations on the lava gull *Larus fuliginosus. Ibis,* **111,** 30–5.

Snow, D.W. (1964). A possible selective factor in the evolution of fruit seasons in a tropical forest. *Oikos,* **15,** 274–81.

Snow, D.W. (1965). The breeding of Audubon's shearwater (*Puffinus lherminieri*) in the Galapagos. *Auk,* **82,** 591–7.

Snow, D.W. & Snow, B.K. (1964). Breeding seasons and annual cycles of Trinidad land-birds. *Zoologica, N.Y.,* **49,** 1–35.

Snow, D.W. & Snow, B.K. (1966). The breeding seasons of the Madeiran storm petrel *Oceanodroma castro* in the Galapagos. *Ibis,* **108,** 283–4.

Snow, D.W. & Snow, B.K. (1967). The breeding cycle of the swallow-tailed gull *Creagus furcatus. Ibis,* **109,** 14–24.

Sousa, W.P. (1979). Disturbance in marine intertidal boulder fields: the non-equilibrium maintenance of species diversity. *Ecology,* **60,** 1225–39.

Sousa, W.P. (1980). The responses of a community to disturbance: the importance of successional age and species life histories. *Oecologia (Berlin),* **45,** 72–81.

Sowls, L.K. (1984). *The Peccaries.* University of Arizona Press, Tucson, Arizona.

Stiles, F.G. (1977). Coadapted competitors: the flowering seasons of hummingbird pollinated plants in a tropical forest. *Science,* **198,** 1177–8.

Stiles, F.G. (1978). Temporal organization of flowering among the hummingbird foodplants of a tropical wet forest. *Biotropica,* **10,** 194–210.

Stiles, F.G. (1980). The annual cycle in a tropical wet forest hummingbird community. *Ibis,* **122,** 322–43.

Stonehouse, B. (1962). The tropic birds (genus *Phaeton*) of Ascension island. *Ibis,* **103b,** 124–61.

Struhsaker, T.T. (1978). Interrelations of red Colobus monkeys and rainforest trees in the Kibale forest, Uganda. *The Ecology of Arboreal Folivores* (Ed. by G.G. Montgomery), pp. 397–422. Smithsonian Institution Press, Washington, DC.

Tanaka, S., Denlinger, D.L. & Wolda, H. (1987). Daylength and humidity as environmental cues for diapause termination in a tropical beetle. *Physiological Entomology* (in press).

Taylor Jr., F.G. (1974). Phenodynamics of production in a mesic deciduous forest. *Phenology and Seasonality Modelling* (Ed. by H. Lieth), pp. 237–54. Springer-Verlag, New York.

Thiele, H.-U. (1969). The control of larval hibernation and of adult aestivation in the carabid beetles *Nebria brevicollis* L. and *Patrobus atrorufus* Stroem. *Oecologia (Berlin),* **2,** 347–61.

Thomson, A.L. (1950). Factors determining the breeding seasons of birds: an introductory review. *Ibis,* **92,** 173–84.

Thresher, R.E. (1984). *Reproduction in Reef Fishes.* T.F.H. Publications, Neptune City, New Jersey.

Turner, A.K. (1983). Food selection and the timing of breeding of the blue-and-white swallow *Notiochelidon cyanoleuca* in Venezuela. *Ibis,* **125,** 450–62.

Voous, K.H. (1950). The breeding season of birds in Indonesia. *Ibis,* **92,** 279–87.

Wellington, G.M. & Victor, B.C. (1985). El Niño mass coral mortality: a test of resource limitation in a coral reef damselfish population. *Oecologia (Berlin),* **68,** 15–19.

White, T.C.R. (1984). The abundance of invertebrate herbivores in relation to the availability of nitrogen in stressed food plants. *Oecologia (Berlin),* **63,** 90–105.

Whitmore, T.C. (1975). *Tropical Rain Forests of the Far East.* Clarendon Press, Oxford.

Wielgolaski, F.E. (1974). Phenological studies in tundra. *Phenology and Seasonality Modelling* (Ed. by H. Lieth). Springer-Verlag, New York.

Wilbur, H.M. & Alford, R.A. (1985). Priority effects in experimental pond communities: responses of *Hyla* to *Bufo* and *Rana. Ecology,* **66,** 1106–14.

Wilkinson, R. (1983). Biannual breeding and moult-breeding overlap of the chestnut-bellied starling *Spreo pulcher. Ibis,* **125,** 353–61.

Wolda, H. (1978). Fluctuations in abundance of tropical insects. *American Naturalist,* **112,** 1017–45.

Wolda, H. (1980). Seasonality of tropical insects. I. Leafhoppers (Homoptera) in Las Cumbres, Panama. *Journal of Animal Ecology,* **49,** 277–90.

Wolda, H. & Denlinger, D.L. (1984). Diapause in a large aggregation of a tropical beetle. *Ecological Entomology,* **9,** 217–30.

Wolda, H. & Fisk, F.W. (1981). Seasonality of tropical insects. II. Blattaria in Panama. *Journal of Animal Ecology,* **50,** 827–38.

Wolda, H. & Galindo, P. (1981). Population fluctuations of mosquitos in the non-seasonal tropics. *Ecological Entomology,* **6,** 99–106.

Wolda, H. (1983). 'Long-term' stability of tropical insect populations. *Researches on Population Ecology, Supplement 3,* 112–26.

Wolf, L.L. (1969). Breeding and molting periods in a Costa Rican population of the Andean Sparrow. *Condor,* **71,** 212–19.

Wong, M. (1983). Understory phenology of the virgin and regenerating habitats in Pasoh forest reserve, Negeri Sembilan, West Malaysia. *Malaysian Forester,* **46,** 197–223.

II
SPATIAL AND TEMPORAL ORGANIZATION IN CONTEMPORARY COMMUNITIES

Terrestrial Assemblages

5. GLOBAL TRENDS IN SPECIES-RICHNESS IN TERRESTRIAL VEGETATION: A VIEW FROM THE NORTHERN HEMISPHERE

P. J. GRUBB

Botany School, University of Cambridge, Downing Street, Cambridge CB2 3EA, UK

INTRODUCTION

There are two basic problems to be tackled in a global analysis. First, we wish to know how it has come about that some plant communities and regions are richer in species than others. Second, we wish to know how species-richness is maintained in different kinds of plant community. During the last decade there have been numerous reviews concerned with the second question, as shown in the chapters by Chesson, Cody, Grubb, Hubbell & Foster, Knoll, Tilman and Yodzis in the book edited by Diamond & Case (1986). This chapter is concerned primarily with the first question, although the two problems are ultimately inseparable, and cross-references are made throughout to the maintenance of species-richness.

Three general issues must be considered at the start. First, a plant 'community' must be defined. I suggest that it is a collection of plant populations found in one habitat-type in one area, and integrated to a degree by competition, complementarity and dependence (cf. Poore 1964; Grubb 1977). This definition deliberately begs the issue of how much environmental heterogeneity can be tolerated in 'a community', and in this article a relatively broad definition is accepted, including in some cases appreciable heterogeneity in the soil. Certainly 'a community' includes all phases of regeneration, whether that process follows death of single plants from old age, or disturbance of various-sized patches by outside factors.

Second, there is the question of the scale on which patterns of richness are considered. For convenience, the various sample-sizes which have been used are divided here into four categories:

(a) micro-scale, i.e. 0.1–10 (rarely 100) m^2, adopted mainly for herbaceous vegetation and used in innumerable reports by 'phytosociologists' as defined by Ellenberg & Mueller-Dombois (1974),

(b) meso-scale, i.e. 0.1–10 (rarely 100) ha, used originally for forest trees, although under the influence of Whittaker (1965) 0.1 ha plots have been used for all life-forms in all sorts of vegetation,

(c) macro-scale, i.e. $10-100\,km^2$, used for whole reserves or study areas, and (d) mega-scale, i.e. $10^3-10^5\,km^2$, commonly the scale of regional studies by phytogeographers.

Third, in thinking about floristic richness, we have to recognize that there is an arbitrariness about our being concerned with the species as the prime unit. In certain contexts it is helpful to consider numbers of genera and families too; this is particularly so in connection with palaeobotanical studies (see Collinson & Scott, Chapter 18) and concepts such as 'harshness for plant growth'. There is also an arbitrariness about ignoring the extent of intraspecific variation in different types of community.

Reviews of global patterns in species-richness of terrestrial vegetation by Whittaker (1965, 1977), Terborgh (1973), Good (1974), Huston (1979) and Gentry (1982) have identified at least eight factors as having played some part in determining the richness of a given system. Three have to do predominantly with proximal mechanistic effects: (i) the heterogeneity of the system, (ii) the abundance of resources, and (iii) the regime of disturbance. The other five factors are believed to have had long-term effects: (iv) age, (v) area, (vi) degree of isolation from sources of suitable plants, (vii) incidence of catastrophes (e.g. glacial periods), and (viii) incidence of potentially damaging factors, whether permanent (e.g. salt) or periodic (e.g. frost). It is not possible to offer a comprehensive review here. I therefore concentrate on just three themes, and restrict myself to vascular plants.

First, I consider herbs in the Northern Cool Temperate Zone, i.e. that area characterized by winter-deciduous forest, steppe and cool semi-desert. Initially I discuss the idea that species-richness in the herb layer peaks at an intermediate level of standing crop, and show that whereas it is common to find a reduction in species-density associated with large standing crop in grasslands, marshes and mires, this situation is relatively unusual in forests. I then point out the need to consider two sorts of historical factor in accounting for present-day patterns of richness in relation to soil-type: 'apparency' and 'harshness'.

Second, I consider trees, climbers and shrubs in the same geographic area. In this case historical factors are taken first, and a tentative explanation is offered for the gradient in richness from eastern Asia through eastern North America to Europe. In each region species-richness in trees and climbers is found to peak with maximum productivity, providing a marked contrast with the herbs. The pattern for shrubs is intermediate. A hypothesis is presented to account for the differing trends in herbs and in trees and climbers.

Third, I deal with the latitudinal gradient in species-richness at the meso-

and mega-scales, and try to disentangle the roles of age and area on the one hand, and 'harshness' of the environment on the other.

Various themes that emerge under the three topics are drawn together in some concluding remarks.

HERBS IN VEGETATION OF THE NORTHERN COOL TEMPERATE ZONE

At the micro- and meso-scales, in any one region, the trends in species-richness in the herb layer can be largely understood in terms of the word model put forward by Grime (1973). He suggested that species density should increase with increasing productivity up to a certain point (as more and more species can tolerate the conditions) and then decrease with further increase in productivity (as more and more species are eliminated by interference from those best able to make use of the resources). In practice, tests of this model have been made using the maximum standing crop of living and dead material above-ground rather than productivity. The results of Grime (1973, 1979), Wheeler & Giller (1982), Vermeer & Berendse (1983) and During & Willems (1984) for grasslands and mires in northern Europe have proved to be consistent with Grime's model.

The build-up of biomass which is associated with species-loss can come about in two different ways: either because there is a lack of disturbance or because there is greater availability of resources—most often mineral nutrients. In the case of managed grasslands and fens lack of disturbance means lack of cutting, grazing and burning, while in natural grasslands (for which we have no directly comparable data on species-density and standing crop) it would imply lack of fire and grazing, and lack of patchy disturbance by mound-building, trampling or scratching animals (cf. Curtis 1959). Whether the standing crop builds up through greater productivity or lack of disturbance, obligately low-growing perennials and short-lived plants are eliminated by a combination of shade and below-ground interference; in the absence of disturbance the physically constraining effects of the accumulated litter may also be important (Grime 1979).

In the moister parts of the Northern Cool Temperate Zone one is constantly finding examples of grasslands in which species-paucity is causally related to dense shading, below-ground interference and/or accumulation of litter, but in the drier parts one may find an increase in species-richness right up to the most productive sites, at least in natural and semi-natural vegetation. For example, in the steppe region of eastern Washington State the mean number of native species in 4 m² samples in stands with minimal numbers of exotics was found to increase steadily from eight in a semi-desert community with 74% mean cover of all life-forms to thirty-five in a dense grasse-steppe with a mean cover of 296% (values based on Appendix D of Daubenmire 1970). Herbaceous communities in which species-density is reduced by interference from one or a few species either do not occur in the region or are rare enough not to have been sampled.

An analogous position is found much more commonly in the herb layer of forests in the Northern Cool Temperate Zone. Impoverishment of the stand by interference from one or a very few species does occur, but only over a tiny fraction of the landscape. In northern Europe the forest herbs most often seen in extensive, almost monospecific stands are *Mercurialis perennis* and *Urtica dioica*. To form such stands both need freedom from disturbance by trampling animals and falling branches, as well as abundant moisture and nitrate, while the *Urtica* also needs a greater irradiance and phosphate (Pigott & Taylor 1964). In a similar way one can find local impoverishment of the herb layer related, at least in part, to accumulation of herb-litter. Good examples are provided by stands of the grass *Deschampsia caespitosa* and the fern *Pteridium aquilinum*, but these species cover extensive areas only where humans have opened up the forest.

In general, peaks of species-richness in forest herbs are found where conditions for that life-form are the most favourable to be found in the area. Two types of site are involved. First, there are the lower slope and gully forests where the extra supply of water and mineral nutrients brought by flushing enables more species to withstand the shade and below-ground interference from the trees, documented for eastern North America by Whittaker (1965) and Peet & Christensen (1980). Second, there are the thin-canopied forests on ridge-tops and upper-slopes where water and mineral nutrients may both be in short supply but the effects of shade and/or below-ground interference are reduced, documented for the western USA by Peet (1978) and del Moral & Fleming (1979). Peaks of herb-richness in both types of forest in a representative area in central Europe are shown in Table 5.1.

Two of the trends in species-richness shown in Table 5.1 cannot be explained in the proximal mechanistic terms set out so far. First, the decline in species-number on strongly acidic soils; this is not confined to forests of beech but is also seen on similar soils under oaks where there may be more light and less below-ground interference (Ellenberg 1982). Yet, on a world scale, there must be more species of terrestrial plants living on soils of pH 4.0–4.5 than of pH 7.0–7.5. Strong acidity, with its associated chemical factors such as a high concentration of aluminium ions, does not constitute an environmental factor against which plants have, in general, a low chance of survival in evolutionary terms. I suggest a historical explanation for the

TABLE 5.1. Mean numbers of species (mostly $100-400$ m^2) of ten major forest-types in southern Germany and nearby northern Switzerland: data from Oberdorfer (1953) for the floodplain forest, from Braun-Blanquet (1932) for the pubescent oak forest, and Lang (1973) for the other eight types

	Mean number of species [1]				Number of stands sampled
	Herbs	Shrubs	Trees	Climbers	
Floodplain forest					
Elm-ash-oak					
(*Querco-Ulmetum* [2])	20	5.9	8.0	2.1	27
Summer-dry forests rich in half-shade species					
Pubescent oak (*Lithospermo-Quercetum*)	32	9.6	8.5	1.3	12
Oak-hornbeam (*Galio-Carpinetum*)	27	3.1	6.5	1.0	8
Scots Pine (*Cytiso-Pinetum* [3])	27	4.9	3.6	0.1	16
Ridge-top and slope forests of beech not so subject to summer drought					
On calcareous soils (*Carici-Fagetum* [4])	22	3.7	5.6	1.1	18
On moderately acidic soils (*Asperulo-Fagetum* [5])	13	0.3	4.4	0.8	19
On strongly acidic soils (*Luzulo-Fagetum* [6])	8	0.1	3.5	0.5	11
Gulley forests with moist, heavy-textured soils not subject to year-round waterlogging					
Ash-maple (*Aceri-Fraxinetum* [7])	26	3.8	5.9	0.7	18
Streamside forests with soils waterlogged much of the year					
Ash-alder (*Carici-remotae-Fraxinetum* [8])	16	1.3	2.8	0.1	14
Swamp forest					
Alder (*Carici-elongatae-Alnetum*)	12	1.4	2.2	0.0	5
Maximum/minimum [9]	4.0	96	3.9	∞	

[1] The delimitation between 'trees' and 'shrubs' is made as in the text, and not as in the three original papers cited, while 'herbs' include shrubs < 1 m tall and scramblers, and hybrids and exotics are omitted;

[2] Oberdorfer gives only constancy classes (V for $80-100\%$, IV for $60-80\%$, etc.) and the values tabulated are estimates based on V $= 25/27$, IV $= 19/27$, III $= 13/27$, II $= 8/27$ and I $= 3/27$;

[3] Records from subassociations *typicum*, *seslerietosum* and *molinietosum* used;

[4] *typicum* and *caricetosum albae*;

[5] *typicum* and *caricetosum umbrosae*;

[6] *typicum*, *luzuletosum sylvaticae* and *caricetosum umbrosae*;

[7] *aruncetosum* and *allietosum*;

[8] *typicum* and *equisetetosum*;

[9] Largest value in column divided by the smallest.

small number of acid-soil-tolerant species in Europe, put forward earlier for grassland plants (Grubb 1986). During the Quaternary period most soils in northern and central Europe have been rejuvenated several times with additions of glacial till or loess, generally base-rich, and there have been only small areas on which acid-tolerant species could evolve. Considering the area of such sites, integrated through time, they have had a low 'apparency' in Feeny's (1976) terminology. In contrast, along the eastern seaboard of North America there has been in the same period a huge coastal plain of sandy soils, mostly strongly acidic; very species-rich communities of acid-soil-tolerant herbs are to be found there (Walker & Peet 1984).

The second trend not readily explained in proximal mechanistic terms is the reduction in richness at waterlogged sites, seen not only in herbs but in other life-forms (Table 5.1), and not only in European forests but in North American too (Curtis 1959). Moreover, the same trend is seen in trees at waterlogged sites in 'ever-wet' tropical forests (Whitmore 1984), in trees and herbs at waterlogged sites in savannas (Eiten 1978), and shrubs and herbs in the Gran Chaco of Argentina (J. P. Lewis, personal communication). It seems that for reasons not yet understood the chance of evolution of a tolerant genotype has not been equal for all potentially limiting environmental factors. An environmental factor for which this chance has been small may be designated 'harsh'. Persistent waterlogging appears to be a prime example. Salt is certainly another factor of this type, although its effects are often confounded with those of waterlogging, as in salt-marshes and mangrove swamps. 'Harshness' should be distinguished from the concepts of 'stress' *sensu* Grime (1974) and 'adversity' *sensu* Whittaker (1975a), both of which are scaled in terms of productivity. The comparatively few species that grow on waterlogged and/or saline sites can be highly productive (Whittaker 1975b). 'Harsh' sites cannot be equated with the 'peripheral' or 'non-mesic' sites of Terborgh (1973); of the latter some are decidedly 'harsher' than others, e.g. waterlogged and salt-rich sites compared with seasonally very dry sites or sites very poor in available nutrients, both of which can be very rich in families, genera and species.

Persistent and extreme cold is another 'harsh' factor. Arctic and Alpine communities of vascular plants, by definition composed of herbs and low shrubs, are relatively poor to very poor in species (Bliss 1971; Larsen 1974), despite the fact that in the Quaternary period tundra covered large areas of the Northern Hemisphere during each glacial period, and the glacials lasted much longer than the warmer interglacials (Deacon 1983). In this case harshness in evolutionary terms is correlated with low productivity (0.1–4 t ha^{-1} $year^{-1}$; Whittaker (1975b).

TREES, CLIMBERS AND SHRUBS IN THE NORTHERN COOL TEMPERATE ZONE

Trees

At the mega-scale, it is well known that the tree flora of Europe is very poor compared with that of corresponding forested regions in eastern North American and eastern Asia. There are only forty-five apparently indigenous species of angiospermous tree > 9 m tall (forty-four deciduous) and five species of gymnosperm (one deciduous) north of the Alps and west of the USSR, based on Tutin *et al.* (1964–80) and excluding 'critical' species of *Sorbus*. In north-eastern North America (from Virginia and Illinois northward) the figures are 148 (140 deciduous) and twenty-three (one deciduous), based on Gleason & Cronquist (1963). In China the figures are higher again; the two most striking features are (i) the presence of almost all the genera found in the eastern USA plus many others, and (ii) the considerable number of allopatric species in some of the genera. For example, there are 111 indigenous deciduous species of *Acer* (Fang 1981), twenty-five of *Carpinus* (Li & Cheng 1979), and at least four of *Fagus* and eleven of *Ulmus* (Anon. 1972–76), compared with ten, one, one and seven in the whole of eastern North America (based on Small 1913 and Gleason & Cronquist 1963). Moreover, although some of the American species are confined to the south, most are wide-ranging, whereas in China only two to five species of *Acer* and *Ulmus* and one to two of *Carpinus* and *Fagus* are encountered in any one area of a few hundred km^2 (Wang 1961).

Many genera found now in North America and Asia did exist in Europe in the Tertiary, and it is generally accepted that the extinction of many genera in Europe during the Pleistocene resulted from lack of access to suitable refugia to the south during the glacial periods (Takhtajan 1969). One explanation is that the major mountain chains run east–west rather than north–south, but perhaps more important was the drying out of the Mediterranean basin during the glacial periods, the development of steppe and the confinement of even sclerophyllous trees to refugia in the mountains (Wright 1977). Analogous drying out occurred in south-eastern North America but seems to have been less severe, with dry woodland covering most of the area rather than steppe (Delcourt & Delcourt 1985). The greater richness in genera in China suggests that the region dried out less in each glacial period, and certainly the temperate deciduous forest there today is continuous with a series of evergreen rain forests going south (warm temperate, subtropical and tropical), which is not true of the eastern USA. The relative uniformity of the eastern American flora, compared with that of

China, is consistent with a scattering of genetically interlinked refugia for
'mixed mesophytic' elements in the south (cf. Bennett 1985). In China, in
successive glacial periods, the refugia were perhaps more numerous and
scattered over a wider area so that genetic isolation and speciation occurred
much as has been hypothesized for various regions with mediterranean or
lowland tropical climates at the present day (cf. compilations of Valentine
1972 and Prance 1982).

At the meso-scale, there can be no doubt that the greatest number of
different tree species is to be found in the broad floodplain forests of the
great rivers such as the Mississippi, Rhine and Danube. Thus, Robertson,
Weaver & Cavanaugh (1978) recorded fifty-three tree species in one reserve
of 64 ha on the Mississippi. In central Europe twenty-seven of the forty-five
angiospermous trees native north of the Alps occur in remnants along the
Rhine (Carbiener 1970). The few remnants of deciduous floodplain forest in
China can also be extremely species-rich (Chu & Cooper 1950). Although
floodplain forests are generally the richest in tree species at a scale of > 1 ha,
they are not always the richest at the scale of 0.1 ha (Table 5.1), probably
mainly as a result of the large mean tree-size at maturity (cf. Curtis 1959).
Peet & Christensen (1980) found a mean of seventeen tree species per 0.1 ha
in floodplain forest in part of North Carolina, and a mean of twenty-one per
0.1 ha in 'dry mesic eutrophic forest'. These values may be compared with
those obtained in the Great Smoky Mountains by Whittaker (1965): ten per
0.1 ha in cove forest, and thirteen in the adjacent slope forest; he exceeded
these values only by sampling in the ecotone (eighteen per 0.1 ha).

Mention of the ecotone leads naturally to the prime problem in assessing
the species-richness of floodplain forests in relation to the concept of 'a
community'. There is no doubt that its richness is to a significant degree a
result of variation in soil conditions over tens of metres. Because of the
nature of alluvial deposits, there is great variation in texture and drainage,
and the topography may rise and fall by up to 3 m. Thus, in the Mississippian
forest Robertson *et al.* (1978) found *Fraxinus pennsylvanica* and *Taxodium
distichum* concentrated at the wettest sites, while *Fagus grandifolia* and *Tilia
americana* were concentrated at the best-drained. However, they also found
that several species had notably wide tolerances. In the simplest case all tree
species in a floodplain forest can withstand some flooding each year, as in the
remnants along the Rhine (Carbiener 1970). At sites including slightly
higher ground, probably including the Mississippian example quoted, the
trees confined to the highest spots die if flooded in summer (cf. Ellenberg
1982, p. 364, on *Fagus sylvatica* in Europe). Overall, it appears likely that
the species-richness in trees is a result of the fact that so many species can
withstand some degree of annual flooding and benefit from the persistent

supply of water in the subsoil, the relatively large amounts of nitrogen in the litter, and the continual renewal of the nutrient-supply by the yearly floods (Carbiener 1970; Trémolières & Carbiener 1985).

The soil conditions suggest that floodplains forests should be among the most productive in their respective areas, and the little evidence available supports this view. In North Carolina Peet (1981) found distinctly higher values for above-ground net primary productivity in alluvial forests than in upland forests (14–21 vs. 7–15 t ha^{-1} year^{-1}). Vyskot (1978) reported a value of 18 t ha^{-1} year^{-1} for above-ground woody parts at a site in Czechoslovakia, and the probable total productivity was *c.* 25 t ha^{-1} year^{-1}, which is at the upper limit of the range for temperate forests (6–25) given by Whittaker (1975b).

In the case of trees we do not see any tendency in the most productive forests for shade and below-ground interference to reduce the stand to a few species best able to use the resources, as is generally found in the case of herbs. In fact, there are certain forest-types where it does appear that shade and below-ground interference do have this effect, but they are among the less productive. Two types are involved: oceanic and montane forests in which species of *Fagus* or *Tsuga* are the main species, and forests on relatively dry sites with abundant *Taxus baccata* or *Tsuga diversifolia* (Ishizuka 1974; Ellenberg 1982).

How is it that trees and herbs behave differently in this respect? I suggest that it is partly a result of the fact that the trees found in the most productive forests do not have the most densely shading crowns (a point made by Carbiener 1970) and partly a result of different relative rates of gap formation and competitive exclusion, the importance of which is spelt out in the general model of Huston (1979). Herbs of nutrient-rich, moist sites commonly form tall stands which extinguish light extremely effectively and are able to shade out both juveniles and adults of competitors; on invading a gap they soon reach mature height. Many of the species concerned have widespreading stolons or rhizomes, and form large clones. Any gaps are quickly filled. In contrast, among trees the species associated with nutrient-rich, moist sites do not cast the greatest shade, and rarely do they form extensive clones by root-budding. They take many years to reach their mature height. Gaps are not filled so quickly. As long as gaps occur by tree-death or tree-fall reasonably often, it is possible to envisage the maintenance of a high degree of species-richness by differentiation in the regeneration niche *sensu* Grubb (1977). In addition, in any one community at any one time, there will be among the sparse species a number that have become established after invasion from sites where they can maintain themselves indefinitely (the 'mass effect' of Shmida & Whittaker 1981). In communities composed of species that have not been forced to migrate over long distances by changes in climate, many species that hardly differ in their niche may 'coexist' regionally for a very long time as they form small, patchy populations that move about locally in a kaleidoscopic pattern, hardly ever meeting each other (Grubb 1986).

Climbers

Like trees, woody climbers and tall herbaceous climbers (growing to > 2 m) are impoverished in species in Europe relative to eastern North America and eastern Asia, with only eight indigenous species north of the Alps (based on Tutin *et al.* 1964–80) compared with fifty-seven in north-eastern North America (based on Gleason & Cronquist 1963), and even more in the relevant parts of China (Anon. 1972–76). The facts are consistent with the historical explanation offered above for the different richness in tree species.

Within each region, at the meso-scale, woody and tall herbaceous climbers, like trees, are most species-rich where their productivity appears to be greatest, that is, in floodplain forest. In central Europe six of the eight indigenous species in this life-form occur in such forest, and there is a mean of just over two per stand (Table 5.1). In Wisconsin the total numbers of climbers in this forest-type is thirteen, higher than in any other type; Curtis (1959) estimated that they account for up to 25% of the canopy leaves, compared with only 0.1% in upper-slope forest. In floodplain forest one tree may support six to seven individual climbers, compared with only one to two in upland forest, and in the latter the climbers rarely grow far up the trunks. In central China, Chu & Cooper (1950) recorded thirteen species of woody climbers in 0.1 ha of floodplain forest.

It appears that tall climbers escape impoverishment in species at the most productive sites through the shade being moderate and through their regeneration in gaps made by tree-falls, just like the trees themselves.

Shrubs

Tall and medium-height shrubs (woody plants > 1 m and ⩽ 9 m tall, usually with several main stems) are also less diverse in Europe north of the Alps and west of the USSR than in north-eastern North America: 105 (103 angiosperms, ninety-four deciduous) versus 193 (190 angiosperms, 179 deciduous), based on Tutin *et al.* (1964–80) and Gleason & Cronquist (1963) respectively. The ratio for shrub species (1.8:1) is less than that for trees (3.4:1) and much less than that for climbers (7.1:1). Perhaps shrubs, as a group, were better able to withstand the desiccation of the glacial periods and/or could survive in smaller and more numerous refugia.

In both Europe and North America the greatest richness at the meso-scale (eight to fifteen spp. per 0.1 ha) is found on dry and/or nutrient-poor sites, where the biomass is often also greatest (Table 5.1; Curtis 1959; Whittaker 1965, 1966). Throughout the eastern USA shrubs tend to take over from herbs on less fertile soils, whether these are wet, as in the swamps of eastern white cedar (*Chamaecyparis thuyoides*) in New Jersey (Little 1951) and the 'pocosins' of the Carolinas (Christensen *et al.* 1981), or dry as in the chestnut-oak (*Quercus prinus*) forests widespread along the Appalachian chain (Braun 1950). In Europe the greatest richness is in oak forests on dry calcareous soils (Table 5.1) but there is no corresponding richness in thin-canopied oak forest on acidic soils (only two to three species of tall shrubs are found in the alliance Quercion robori-petraeae of oceanic Europe, Ellenberg 1982), and I suggest that the most probable explanation is a historical one, as set out above for herbs.

In Europe the species-density of shrubs is also notably high in floodplain forest (Table 5.1 and Carbiener 1970), and the same can be true in China (Chu & Cooper 1950) and at some North American sites, e.g. on the Piedmont of North Carolina (Peet & Christensen 1980). The density of the shrub layer (number of individuals or mass of foliage per unit area) is very variable, and appears to be related to past disturbance-events. All of the shrubs concerned in Europe are believed to be relatively demanding of light for establishment (rated 6–8 on a scale of 1–9 by Ellenberg 1982), although most can persist in shade. The shrubs are highly productive, and species usually seen on the uplands as short multi-stemmed individuals grow much taller and have single, thick trunks (Carbiener 1970). It is unclear how the balance is maintained between shrub species in some floodplain forests, while in others one or two species of shrub or short tree oust most of the shrubs, as recorded on the Mississippi (Robertson *et al.* 1978) and in Texas (Marks & Harcombe 1981).

Shrubs appear to be intermediate between trees and herbs in terms of their potential for competitive exclusion relative to the rate of gap-formation. They are clearly like the herbs in grasslands and marshes in reaching their highest species-densities at sites with slow growth, i.e. on nutrient-poor and/or dry soils. However, their intermediate nature is emphasized by the fact that there are certain communities in which a low species-density for shrubs is associated with low productivity and apparently caused by competitive exclusion through shading and/or below-ground interference. As in the case of trees, two types of stand are involved: one type is associated with persistent moisture and low nutrient-status, e.g. the

3 m tall stands of evergreen *Kalmia* and *Rhododendron* found on north-facing bluffs near streams in the Piedmont of North Carolina (Peet & Christensen 1980), and the other is associated with dry nutrient-poor sites, e.g. thickets of evergreen *Buxux sempervirens* in central Europe (Ellenberg 1982).

A clear case of the parallel between shrubs and well-lit herbs in their association of high productivity with low species-density is seen in the contrast between the mediterranean-climate shrublands of the Mediterranean basin, central California and Chile on the one hand, and those of the Cape Province in South Africa and of south-western Australia on the other. The two latter regions are very much richer in shrub species at not only the macro-scale but also the meso- and micro-scales (Bond 1983), and this fact is clearly related to the extremely low availability of nutrients in the soils of these regions and the relatively low productivity of the plants (Specht & Moll 1983).

THE LATITUDINAL GRADIENT IN SPECIES-RICHNESS

There is no simple or neat relation between latitude and species-richness at the meso- or macro-scale because of the uneven distribution of land masses, mountain ranges and climatic types, as evidenced by any map of the world's major vegetation-types. Nevertheless, if one broadly defined vegetation-type, say lowland forest, is followed from the Equator to the northern and southern limits, a marked decline in richness is evident. Few cool temperate forests have more than eighty species of vascular plant per 0.1 ha and many have less than sixty (Peet & Christensen 1980), and the same is probably true of both warm temperate rain forests (Wang 1961) and mediterranean-climate sclerophyll forests not degraded by humans and excessively enriched with exotic annuals (Naveh & Whittaker 1979), while tropical forests have 100–300, possibly even 500 (Table 5.2). The overall trend is reflected in the individual trends for trees, climbers and epiphytes (Richards 1952), but may be reversed in ground-dwelling herbs, which are advantaged in the cool temperate forests where the trees are leafless in spring. There is little doubt that, in general, the increase in species-richness toward the Equator is correlated with an increase in total forest productivity (Jordan 1983), much as trees and climbers in the Northern Cool Temperate Zone have been seen to reach their maximum richness in those habitats where they are most productive. Within the lowland tropics the same correlation of high species-richness and high productivity is found; richness in trees and climbers is generally greater in lowland rain forests on average soils than in lowland rain forests on infertile soils or forests at seasonally dry sites (Table 5.2). In a similar way the richness in epiphytic species generally increases from lowland to lower montane rain forests, where the standing crop is greater (Grubb *et al.* 1963; Edwards & Grubb 1977; Grubb & Stevens 1985), but some lowland rain forests with very high rainfall ($\geqslant 4000$ mm year^{-1}) also have large standing crops and high species-density (Whitmore, Peralta & Brown 1985). The evidence presented by Ashton (1977) and Huston (1980) for lower richness in species of large trees in lowland forests on what are believed to be some of the most fertile soils in Borneo and Costa Rica respectively runs counter to the general trend.

How have the cool, temperate regions come to be comparatively poor in species, especially eastern Asia, which suffered fewest extinctions in the Pleistocene? Virtually all the genera of trees found in the Northern Cool Temperate Zone today are known to have evolved by the beginning of the Tertiary, that is, by 60 million years ago, and within 25 million years of the angiosperms becoming the dominant group on earth (Takhtajan 1969; Collinson & Scott, Chapter 18). Furthermore, the 'Arcto-Tertiary' deciduous forest flora covered a huge area, including most of the land under Boreal forest today (Fig. 30 in Takhtajan 1969). It is hardly reasonable to suggest that the relative paucity in species is a result of insufficient age or area. I suggest that the harshness of winter cold has been responsible for the development of a relatively poor cool, temperate flora, at least at the generic level, and that it acted effectively 85–60 million years ago. Certain genera have speciated much more than others, notably *Quercus* in eastern North America (Gleason & Cronquist 1963) and *Acer, Carpinus* and *Quercus* in China (Wang 1961; Anon. 1972–76). The same phenomenon of differing potential for speciation is seen in the tropics (Richards 1969), and is equally difficult to explain.

CONCLUDING REMARKS

Four points of general significance may be made at this stage. First, a paradox is posed by the association of high productivity with species-richness on a world scale, and with species-paucity in the much-studied herbaceous vegetation of the Northern Cool Temperate Zone. I have shown that this paradox does not result from a fundamental difference between tropical and temperate systems, but from a difference between trees and climbers on the one hand and herbs on the other. The potential for competitive exclusion relative to the rate of gap formation and establishment of new individuals is inherently greater in herbs than in trees or tall climbers. Shrubs are intermediate in this respect.

TABLE 5.2. The number of species of vascular plants recorded or estimated in samples of particular tropical communities and in regional floras based largely on these communities

Vegetation-type	Physical conditions	Species-richness			Life-forms especially rich in species
		Meso-scale	Macro-scale	Mega-scale	
Lowland rain forest	Rainfall 1800 to > 10000 mm year^{-1}. 0–1 dry months. Mean screen temperature 20–28°C. Typically large nitrogen supply.	200–?2500 per 0.1 ha and ?2400–600 per ha[1] (ref. 1)	> 1300 in 16 km^2 (ref. 2)	7900 in 1.3×10^5 km^2 (Malaya) 9000 in 6.9×10^5 km^2 (New Guinea)	Trees, climbers
Lower montane rain forest	Rainfall 1500–5000 mm yr^{-1}. 0 dry months. Frequent cloud. Mean screen temperature 11–23°C. Soils more organic; nitrogen availability probably reduced.	100–?350 per 0.1 ha and ?2200–400 per ha (ref. 4)		25000 in 31 ×10^5 km^2 (Malesia) (ref. 3)	Trees and epiphytes
Heath forest ('kerangas' and 'caatinga')	Rainfall 2000–4000 mm year^{-1}. 0–1 dry months. Mean screen temperature 20–28°C. Soils sandy, exceedingly poor in major nutrients.	200–?300 per ha (ref. 5)		> 850 in c. 0.5×10^4 km^2 in Brunei and Sarawak (ref. 6)	Trees
Semi-deciduous seasonal forest ('cerrado')	Rainfall 750–1500 (–2000) mm year^{-1}. 3–4 dry months. Mean screen temperature 20–27°C. Soils acidic, ancient, deep, very low in available nutrients.	Up to 230 per 0.1 ha > 300 per ha (ref. 7)	700–800 in 20–40 km^2 (ref. 7)	About 1000 in 15×10^5 km^2 in Brazil (ref. 8)	Trees, shrubs and herbs

| Deciduous forest | Rainfall 750 mm year^{-1}. 8 dry months. Mean screen temperature 20–25°C. Soils various, not especially old or poor in available nutrients. | > 70 per 0.1 ha and > 100 per ha (ref. 9) | 750–800 in 16 km^2 (ref. 10) | Trees, shrubs, herbs and climbers |

(1) At least five studies have recorded 100–200 species of tree of diameter at breast height (dbh) $\geqslant$ 10 cm in 1 ha (Whitmore 1984). In complete inventories of all life-forms Hall & Swaine (1981) recorded 350 spp. in 0.5 ha in Ghana, and Whitmore, Peralta & Brown (1985) found 233 spp. on 100 m^2 in Costa Rica.

(2) Croat (1978) recorded 1318 spp. from 15.6 km^2 in Panama; in a much wetter area Dodson & Gentry (1978) found > 800 spp. in a mere 0.87 km^2.

(3) Good (1974); Whitmore (1984).

(4) At 2500 m in Papua New Guinea an attempt at a complete inventory on 35–40 ha yielded 348 spp. (Grubb & Stevens 1985); an incomplete inventory of 400 m^2 yielded 80 spp. (Edwards & Grubb 1977). At 1500 m in Ecuador an incomplete inventory yielded 285 spp. on 465 m^2 (Grubb et al. 1963).

(5) About 80–100 spp. of tree of dbh $\geqslant$ 10 cm have been recorded on 1 ha in three studies (Whitmore 1984); a complete inventory is probably 2–3 times as large.

(6) Brünig (1974) and personal communication (1987).

(7) Eiten (1978), quoting results of his own and those of E.P. Heringer and I. Silberbauer-Gottsberger for 0.1–1.0 ha, and those of himself and E. Warming for larger areas; see also Goodland (1970).

(8) C.T. Rizzini, cited by Goodland (1970), lists 600 species of trees and shrubs for the entire cerrado area, taken to be 15×10^5 km^2 by P. de T. Alvim & W.A. Araujo, also cited by Goodland (1970); the list for all life-forms is likely to be 1.5–2.0 times as long.

(9) Gentry (1982) recorded 53–69 spp. per 0.1 ha for trees, shrubs and climbers of dbh > 2.5 cm at five sites in Costa Rica and Venezuela; E.J. Lott (unpublished) found > 100 spp., including all life-forms, in 1 ha in Mexico (quoted by Bullock 1985).

(10) Bullock (1985).

The second general point is that evidence from the fossil record proves that present-day species-richness is strongly dependent on the degree of exposure to past climatic catastrophes. Very severe and all-pervading catastrophes have led to the impoverishment of floras, while moderately severe catastrophes (which have left more refugia) have led to enrichment of floras through speciation in the genetically isolated refugia ('species-pumping' in the sense of Robinson & Gibbs Russell 1982). The first case is illustrated by the European forest flora, and the second is probably illustrated by the Chinese Cool Temperate forest flora as well as the mediterranean-climate floras and tropical rain forest considered in detail by other authors (Valentine 1972; Raven 1977; Hopper 1979; Prance 1982). I have suggested that a second historically important factor has been the 'apparency' *sensu* Feeny (1976) of certain habitat types during the Quaternary.

The third point is that the chance of formation of a tolerant genotype has not been equal for all potentially limiting environmental factors. Many more families, genera and species are tolerant of relatively severe drought or soils very low in available nutrients than are tolerant of waterlogging, salt or winter cold—even fewer tolerate persistent and extreme cold. It would be fascinating to know the reasons at the cellular and molecular level. Meanwhile, it is important not to confuse the concept of 'harshness' here advocated with that of 'stress' or 'adversity' used by other authors and scaled in terms of productivity.

The final point is that the mechanisms of speciation in different types of vegetation, dealt with by many authors in the symposia edited by Valentine (1972) and Prance (1982) are still imperfectly understood. In particular, the tendency of certain genera to speciate more than others is a major puzzle.

All these points need to be explored imaginatively in the future if the global patterns of species-richness in terrestrial vegetation are truly to be explained.

ACKNOWLEDGMENTS

I thank very warmly the many ecologists who have introduced me to the various vegetation-types discussed in this article, Drs H. J. During, P. L. Marks, E. I. Newman and R. K. Peet who constructively criticized the first draft, and Drs D. H. Walton, J. White and T. C. Whitmore who provided valuable references.

REFERENCES

Anonymous (1972−76). *Iconographia Cormophytorum Sinicorum,* 5 Vols (in Chinese). Science Press, Beijing.

Ashton, P.S. (1977). A contribution of rain forest research to evolutionary theory. *Annals of the Missouri Botanical Garden,* **64,** 694−705.

Bennett, K.D. (1985). The spread of *Fagus grandifolia* across eastern North America during the last 18 000 years. *Journal of Biogeography,* **12,** 147−84.

Bliss, L.C. (1971). Arctic and alpine plant life cycles. *Annual Review of Ecology and Systematics,* **2,** 405−38.

Bond, W.J. (1983). On alpha diversity and the richness of the Cape flora: a study of Southern Cape fynbos. *Mediterranean-Type Ecosystems: the Role of Nutrients* (Ed. by F.J. Kruger, D.T. Mitchell & J.U.M. Jarvis), pp. 337−56. Springer-Verlag, Berlin.

Braun, E.L. (1950). *Deciduous Forests of Eastern North America.* Blakiston, Philadelphia.

Braun-Blanquet, J. (1932). Zur Kenntnis nordschweizerischer Waldgesellschaften. *Beihefte zum Botanischen Zentralblatt,* **49,** 7−42.

Brünig, E.F. (1974). *Ecological Studies in the Kerangas Forests of Sarawak and Brunei.* Borneo Literature Bureau, Kuching.

Bullock, S.H. (1985). Breeding systems in the flora of a tropical deciduous forest in Mexico. *Biotropica,* **17,** 287−301.

Carbiener, R. (1970). Un exemple de type forestier exceptionnel pour l'Europe Occidentale: la forêt du lit majeur du Rhin au niveau du fossé Rhénan (*Fraxino-Ulmetum* Oberd. 53). Interêt écologique. Comparison à d'autres forêts thermophiles. *Vegetatio,* **20,** 97−148.

Christensen, N.L., Burchell, R.B., Liggett, A. & Simms, E.L. (1981). The structure and development of pocosin vegetation. *Pocosin Wetlands* (Ed. by C.J. Richardson), pp. 43−61. Hutchinson Rose, Stroudsberg, Pennsylvania.

Chu, C.-H. & Cooper, W.S. (1950). An ecological reconnaisance in the native home of *Metasequoia glyptostroboides. Ecology,* **31,** 260−78.

Croat, J.B. (1978). *The Flora of Barro Colorado Island.* Stanford University Press.

Curtis, J.T. (1959). *The Vegetation of Wisconsin.* University of Wisconsin Press, Madison.

Daubenmire, R. (1970). Steppe vegetation of Washington. *Technical Bulletin of the Agricultural Experimental Station, Washington State,* **62,** 1−131.

Deacon, H.J. (1983). The comparative evolution of mediterranean-type ecosystems: a southern perspective. *Mediterranean-Type Ecosystems: the Role of Nutrients* (Ed. by F.J. Kruger, D.T. Mitchell & J.U.M. Jarvis), pp. 3−40. Springer-Verlag, Berlin.

Delcourt, H.R. & Delcourt, P.A. (1985). Quaternary palynology and vegetational history of the southeastern United States. *Pollen Records of Late-Quaternary North American Sediments* (Ed. by V.M. Bryant & R.G. Holloway), pp. 1−37. American Association of Stratigraphic Palynologists Foundation.

Diamond, J. & Case, T.J. (Eds) (1986). *Community Ecology.* Harper & Row, New York.

Dodson, C. & Gentry, A.H. (1978). Flora of the Rio Palenque Science Center. *Selbyana,* **4,** 1−628.

During, H.J. & Willems, J.H. (1984). Diversity models applied to a chalk grassland. *Vegetatio,* **57,** 103−14.

Edwards, P.J. & Grubb, P.J. (1977). Studies of mineral cycling in a montane rain forest in New Guinea. I. The distribution of organic matter in the vegetation and the soil. *Journal of Ecology,* **65,** 943−69.

Eiten, G. (1978). Delimitation of the cerrado concept. *Vegetatio,* **36,** 169−78.

Ellenberg, H. (1982). *Vegetation Mitteleuropas mit den Alpen in ökologischer Sicht,* 3rd edn. Ulmer, Stuttgart.

Ellenberg, H. & Mueller-Dombois, D. (1974). *Aims and Methods of Vegetation Ecology.* Wiley, New York.

Fang, W. (1981). Aceraceae. *Flora Reipublicae Popularis Sinicae, Vol. 45* (Ed. by W. Fang), pp. 66–273. Science Press, Beijing.

Feeny, P. (1976). Plant apparency and chemical defense. *Recent Advances in Phytochemistry,* **10,** 1–40.

Gentry, A.H. (1982). Patterns of neotropical plant species diversity. *Evolutionary Biology,* **15,** 1–84.

Gleason, H.A. & Cronquist, A. (1963). *Manual of Vascular Plants of North-Eastern United States, and Adjacent Canada.* Van Nostrand, Princeton.

Good, R. (1974). *The Geography of the Flowering Plants,* 4th edn. Longman, London.

Goodland, R.J.A. (1970). Plants of the cerrado vegetation of Brasil. *Phytologia,* **20,** 57–78.

Grime, J.P. (1973). Control of species density in herbaceous vegetation. *Journal of Environmental Management,* **1,** 151–67.

Grime, J.P. (1974). Vegetation classification by reference to strategies. *Nature,* **250,** 26–31.

Grime, J.P. (1979). *Plant Strategies and Vegetation Processes.* Wiley, Chichester.

Grubb, P.J. (1977). The maintenance of species-richness in plant communities: the importance of the regeneration niche. *Biological Reviews,* **52,** 107–45.

Grubb, P.J. (1986). Problems posed by sparse and patchily distributed species in species-rich plant communities. *Community Ecology* (Ed. by J. Diamond & T.J. Case), pp. 207–25. Harper & Row, New York.

Grubb, P.J., Lloyd, J.R. Pennington, T.D. & Whitmore, T.C. (1963). A comparison of montane and lowland rain forest in Ecuador. I. The forest structure, physiognomy and floristics. *Journal of Ecology,* **51,** 567–601.

Grubb, P.J. & Stevens, P.F. (1985). *The Forests of the Fatima Basin and Mt Kerigomna, Papua New Guinea with a Review of Montane and Subalpine Rainforests in Papuasia.* Department of Biogeography and Geomorphology Publication BG/5. Australian National University Press, Canberra.

Hall, J.B. & Swaine, M.D. (1981). *Distribution and Ecology of Vascular Plants in a Tropical Rain Forest: Forest Vegetation in Ghana.* Junk, The Hague.

Hopper, S.D. (1979). Biogeographical aspects of speciation in the southwest Australian flora. *Annual Review of Ecology & Systematics,* **10,** 399–422.

Huston, M. (1979). A general hypothesis of species diversity. *American Naturalist,* **113,** 81–101.

Huston, M. (1980). Soil nutrients and tree species in Costa Rican forests. *Journal of Biogeography,* **7,** 147–57.

Ishizuka, K. (1974). Mountain vegetation. *The Flora and Vegetation of Japan* (Ed. by M. Numata), pp. 173–210. Kodansha, Tokyo.

Jordan, C.F. (1983). Productivity of tropical rain forest ecosystems and the implications for their use as future wood and energy resources. *Ecosystems of the World, 14A. Tropical Rain Forest Ecosystems: Structure and Function* (Ed. by F.B. Golley), pp. 117–36. Elsevier, Amsterdam.

Lang, G. (1973). Die Vegetation des westlichen Bodenseegebietes. *Pflanzensoziologie,* **17,** 1–451.

Larsen, J.A. (1974). Ecology of the northern continental forest border. *Arctic and Alpine Environments* (Ed. by J.D. Ives & R.G. Barry), pp. 341–69. Methuen, London.

Li, P. & Cheng, S. (1979). Betulaceae. *Flora Reipublicae Popularis Sinicae, Vol. 21* (Ed. by K. Kuang & P. Li), pp. 44–137. Science Press, Beijing.

Little, S. (1951). Observations on the minor vegetation of the Pine Barren Swamps in southern New Jersey. *Bulletin of the Torrey Botanical Club,* **78,** 153–60.

Marks, P.L. & Harcombe, P.A. (1981). Forest vegetation of the Big Thicket, southeast Texas. *Ecological Monographs,* **51,** 287–305.

Moral, R. del & Fleming, R.S. (1979). Structure of coniferous forest communities in western Washington: diversity and ecotope properties. *Vegetatio,* **41,** 143–54.

Naveh, Z. & Whittaker, R.H. (1979). Structural and floristic diversity of shrublands and woodlands in northern Israel and other mediterranean areas. *Vegetatio,* **41,** 171–90.

Oberdorfer, E. (1953). Der europäische Auenwald. *Beiträge zur naturkundlichen Forschung in Südwestdeutschland,* **12,** 23–70.

Peet, R.K. (1978). Forest vegetation of the Colorado Front Range: patterns of species diversity. *Vegetatio,* **37,** 65–78.

Peet, R.K. (1981). Changes in biomass and production during secondary forest succession. *Forest Succession: Concepts and Application* (Ed. by D.C. West, H.H. Shugart & D.E. Botkin), pp. 324–38. Springer-Verlag, New York.

Peet, R.K. & Christensen, N.L. (1980). Hardwood forest vegetation of the North Carolina Piedmont. *Veröffentlichungen des geobotanischen Institutes, Zürich,* **69,** 14–39.

Pigott, C.D. & Taylor, K. (1964). The distribution of some woodland herbs in relation to the supply of nitrogen and phosphorus in soil. *Supplement to Journal of Ecology,* **52,** 175–85.

Poore, M.E.D. (1964). Integration in the plant community. *Supplement to Journal of Ecology,* **52,** 213–26.

Prance, G.T. (Ed.) (1982). *Biological Diversification in the Tropics.* Columbia University Press, New York.

Raven, P.H. (1977). The California flora. *Terrestrial Vegetation of California* (Ed. by M.G. Barbour & J. Major), pp. 109–37. Wiley, New York.

Richards, P.W. (1952). *The Tropical Rain Forest.* Cambridge University Press, Cambridge.

Richards, P.W. (1969). Speciation in tropical rain forest and the concept of the niche. *Biological Journal of the Linnean Society,* **1,** 149–53.

Robertson, P.A., Weaver, G.T. & Cavanaugh, J.A. (1978). Vegetation and tree species patterns near the northern terminus of the southern floodplain forest. *Ecological Monographs,* **48,** 249–67.

Robinson, E.R. & Gibbs Russell, G.E. (1982). Speciation environments and centres of diversity in southern Africa. I. Conceptual framework. *Bothalia,* **14,** 83–88.

Shmida, A. & Whittaker, R.H. (1981). Pattern and biological microsite effects in two shrub communities, southern California. *Ecology,* **62,** 234–51.

Small, J.K. (1913). *Flora of the Southeastern United States,* 2nd edn. Published by the author, New York.

Specht, R.L. & Moll, E.J. (1983). Mediterranean-type heathlands and sclerophyllous shrublands of the world: an overview. *Mediterranean-Type Ecosystems: the Role of Nutrients* (Ed. by F.J. Kruger, D.T. Mitchell & J.U.M. Jarvis), pp. 41–65. Springer-Verlag, Berlin.

Takhtajan, A.L. (1969). *Flowering Plants—Origin and Dispersal.* Oliver & Boyd, Edinburgh.

Terborgh, J. (1973). On the notion of favourableness in plant ecology. *American Naturalist,* **107,** 481–501.

Trémolières, M. & Carbiener, R. (1985). Quelques aspects des interactions entre litière forestières et écosystemes aquatiques ou terrestres. *Revue d'Écologie (Tierre et Vie),* **40,** 435–49.

Tutin, T.G., Heywood, V.H., Burges, N.A., Valentine, D.H., Walters, S.M. & Webb, D.A. (1964–80). *Flora Europaea,* Vols. *1–5.* Cambridge University Press, Cambridge.

Valentine, D.H. (ed.) (1972). *Taxonomy, Phytogeography and Evolution.* Academic Press, London.

Vermeer, J.G. & Berendse, F. (1983). The relationship between nutrient availability, shoot biomass and species richness in grassland and wetland communities. *Vegetatio,* **53,** 121–36.

Vyskot, M. (1978). Tchechoslowakische Urwaldreservate als Lehrobjekte. *Allgemeine Forstzeitschrift,* **24,** 696–7.

Walker, J. & Peet, R.K. (1984). Composition and species diversity of pine-wiregrass savannas of the Green Swamp, North Carolina. *Vegetatio*, **55**, 163–79.

Wang, C.-W. (1961). The forests of China. *Maria Moors Cabot Foundation (Cambridge, Massachusetts) Publication*, **15**, 1–313.

Wheeler, B.D. & Giller, K.E. (1982). Species richness of herbaceous fen vegetation on Broadland, Norfolk in relation to the quantity of above-ground plant material. *Journal of Ecology*, **70**, 179–200.

Whitmore, T.C. (1984). *Tropical Rain Forests of the Far East*, 2nd edn. Clarendon Press, Oxford.

Whitmore, T.C., Peralta, R. & Brown, K. (1985). Total species count in a Costa Rican tropical rain forest. *Journal of Tropical Ecology*, **1**, 375–8.

Whittaker, R.H. (1965). Dominance and diversity in land plant communities. *Science*, **147**, 250–60.

Whittaker, R.H. (1966). Forest dimensions and production in the Great Smoky Mountains. *Ecology*, **47**, 103–21.

Whittaker, R.H. (1975a). The design and stability of plant communities. *Unifying Concepts in Ecology* (Ed. by W.H. van Dobben & R.H. Lowe-McConnell), pp. 169–81. Junk, The Hague.

Whittaker, R.H. (1975b). *Communities and Ecosystems,* 2nd edn. Macmillan, New York.

Whittaker, R.H. (1977). Evolution and species diversity in land communities. *Evolutionary Biology*, **10**, 1–67.

Wright, H.E. (1977). Environmental change and the origin of agriculture in the Old and New Worlds. *Origins of Agriculture* (Ed. by C.A. Reed), pp. 281–318. Houton, The Hague.

6. SPATIAL AND TEMPORAL SEPARATION OF ACTIVITY IN PLANT COMMUNITIES: PREREQUISITE OR CONSEQUENCE OF COEXISTENCE?

A. H. FITTER

Department of Biology, University of York, Heslington, York YO1 5DD, UK

INTRODUCTION

A central problem of community ecology is to explain the coexistence of plant species within communities. In this paper I shall examine the role of differentiation in the physical niche—the ability of coexisting species to reduce overlap in their utilization of resources—in the organization of plant communities. It is often assumed, following Gause (1934), that coexistence implies niche differentiation, and much of the ecological literature on the control of species-richness or alpha diversity is in this vein, for example the analysis of dominance–diversity curves (Whittaker 1972). Where the community is unstable or is subject to repeated disturbance, this is not the case, and powerful external forces such as grazing can also modify species' behaviour patterns so as to render niche differentiation unnecessary by suppressing competitive interactions.

Where the component species of a community do have the potential to interact strongly, however, differentiation seems intuitively a likely explanation for coexistence. There is, of course, no plant equivalent of the animal 'food niche' because of the identity of the resource set utilized by green plants. When plants compete for these resources they may do so consumptively, reducing the rate of supply of the resource to a competitor, or by pre-empting space and the resources contained in it (Nicholson 1954; Yodzis 1986). Niche differentiation could therefore occur in respect of either of these modes of competition: differences in the relative demands for various resources (Williamson 1957; Tilman 1982) would alleviate consumptive competition, while avoidance in either space (e.g. different soil horizons) or time of the activity of a competitor would influence space competition.

It is this last possibility, which is often assumed but less frequently documented, that I wish to explore in detail. If plant species do compete for space, then selection is presumed to lead to such avoidance—Connell's (1980) 'ghost of competition past'. Do plant communities therefore comprise co-evolved sets of species? (Turkington & Aarssen 1985)—a result that would be startlingly Clementsian—or, as Connell (1980) believes more probable, do they consist of species that happen to have complementary behaviour patterns and so find it possible to coexist?—a more Gleasonian view. I shall also explore a third possibility, consistent with the last, but in which complementary patterns of behaviour arise secondarily, as a result of interference interactions which result in some species pre-empting particular spatial or temporal domains.

My approach here will necessarily be more mechanistic than is often the case, since it will be insufficient to demonstrate that species coexist, without explaining the mechanism of coexistence. In particular I want to draw attention to a part, arguably the most important part, of the plant environment in which competitive interactions occur: the soil. The existence of competition between roots has been known for many years (Watt & Fraser 1933; Donald 1958) and is well documented in agricultural systems (Aspinall 1960; Idris & Milthorpe 1966), but there are very few clear demonstrations of the role of root competition in natural communities. This is surprising, since I hope to show that the possibilities for niche differentiation are much greater below-ground than in the aerial environment.

NICHE DIFFERENTIATION AND COEXISTENCE

I take coexistence here to mean that individuals of two species grow in sufficient proximity to have the potential to interfere with one another. Where there is spatial heterogeneity of environmental factors or resources, species may be distributed differentially with respect to those factors, so that their overlap is minimized. In the extreme, this may lead to two species occurring within the same area, but in mutually exclusive parts of it, and so not coexisting, as might occur with the vegetation of wet or saline depressions in grassland.

The spatial pattern of such differentiation may be large and typically along environmental gradients, as in Whittaker's (1956) study of tree species distributions in relation to elevation and topography in the Great Smoky Mountains, or it may be on a smaller scale. Differential response to moisture gradients commonly leads to species being found in non-coincident sites within an area (Platt & Weiss 1977; Fekete & Précsényi 1981), and other factors such as topography (Cody 1978) are also frequently involved in delineating niche boundaries. In all these cases, therefore, differentiation clearly reduces the spatial (horizontal) overlap of the distributions, and, while this is a satisfactory correlate of beta diversity, it remains to be seen whether equivalent mechanisms can explain the coexistence of species within habitats, where they apparently are sharing space.

Précsényi *et al.* (1979) measured niche parameters of the main species of a Hungarian grassland, and found that they formed two clear groups with respect to soil moisture and root depth, those found in the driest sites being more shallow-rooted (3–5 as against 7–9 cm). Of the two axes used there, one (soil moisture) clearly leads to horizontal spatial separation, while the other has the potential to allow coexistence within a site.

Above ground

Because the main above-ground resource, photons, is directional, spatial differentiation of the above-ground environment with respect to irradiance is not possible. This is not to suggest that competition for light cannot occur, as it clearly does where individuals within a canopy have overlapping leaf distributions, but only that this cannot lead to stable coexistence on its own. Instead, there is physiological differentiation, such that lower leaves in a canopy are able to operate at lower resource levels. This may be a phenotypic response, shown in the sun and shade forms of leaves of forest trees (Haberlandt 1884), or a genotypic difference exemplified by shade races of plants such as *Solidago virgaurea* (Björkman 1968) and by the adaptation to deep shade of rain-forest floor species such as *Cordyline rubra* (Björkman, Ludlow & Morrow 1972). There is no sense here in which canopy and forest-floor individuals are competing for light; the canopy acts simply to create an environment exploited by the other species. To allow such one-sided interactions to be termed competitive, would mean that inanimate objects such as stones could be held to compete with plants.

This is more than a semantic point, since there are numerous published experiments in which species have been removed from communities and increased growth of some remaining species recorded (e.g. Allen & Forman 1976; Fowler 1981). However, these responses are usually diffused through the community and are rarely reciprocal; only Silander & Antonovics (1982) appear to have recorded a reciprocal response, and that because the two species were the co-dominants in a species-poor community. This strongly suggests that these competitive interactions are not perceived by the plant in terms of the effects of particular competitors, but in terms of the diminution of resource levels, and that selection acts accordingly.

In contrast, the separation of above-ground growth and flowering times has been interpreted as a mechanism that allows coexisting species to avoid competition. It is normal to find that species within a community display non-synchronous activity, often but by no means always in such a way as to permit the identification of guilds (Folwer & Antonovics 1981; Fitter 1986a; and see Wolda, p. 83). A number of explanations are available for this pattern, including avoidance of competition for nutritional resources (water, light, mineral nutrient ions) or for pollinators (Stiles 1977; Rathcke & Lacey 1985), or avoidance of hybrid production in closely-related species (Rogers & Westman 1979). The first of these explanations seems intuitively to be the most general, as it can apply to any group of species, not just to closely-related species or those that share pollinators, and indeed its truth is often assumed. Nevertheless there is little clear evidence to demonstrate it in plant communities. Indeed, if the productivity of a group of coexisting but temporally staggered species is plotted against their time of activity, it will generally be found that the early active species, in a temperate grassland for instance, are less productive than those reaching maximum productivity later (Fig. 6.1). This could imply that there is an optimum time for activity,

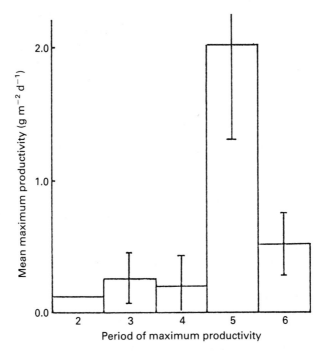

Fig. 6.1. Mean maximum productivity of groups of species that reach their maximum productivity during different time periods within the growing season. Numbers 2–6 refer to the following periods (no species reached maximum productivity during period 1, before 3 April): 2: 3–19 April; 3: 19 April–2 May; 4: 2–30 May; 5: 30 May–11 June: 6: 11–26 June. Data are from an alluvial grassland near York (from Fitter 1986a).

occupied by the most productive species, and that other species then fill in the remaining time periods, which are markedly less favourable. I would interpret this as competition for 'time' in the same way that pre-emption of spatial parts of the environment is competition for space (de Wit 1960; Yodzis 1986). The relevant test is a reciprocal removal experiment, in which the prediction would be that the early-active species would encroach more upon the time left vacant by the dominant than vice versa; to my knowledge this has not been done.

Such pre-emption might of course produce a number of responses in the escaping species. In deciduous woodland, some forest-floor species are active early before the canopy expands (e.g. *Anemone nemorosa, Erythronium americanum*) and have an appropriately modified physiology and life-history, whereas others persist at low rates of activity throughout the summer (e.g. *Mercurialis perennis, Aster acuminatus*). Equally, early-season activity may for some grassland species be their fundamental niche (e.g. *Ranunculus ficaria, Pulsatilla* spp.), whereas others might have a broader fundamental niche, but be forced into a limited (realized) part of it by the dominant species (*Poa pratensis*: Veresoglou & Fitter 1984). An excellent system in which to test these ideas would be continuously-moist tropical rainforest, where there should be no clear optimum period available for pre-emption.

Below ground

Above ground in terrestrial plant communities there appears, therefore, to be little opportunity for niche differentiation within a community. Indeed it is hard to maintain the distinction between competition for space and consumptive competition in this area because to achieve one involves the other. Below ground, in contrast, there are both more resources (water and the various ions) to be competed for and a diffuse supply pattern which means that occupation of space does not automatically confer title to the resources in that space. In other words the resources that lie between two roots may be obtained by either, depending upon the physiological characteristics of the two roots and the plants to which they are attached (Fitter 1986b), whereas the relative amounts of radiation absorbed by two leaves are determined almost entirely by their positions.

There is, however, one major problem in assessing the importance of below-ground niche differentiation. Despite the existence of many diagrams that show roots of coexisting plants apparently occupying different soil layers, and therefore suggesting that spatial separation of roots may be important in many communities, there is remarkably little quantitative data with which one can test such a hypothesis. This is due partly to inadequate techniques and partly to a failure to recognize the importance of below-ground phenomena in determining community structure.

Methodology

Root ecology got off to an excellent start with the pioneering studies of Cannon (1911), Weaver (1919) and others. The excavation methods they used were extremely laborious and by modern standards prohibitively expensive. Unfortunately, too, they reveal a typological viewpoint which is unacceptable to modern ecologists. The characterization, in fine cartographic detail and in two dimensions, of the root system of individual plants cannot be used as a basis for generalization about root systems, since the sample sizes are necessarily small. Important deductions about root distribution, particularly in arid and semi-arid communities, did come from these early studies, but for a long time afterwards little progress was made, though various methods were developed (Böhm 1979).

One such is direct observation through root observation windows. Although giving continuous records of root growth *in situ,* this method suffers from its restrictions to two dimensions and again from the enormous variability in the data it produces (Atkinson 1985). More sophisticated versions, using endoscopes and fibre optic systems (Vos & Groenwold 1983) have been developed recently. Though they involve less disturbance, they suffer from many of the same limitations.

The most fruitful methods involve tracers. These can be fed to leaves, for example as $^{14}CO_2$ (e.g. Litav & Harper 1967; Baldwin & Tinker 1972), and recovered from soil cores to give an indication of root sink strength in different locations, or they can be injected into soil and recovered in the shoots. The latter method has the advantage that it measures uptake activity, which in studies of community function is the most important variable; it has been widely used with ^{32}P (Nye & Foster 1961; Newbould 1969; Syers, Ryden & Garwood 1984). Non-radioactive tracers, such as strontium (Fox & Lipps 1964; Soileau 1973; Veresoglou & Fitter 1984) and lithium (Sayre & Morris 1940) may be more versatile for field use. Combinations of tracers give much greater precision for field work: Fitter (1986a) used Li, Rb and Sr simultaneously, and Caldwell *et al.* (1985), in an elegant study, used ^{32}P and ^{33}P together.

The techniques for studying root activity in the field are now, therefore, well developed and, although these tracer techniques are all indirect, they are capable of providing the necessary data to test hypotheses about root activity in natural communities.

Temporal variation

One obvious question is whether there is separation of root activity in time. Paradoxically, in one case where this has been unequivocally demonstrated, the result is exclusion rather than coexistence. Harris (1967), in one of the best investigated cases of interspecific competition, showed that the introduced Eurasian annual grass, *Bromus tectorum*, replaces the native perennial *Agropyron spicatum* over much of the inter-mountain region of North America by its earlier root elongation (Fig. 6.2). In a habitat of low summer rainfall, this results in *B. tectorum* pre-empting stored soil water which is therefore never accessible to *A. spicatum*. Importantly, in areas where summer precipitation occurs regularly, such as the Great Plains and south-eastern Washington, *B. tectorum* is a much poorer competitor, confined to disturbed areas (Daubenmire 1942; Harris 1967).

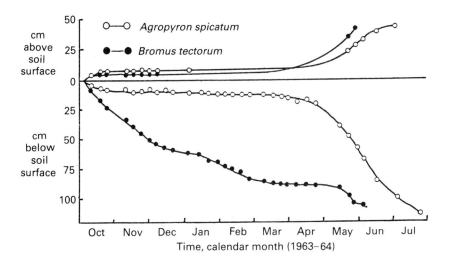

FIG. 6.2. Root penetration and shoot elongation of co-occurring *Agropyron spicatum* and *Bromus tectorum* (from Harris 1967).

In this example, as with that of competition for light between shoots above, it is the particular distribution of the resource which permits its pre-emption by one species. Veresoglou & Fitter (1984) examined time of root activity, using stable Sr as a tracer, in a species-poor acid grassland. In an area with only three species, there was a clear temporal pattern of root activity: *Poa pratensis* followed by *Holcus lanatus* and then *Agrostis capillaris*. This series was, however, linked with a spatial heterogeneity: the order of depth of root activity was exactly the same, with *P. pratensis* shallowest and *A. capillaris* deepest, and this in turn was linked to the ability of the plants to grow (in growth room experiments) at low temperature or soil moisture. In other words there were differences in fundamental as well as realized niches: the ability of *P. pratensis* to tolerate low temperatures suited it to growth early in the season and that of *A. capillaris* to withstand drought suited it to later activity. The same link between the spatial and temporal patterns was found by Fitter (1986a).

Spatial variation

Several studies have looked at root distribution or activity at different depths in soil, and some indicate that coexisting species may be well separated. Kummerow & Mangan (1981), for instance, extracted roots of five species in a *Quercus dumosa*-dominated chaparral in southern California. Only one species, *Q. dumosa* itself, showed significant root development below 60 cm, while the patterns for all five in the top 60 cm departed significantly from random (Table 6.1). The largest contribution to χ^2 was from *Ceanothus greggii* which had a large excess of roots in the top 20 cm, and from *Q. dumosa* which had fewer than expected in that layer.

TABLE 6.1. Distribution of fine roots (< 5 mm diameter) of five coexisting shrubs in *Quercus dumosa*-dominated chaparral. Values are in g per 36 m² plot. (Data from Kummerow & Mangan 1981)

Species	Depth (cm)			Contribution to χ^2
	0–20	20–40	40–60	
Eriogonum fasciculatum	100*	339	208	5.7
Quercus dumosa	271†	2136	1746	22.0
Adenophora fasciculatum	185	517	409*	12.5
Ceanothus greggii	118*	207	189	102.8
Cercocarpus betulifolius	11	156	240*	12.2
			Total χ^2	155.2

*Greatest contribution to χ^2 for that species; roots more abundant than expected.
†Greatest contribution to χ^2 for that species; roots less abundant than expected.

The step from such a level of description to one that encompasses niche differentiation is large. We need to know the activity of the roots and the

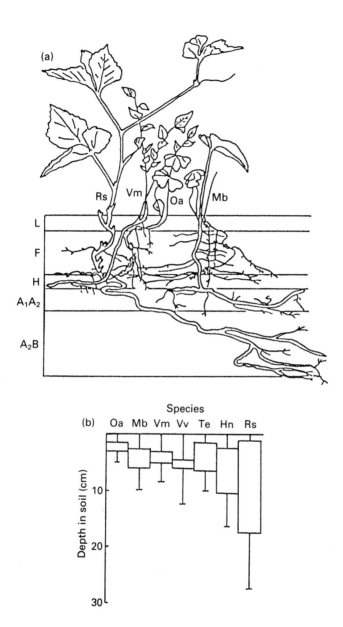

FIG. 6.3. (a) Profile diagram of root and shoot systems of four coexisting spruce-forest species (Kirikova 1970): Rs = *Rubus saxatilis*; Vm = *Vaccinium myrtillus*; Oa = *Oxalis acetosella*; Mb = *Maianthemum bifolium*. (b) Upper and lower bounds of root zone of seven coexisting species in a spruce-forest, with 95% confidence limits (from Kirikova 1970). Abbreviations as in Fig. 6.3a, plus: Vv = *Vaccinium vitis-idaea*; Te = *Trientalis europaea*; Hn = *Hepatica nobilis*.

effect of varying degrees of overlap. There is certainly little evidence for completely exclusive root distributions, even in forest systems where profile diagrams seem to suggest it. Kirikova (1970), for example, presents a profile diagram of a spruce forest floor (Fig. 6.3a) which appears to show mutually exclusive rooting patterns, but the accompanying data show that rooting zones of component species overlap strongly (Fig. 6.3b), although there were significant differences in rhizome depth.

I have attempted to approach this problem by measuring root activity with tracers, rather than actual root distribution. In a species-rich alluvial grassland I could find little evidence of spatial separation of root activity with depth, beyond a general tendency for early-active species to be shallow-rooting and vice versa (Fitter 1986a). This, however, was apparently merely a facet of a general trend to deeper rooting as the season progressed, displayed by all species (Fig. 6.4).

If separation of root activity in distinct soil layers is by no means a universal phenomenon, it is possible that spatial differentiation takes place in a more subtle sense. Sheikh & Rutter (1969) found that *Molinia caerulea* had more root mass in a series of mire sites than *Erica tetralix* with which it was associated, and that it had a smaller proportion of its roots at the surface (around 10 as opposed to 30%), but more at 19–27 cm depth (12 as opposed to 2%). In addition to these differences, the modal diameter of *E. tetralix* roots at their deepest penetration (15 cm) was 200–250 μm, as compared to 100–150 μm for *M. caerulea* roots, and this was reflected in the different sizes of soil pores occupied most frequently by the two species: 600–1200 μm for *E. tetralix* and 300–600 μm for *M. caerulea*. The possibility exists, therefore, that plant species might express niche differentiation in relation to the pore size classes that their roots exploited.

To test this idea, Martin (1979) performed de Wit replacement series experiments using species with different mean root diameters, and in soil prepared by sieving to offer different dominant pore sizes. He grew three species, *Holcus lanatus, Poa annua* and *Silene alba*, in two-species mixtures and in monoculture in three sand types sieved to give a range of 300–425 μm (fine sand) and 425–600 μm (medium sand). These, when packed, should give minimum pore sizes of 50–70 μm and 70–90 μm respectively (Goss 1977); in practice, since they were not tightly packed, these figures would be a little larger. The third type was a mixture of these two fractions, which should offer a wide range of pore sizes. *H. lanatus* and *P. annua* had very similar mean root diameters (126 and 121 μm respectively), but roots of *S. alba* were significantly larger at 147 μm. These values were not altered by growth conditions. The ability of plants of these three species to coexist, as indicated by the relative yield total (RYT) in a 1:1 mixture, depends upon

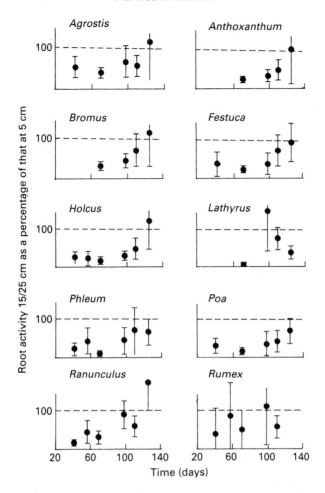

FIG. 6.4. Root activity at 15/25 cm depth (mean separate measurements at 15 and 25 cm) expressed as a percentage of that at 5 cm for ten coexisting species in an alluvial grassland near York over the period March–June 1984 (from Fitter 1986a)

sand type (Table 6.2). In all cases but one where a uniform medium was used, RYT values were not significantly different from 1; in the exception (*H. lanatus* vs *P. annua* in medium sand), the RYT was only trivially, though significantly ($P < 0.05$), less than 1. In all three cases where mixed sand was used, however, RYT values were significantly greater than 1. It is tempting to relate this simply to the occupation of different sized pores by different sized roots, but this cannot be a wholly adequate explanation, since *H. lanatus* and *P. annua,* with roots of similar diameter, also produced an RYT

TABLE 6.2. Relative yield totals (RYT) of mixtures of three species grown for 76 days in 1:1 mixtures in three sand types. Values greater than 1.00 indicate coexistence. Data are means ±95% confidence limit. (From Martin 1979)

Sand type	Species contribution		
	Holcus lanatus and *Poa annua*	*Holcus lanatus* and *Silene alba*	*Poa annua* and *Silene alba*
Fine	0.97±0.13	0.90±0.77	0.80±0.60
Medium	0.97±0.02	1.18±0.65	1.22±1.08
Mixed	1.44±0.18	1.55±0.36	1.45±0.43

> 1 in the mixed sand. Nevertheless, these results suggest that soil pore size may be an important and largely unexplored resource axis for plant roots.

Aquatic habitats

In terrestrial habitats the soil is potentially a suitable environment for the development of niche differentiation, as described above. Aquatic plants, however, absorb mineral nutrient ions directly from the water through their leaves and the root system appears to act primarily as an anchor. In aquatic communities, therefore, there should be less opportunity for niche differentiation and consequently lower species diversity. Testing this prediction is complicated by the problem of identifying appropriate terrestrial comparisons where grazing and other extrinsic forces exert similar effects, but the very short species lists for vascular plants in most aquatic habitats suggest that they have low species diversity (Tansley 1949). This represents a worthwhile area for experimental study.

COMPLEMENTARY PATTERNS

In terrestrial communities, there is as yet no clear picture of the importance of below-ground niche differentiation. In cases where environmental heterogeneity is marked, as in arid environments with seasonal rainfall, a pronounced discontinuity in root distribution in space and time may appear between deep-rooting perennials and shallow-rooted ephemerals. In situations where apparent true coexistence can be found, we await a fully quantitative demonstration of below-ground relationships.

Nevertheless, it seems that complementary patterns in both time and space between coexisting species can often be observed. There are two possible explanations for this: the first assumes that species are inherently complementary, either because of an evolutionary response by species that frequently coexist (Connell's (1980) 'ghost of competition past') or because for quite other reasons the species have distinct ecologies. Connell (1980) cogently argues in favour of the latter, but an alternative view is that this is the result of the displacement of some less aggressive species from favourable volumes of the environment or time-periods, in other words as a consequence rather than a prerequisite of coexistence. If the former explanation were valid, one might argue that communities were composed of species subtly co-adapted or otherwise suited for coexistence. For this to have arisen, two conditions would be necessary:

1 Neighbours would have to be predictable in space or time (where neighbours might represent a single genotype or species or any set of members of a guild). This also means that differentiation is only likely between abundant species (Grubb 1986).

2 There would have to be sufficient environmental heterogeneity for it to be possible for distinct response patterns to be adopted by coexisting species.

Most studies of small-scale pattern in vegetation suggest that the first condition is unlikely to be sufficiently met (e.g. Fowler & Antonovics 1981), although Turkington & Harper's (1979) studies on *Trifolium repens* in association with different grass species might suggest that particular genotypes can adopt behaviours which lead to best growth with particular associates. This study was not reciprocal. More recently Aarssen & Turkington (1985a) have found that whereas *T. repens* genotypes tend to perform best in association with their 'usual' *Lolium perenne* genotype (i.e. that with which they were found in the field), there is no reciprocal response by the *L. perenne* genotypes. Indeed they tended to perform badly in the 'usual' combination, and Aarssen & Turkington (1985b) suggest that the explanation for these phenomena does not lie in traditional niche differentiation.

Evidence for the first condition is therefore equivocal; the second condition is inherently more likely and is of course equally necessary for the development of coexistence by the pre-emption of favoured space or time by dominant species.

There are a number of experimental studies in which the effect of one species on the pattern of resource acquisition in soil by another has been monitored. O'Brien, Moorby & Whittington (1967) showed that x *Festulolium loliaceum*, the triploid hybrid of *Festuca arundinacea* and *Lolium perenne*, was deep-rooting and competitively dominant to both parents and effectively excluded all root activity by the other two species in deep soil layers at around 60 cm depth, even though these layers were important phosphorus sources to both parents in monoculture. Similarly Fitter (1976) showed that *Lolium perenne* preferentially developed roots in soil volumes not occupied by *Plantago lanceolata*, and Bookman & Mack (1982) explained the supplanting of *Bromus tectorum* by *Poa pratensis* as a consequence of the restriction imposed by the latter on the lateral root development of *B. tectorum*.

All these studies suggest that complementary patterns observed in the field are likely to be the result of an interaction. This is particularly clear in the species-poor, acid grasslands studied by Veresoglou & Fitter (1984). On one site, where only *Poa pratensis*, *Holcus lanatus* and *Agrostis capillaris* occurred, *H. lanatus* occupied the favourable growth period in early summer, achieving a peak productivity of around 6 g m^{-2} d^{-1}. Nearby, *Arrhenatherum elatius* and *Deschampsia cespitosa* also occurred, and had peak productivities simultaneous with *H. lanatus* and together totalling about 7.5 g m^{-2} d^{-1}; here *H. lanatus* achieved only 1.6 g m^{-2} d^{-1}. Most significantly, although peak productivities coincided in time (3 June: Fig. 6.5a), peak nutrient (P and K) uptake was staggered, with *H. lanatus* reaching peak uptake on 20 April as opposed to 10 May where it was the dominant species, *D. cespitosa* on 10 May and *A. elatius* on 3 June (Fig. 6.5b).

This is clear evidence that the introduction of two dominant species had caused a shift in the time of resource acquisition in a suppressed species. Similarly Fitter (1986a), in a species-rich grassland, has found evidence for a comparable spatial shift. Here patches of *Sanguisorba officinalis* occurred in which most other species were suppressed. Two that managed to persist in *Sanguisorba* patches were *Ranunculus acris* and *Agrostis capillaris*, the first of which was apparently excluded from deep (25 cm) soil layers by *S. officinalis*, whereas the pattern of uptake of the latter was unaffected (Fig. 6.6).

CONCLUSIONS

Differentiation certainly occurs in the timing and location of root activity between coexisting species. Even where above-ground productivity seems to be coincident, below-ground events may be staggered. Temporal and spatial patterns may, however, be strongly linked: in many temperate systems the surface soil layers are the first to become warm enough for root growth in spring, but may dry out in summer, so that a progression from early-growing, shallow-rooting to later, deeper-rooting species may occur (Veresoglou & Fitter 1984). This may give the appearance of a mechanism that promotes coexistence, but the introduction of a species able to continue

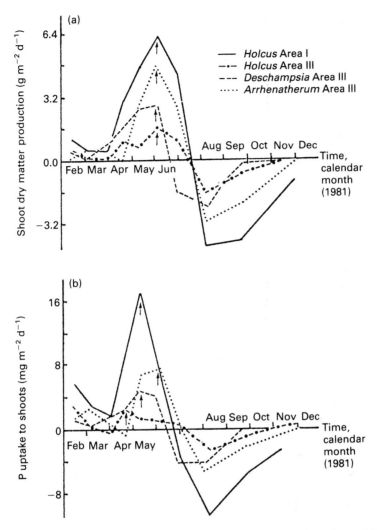

FIG. 6.5. Distribution of (a) shoot dry matter production and (b) P uptake to shoots of three grass species in an acid, sandy grassland near York (from Veresoglou & Fitter 1984). In Area II *Holcus lanatus* was dominant and the other two species absent; in Area III *Holcus* was subordinate to the other two.

root growth at lower temperatures may lead to competitive exclusion, as shown by Harris (1967) for *Bromus tectorum*.

This correlation of spatial and temporal activity means that if we wish to demonstrate niche differentiation below-ground, it is insufficient to sample root distributions on one occasion. The data required are measurements of

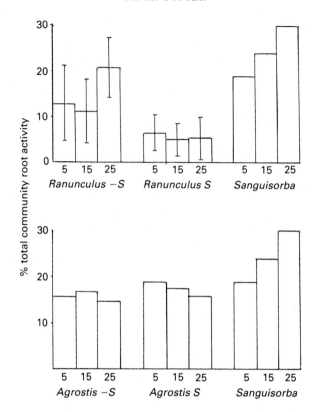

FIG. 6.6. Mean percentage of total community root activity (as measured by uptake of Li, Rb and Sr tracers) at three depths (5, 15 and 25 cm) for three species in an alluvial grassland near York (Fitter 1986a). Values are means of three separate harvests in May and June 1984. Data for *Ranunculus acris* and *Agrostis capillaris* have been separated into plots with (S) and without (−S) *Sanguisorba officinalis*. Vertical bars represent 95% confidence intervals for *Ranunculus* (no differences existed for *Agrostis*).

root activity in different locations at different times. Where this has been done, for example by Bookman & Mack (1982), Veresoglou & Fitter (1984) and Fitter (1986a), it is common to find pre-emption of favourable parts of the soil volume–time domain by dominant species, with others occupying the remaining parts.

This is not universally true. I found this effect of the dominant *Sanguisorba officinalis* (peak productivity 4.86 g m^{-2} d^{-1}) on *Ranunculus acris* (0.55 g m^{-2} d^{-2}) but not on *Agrostis capillaris* (2.13 g m^{-2} d^{-2}) (Fitter 1986a). For *Sanguisorba* and *Agrostis*, therefore, it seems reasonable to postulate a high degree of competitive interaction—consumptive competition *sensu* Yodzis (1986)—which might imply a high competitive combining ability (Aarssen 1983) or differentiation on some other niche axis.

The bulk of the evidence seems then to suggest that there is less spatio-temporal niche differentiation than has generally been assumed to be the case. Where near-exclusive distributions of root activity are found, these are often due to pre-emption by dominant species, with subsidiary species escaping competition by activity in less favoured zones, a situation comparable to that of spring-active species above-ground in deciduous forests and to the productive sub-alpine plant communities studied by Del Moral, Crampitt & Wood (1985). Many plant species, nevertheless, appear to coexist with strongly overlapping root activity distributions. Theory predicts, therefore, that they would compete strongly for soil-based resources. Factors mitigating this might include:

1 Low demand for soil resources or very low root densities such that root exploitation zones do not overlap: this may occur where powerful external forces such as grazing are active, and in special cases such as rock faces and some arid communities;

2 Interactions with soil micro-organisms: nitrogen-fixing plant species can coexist with others (Hall 1974) and mycorrhizal infection can certainly alter the competitive balance between species (Fitter 1977). Since rhizosphere microbial populations are affected by neighbours (Newman *et al.* 1979), many competitive interactions might be expressed in this way;

3 Fine-scale differentiation: Litav & Wolowitch (1971) showed that a suppressed species could coexist with a dominant if a small part of its root system was placed in a protected soil volume. The experiments of Martin (1979) and Sheikh & Rutter (1969), described above, suggest that this may occur in soil in relation to pore size.

There is little support here, therefore, for the idea that plant communities are composed of complementary species, co-adapted by past interactions. Indeed it is hard to see how this could come about. When plants interact, the effect of that interaction is, in the vast majority of cases, a change in the concentration of some physico-chemical variable (irradiance, soil water potential, ion concentration, etc.). The target plant cannot experience the interaction in any other way, and if selection acts to increase an ability to withstand that effect, it will do so just as it might were the cause of the change in concentration wholly abiotic. Only if neighbours are sufficiently predictable and specific in their effects on the environment, would there be selection for the ability to coexist with particular competitors. The concept of coexistence between plants thus has limited meaning in an evolutionary sense, except in so far as common sets of physico-chemical parameters will recur and so species or genotypes well

136   A. H. FITTER

suited to growing in those conditions will occur in the same community, but without the predictability of juxtaposition required if co-adaptation is to be invoked as an explanation.

Such apparent niche differentiation as is seen in plant communities seems, therefore, to arise as a result of the pre-emption of the most favourable domains by dominant species, with subordinate species occupying peripheral regions, which are presumably not utilized by the dominant because of inadequate returns on the required investment. Niche occupation is thus apparently very plastic and such structure as plant communities possess probably derives largely from the dominants.

ACKNOWLEDGMENTS

I am grateful to Richard Law, John Lawton and Mark Williamson for their perceptive comments on the draft of this paper, and to Ruth Nichols who drew the figures.

REFERENCES

Aarssen, L.W. (1983). Ecological combining ability and competitive combining ability in plants: toward a general evolutionary theory of coexistence in systems of competition. *American Naturalist,* **122,** 707–31.

Aarssen, L.W. & Turkington, R. (1985a). Biotic specialisation between neighbouring genotypes in *Lolium perenne* and *Trifolium repens* from a permanent pasture. *Journal of Ecology,* **73,** 605–14.

Aarssen, L.W. & Turkington, R. (1985b). Competitive relations among species from pastures of different ages. *Canadian Journal of Botany,* **63,** 2319–25.

Allen, F.B. & Forman, R.T. (1976). Plant species removals and old field community structure and stability. *Ecology,* **57,** 1233–43.

Aspinall, D. (1960). An analysis of competition between barley and white persicaria II. Factors determining the course of competition. *Annals of Applied Biology,* **48,** 637–54.

Atkinson, D. (1985). Spatial and temporal aspects of root distribution as indicated by the use of a root observation laboratory. *Ecological Interactions in Soil* (Ed. by A.H. Fitter, D. Atkinson, D.J. Read & M.B. Usher), pp. 43–66. Blackwell Scientific Publications, Oxford.

Baldwin, J.P. & Tinker, P.B. (1972). A method for estimating the lengths and spatial pattern of two interpenetrating root systems. *Plant and Soil,* **37,** 209–13.

Björkman, O. (1968). Further studies on differentiation of photosynthetic properties of sun and shade ecotypes of *Solidago virgaurea. Physiologia Plantarum,* **21,** 1–10.

Björkman, O., Ludlow, M.M. & Morrow, P.A. (1972). Photosynthetic performance of two rainforest species in their native habitat and analysis of their gas exchange. *Carnegie Institute of Washington Yearbook,* **71,** 94–102.

Böhm, W. (1979). *Methods for Analysing Root Systems.* Ecological Studies. Springer-Verlag, Berlin.

Bookman, P.A. & Mack, R.N. (1982). Root interaction between *Bromus tectorum* and *Poa pratensis*: a three-dimensional analysis. *Ecology,* **63,** 640–6.

Cannon, W.A. (1911). *The Root Habits of Desert Plants*. Carnegie Institute of Washington, Publication No. 131.

Caldwell, M.M., Eissenstat, D.M., Richards, J.H. & Allen, M.F. (1985). Competition for phosphorus: differential uptake from dual-isotope-labelled soil interspaces between shrub and grass. *Science*, 229, 384–6.

Cody, M.L. (1978). Distribution ecology of *Haplopappus* and *Chrysothamnus* in the Mojave desert. I. Niche position and niche shifts on north-facing granite slopes. *American Journal of Botany*, 65, 1107–16.

Connell, J.H. (1980). Diversity and the coevolution of competitors, or the ghost of competition past. *Oikos*, 35, 131–8.

Daubenmire, R.F. (1942). An ecological study of the vegetation of south-western Washington and adjacent Idaho. *Ecological Monographs*, 12, 53–79.

Del Moral, R., Crampitt, C.A. & Wood, D.M. (1985). Does interference cause niche differentiation? Evidence from subalpine plant communities. *American Journal of Botany*, 72, 1891–901.

Donald, C.M. (1958). The interaction of competition for light and nutrients. *American Journal of Agricultural Research*, 9, 421–35.

Fekete, G. & Précsényi, I. (1981). Niche structure of a perennial sandy grassland. *Man and the Biosphere Programme: Survey of 10 years activity in Hungary* (Ed. by P. Stefanovits, A. Berczik, G. Fekete & M. Seidl), Budapest.

Fitter, A.H. (1976). Effects of nutrient supply and competition from other species on root growth of *Lolium perenne* in soil. *Plant and soil*, 45, 177–89.

Fitter, A.H. (1977). Influence of mycorrhizal infection on competition for phosphorus and potassium by two grasses. *New Phytologist*, 79, 119–25.

Fitter, A.H. (1986a). Spatial and temporal patterns of root activity in a species-rich alluvial grassland. *Oecologia*, 69, 594–9.

Fitter, A.H. (1986b). Acquisition and utilisation of resources. *Plant Ecology* (Ed. by M.J. Crawley) pp. 375–405. Blackwell Scientific Publications, Oxford.

Fowler, N. (1981). Competition and co-existence in a North Carolina grassland. II. The effects of the experimental removal of species. *Journal of Ecology*, 69, 843–54.

Fowler, N. & Antonovics, J. (1981). Competition and co-existence in a North Carolina grassland. I. Patterns in undisturbed vegetation. *Journal of Ecology*, 69, 825–42.

Fox, R.L. & Lipps, R.C. (1964). A comparison of stable strontium and P32 as tracers for estimating alfalfa root activity. *Plant and Soil*, 20, 337–50.

Gause, G.F. (1934). *The Struggle for Existence*. Reprinted 1964. Hafner, New York.

Goss, M.J. (1977). Effect of mechanical independence on growth of seedlings. *Journal of Experimental Botany*, 28, 96–111.

Grubb, P.J. (1986). Problems posed by sparse and patchily distributed species in species-rich plant communities. *Community Ecology* (Ed. by J. Diamond & T.J. Case), pp. 207–22. Harper & Row, New York.

Haberlandt, G. (1884). *Physiologische Pflanzenanatomie*. Engelmann, Leipzig.

Hall, R. (1974). Analysis of the nature of interference between plants of different species. II. Nutrient relations in a Nandi *Setaria* and Greenleaf *Desmodium* association with particular reference to potassium. *Australian Journal of Agricultural Research*, 25, 749–56.

Harris, G.A. (1967). Some competitive relationships between *Agropyron spicatum* and *Bromus tectorum*. *Ecological Monographs*, 37, 89–111.

Idris, H. & Milthorpe, F.L. (1966). Light and nutrient supplies in the competition between barley and charlock. *Oecologia Plantarum*, 1, 143–64.

Kirikova, L.A. (1970). Razmeshchenie podzemnykh chastei nokotorhky vidov travyanokustarnikovogo yarusa yelovogo lesa (The distribution of underground parts of certain species of the herbaceous/dwaft-shrub layer in a spruce forest). *Botanicheskii Zhurnal*, 55, 1290–1300.

Kummerow, J. & Mangan, R. (1981). Root systems in *Quercus dumosa* Nutt. dominated chaparral in southern California. *Oecologia Plantarum*, 2, 177–88.

Litav, M. & Harper, J.L. (1967). A method for studying the spatial relationship between the root systems of two neighbouring plants. *Plant and Soil*, 26, 389–92.

Litav, M. & Wolowitch, S. (1971). Partial separation of roots as a means of reducing the effect of competition between two grasses. *Annals of Botany*, 35, 1163–78.

Martin, S. (1979). *The effects of pore sizes on root diameters and plant interference.* Unpublished B.Sc. Thesis, University of York.

Newbould, P. (1969). The adsorption of nutrients by plants from different zones in the soil. *Ecological Aspects of the Mineral Nutrition of Plants* (Ed. by I.H. Rorison), Symposia of the British Ecological Society, 9, pp. 177–90. Blackwell Scientific Publications, Oxford.

Newman, E.I., Campbell, R., Christie, P., Heaps, A.J. & Lawley, R.A. (1979). Root micro-organisms in mixtures and monocultures of grassland plants. *The Soil-Root Interface* (Ed. by J.L. Harley & R. Scott Russell), pp. 161–73. Academic Press, London.

Nicholson, A.J. (1954). An outline of the dynamics of animal populations. *Australian Journal of Zoology*, 2, 9–65.

Nye, P.H. & Foster, W.N.M. (1961). The relative uptake of phosphorus by crops and natural fallow from different parts of their root zone. *Journal of Agricultural Science*, 56, 299–306.

O'Brien, T.A., Moorby, J. & Whittington, W.J. (1967). The effect of management and competition on the uptake of ^{32}phosphorus by ryegrass, meadow fescue, and their natural hybrid. *Journal of Applied Ecology*, 4, 513–20.

Platt, W.J. & Weiss, I.M. (1977). Resource partitioning and competition within a guild of fugitive prairie plants. *American Naturalist*, 11, 479–513.

Précsényi, I., Fekete, G., Molnar, E., Melko, E. & Viragh, K. (1979). Niche studies on some plant species of a grassland community. V. The position of the species in the three-dimensional niche space. *Acta Botanica Academiae Scientiarum Hungaricae*, 25, 131–8.

Rathcke, B. & Lacey, E. (1985). Phenological patterns of terrestrial plants. *Annual Review of Ecology and Systematics*, 16, 179–214.

Rogers, R.W. & Westman, W.E. (1979). Niche differentiation and maintenance of identity in cohabiting *Eucalyptus* species. *Australian Journal of Ecology*, 4, 429–39.

Sayre, J.D. & Morris, V.H. (1940). The lithium method of measuring the extent of corn root systems. *Plant Physiology*, 15, 761–4.

Sheikh, K.H. & Rutter, A.J. (1969). The responses of *Molinia caerulea* and *Erica tetralix* to soil aeration and related factors. I. Root distribution in relation to soil porosity. *Journal of Ecology*, 57, 713–26.

Silander, J.A. & Antonovics, J. (1982). Analysis of interspecific interactions in a coastal plant community—a perturbation approach. *Nature*, 298, 557–60.

Soileau, J.M. (1973). Activity of barley seedling roots as measured by strontium uptake. *Agronomy Journal*, 65, 625–8.

Stiles, F.G. (1977). Coadapted competitors: the flowering seasons of hummingbird pollinated plants in a tropical forest. *Science*, 198, 1177–8.

Syers, J.K., Ryden, J.C. & Garwood, E.A. (1984). Assessment of root activity of perennial ryegrass and white clover using ^{32}phosphorus as influenced by method of isotope placement, irrigation and method of defoliation. *Journal of the Science of Food and Agriculture*, 35, 959–69.

Tansley, A.G. (1949). *The British Isles and their Vegetation.* Cambridge University Press, Cambridge.

Tilman, D. (1982). *Resource Competition and Community Structure.* Princeton University Press, Princeton, New Jersey.

Turkington, R. & Aarssen, L.W. (1985). Local-scale differentiation as a result of competitive interactions. *Perspectives on Plant Population Ecology* (Ed. by R. Dirzo & J. Sarukhan), pp. 107–29. Sinauer, Sunderland, Massachusetts.

Turkington, R. & Harper, J.L. (1979). The growth, distribution and neighbour relationships of *Trifolium repens* in a permanent pasture. IV. Fine-scale biotic differentiation. *Journal of Ecology*, **67**, 245–54.

Veresoglou, D.S. & Fitter, A.H. (1984). Spatial and temporal patterns of growth and nutrient uptake of five co-existing grasses. *Journal of Ecology*, **72**, 259–72.

Vos, J. & Groenwold, J. (1983). Estimation of root densities by observation tubes and endoscope. *Plant and Soil*, **74**, 295–300.

Watt, A.S. & Fraser, G.K. (1933). Tree roots in the field layer. *Journal of Ecology*, **21**, 404–14.

Weaver, J.E. (1919). *The Ecological Relation of Roots.* Publication No. 286, Carnegie Institute, Washington.

Whittaker, R.H. (1956). Vegetation of the Great Smoky Mountains. *Ecological Monographs*, **26**, 1–80.

Whittaker, R.H. (1972). Evolution and measurement of species diversity. *Taxon*, **21**, 213–51.

Williamson, M.H. (1957). An elementary theory of interspecific competition. *Nature*, **180**, 422–5.

Wit, C.T. de (1960). *On Competition.* Versl. Landbouwk. Onderzoek No. 66.8, Wageningen.

Yodzis, P. (1986). Competition, mutuality and community structure. *Community Ecology* (Ed. by J. Diamond & T.J. Case), pp. 480–91. Harper & Row, New York.

7. INSECT ASSEMBLAGES— DIVERSITY, ORGANIZATION, AND EVOLUTION

M. F. CLARIDGE

Department of Zoology, University College of Wales,
P.O. Box 78, Cardiff CF1 1XL, Wales, UK

INTRODUCTION

'Ecological communities are groups of species living closely enough together for the potential of interaction.' (Strong *et al.* 1984). This broad definition embraces two important concepts: (i) communities consist of assemblages of species, and (ii) the species of any community may interact (see Southwood, Chapter 1). Among the possible interactions are the familiar processes of competition, predation and parasitism, but also commensalism, amensalism and mutualism. All may be important in the organization of communities, but until recently ecologists have concentrated mainly on the roles of competition and predation/parasitism. Indeed, much of the current controversy about the development of community structure has concerned the relative importance of these (Strong *et al.* 1984; Diamond & Case 1986; Southwood, Chapter 1).

The processes that determine the nature and structure of insect dominated communities form an area of current interest in ecology and a number of recent reviews are available (e.g. Price 1980; Lawton & Strong 1981; Lawton 1984; Lawton & Hassell 1984; Price 1984; Strong 1984; Strong, Lawton & Southwood 1984; Lawton 1986). In this chapter, therefore, I shall be selective and consider only terrestrial insects (see Hildrew & Townsend, Chapter 16, for freshwater insect assemblages), and among these, particularly herbivores and their parasites and predators. My aim is to highlight some of the types of associations and interactions that may influence the diversity of such insect assemblages and the organizing processes that give structure to the natural communities of which they are part. In addition I shall draw attention to another part of the community concept which is usually either ignored or taken for granted by most ecologists. That is the nature of species and speciation: the ultimate process upon which all ecological diversity depends. Unless we can reliably recognize species, then our analyses of processes may be incomplete or even totally wrong.

Why insects?

It may seem perverse to select just one class from one of the animal phyla for special attention, but to the terrestrial ecologist the justification is obvious. About 57% of all living species of organisms (excluding fungi, algae and 'microbes') are insects (Southwood 1978). They are almost absent from marine ecosystems, but they are abundant in most terrestrial and many freshwater ones. Of the remaining 43% of living organisms, green plants make up 22% of the total species-richness. Thus, insects and green plants together dominate the diversity of most terrestrial communities and the interactions within and between these two groups are major processes contributing to the organization of such communities. Strong, Lawton & Southwood (1984) estimate that almost half of the known species of modern insects, some 361 000, are herbivores (or phytophages).

Among insects a special feeding category is often recognized which combines some of the attributes of both predators and parasites. Such insects are known as parasitoids and are usually very abundant in terrestrial communities. They attack a single individual of a host organism on or in which their progeny complete development before killing it. Parasites, parasitoids and predators are useful practical categories, but intermediates are frequently found so that there is effectively a biological continuum between them. Very large numbers of insects, mostly from the Hymenoptera and Diptera, exhibit the parasitoid mode of life and use as hosts mostly other insect species. A recent review of the biology of parasitoids is given in Waage & Greathead (1986).

HERBIVORE ASSEMBLAGES

The complex communities of green plants with insect herbivores and their predators and parasitoids are sometimes termed compound communities (Root 1973). These in turn then include a series of partially interacting component communities (here termed assemblages) usually associated with some special food resource in the form of a plant species, group of species, or even a particular plant structure. Food specialization is thus usually extreme and most insect herbivores take only a limited range of host plants: many may be monophagous (Figs 7.1a,b).

One of the questions that has received much attention from insect ecologists in recent years has been why the diversity of herbivores is different on different plant species. Unfortunately, because of the complexity of the data required and the very large numbers of insect species usually involved, few good data sets have been published and most can only be

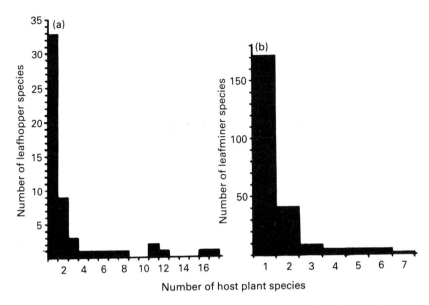

FIG. 7.1. Histograms to show spectrum of host specificity of two guilds of insect herbivores associated with the foliage of trees in Britain. (a) Mesophyll feeding leafhoppers (after Claridge & Wilson 1981). (b) Leafminers (after Claridge & Wilson 1982b).

analysed in terms of species-richness. Additionally, available studies are dominated by works from temperate regions, mainly because of the daunting taxonomic problems with almost all groups of tropical insects.

Species-richness of insect herbivores

The British Isles are floristically and faunistically one of the best known regions in the world. It is therefore not surprising that special attention has been paid to analyses of data from Britain. Southwood (1961), in his now classic study, accumulated from a literature survey a list of insect herbivores associated with most British trees. He attempted also to obtain a measure of the abundance of the host plants in geologically recent time by estimating the frequency of fossil remains, mostly of pollen, during the post-glacial period (from Godwin 1956). He obtained a significant positive regression between these two data sets ($r = 0.85$, $P < 0.001$). Thus 72% of the variance was explained by the relationship. Some subsequent authors interpreted this as evidence for what Opler (1974) called a 'geological time theory', where it is suggested that numbers of associated herbivore species depend on the length of time during which a plant has been present in a region. Strong (1974) obtained a new set of recent abundance data for British trees by counting the 10 km squares recorded for each in the Botanical Atlas of the British Isles (Perring & Walters 1962). He showed a very significant relationship between his new abundance data and the original insect data from Southwood ($r = 0.78$, $P < 0.001$). Subsequently, many authors have used species–area relationships in analysing a variety of insect–plant assemblages (reviewed by Strong 1979; Strong, Lawton & Southwood 1984). However, the species–area relationship is widely associated with the equilibrium theory of island biogeography of MacArthur & Wilson (1967) and therefore may be taken to imply dynamic equilibrium levels maintained by balanced rates of migration and extinction. In the present context, such an interpretation has been severely criticized (e.g. Connor & McCoy 1980; Gilbert 1980) and it is now obvious that a significant species–area relationship in herbivore–plant studies should be regarded as purely empirical and certainly not as demonstrating the existence of balanced equilibria. Indeed, as Connor & McCoy (1980) clearly showed, such relationships are of most value as practical means for comparing species-richness between different geographical regions or between habitats (which may often be different host plants). Techniques of multiple regression analysis may then be used to evaluate factors other than area in contributing to the relationship. Strong (1974) first used this approach to show that cumulative ancient abundance in Southwood's 1961 study was not significant after taking into account recent abundance.

More recently Lawton & Schröder (1977) and Lawton (1978) suggested that structural complexity or 'architecture' might often be an important partial determinant of species-richness. Increased complexity of plant structure has since been implicated in various species–area relationships (e.g. Strong & Levin 1979; Rigby & Lawton 1981).

A further important factor that has been analysed in these types of total faunal studies on plants is the effect of 'taxonomic isolation' of host plants. Thus, it may be expected that insect herbivores may more easily colonize new hosts that are related to their existing one than hosts that are taxonomically isolated. Southwood (1961) used this argument to account for the relative paucity of insect herbivores on ash (*Fraxinus excelsior*) in Britain. Connor *et al.* (1980), using residuals from Strong's (1974) analysis, showed that introduced trees with native relatives in Britain had accumulated significantly more herbivores than other taxonomically more isolated species, after accounting for the species–area relationship itself.

Thus, most recent studies on the species-richness of large assemblages of insect herbivores have demonstrated the importance to varying degrees of (i) area or abundance, (ii) structural complexity or architecture, and (iii) taxonomic relatedness, of host plants. Kennedy & Southwood (1984), in an analysis of some newly abstracted data on insects of British trees, found evidence also of a contribution from a geological time factor.

Herbivore guilds: smaller assemblages

Total assemblages of insect herbivores consist of groups of species, or guilds (Root 1967), that exploit the same types of resources in similar ways, e.g. leaf-chewers, stem-borers, leaf-miners, gall-formers, sap-suckers, etc. These guilds are not absolute divisions, and the level at which they may be used will depend on the purpose of any study.

From a purely practical point of view it is easier to study restricted feeding guilds than complete herbivore assemblages for groups of host plants, so that reliability and comparability of data are likely to be good. Even the most recent complete British herbivore–tree data used by Kennedy & Southwood (1984) are not of uniform quality. It is virtually impossible for any one group of workers themselves to work effectively on such enormous numbers of species.

Claridge & Wilson (1976, 1981, 1982b) studied two taxonomically different guilds of insect herbivores, both of which exploit primarily the mesophyll tissues of the leaves of trees and larger shrubs in Britain, but in quite different ways. Our first study was on mesophyll sap-suckers. Feeding adults and nymphs cause characteristic pale, stippling marks on leaves where cells have been emptied of their green contents. These are overwhelmingly dominated in numbers of species and individuals by leafhoppers (Cicadellidae, Typhlocybinae). We first studied thirty-four species of leafhoppers and their associations with twelve species of trees and shrubs (Claridge & Wilson 1976). In our later study (1981) we extended the work to include fifty-three species of leafhoppers from thirty-four species of trees. The latter included all of the major native trees, but also some species introduced during historic times, such as *Acer pseudoplatanus* and *Aesculus hippocastanum*, together with the recently introduced *Nothofagus* species. No relationship between antiquity in the British flora and numbers of associated leafhoppers was evident. Indeed some of the recently introduced trees carried relatively large faunas. Indices of similarity showed that most tree species, including some close relatives, had very distinctive leafhopper faunas.

Species–area relationships were calculated using our own two sets of leafhopper host records and newly estimated measures of area or abundance of trees, following the method of Strong (1974, see above). The relationship for the first more limited data set was not significant (Claridge & Wilson

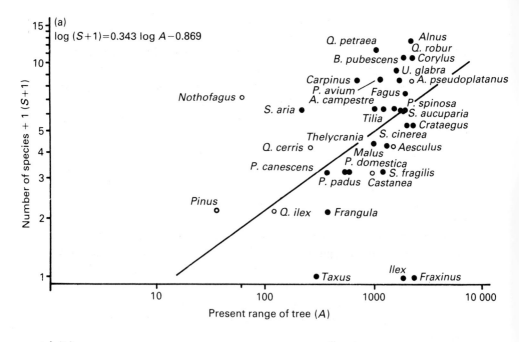

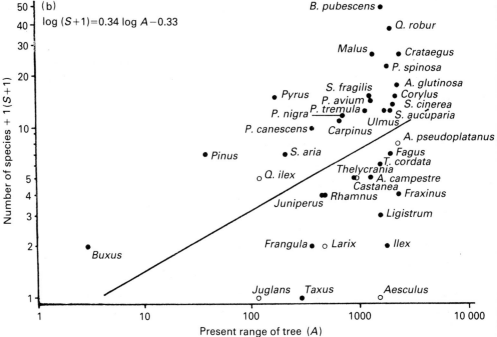

FIG. 7.2. The relationships between numbers of insect species (S) associated with trees and the present range (A) of each tree in Britain (computed as the number of 10 km squares from which each has been recorded) for two guilds of herbivores. Open circles denote trees introduced to Britain in recent and historical times. (a) Mesophyll feeding leafhoppers (after Claridge & Wilson 1981). (b) Leafminers (after Claridge & Wilson 1982b).

1978), but the later larger set was (Fig. 7.2a). However, the predictive value of host plant geographical range was very low and accounted for only 16% of the variance about the regression. Rey, McCoy & Strong (1981) suggested that our failure to obtain a significant relationship with the first data set was due to undercollecting, and to using leafhopper host records for only a small part of Britain but plant data for the whole country. Neither of these suggestions was true (Claridge & Wilson 1982a). In our view the critical difference between our two data sets was the small number of trees and therefore insects in the earlier study. Inevitably it was more difficult to obtain statistically significant results for smaller data sets than for bigger ones.

For comparison with our leafhopper study, species–area analyses were made on the leaf-mining guild of insect larvae associated with a similar array of trees in Britain (Claridge & Wilson 1982b). This guild also exploits the mesophyll tissue of leaves, but is taxonomically more diverse, being dominated by Lepidoptera, together with some Diptera, Hymenoptera and Coleoptera. Host records for 239 leafmining species were obtained for thirty-seven species of trees by a detailed survey of recent literature. A species–area relationship using the previous area data was computed (Fig. 7.2b). A significant relationship was obtained, but it explained only about 19% of the variation in the regression.

Thus we concluded that when insect feeding guilds associated with a wide taxonomic array of host plants are studied, species–area relationships by themselves seem to explain little of the variation in species–richness. This contrasts with the study of Opler (1974) on leafmining insects on oaks— mostly species of *Quercus* together with some closely-related genera—in California, which gave a significant species–area relationship with 90% of the variance explained by area alone. No analyses of other insect/plant assemblages have produced such a significant species–area relationship. We speculated (Claridge & Wilson 1982b) that the major difference between these studies lay in the close taxonomic relatedness of the Californian hosts in contrast to the taxonomically diverse tree flora of Britain. This view is supported by the studies of Cornell & Washburn (1979) on gall-forming insects on oaks and related trees in California. They obtained a species–area relationship which could explain 33% of the variation in species-richness when all hosts were included. However, when only species of *Quercus* were considered, a more significant regression was obtained which accounted for 72% of the variation.

It is thus obvious that taxonomic relationship of host plants is an important element in these studies. Using simple techniques for testing for the importance of such factors in the accumulation of insect herbivores

(Lawton & Schröder 1977; Connor *et al.* 1980), we were able to show a distinct influence of taxonomic isolation in our leafminer data which accounted for 23% of the unexplained variation in our species–area relationship (Claridge & Wilson 1982b). We could find no such relationship for leafhoppers. However, using a slightly different method, Neuvonen & Niemela (1981) reanalysed our earlier leafhopper data (Claridge & Wilson 1978) and were able to show a significant relationship with their measure of taxonomic isolation.

Godfray (1984) made an independent study of leafmining insects associated with British trees using tree genera instead of species as host plant groupings. A multiple regression analysis was undertaken using four plant properties as independent variables. The regression explained 69% of the variation in the relationship, with taxonomic isolation and plant species per genus having significant effects, but most interestingly, plant range had no demonstrable effect. In addition to the difference in the taxonomic range of hosts, Godfray noted that Opler's (1974) data derived from the complete geographic range of oaks studied: these species are endemic to California. All studies of trees in Britain alone are inevitably restricted to a small part of their natural ranges since they also extend widely over mainland Europe. It is certainly a valid point that few of the insect–plant host relationships observed in Britain may have evolved here. A thorough understanding of relationships can only be obtained by studying complete areas of distribution. Such criticism, of course, must apply equally to total herbivore assemblages as to specialist feeding guilds.

It is perhaps worth speculating on the nature of the factors that have been identified as 'taxonomic isolation' in species–area studies by many of the authors cited above. It seems unlikely that taxonomic isolation *per se* is an important factor but rather that it reflects other differences between the plants under study. It is axiomatic that plants that are closely related in taxonomy resemble each other more than do distantly related ones. It is then probable that such resemblances will also extend to chemical characteristics, which are known to be important in host plant recognition and utilization by insect herbivores. Thus it is possible that chemical differences between plants will primarily determine the likelihood of colonization by insect herbivores. However, not all studies show such obvious effects of taxonomic isolation and relationships of hosts. For example, Futuyma & Gould (1979) in a study of the leaf-chewing guild of insect larvae associated with common trees and shrubs in the north-eastern USA, could find no such distinct patterns.

We may conclude therefore that where detailed information is available on assemblages of insect herbivores, species–area relationships may be useful means for evaluating a variety of other factors that could explain some of the variation in such regressions, but there is little evidence for area as such playing a major role in determining species-richness. Many complex factors are probably involved which will be intransigent to such analyses. New approaches and detailed field studies on particular guilds are needed for a more complete understanding of the apparently simple question: 'Why do different plant species have different numbers of associated insect herbivores?'

HERBIVORE–PREDATOR/PARASITOID COMMUNITIES

Adequate analyses of particular plants and their associated insect herbivores in natural ecosystems are relatively few. To extend such studies to include further trophic levels is much more difficult and even fewer examples are available. Many individual records of particular predator/prey, or parasitoid/host interactions are known, especially for agricultural and forest pests. Askew & Shaw (1986) have recently reviewed work on parasitoid assemblages.

Generally, predators may be defined as requiring more than one individual prey, while parasitoids require only one host, in order to complete a life cycle. Predators usually take a wide array of prey species whereas many parasitoids are very specialized and attack a limited range of hosts. Of course there are many exceptions. However, adequate field data on prey and host ranges are difficult to obtain. It is relatively easy to determine what species attack a particular herbivore in a complex community, but much more difficult to be sure that all prey or host records for any predator or parasitoid have been collected.

The best examples of detailed food web studies on terrestrial insect assemblages involve insect herbivores that live either within the tissues of their host plants or at least concealed within particular plant structures. Such endophytic herbivores are generally easier to sample and to study than free-living forms since they are at least partly confined by the plant itself. Thus, assemblages associated with plant gall-formers, leafminers, and flower head and fruit-feeders are among the best studied and analysed.

Askew studied the assemblages of parasitoids associated with gall wasps (Hymenoptera, Cynipidae), especially those that attack oaks in Britain (for recent review see Askew 1984). All species of the subfamily Cynipinae either form galls on their food plants in which larval development is completed, or they live in and feed on the gall tissues of other species (inquilines). These insects are usually host plant specific. Remarkably, more than 75% of all known species attack *Quercus* species. In Britain more than fifty species are confined to oaks and most produce galls. Askew (1961) published food webs for many of the commoner British species of oak-feeding gall wasps. Associated with each is generally a large complex of parasitoids—mostly of chalcid wasps. Each gall is typically attacked by a set of polyphagous parasitoids together with a few monophagous species. However, the parasitoid assemblage of oak galls differs markedly from that in galls caused by related wasps on roses (*Diplolepis* and *Periclistus* on *Rosa* species) (Askew 1960). Thus, even the more polyphagous parasitoids are generally restricted to guilds of gall wasps associated with groups of related hosts. Host-specificity is relative, and depends on the availability of hosts: taxonomically isolated species of gall wasps associated with different host plants have mostly very specific associated parasitoids.

A similar large complex of parasitoids was found attacking the leaf-mining insect larvae associated with trees in Britain (Askew & Shaw 1974). Most are polyphagous, but some are specific, usually to the tree rather than to a particular species of leafminer.

Thus, not surprisingly, the diversities of parasitoid assemblages associated with plant galls and leafminers are highest where most species of hosts occur on a single plant species—that is on trees for both galls and leafminers (Askew 1980). The large numbers of coexisting distinctive galls on oaks support the most complex assemblages with some subdivision of resources by more specific parasitoids.

Similar long-term and detailed community studies have been made on the insect inhabitants of flower heads of thistles, knapweeds and related Compositae by Zwölfer (1968, 1979, 1980). Unlike most of the parasitoids above, those associated with flower-heads are often quite specific to particular herbivores or groups of related herbivores. Within one species of flowerhead most parasitoids are monophagous. Thus even within these component communities partitioning of hosts by parasitoids is extreme.

It is clear then that different assemblages of insect herbivores and parasitoids may be organized in different ways and differ at least in the relative importance of monophagous and polyphagous species. Host plants always play an important role in determining the specific composition of the parasitoid complexes of insect herbivores.

ORGANIZING PROCESSES IN INSECT HERBIVORE AND PARASITOID ASSEMBLAGES

Many of the concepts and much of the terminology of community ecology, such as 'resource partitioning', 'species packing', etc., implicitly rest on the assumption that interspecific competition is a major organizing process in animal communities. The best evidence for the dominating influence of competition comes from vertebrate studies (see Brown, Chapter 9; O'Connor, Chapter 8; Rosenzweig, Chapter 21). However, in recent years more and more workers have questioned the universality of such processes in natural communities (e.g. Strong *et al.* 1984). Lawton & Strong (1981) reviewed the literature on insect herbivores and concluded that there are few studies that unequivocally demonstrate competition in the field. Price (1980) came to a similar conclusion in his wide-ranging review of parasite biology. It is suggested that insect herbivore populations rarely achieve sufficiently high densities for classical resource-based competition to be important. The glib interpretation of differences in distribution and resource utilization by coexisting species as the result of interspecific competition has been rightly criticized (Strong 1984). However, it may be equally glib to attribute such differences to other causes without experimental evidence. Indeed, as Southwood (Chapter 1) points out, subtle competitive interactions may be more widespread, even amongst phytophagous insects, than is often currently thought.

The dominating role of predation and parasitism in insect assemblages, in contrast to competition, has been emphasized by Jeffries & Lawton (1984) and Lawton (1986). These authors suggested that many of the types of species differences that traditionally have been attributed to classical interspecific competition for food resources could more easily be attributed to indirect competition for 'enemy free space'. Thus pressure of predation and parasitism may have profound effects on the species composition and evolution of many insect communities. The extensive recent literature makes it unnecessary to repeat the arguments here. Instead I wish to consider some particular examples in which field experiments have been made and in which more definite conclusions concerning ecological processes may follow.

The possibility for massive field experiments and the introduction of new species to insect communities on a scale not usually possible for an ecological experiment is the normal procedure in the practice of biological pest control. Unfortunately, until recently, the enormous literature resulting from such work had been largely ignored by community ecologists. For example, more than 20 years ago workers in biological control were concerned with the possibility that the introduction of an 'ecological homologue' of a pest species into a new environment might result in the competitive displacement and therefore elimination of the pest (DeBach & Sundby 1963; DeBach 1966).

A number of examples attributed to competitive displacement among insect herbivores derive from experimental introductions in attempts at the biological control of weeds. One of the most thoroughly documented concerns the efforts to control the St. John's-wort of European origin, *Hypericum perforatum,* in California during the 1940s (Huffaker & Kennett 1969). A series of insect herbivores was introduced from Europe including a gall midge, a root borer and two species of leaf-feeding beetles, *Chrysolina hyperici* and *C. quadrigemina.* At first both species of *Chrysolina* spread rapidly after introduction and coexisted widely when the weed was still abundant. However, after several years Huffaker & Kennett (1969) reported the total extinction of *C. hyperici* and the elimination of the weed from open areas. Both host and beetle continued to survive at low densities, but only in shaded habitats. The field data were supported by laboratory experiments which showed *C. quadrigemina* to be competitively superior to *C. hyperici* under Californian conditions.

It seems reasonable to conclude that this is an example of true competitive displacement. Other examples include the apparent restriction of the Mediterranean fruit fly, *Ceratitis capitata,* previously widespread throughout Hawaii, to littoral areas only, following the establishment of the Oriental fruit fly, *Dacus dorsalis* (Christenson & Foote 1960); and the replacement of the Yellow Scale, *Aonidiella citrina* by the Californian Red Scale, *A. aurantii,* in southern California (DeBach & Sundby 1963; DeBach, Hendrickson & Rose 1978). Such large scale experiments suggest that under some circumstances, especially of high initial host plant densities, a number of insect herbivores may coexist, only to suffer intense competition when the total herbivore pressure reduces host abundance. Competition may be a dominating process for short, but critical, periods of time when food resources are limiting and herbivore numbers are high. The 'ghost of competition past' (Strong 1984) may be difficult to detect, but nevertheless short periods of intense competition may be important in establishing species distribution patterns during evolution.

Among parasitoid species there is more evidence for major competitive effects. The best known and most widely studied example derives from attempts at the biological control of the California Red Scale, *Aonidiella aurantii,* by the introduction of a series of closely-related species of chalcid wasps of the genus *Aphytis* (DeBach & Sundby 1963; DeBach, Rosen & Kennett 1969).

In 1948 *A. lingnanensis* was introduced to California where *A. chrysomphali* was already present as a major parasitoid of *A. aurantii.* Within 10 years *A. chrysomphali* had been completely replaced by *A. lingnanensis* over an area of some 4000 square miles (6440 km^2). Between 1956 and 1957 another very closely-related species, *A. melinus,* was introduced from India.

It in turn replaced *A. lingnanensis,* particularly in inland citrus growing areas. By 1965 *A. lingnanensis* was present only in coastal regions.

The interpretation of these changes of distribution as the result of competitive exclusion was supported by laboratory experiments which showed each species to be completely superior to the other under particular conditions of temperature and humidity. More recently some authors have questioned some details of the interpretation of this example (Ehler & Hall 1982; Keller 1984; Luck & Podoler 1985), but the evidence appears to be so strong that it is hard not to accept a considerable role for interspecific competition.

Numerous other examples of apparent competitive interactions between introduced biological control agents have been well documented by Huffaker, Messenger & DeBach (1969). However, a weakness of most of these studies is that it is not usually possible to rule out the importance of factors other than interspecific competition, such as differential predation/ parasitism. One of the best series of studies that do allow competition and parasitism to be evaluated is that by Zwölfer on the interactions between insect herbivores and their parasitoids in flower-heads (Zwölfer 1968, 1979, 1980). This work was originally stimulated by the need to discover possible biological control agents for weeds of the genera *Carduus, Cirsium* and *Centaurea* of Palearctic origin introduced into North America. Flower-heads of these plants are complicated structures with numerous florets and seeds, which develop on a large receptacle and are surrounded by groups of often complex bracts. These structures form a specialized habitat for a variety of insect herbivores and associated parasitoids. A typical assemblage of species in the flower-head of one plant species may include such diverse herbivores as gall-forming and seed-feeding tephritid flies, gall-forming Cynipidae, seed and receptacle-feeding weevils (Curculionidae) and lepidopterous larvae. Some of the lepidopterous larvae feed not only on plant tissues but often also prey on other herbivores and thus function as omnivores. A maximum of ten different coexisting herbivore species was found in one population of *Centaurea* flower-heads (Zwölfer 1979).

All of Zwölfer's studies show that parasitoid complexes have a major impact on the density of hosts, but in most flower-head populations this was not sufficient to prevent intense competition between herbivorous species. In a population of the thistle, *Carduus nutans,* in France, Zwölfer (1979) found five major flower-head herbivores (Fig. 7.3). Solitary larvae of the Lepidoptera, *Homoeosamia* and *Eucosma* species, feed on plant tissue as well as on other insect species. The fly, *Urophora solstitialis,* causes a compact gall which soon becomes hardened and unavailable as food for other species. In addition, larvae of two weevils also inhabit the same heads,

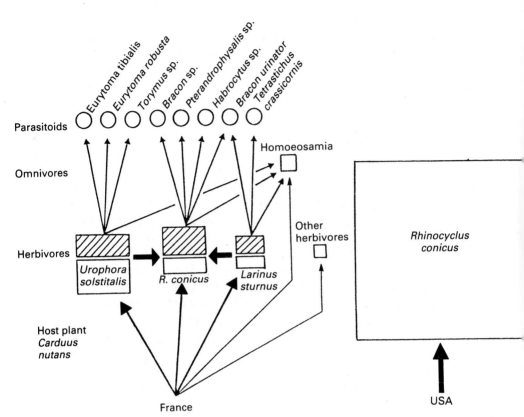

Fig. 7.3. Food webs associated with insect herbivores in the flower-heads of *Cardus nutans* at Mulhouse, France, in 1971, where the plant is native, and in Montana, USA, in 1977, where it is an introduced weed with only one insect herbivore and no parasitoids. Thickness of linking arrows indicates relative importance of individual interactions. Areas of rectangles for each herbivore are proportional to the biomass per flower-head of each at each locality, with cross-hatched subdivisions representing proportions of each removed by parasitoids (after Zwölfer 1980).

Larinus sturnus and *Rhinocyllus conicus*. Zwölfer found evidence of competition between *Urophora*, *Larinus* and *Rhinocyllus* in about 50% of flower-heads, but showed that differences in life cycles and behaviour enabled them to coexist.

C. nutans is a major introduced weed in North America. The weevil, *R. conicus* was introduced in the late 1960s as a control agent. In the absence of competitors (competitive release) and natural enemies, *R. conicus* spread rapidly and became more abundant than in any European populations, often attacking over 90% of the flower-heads and dramatically reducing seed production (Fig. 7.3) (Zwölfer & Harris 1984).

Studies on weevils of the genus *Larinus* also give a good example of competition and apparent ecological character displacement (Zwölfer 1979, 1980). Sampling of *Carduus nutans* in south-eastern France, Switzerland, Austria and southern Germany showed that *L. sturnus* and *L. jaceae* both attack the flower-heads, but never in the same locality. In twenty-one study sites the two species always fed on different host plants when they occurred together. In competition experiments on caged thistles, *L. sturnus* had lower breeding success in the presence of *L. jaceae*. In field experiments in 1965 and 1969 populations of *L. sturnus* from *C. nutans* in the Rhine valley were introduced to an area of Switzerland where *L. jaceae* alone was naturally present on *C. nutans*. The introduced populations survived for 1 or 2 years only, but *L. jaceae* was still present at least until 1974. It is thus probable that competitive interactions are responsible for the non-overlapping ranges of the *C. nutans* feeding populations of these two weevils.

The effects of parasitoid assemblages are also demonstrated in Zwölfer's studies of the gall-fly, *Urophora cardui*. Some populations were even driven to extinction by pressure of parasitoids (Zwölfer 1979). Unlike other species of *Urophora*, *U. cardui* does not make flower-head galls but forms swellings on the flowering stems of creeping thistle, *Cirsium arvense*. The fly larvae are attacked particularly by two chalcid wasps, *Eurytoma serratulae*, an internal parasitoid, and *E. robusta* Mayr, an external parasitoid (Claridge 1961). *E. serratulae* attacks the galls early before swelling is complete and is able potentially to reach all individual host larvae. *E. robusta* attacks mature galls and is often not able to reach some of the host larvae because of the thickness of the gall (Zwölfer 1979). *E. robusta* is inherently superior to *E. serratulae*, since it is able to attack hosts already containing larvae of the latter. It also attacks other species of *Urophora* galls on other host plants, whereas *E. serratulae* is specific to *U. cardui*. Coexistence is possible because *E. robusta* is only efficient in exploiting small to medium-sized galls, whereas *E. serratulae* can exploit all sizes of host galls. Thus both species of *Eurytoma* coexist in the field despite often intense competition.

These examples of detailed studies of herbivore–parasitoid interactions are illuminating. It is clear that, where adequately studied, both competitive effects within herbivore and parasitoid guilds and predator/parasitoid mediated effects on herbivore assemblages may be extremely complex and frequently interdependent. It is almost certainly too simple-minded to imagine that just one type of interaction will account for most of what we see as structure and organization in natural insect communities, but we certainly need more detailed studies of different communities.

SPECIES AND SPECIATION: THE SOURCE OF
DIVERSITY

It is surprising that ecologists generally give so much attention to evaluating the importance of different interspecific interactions, but so little to considering the nature of species and their evolution. They have generally been content to use a primarily morphological approach to the species concept, though above all we are interested in biological interactions.

The biological species is the basic unit of evolution, and the diversification of species—speciation—is the ultimate source of organic diversity. The species concept is a complex and multifaceted one which cannot be discussed in detail here. However, it is generally agreed that species are groups of populations genetically isolated from other such groups by barriers to interbreeding (e.g. Mayr 1942, Cain 1954, but see Paterson 1985 for a radical new approach). In the present context most workers would agree that at any one place and time biological species either do not interbreed, or interbreed so infrequently as to retain their genetic integrities. They are therefore the units from which ecological communities are built and the nature of the species concept may affect the ways in which we view community processes (e.g. Walter, Hulley & Craig 1984).

In order to recognize species in field samples it is usual to rely on morphological differences. The problems are specially great in many groups of insects because of the enormous numbers of species. Often, even in well-known areas of the world, basic taxonomic work has not been done on major groups. More importantly, there are now numerous examples of insects in which morphological differences between biological species are very slight or even non-existent: these are so-called sibling or cryptic species. Thus morphologically similar populations from different hosts or habitats may be mistakenly identified as the same species and consequently ecological interactions and food web structure may be misinterpreted. For example, detailed studies revealed that *Eurytoma rosae,* a polyphagous chalcid parasitoid of gall wasps on different plants, was in reality a group of biological species with much narrower host ranges (Claridge & Askew 1960). *E. rosae* was previously thought to link the food webs associated with galls on oaks and roses, but in reality two biological species are involved: *E. rosae* on rose galls and *E. brunniventris* on oak. It is still not possible to identify these species from dead adults alone.

Long-term studies on leafhoppers of the genus *Oncopsis* clearly illustrate the difficulties that may arise (Claridge & Nixon 1986). These insects are all phloem-feeders associated with trees and are very restricted in host plant preferences. Two morphologically similar species, previously of doubtful

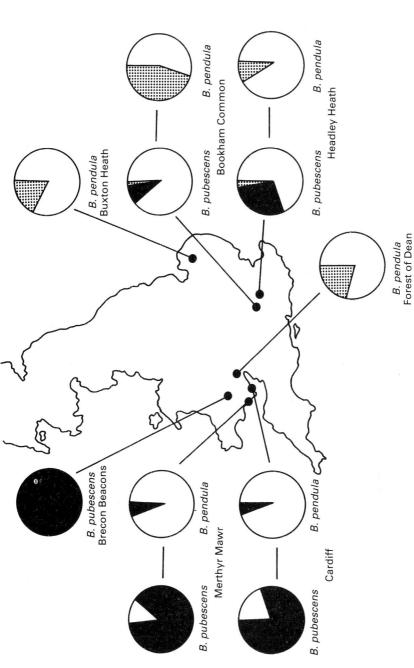

FIG. 7.4. Sketch map of southern mainland Britain to show proportions of three sibling species of leafhoppers presently collectively known as *Oncopsis flavicollis* from the two species of birch trees, *Betula pendula* and *B. pubescens*. Pie diagrams show proportions of species 1, shaded black, species 2, unshaded and species 3, stippled, from samples of adult males from identified host plants at seven localities (after Claridge & Nixon 1986).

status, *O. flavicollis* and *O. subangulata,* are associated with the two closely-related species of tree birches, *Betula pendula* and *B. pubescens,* in Britain. Studies on acoustic mate recognition signals revealed that the two species are indeed biologically distinct. *O. subangulata* is widely distributed on both species of *Betula.* However, *O. flavicollis* itself consists of three distinct, but as yet unnamed, biological species with different, but widely overlapping distributions and patterns of host plant utilization (Fig. 7.4). Only with this knowledge is it now possible to ask meaningful questions about the roles of interspecific competition and other factors in the establishment of assemblages of these insects on birch trees.

Other similar examples could be cited, but it is perhaps particularly interesting to note that the introduction of the parasitoid *Aphytis lingnanensis* to California (p. 151 above) was long delayed, as it was for many years thought to be the same as the already resident and morphologically similar. *A. chrysomphali.* It is worrying that in most insect groups where detailed studies are made, examples of such sibling species are regularly revealed. It is no coincidence that the phenomenon is best known in groups of biting flies, intensively worked because of their medical and veterinary importance, and species of *Drosophila* widely studied by geneticists and evolutionists. Where insect herbivores have been investigated in such detail, usually because they are pests, similar arrays of sibling species have been suggested (e.g. Jaenike & Selander 1980; Claridge, Den Hollander & Morgan 1984).

Clearly if, as seems likely, such problems reflect a widespread inadequacy of morphological taxonomy in determining biological species limits of insects, many ecological ideas concerning compartmentation and food web structure and the relative importance of interspecific competition and other population processes need to be treated with caution and possibly re-evaluated.

The process of speciation in specialist insect feeders is little understood and we do not have sufficient evidence to test several competing hypotheses. However, it is certainly too soon to conclude that sympatric processes of speciation necessarily dominate such assemblages of insects as those reviewed here, as suggested by Price (1980) and Strong, Lawton & Southwood (1984). There is little good genetic evidence to support such ideas (Futuyma & Mayer 1980; Jaenike 1981). Just because assemblages of insect herbivores and parasitoids often differ most obviously in host species, it cannot be concluded that such differentiation was the major factor in their speciation. The limited evidence suggests that some process of allopatric speciation, requiring a period of spatial separation of incipient species populations to allow some genetic divergence, is likely to be a necessary prelude to speciation.

ACKNOWLEDGMENTS

I thank Simon Fowler, Mark Jervis, Neil Kidd, Alan Stewart and an anonymous reviewer for helpful comments on drafts of this paper, Anne Weldon for preparing the typescript, and my wife, Clare, for help in the preparation of the illustrations. I am also grateful to Dick Askew and John Lawton for sending me copies of their then unpublished manuscripts.

REFERENCES

Askew, R.R. (1960). Some observations on *Diplolepis rosae* (L.) (Hym., Cynipidae) and its parasites. *Entomologists Monthly Magazine,* **95,** 191–2.

Askew, R.R. (1961). On the biology of the inhabitants of oak galls of Cynipidae in Britain. *Transactions of the Society for British Entomology,* **14,** 237–68.

Askew, R.R. (1980). The diversity of insect communities in leaf-miners and plant galls, *Journal of Animal Ecology,* **49,** 817–29.

Askew, R.R. (1984). The biology of gall wasps. *Biology of Gall Insects* (Ed. by T.N. Ananthakrishnan), pp. 223–71. Edward Arnold, London.

Askew, R.R. & Shaw, M.R. (1974). An account of the Chalcidoidea (Hymenoptera) parasitising leaf-mining insects of deciduous trees in Britain. *Biological Journal of the Linnean Society,* **6,** 289–335.

Askew, R.R. & Shaw, M.R. (1986). Parasitoid communities: their size, structure and development. *Insect Parasitoids* (Ed. by J. Waage & D. Greathead), pp. 225–64. Symposia of the Royal Entomological Society of London, 13. Academic Press, New York.

Cain, A.J. (1954). *Animal Species and their Evolution.* Hutchinson, London.

Christenson, L.D. & Foote, R.H. (1960). Biology of fruit flies. *Annual Review of Entomology,* **5,** 171–92.

Claridge, M.F. (1961). Biological observations on some Eurytomid (Hym., Chalcidoidea) parasites associated with Compositae, and some taxonomic implications. *Proceedings of the Royal Entomological Society of London, Series A,* **36,** 153–8.

Claridge, M.F. & Askew, R.R. (1960). Sibling species in the *Eurytoma rosae* group (Hym., Eurytomidae). *Entomophaga,* **5,** 141–53.

Claridge, M.F., Den Hollander, J. & Morgan, J.C. (1984). The status of weed-associated populations of the brown-planthopper, *Nilaparvata lugens* (Stål)—host race or biological species? *Zoological Journal of the Linnean Society,* **84,** 77–90.

Claridge, M.F. & Nixon, G.A. (1986). *Oncopsis flavicollis* (L.) associated with tree birches (*Betula*): a complex of biological species or a host plant utilization polymorphism? *Biological Journal of the Linnean Society,* **27,** 381–97.

Claridge, M.F. & Wilson, M.R. (1976). Diversity and distribution patterns of some mesophyll-feeding leafhoppers of temperate woodland canopy. *Ecological Entomology,* **1,** 231–50.

Claridge, M.F. & Wilson, M.R. (1978). British insects and trees: a study in island biogeography or insect/plant coevolution? *American Naturalist,* **112,** 451–6.

Claridge, M.F. & Wilson, M.R. (1981). Host plant associations, diversity and species-area relationships of mesophyll-feeding leafhoppers of trees and shrubs in Britain. *Ecological Entomology,* **6,** 217–38.

Claridge, M.F. & Wilson, M.R. (1982a). Species–area effects for leafhoppers on British trees: comments on the paper by Rey *et al. American Naturalist,* **119,** 573–5.

Claridge, M.F. & Wilson, M.R. (1982b). Insect herbivore guilds and species–area relationships: leafminers on British trees. *Ecological Entomology,* **7,** 19–30.

160   M. F. CLARIDGE

Connor, E.F., Faeth, S.H., Simberloff, D. & Opler, P.A. (1980). Taxonomic isolation and the accumulation of herbivorous insects: a comparison of introduced and native trees. *Ecological Entomology*, **5**, 205–11.

Connor, E.F. & McCoy, E.D. (1980). The statistics and biology of the species–area relationship. *American Naturalist*, **113**, 791–833.

Cornell, H.V. & Washburn, J.O. (1979). Evolution of the richness–area correlation for cynipid gall wasps on oak trees: a comparison of two geographic areas. *Evolution*, **33**, 257–74.

DeBach, P. (1966). The competitive displacement and coexistence principles. *Annual Review of Entomology*, **11**, 183–212.

DeBach, P., Hendrickson, R.M. & Rose, M. (1978). Competitive displacement: extinction of the Yellow Scale, *Aonidiella citrina* (Coq.) (Homoptera: Diaspididae), by its ecological homologue, the California Red Scale, *Aonidiella aurantii* (Mask.) in Southern California. *Hilgardia*, **46**, 1–35.

DeBach, P., Rosen, D. & Kennett, C.E. (1969). Biological control of coccids by introduced natural enemies. *Biological Control* (Ed. by C.B. Huffaker), pp. 165–94. Plenum, New York.

DeBach, P. & Sundby, R.A. (1963). Competitive displacement between ecological homologues. *Hilgardia*, **34**, 105–66.

Diamond, J.M. & Case, T.J. (Eds) (1986). *Community Ecology*. Harper & Row, New York.

Ehler, L.E. & Hall, R.W. (1982). Evidence for competitive exclusion of introduced natural enemies in biological control. *Environmental Entomology*, **11**, 1–4.

Futuyma, D.J. & Gould, F. (1979). Associations of plants and insects in a deciduous forest. *Ecological Monographs*, **49**, 33–50.

Futuyma, D.J. & Mayer, G.C. (1980). Non-allopatric speciation in animals. *Systematic Zoology*, **29**, 254–71.

Gilbert, F.S. (1980). The equilibrium theory of island biogeography: fact or fiction? *Journal of Biogeography*, **7**, 209–35.

Godwin, H. (1956). *The History of the British Flora*. Cambridge University Press, Cambridge.

Godfray, H.C.J. (1984). Patterns in the distribution of leaf-miners on British trees. *Ecological Entomology*, **9**, 163–8.

Huffaker, C.B. & Kennett, C.E. (1969). Some aspects of assessing efficiency of natural enemies. *Canadian Entomologist*, **101**, 425–47.

Huffaker, C.B., Messenger, P.S., DeBach, P. (1969). The natural enemy component. *Biological Control* (Ed. by C.B. Huffaker), pp. 16–67. Plenum, New York.

Jaenike, J. (1981). Criteria for ascertaining the existence of host races. *American Naturalist*, **117**, 830–4.

Jaenike, J. & Selander, R.K. (1980). On the question of host races in the fall webworm, *Hyphantria cunea*. *Entomologia Experimentalis et Applicata*, **27**, 31–7.

Jeffries, M.J. & Lawton, J.H. (1984). Enemy free space and the structure of ecological communities. *Biological Journal of the Linnean Society*, **23**, 269–86.

Keller, M.A. (1984). Reassessing evidence for competitive exclusion of introduced natural enemies. *Environmental Entomology*, **13**, 192–5.

Kennedy, C.E.J. & Southwood, T.R.E. (1984). The numbers of species of insects associated with British trees: a re-analysis. *Journal of Animal Ecology*, **53**, 455–78.

Lawton, J.H. (1978). Host-plant influences on insect diversity: the effects of space and time. *Diversity of Insect Faunas* (Ed. by L.A. Mound & N. Waloff), pp. 105–25. Symposia of the Royal Entomological Society of London, 9, Blackwell Scientific Publications, Oxford.

Lawton, J.H. (1984). Non-competitive populations, non-convergent communities and vacant niches: the herbivores of bracken. *Ecological Communities: Conceptual Issues and the Evidence* (Ed. by D.R. Strong, D. Simberloff, L.G. Abele & A.B. Thistle), pp. 67–100. Princeton University Press, Princeton, New Jersey.

Lawton, J.H. (1986). The effect of parasitoids on phytophagous insect communities. *Insect Parasitoids* (Ed. by J. Waage & D. Greathead), pp. 265–87. Symposia of the Royal Entomological Society of London, 13. Academic Press, New York.

Lawton, J.H. & Hassell, M.P. (1984). Interspecific competition in insects. *Ecological Entomology* (Ed. by C.B. Huffaker & R.L. Rabb), pp. 451–95. Wiley, New York.

Lawton, J.H. & Schröder, D. (1977). Effects of plant type, size of geographical range and taxonomic isolation on number of insect species associated with British plants. *Nature (London),* **265,** 137–40.

Lawton, J.H. & Strong, D.R. Jr. (1981). Community patterns and competition in folivorous insects. *American Naturalist,* **118,** 317–38.

Luck, R.F. & Podoler, H. (1985). Competitive exclusion of *Aphytis lingnanensis* by *A. melinus*: potential role of host size. *Ecology,* **66,** 904–13.

MacArthur, R.H. & Wilson, E.O. (1967). *The Theory of Island Biogeography.* Princeton University Press, Princeton, New Jersey.

Mayr, E. (1942). *Systematics and the Origin of Species from the Viewpoint of a Zoologist.* Columbia University Press, New York.

Neuvonen, S. & Niemela, P. (1981). Species richness of Macrolepidoptera on Finnish deciduous trees. *Oecologia (Berlin),* **51,** 364–70.

Opler, P. (1974). Oaks as evolutionary islands for leaf-mining insects. *American Scientist,* **62,** 67–73.

Paterson, H.E.H. (1985). The recognition concept of species. *Species and Speciation* (Ed. by E.S. Vrba), pp. 21–9. Transvaal Museum Monograph No. 4, Transvaal Museum, Pretoria.

Perring, F.H. & Walters, S.M. (1962). *Atlas of the British Flora.* Nelson, London.

Price, P.W. (1980). *Evolutionary Biology of Parasites.* Princeton University Press, Princeton, New Jersey.

Price, P.W. (1984). *Insect Ecology,* 2nd edn. Wiley, New York.

Rey, J.R., McCoy, E.D. & Strong, D.R. (1981). Herbivore pests, habitat islands and the species–area relationship. *American Naturalist,* **117,** 611–22.

Rigby, C. & Lawton, J.H. (1981). Species–area relationships of arthropods on host plants: herbivores on bracken. *Journal of Biogeography,* **8,** 125–33.

Root, R.B. (1967). The niche exploitation pattern of the blue-grey gnatcatcher. *Ecological Monographs,* **37,** 317–50.

Root, R.B. (1973). Organization of a plant–arthropod association in simple and diverse habitats: the fauna of collards (*Brassica oleracea*). *Ecological Monographs,* **43,** 95–124.

Southwood, T.R.E. (1961). The number of species of insect associated with various trees. *Journal of Animal Ecology,* **30,** 1–8.

Southwood, T.R.E. (1978). The components of diversity. *Diversity of Insect Faunas* (Ed. by L.A. Mound & N. Waloff), pp. 19–40. Symposia of the Royal Entomological Society of London, 9. Blackwell Scientific Publications, Oxford.

Strong, D.R. Jr. (1974). Nonasymptotic species richness models and the insects of British trees. *Proceedings of the National Academy of Sciences, USA,* **71,** 2766–9.

Strong, D.R. Jr. (1979). Biogeographic dynamics of insect–host plant communities. *Annual Review of Entomology,* **24,** 89–119.

Strong, D.R. Jr. (1984). Exorcising the ghost of competition past from insect communities. *Ecological Communities: Conceptual Issues and the Evidence* (Ed. by D.R. Strong, D. Simberloff, G. Abele & A.B. Thistle), pp. 28–41. Princeton University Press, Princeton, New Jersey.

Strong, D.R., Lawton, J.H. & Southwood, R. (1984). *Insects on Plants, Community Patterns and Mechanisms.* Blackwell Scientific Publications, Oxford.

Strong, D.R. Jr. & Levin, D.A. (1979). Species richness of plant parasites and growth form of their hosts. *American Naturalist,* **114,** 1–22.

Strong, D.R. Jr., Simberloff, D., Abele, L.G. & Thistle, A.B. (Eds) (**1984**). *Ecological Communities: Conceptual Issues and the Evidence.* Princeton University Press, Princeton, New Jersey.

Waage, J. & Greathead, D. (Eds) (**1986**). *Insect Parasitoids.* Symposia of the Royal Entomological Society, 13. Academic Press, New York.

Walter, G.H., Hulley, P.E. & Craig, A.J.F.K. (**1984**). Speciation, adaptation and interspecific competition. *Oikos,* **43**, 246–8.

Zwölfer, H. (**1968**). Untersuchungen zur biologischen Bekämpfung von *Centaurea solstitialis* L.—Strukturmerkmale der Wirtspflanze als Auslöser des Eiablageverhaltens bei *Urophora siruna-seva* (Hg.) (Dipt., Trypetidae). *Zeitschrift für angewandte Entomologie,* **61**, 119–30.

Zwölfer, H. (**1979**). Strategies and counterstrategies in insect population systems competing for space and food in flower heads and plant galls. *Fortschritte der Zoologie,* **25**, 331–53.

Zwölfer, H. (**1980**). Distelblütenkopfe als ökologische Kleinsysteme: Konkurrenz und Koexistenz in Phytophagenkomplexen. *Mitteilungen der deutschen Gesellschaft für allgemeine und angewandte Entomologie,* **2**, 21–37.

Zwölfer, H. & Harris, P. (**1984**). Biology and host specificity of *Rhinocyllus conicus* (Froel.) (Col., Curculionidae), a successful agent for biocontrol of the thistle, *Carduus nutans* L. *Zeitschrift für angewandte Entomologie,* **97**, 36–62.

8. ORGANIZATION OF AVIAN ASSEMBLAGES—THE INFLUENCE OF INTRASPECIFIC HABITAT DYNAMICS

RAYMOND J. O'CONNOR

British Trust for Ornithology, Beech Grove, Tring,
Hertfordshire HP23 5NR, UK

INTRODUCTION

Two distinct schools of thought exist as to how communities are organized. On the one hand is the idea that communities are random assemblages of individual species that come together because environmental conditions at the site concerned are locally suitable for those rather than for other species (Beals 1960). On the other hand is the idea that communities are highly self-organized entities with definite structure resulting from species inter-actions, often competitive (MacArthur 1958; Lack 1971). This latter school of thought has dominated the study of avian communities until recently, when the work of Wiens (Wiens & Rotenberry 1981;Wiens 1983, 1984) in particular challenged the general validity of competition as an organizing force in avian community ecology. Wiens stresses the variability of patterns observed in bird communities, a feature that suggests greater dynamism within these assemblages than is normally recognized.

A factor contributing to such dynamism and mitigating or even abolishing the effects of interspecific competition is the effect of population pressure on the use of different habitats by birds. As bird populations vary in density in response to environmental conditions (Marchant & Hyde 1980), they vary also in their use of the available habitats (Fretwell & Lucas 1969; O'Connor 1985). Since differential habitat use has long been regarded as an effective means of ecological isolation among birds (MacArthur 1958; Lack 1971), a change in the spectrum of habitats occupied in the short-term should also result in changed competitive relationships, thereby blurring any competitive organization of the communities in question. The role of avian habitat dynamics in thus generating non-equilibrium communities has, however, hardly been explored, with the work of Rosenzweig (1979, 1981 and this volume, Chapter 21) the major exception.

The idea that autecological processes may have as large a role to play in shaping ecological communities as have synecological ones has recently

regained favour (Wiens 1983; Strong *et al.* 1984). Wiens' arguments in
favour of non-equilibrium communities derive in large part from his work on
the birds of shrub-steppe communities of cold deserts, in which restricted
vertical scope results in most of the spatial pattern apparent at any one time
being horizontal and the year to year changes in patterning appear to have a
large component of chance event in them (Rotenberry & Wiens 1980a,b). It
is arguable that his results (and views based on them) are largely the
outcome of work on communities lacking the vertical structure that might
otherwise facilitate the emergence of the classic correlations of bird species
diversity (BSD) with foliage height diversity (FHD) (see MacArthur &
MacArthur 1961). In the present paper I examine some data on the structure
of some British bird communities, principally of agricultural land, in which
strong relationships with structural diversity are apparent, and show that
dynamical aspects of intraspecific competition may significantly influence
the assemblages observed even in such communities. As not all birds
recorded in farmland censuses (the major source of data used here) are
integrated ecologically into the farmland bird *community*, I will reserve the
latter term for contexts where such integration is to be implied and use the
term *assemblage* where a neutral context, e.g. observations without proof of
competition, is intended.

SOME THEORETICAL ASPECTS OF AVIAN HABITAT DYNAMICS

The idea of competition-driven habitat dynamics in bird communities
originated with Brown (1969) and Fretwell & Lucas (1969). Consider a
species which does best in (and therefore prefers) some habitat A and whose
members defend exclusive incompressible territories. At low population
densities all individuals can establish territories in that habitat and breed
there. Above some critical density, however, more individuals are seeking
territories than can be accommodated in the preferred habitat and these
must either remain as non-breeders or establish themselves in some habitat
B that is less preferred because breeding success (or survival) is poorer
there. This secondary habitat will then fill up as the population continues to
increase until it in turn is saturated. At this point a third habitat C may be
colonized and so on. The process continues through a habitat hierarchy until
either no further acceptable habitats remain (so that any remaining surplus
birds can stay only as non-breeding 'floaters') or a habitat is used which is so
poor in quality that birds using it contribute little to population recruitment,
and expansion ceases. Assuming for simplicity that habitat quality is determined only by breeding performance (although somewhat similar argu-

ments apply if survival is the critical factor), three predictions may be made from this model. First, breeding success within each habitat should be independent of density. Second, in an expanding population breeding performance within each habitat should decline with order of use of habitat, and third, in a decreasing population the poorer habitats should be abandoned earlier. Corollaries of these last two predictions are that average breeding performance should vary in a density-dependent way purely in consequence of the ebb and flow through the habitat hierarchy and that the number of habitats in use should vary with population density.

A variant of this model holds if territories within a habitat are compressible, but at a cost to breeding performance, over the full range of population densities experienced by the species. Now habitat B will not come into use as long as the cost of further territory compression within A is less than the difference between the current quality of A and the intrinsic (uncrowded) quality of B. Once this point is reached, however, further population expansion will result in birds entering either A or B in such a way as to equalize the resulting fitnesses within the two habitats. Further population expansion then drives fitnesses in both habitats ever lower until the level of the intrinsic quality of habitat C is reached, whereupon it in turn comes into use, with further density increases now being accommodated over the three habitats so as to equalize density-dependent fitnesses among them. With this model, breeding success *within each habitat* should be density-dependent whilst success *across habitats* at any one time should be fairly similar; other properties of the model resemble those discussed above. Rozenzweig (Chapter 21) presents a theoretical treatment of a similar, density-dependent model of habitat choice.

These habitat dynamics models thus make some specific predictions about the relationships between breeding performance, population density and nest habitat diversity. There are, though, a number of modifying factors (O'Connor 1985; Rosenzweig 1985). First, once an individual has established itself in a particular territory there may accrue to it various advantages (in terms of breeding performance) associated with site familiarity. Such benefits may favour site fidelity (Fretwell 1981) by raising the threshold for a change of territory. If the population decreases somewhat, so that a vacancy now arises in an intrinsically more favoured habitat, the gain in moving to take up that vacancy may be partially or wholly offset by the loss of the benefits of site familiarity. In a decreasing population, therefore, a greater variety of habitats may be in use than at the same density during population expansion. A further complication may arise through interspecific competition: the presence of competitors in an otherwise invasible habitat may reduce the value of that habitat below that of other alternatives or even below that acceptable at all, thus restricting the expanding species (Rosenzweig 1985). Empirical evidence in support of this model has been obtained for hummingbirds by Pimm, Rosenzweig & Mitchell (1985).

In this paper I will show that the success of widespread abundant farmland species is associated more with breadth of nesting habitat than with their demographic characteristics. Population changes over two decades will be used to show that this habitat breadth is dynamically linked to population density but that site fidelity may be a significant modifier of such relationships. Finally, I will discuss the implications of density-dependence in habitat use for the organization of bird communities.

MATERIALS AND METHODS

This paper is based largely on data from two sources, the Common Birds Census (CBC) and the Nest Records Scheme (NRS) of the British Trust for Ornithology. The CBC is a territory-mapping census of breeding birds in Britain, conducted annually on farmland since 1962 and in woodland since 1964. Participants (largely amateur ornithologists) map the distribution of all bird registrations made in the course of 8–12 breeding season visits to a *c.* 60 ha census plot (for details see Marchant 1983). The resulting maps are collated to form species maps on which clusters of registrations, when interpreted consistently by trained professional analysts, closely approximate the locations of individual territories. Provided that each observer is consistent in fieldwork effort from year to year, it is possible to eliminate observer bias by pairing results from individual plots across years to estimate relative changes in population levels. By pro-rating these changes on a 1966 reference index of 100, an index of population level for each species is obtained. Separate indices are calculated for farmland and for woodland. About 100 plots in each habitat are used in the national index calculations each year, with additional censuses by observers entering the scheme but yet lacking a pairing census from the previous year. A few, mostly colonial, species such as reed warbler *Acrocephalus scirpaceus* and rook *Corvus frugilegus* are frequent on farmland but are not amenable to territory mapping and are omitted from CBC index coverage. Detailed instructions for the CBC are given in Marchant (1983), whilst recent reviews of possible biases and limitations of the scheme are provided by O'Connor & Marchant (1981) and by Fuller, Marchant & Morgan (1985).

The Nest Records Scheme began in 1939 and comprises a file of nest histories. Participants, again mostly amateur ornithologists, complete standard cards for each nest they find. Each card records details of the locality of a nest, its environment and habitat, the nest site, and details of its contents on one or more visits by the observer. With a well-timed sequence of visits this information allows computation of laying date, clutch size, incubation period, number of eggs hatched, nestling period, and chicks fledged. The card may also record the cause of nest failure, where this occurs. In practice, only a proportion of the cards are this detailed but cards yielding only partial information can be pooled by region, habitat, date, and so on, to yield good estimates of mean clutch size, etc. for that pooling. Frequencies of hatching and fledging success estimated from partial data (where some nests are found after laying has started) are biased since some nest losses go unrecorded, but the rates of such losses could be computed in an unbiased manner using a procedure devised by Mayfield (1961, 1975) and subsequently validated by Johnson (1979).

Analyses of nest records conducted here were based on computerized sample files of nest records for the years 1962–80 (or, for some species, 1981) which had been screened for transcription and other errors. It was impractical, given the resources available, to process all the records for the more numerous species, there being, for example, some 13 000 nest records for lapwing *Vanellus vanellus*. For such species the files contained annual samples of approximately 100 cards for each species, based as far as possible on cards recording three or more visits, corrected to ensure adequate regional representation.

Nest habitats are poorly classified by observers, who vary greatly in the extent of their ecological knowledge. However, much descriptive information about the nest site and its environs may be written on each card, though not in a standardized form. This problem was dealt with here by each card being coded in a consistent manner by a technical assistant, with the coding based on the total information provided by the observer about the nest, irrespective of how loosely described it was. Habitat categories used in coding were rather crude, with agricultural land being sub-classified only to the level of arable/pastoral/mixed/rough grazing, woodland to deciduous/coniferous/mixed, and so on.

Nevertheless, even at this level between habitat differences in breeding biology have been previously established (see the papers by O'Connor and colleagues cited below), so the classification suffices in testing the models of interest here. For the present paper I adopted a standard species diversity index to compute an index of nest habitat diversity (NHD) as

$$\text{NHD} = \Sigma 1/p_i^2$$

where p_i is the proportion of nests in the sample that were in the ith habitat. Such indices were calculated for each annual sample and the median of these used as a species-specific value for each species.

Two points need to be made about the uses of CBC and NRS data in the present paper. First, although the two sets of data are largely independent of each other, some nest records may have come from CBC plots. It was impractical to check for and exclude such cases but I estimate that such nest records must have constituted fewer than 0.1% of those considered here. Secondly, although the choice of species for the various analyses described below was arbitrary, in that it was based on administrative decisions made in the context of the Nest Record Scheme, the species concerned do constitute what most British ornithologists would recognize as a reasonable cross-section of farmland bird species. Systematic bias within the sample of species can therefore be discounted.

RESULTS AND DISCUSSION

Figure 8.1 shows a clear relationship between the average abundance of individual species on farmland and the frequency with which they occurred in farmland censuses. Species recorded in fewer than 20% of the censuses have been omitted from the plot. As a broad generalization, farmland bird

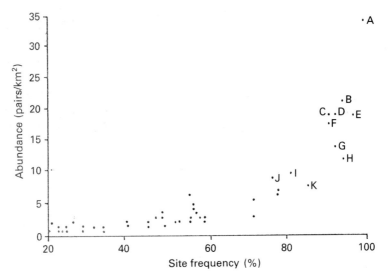

FIG. 8.1. The average abundance (pairs/km^2) of individual species on farmland Common Birds Census plots for 1962–80 in relation to the percentage of returns with that species present (= site frequency). The most abundant species are A, blackbird *Turdus merula*; B, chaffinch *Fringilla coelebs*; C, wren *Troglodytes troglodytes*; D, robin *Erithaculus rubecula*; E, dunnock *Prunella modularis*; F, skylark *Alauda arvensis*; G, blue tit *Parus caeruleus*; H, song thrush *Turdus philomelos*; I, yellowhammer *Emberiza citrinella*; J, willow warbler *Phylloscopus trochilus*; K, great tit *Parus major*.

assemblages in Britain comprise a rather small group of very widespread and abundant species, together with a larger group of less abundant species. Only eight of the 107 species considered were found in 90% or more of the censuses. These widespread species are, by and large, generalist passerines that also dominate the bird communities of other habitats. Thus of the very abundant species in Fig. 8.1 (see legend for species identity) only the skylark—a species of open fields and moorland—was not in the equivalent species list for the woodland communities studied by Fuller (1982), and of Fuller's list only the treecreeper *Certhia familiaris* was scarce on farmland. Even with these differences between the two habitats, the densities in farmland and in woodland of the twelve species named are correlated across species ($r = 0.628$, $P < 0.05$, based on data in Hickling (1983)). Fuller (1982) provides abundance–frequency data for a variety of other habitats and these show broadly the same kind of relationships as is apparent in Fig. 8.1; only a few colonial species present in large colonies at a very few sites introduce discrepancies.

Demographic factors might underlie this link between wide distribution and numerical abundance, i.e. abundant species may have high reproductive rates or be particularly long-lived. O'Connor (1981) found, for example, that migrant species in Britain were more widespread the greater their annual production of eggs, whilst amongst resident species high annual survivorship was correlated with ubiquity. Table 8.1, however, shows for the present data that the frequencies with which the species analysed in O'Connor (1981) were found on farmland census plots were poorly correlated with clutch size, with number of broods, with seasonal egg production, and with adult survivorship, irrespective of whether the analysis considered only residents or only migrants or whether all species were pooled. These differences from the earlier analysis indicate that the different scales of distribution—over all 10 km OS grid squares of Britain and Ireland in the earlier analysis but over smaller and geographically more restricted areas in the present one—involve different levels of explanation, demographic at the regional level, perhaps habitat-oriented (see below) at the farm census plot level. These findings thus echo the different concepts of diversity suggested by Cody (1975), with alpha (α)-diversity at the site level, beta (β)-diversity at the regional level, and gamma (γ)-diversity at the continental level.

If demographic explanations are discarded on the basis of Table 8.1, an explanation for the apparent regularity of the relationship between abundance and site frequency in Fig. 8.1 must be sought in some other general ecological factor, for which habitat availability must, for birds, be a leading candidate (MacArthur & MacArthur 1961, Cody 1975). Figure 8.2

TABLE 8.1. Correlation between (log-transformed) relative frequency of farmland occurrence of various species and their demographic characteristics. Based on data from O'Connor (1981) and O'Connor & Shrubb (1986a)

Variable	Residents (N = 32)	Migrants (N = 11)	Pooled (N = 43)
Log (clutch size)	−0.200	0.258	0.093
Log (number of clutches)	0.208	−0.001	0.272
Log (seasonal egg production)	−0.003	0.287	0.351
Adult survival	−0.004*	−0.363	0.133

*Data available for only twenty-nine species.

shows, for a subset of species, that site frequency (and therefore, on the basis of Fig. 8.1, of density also) was closely related to the diversity of nesting habitat used by these species ($r = 0.591$, $P < 0.01$). The more widespread species were those using the greater diversity of nesting habitats and the more restricted species were those with a narrower range of nesting habitats. Three species (blackbird, dunnock, and blue tit) lie well above the general trend of Fig. 8.2, i.e. they are more ubiquitous than is typical of

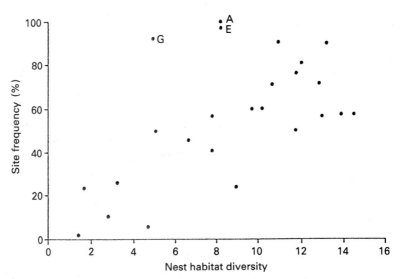

FIG. 8.2. The relationship between nest habitat diversity (median of annual values for the years 1962–81) for various species and site frequency (= the percentage of farmland Common Birds Census returns recording that species). Species identified here are A, blackbird; E, dunnock; G, blue tit.

other species with similar levels of nest habitat diversity. In the case of the blue tit this may well be a reflection of its hole-nesting habits, which perhaps limit the number of distinct habitats in which it is likely to be recorded; the only other hole-nesting species in the sample, the redstart *Phoenicurus phoenicurus*, is also stenotopic, with the third lowest diversity level of the twenty-five calculated. However, no such explanation is readily available for the apparently anomalous positions of the blackbird and dunnock.

The greater breadth of nesting habitat just established for the more ubiquitous, abundant species is not immediately consistent with the absence of demographic differences between abundant and scarcer species established in Table 8.1. Studies such as that of Best & Stauffer (1980) have shown nest predation losses to be greater among species utilizing a broad range of nest sites than among nest site specialists. I have shown a similar relationship between nesting success and nest habitat diversity over a range of British breeding birds, with egg losses and nestling losses (to unknown causes) both increasing with habitat diversity (O'Connor 1985). On the basis of such studies one might expect the habitat generalists to produce more eggs per season (or for the adults to survive better) in compensation for these increased losses. Yet Table 8.1 shows that the abundant, habitat-generalist species are neither more likely to lay more eggs per season nor more likely to live longer than are the scarcer habitat specialists. On simple demographic grounds, then, one would expect the specialists to increase relative to the more abundant generalists, for they start with similar clutches but enjoy lower predation than do the generalists.

The habitat dynamics model described above offers an explanation of this anomaly if the sheer density of abundant habitat generalists in fact creates intraspecific density-dependent competition for good nest sites. Species-typical clutch sizes and arrival rates may merely reflect typical levels of competition for habitat rather than intrinsic life-history characteristics. On this argument, the findings of Best & Stauffer (1980) and of O'Connor (1985) might be due to different levels of intraspecific competition for nest habitat among different species rather than to interspecific differences in demography. On this model, therefore, increased intraspecific competition should lead to a species using a greater variety of habitats at high densities than at low densities, so a more populous species might be expected, on average, to occupy a great variety of habitats of varied intrinsic quality. It was possible to examine the intraspecific effects of population change for farmland birds in Britain because many species suffered very badly during the severe winter of 1962–63, when populations of some species were reduced to as little as a sixth of their national level. Numbers recovered during the subsequent run of mild winters, with the rate of recovery varying from species to species, largely determined by seasonal egg production rates (O'Connor 1981). Figure 8.3 illustrates the time course of species-richness and total bird density (based on species censused within the CBC). Note that the low abundance levels in 1962 and 1979 were also each associated with a preceding severe winter.

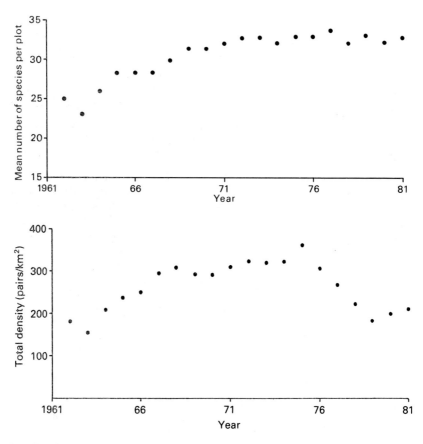

FIG. 8.3. Changes in mean number of species (top) and total density of birds censused by the CBC scheme (bottom) recorded on farmland CBC plots over the period 1962–80.

Figure 8.3 shows that total bird abundance peaked in 1975, declining sharply thereafter. Because farmland is a managed habitat, it is necessary to digress briefly to consider whether changes in farming practice might have changed the carrying capacity of the land for the species involved. Were this the case, it could preclude interpretation of the data within the theory of avian habitat dynamics. Farmland populations peaking in 1975 were those of pied wagtail *Motacilla alba,* wren, dunnock, robin, song thrush, long-tailed tit *Aegithalos caudatus,* blue tit, greenfinch *Carduelis chloris* and reed bunting *Emberiza schoeniclus.* Woodland populations affected were green woodpecker *Picus viridis,* wren, blackbird, goldcrest *Regulus regulus,* bullfinch *Pyrrhula pyrrhula* and yellowhammer. Finally, the numbers of meadow pipits *Anthus pratensis,* a species monitored only through a pooled index based on CBC plots in all available habitats, also peaked in 1975. For most of these populations the CBC indices showed gradual build-ups to the 1975 level, thereafter declining steadily. As all the populations affected were of resident species rather than of migrants, a change in the carrying capacity of farmland in winter or early spring could account for the post-1975 drop in numbers shown by Fig. 8.3. (In contrast, the large-scale collapse after 1968 of the populations of those migrant species that winter in the drought-stricken Sahelian zone of Africa (Winstanley, Spencer & Williamson 1974) is hardly apparent in Fig. 8.3.)

One striking change that has the right scale and timing in farmland to affect carrying capacities in winter or early spring has been the change from the growing of spring-sown to autumn-sown cereals. This change swept through Britain between about 1974 and 1977 (O'Connor & Shrubb 1986a), after which the bulk of cereal sowings were of winter cereals. The change

TABLE 8.2. Percentages of nest record cards from certain agricultural habitats between 1962–75 and 1976–81, for eight species that declined in population density after 1975. See text for details

Species	Percentage of nest record cards involving agricultural habitats*		χ^2
	1962–75	1976–81	
Yellowhammer	40.6	38.2	0.96
Dunnock	19.4	21.7	1.31
Blackbird	14.7	16.7	1.44
Pied wagtail	13.7	15.6	1.13
Reed bunting	12.4	9.0	5.08†
Bullfinch	9.5	8.7	0.35
Wren	10.5	8.2	2.58
Blue tit	3.3	2.0	0.84

*Habitats coded 70–74 in standard BTO habitats classification, as yet unpublished. A listing of the habitat codes used will be provided on request.
†$P < 0.05$.

affects farmland carrying capacity for ground-feeding birds partly by
removing the spring cultivations that used to expose freshly worked soil and
its complement of invertebrates just when females needed extra energy and
nutrients for egg formation, and partly because the more intensive
management of winter cereals is unfavourable to birds. Thus song thrush
numbers in areas switching from spring to winter barley decreased by as
much as 60 per cent, due both to reduced settlement and to lower breeding
success by settlers (O'Connor & Shrubb 1986a). Other ground-feeding
species also decreased in these areas but arboreal species did not (O'Connor
& Shrubb 1986b). If the post-1975 changes in the species listed above were
due to similar changes in carrying capacity, the nest record cards should
show a decrease in the relative use of agricultural habitats. Table 8.2
presents a comparison of such relative use for the eight species that
decreased nationally after 1975 and for which nest record cards were
available for analysis here. For five species (blue tit, yellowhammer, reed
bunting, bullfinch and wren) farmland habitats were less frequently used
from 1976 onwards than before whilst for three species (pied wagtail,
dunnock and blackbird) their use increased. Only for reed bunting was the
change statistically significant. One must conclude, therefore, that a change

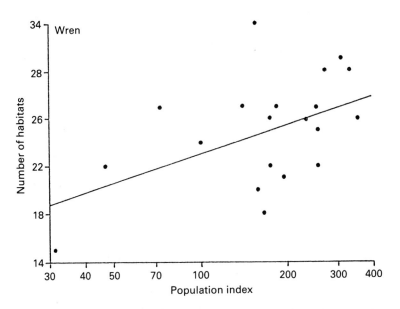

FIG. 8.4. The number of nesting habitats used by wrens (*Troglodytes troglodytes*) each year in
relation to its farmland population density (as farmland Common Birds Census index) in that
year. Line fitted by least squares regression.

in the carrying capacity of farmland cannot explain for all species the pattern of declines observed, though it is probably of major importance for some.

Figure 8.4 therefore returns to the idea of dynamical control of nest habitat diversity by illustrating for the case of the wren the relationship between number of nesting habitats in use each year and the CBC population index of the species in that year. As expected on the basis of the habitat model presented earlier, the two variables are correlated ($r = 0.485$, $P < 0.05$ with logarithmic transformation of CBC index values), though the correlation is clearly poorer at high densities than at low. This latter feature would be expected on the basis of within habitat density-dependence (Rosenzweig 1985). Where density-dependence is present, the distribution of a species across the available habitats is determined both by the intrinsic quality of the habitats concerned and by the population pressure experienced within them. For example, yellowhammers nesting in farmland at low densities lay almost half as many eggs again as do their conspecifics nesting contemporaneously in woodland. As population density increases, however, clutch size decreases both in farmland and in woodland but at a faster rate in farmland, so that clutch sizes are eventually equal both in farmland and in woodland, and the ratio of farmland density to woodland density is proportional to the ratio of farmland to woodland clutch sizes over a wide range of densities (O'Connor 1980). Wrens are also a species subject to density dependence in breeding performance, for in woodland, at least, their clutch sizes decrease with population density (O'Connor & Fuller 1985). In general, this type of density-dependent regulation of within habitat densities can be expected to conceal the simple relationship of habitat frequency and population density (Rosenzweig 1985), as is apparent in Fig. 8.4.

Nest record data allowing examination of the relationship between habitats occupied and population density were available for eight of the species subject to post-1975 population decreases. Of these, density-dependence in reproductive performance has already been established for three species, the wren and yellowhammer already mentioned and the blackbird (Batten 1977), and is suspected for the blue tit (Perrins 1979). I restricted the testing for habitat–density relationships to the first few years of the Common Birds Census (CBC) when populations were almost uniformly expanding (Fig. 8.3), as done by Williamson (1969) in his detailed study of wren habitat use. Williamson documented wren territory placement within preferred and less preferred habitats between 1964 and 1967, as the wren population recovered from the effects of the 1962–63 winter. Table 8.3 examines habitat–density correlation for eight species over the period 1962–67 and shows that seven of the eight species displayed positive correlations between the number of habitats in use and the prevailing density. None of the correlations are statistically significant but the preponderance of positive results is (Signs test, one-tailed $P = 0.035$). The blue tit was the only species with a negative correlation, possibly the result of the large proportion of reported nests in nest boxes: if blue tits were scarce after the 1962–63 winter, one might expect biased reporting of those nests in regularly checked boxes. The overall result thus suggests a tendency for species declining after 1975 to use habitats in accordance with the habitat dynamical model above, at least when the species are at low densities.

TABLE 8.3. Correlations between number of habitats in use for nesting each year and the Common Birds Census index for that year for each of eight species over 1962–67. See text for details

Species	Pearson correlation
Yellowhammer	0.019
Dunnock	0.457
Blackbird	0.099
Pied wagail	0.322
Reed bunting	0.639
Bullfinch	0.352
Wren	0.393
Blue tit	−0.266

Figure 8.5 compares nest habitat diversity in the species of Table 8.3 against the data available for a variety of other species that did not decline after 1975. The figure shows that the former species had significantly greater nest habitat diversity than had the other species. It seems reasonable to conclude that habitat dynamics considerations drive these species into a greater variety of nesting habitats than is true of species at large.

Summarizing the results to this point, the more ubiquitous species in farmland bird communities are neither more productive nor favoured with higher survival than less abundant species. They do, however, use a greater diversity of nesting habitats than do other species, with a bias towards expanding their range of nesting habitats as their populations recovered from a weather-induced population crash. Many of these species decreased synchronously in 1975, with the species affected having rather larger nest habitat diversity than other species and with evidence of a link between farming practices and population density in some but not all of these species. Figure 8.3 shows that the resulting reduction in overall density was not

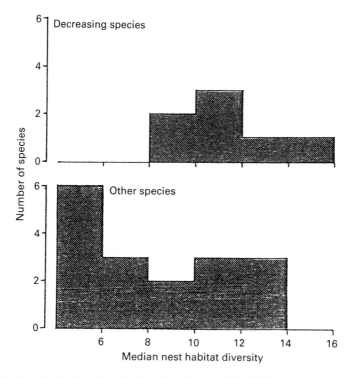

Fig. 8.5. The distribution of nest habitat diversity (as for Fig. 8.2) in samples of species (top) that began to decrease between 1975 and 1976 and (bottom) other species.

accompanied by a reduction in the number of species supported within the community, despite there having been fewer species at those same density levels during the expansion phase. The data suggest that, on the reduction of population levels that took place after 1975, some inhibition by local effects hindered a fall-back into those habitats that are intrinsically best.

The effects of site fidelity would neatly explain this inhibition if knowledge of particular territories conferred a greater advantage on birds surviving there than would transfer to a vacant territory in intrinsically better habitat (Fretwell 1981; Rosenzweig 1985). If such a process operated within the assemblages examined here, a greater variety of nesting habitats should be in use during the post-1975 decrease than in previous years. I tested this idea for the species of Table 8.3 by regressing the number of breeding habitats in use each year on the (logarithmically-transformed) CBC index values, with the 1962–75 increase phase and 1976–81 decrease phase as a dummy variable. Only three of the regressions—those for blue tit, dunnock and wren—were statistically significant and only the first two also have statistically significant dummy coefficients. However, the three dummy coefficients, together with those for reed bunting, bullfinch and yellow-hammer, were all positive, indicating a tendency towards a greater variety of nesting habitats in use (at a given population density) when the population was decreasing. Neither blackbird nor pied wagtail showed this dependence, nor did they show any dependence on the CBC index overall. The blackbird is, however, very strongly sensitive to population density and at any one time has a sizeable non-breeding component to its population (Batten 1977). Such individuals could effectively buffer the population recorded within the Nest Record Scheme, keeping the numbers recorded within the population census and in the breeding surveys stable at the cost of fluctuations in the population of 'floaters' (Kluijver & Tinbergen 1953; Brown 1969; Edwards 1977).

If intraspecific competition is particularly significant among the more ubiquitous of the species breeding on farmland, one might expect to find such species breeding in association with habitat elements found in a wide range of habitats. I examined this idea here by using information on the habitat preferences of some fifty-seven species found breeding on farmland (for a detailed example see Morgan & O'Connor (1980) and for comparative analyses see O'Connor & Shrubb (1986a)). The abundance of each of these species across sixty-five farms was tested for correlation against a spectrum of landscape elements on the farms. Table 8.4 compares the spectrum of maximum and of statistically significant correlations within the group of twelve generalists already discussed above (those named in Fig. 8.1, plus treecreeper) against the corresponding frequencies for the remaining forty-five species. The table shows that the twelve common species were significantly biased towards habitats with marked habitat structure, favouring woodlands, hedges with trees, and scrub (including that found in the gardens and shelterbelts around farmsteads and that around ponds). These are all habitats commonly present in British farmland, except where intensively managed for arable crops. Only the skylark showed a preference for arable habitats. Surprisingly, perhaps, lines of trees without an underlying hedge were not particularly favoured by the abundant species, despite such habitat being favoured by several other species (O'Connor & Shrubb 1986a). These results might suggest that the common generalists are more responsive to habitat structure than are other, less numerous species, and that the well-known relationships between foliage height diversity (FHD) and bird species diversity (BSD) (MacArthur & MacArthur 1961) might be set more by the influence of the commoner species than by the diversity of the community as a whole. In particular, intraspecific rather than interspecific competition may underlie these relationships.

TABLE 8.4. Habitat correlates of farmland density in twelve generalist species (see text) and in a sample of forty-five other species found on farmland

Habitat variable	Species (%) for which this variable is the species maximum*		Number (%) of species with densities correlated with the variable stated†	
	Generalists (N = 12)	Others (N = 45)	Generalists (N = 12)	Others (N = 45)
Hedges with trees	6 (50.0)	9 (20.0)	10 (83.3)	33 (73.3)
Woodland	1 (8.3)	7 (15.6)	10 (83.3)	22 (48.9)
3D-structure index	0 (0.0)	1 (2.2)	10 (83.3)	15 (33.3)
Total hedge density	1 (8.3)	3 (6.7)	8 (66.7)	22 (48.9)
Ponds	2 (16.7)	(3 (6.7)	7 (58.3)	33 (57.9)
Farmsteads	0 (0.0)	3 (6.7)	6 (50.0)	9 (20.0)
Scrub	1 (8.3)	4 (8.9)	5 (41.7)	15 (33.3)
Hedge without trees	0 (0.0)	0 (0.0)	1 (16.7)	7 (15.6)
Arable fraction	1 (8.3)	5 (11.1)	1 (8.3)	17 (37.8)
Lines of trees	0 (0.0)	3 (6.7)	0 (0.0)	20 (44.4)
Linear water	0 (0.0)	3 (6.7)	1 (8.3)	16 (28.1)
Other habitats	0 (0.0)	4 (8.9)	—	—
	$\chi^2 = 5.3$, df = 5, ns. ‡		$\chi^2 = 19.70$, df = 10, $P < 0.05$‡	

*To obtain these data the correlations of each species density with each of the habitat variables listed were computed and the strongest relationship (maximum correlation irrespective of sign) identified for each species. These were then tallied by habitat.
†This column lists the number of species showing statistically significant correlation of density with the extent of the habitat stated on each farm, irrespective of the presence of other habitat correlations for that same species.
‡Comparison of the distributions over habitats for generalists and for other species (see text for detail).

Few of the analyses presented here are conclusive; rather they provide circumstantial evidence that intraspecific competition may operate to drive some of the species through a spectrum of breeding habitats in ways that are not immediately consistent with the classical emphasis on interspecific competition. A highly dynamic system in which intraspecific forces simultaneously operate within each of a large number of abundant species cannot be treated simplistically as an equilibrium community. It is of course possible that interspecific competition operates through some limiting factor other than habitat competition, but so many of Lack's (1971) conclusions are in favour of habitat segregation as the isolating mechanism for competing species that such an argument is weak. It is equally possible that interspecific competition for habitat operates in a manner complementary to intraspecific factors, with combined intra- and interspecific pressure closely tracking available resources, but Wiens (1983) has dismissed this argument because of the tight coupling involved. In the present context such an explanation would necessitate an improbably close equivalence of members of different species over wide ranges of population densities and within different preferred habitats. The idea is challenged, moreover, by the present finding that the most generalist and numerous species each have wide diversity of nesting habitat rather than each specializing within a narrow spectrum of individually abundant habitats.

Some competition for resources undoubtedly occurs within bird communities. Ashmole's (1963) hypothesis of a close association between the proportion of migrant birds in a given area and the seasonality of that area predicates resource limitation of avian communities and has been supported by the findings of Herrera (1978) and Ricklefs (1980). It might be thought that farmland in Britain might, as a highly managed habitat, lie outside the scope of such generalizations, with Williamson (1971) and Bull, Mead & Williamson (1976), for example, suggesting that farmland habitat is rarely saturated with birds. The massive decrease in numbers of many species of birds supported on farmland following the wide-scale introduction of winter cereals—effectively a change in the seasonality of agricultural land—suggests that this is not so. Edwards (1977) also provides relevant experience, showing that in several abundant bird species on agricultural land a local population may contain a surprising number of non-breeding individuals unable to obtain territories, whilst Parr (1979) also documents cases of golden plovers *Pluvialis apricaria* waiting on the vacation of early territories by their owners before they themselves breed there. Nor is farmland a poorly-structured habitat akin to the shrub-steppe habitats studied by Rotenberry & Wiens (1980a,b), for three-dimensional structure is a major predictor of species and bird numbers on farmland (O'Connor & Shrubb 1986a), thus providing options for the classic processes of habitat segregation. Note, though, that resource limitation is not synonymous with community equilibrium. For example, birds benefiting from site fidelity may effectively alter the value of the original resource (e.g. a territory in suboptimal habitat) as a result of their use of it, thereby dynamically altering the carrying capacity of the habitat and the size of the community it can support yet leaving it at all times limited by the resource concerned. It seems clear that intraspecific dynamics do take place in a competitive milieu but that their very dynamism constitutes a significant questioning of the global validity of present theories of competitive coexistence. Farmland bird communities are not rigidly organized by a framework of immutably fixed species interactions, as classic concepts would suggest. Instead, some of the numerically dominant species may ebb and flow into and out of particular habitats in response to their own population levels and what we observe as interspecific interactions depends on the confluence of separate population trends of the potentially competing species. For most birds, after all, conspecifics are their most effective competitors.

ACKNOWLEDGMENTS

I thank the staff of the BTO Nest Records Scheme, particularly Alan Eardley, Mercedes Tourle, David Glue, and Margaret Phillips for their help in processing nest records. John Marchant and Phil Whittington were ever helpful in accessing CBC data and Elizabeth Murray once again prepared artwork for publication. The paper benefited considerably from review by the two editors and an anonymous referee, which help is gratefully acknowledged. This paper was written whilst holding a post funded by the Nature Conservancy Council who also fund the Common Birds Census and part-fund the Nest Records Scheme. Computerization of past nest records was assisted by a grant from the Natural Environment Research Council. I thank NCC and NERC for their support.

REFERENCES

Ashmole, N.P. (1963). The regulation of numbers of tropical oceanic birds. *Ibis*, **103**, 458–73.

Batten, L.A. (1977). *Studies on the population dynamics and energetics of Blackbirds, Turdus merula Linnaeus.* Unpublished Ph.D. thesis, University of London.

Beals, E.W. (1960). Forest bird communities in the Apostle Islands of Wisconsin. *Wilson Bulletin*, **72**, 156–81.

Best, L.B. & Stauffer, D.F. (1980). Factors affecting nesting success in riparian bird communities. *Condor*, **82**, 149–58.

Brown, J.L. (1969). The buffer effect and productivity in tit populations. *American Naturalist*, **103**, 347–54.

Bull, A.L., Mead, C.J. & Williamson, K. (1976). Bird-life on a Norfolk farm in relation to agricultural changes. *Bird Study*, **23**, 203–18.

Cody, M.L. (1975). Towards a theory of continental species diversities: bird distributions over Mediterranean habitat gradients. *Ecology and Evolution of Communities* (Ed. by M.L. Cody & J.M. Diamond), pp. 214–57. Belknap Press, Cambridge, Massachusetts.

Edwards, P.J. (1977). 'Re-invasion' by some farmland bird species following capture and removal. *Polish Ecological Studies*, **3**, 53–70.

Fretwell, S.D. (1981). Evolution of migration in relation to factors regulating bird numbers. *Migrant Birds in the Neotropics* (Ed. by A. Keast & E.S. Morton), pp. 517–27. Smithsonian Institution Press, Washington, DC.

Fretwell, S.D. & Lucas, H.L. (1969). On territorial behaviour and other factors influencing habitat distribution in birds. I. Theoretical development. *Acta Biotheoretica*, **19**, 16–36.

Fuller, R.J. (1982). *Bird Habitats in Britain.* Poyser, Calton, Stoke-on-Trent.

Fuller, R.J., Marchant, J.H. & Morgan, R.A. (1985). How representative of agricultural practice in Britain are Common Birds Census farmland plots? *Bird Study,* **32,** 60–74.

Herrera, C.M. (1978). On the breeding distribution pattern of European migrant birds: MacArthur's theme reexamined. *Auk,* **95,** 496–509.

Hickling, R.A.O. (1983). *Enjoying Ornithology.* Poyser, Calton, Stoke-on-Trent.

Johnson, D.H. (1979). Estimating nest success: the Mayfield method and an alternative. *Auk,* **96,** 651–61.

Kluijver, H.N. & Tinbergen, L. (1953). Territory and the regulation of density in titmice. *Archives neerlandisches Zoologie,* **10,** 265–89.

Lack, D. (1971). *Ecological Isolation in Birds.* Harvard University Press, Cambridge, Massachussetts.

MacArthur, R.H. (1958). Population ecology of some warblers of northeastern coniferous forests. *Ecology,* **39,** 599–619.

MacArthur, R.H. & MacArthur, J.W. (1961). On bird species diversity. *Ecology,* **42,** 594–8.

Marchant, J.H. (1983). *Common Birds Census Instructions.* British Trust for Ornithology, Tring.

Marchant, J.H. & Hyde, P.A. (1980). Bird population changes for the years 1978–79. *Bird Study,* **27,** 173–8.

Mayfield, H. (1961). Nesting success calculated from exposure. *Wilson Bulletin,* **73,** 255–61.

Mayfield, H. (1975). Suggestions for calculating nest success. *Wilson Bulletin,* **87,** 456–66.

Morgan, R.A. & O'Connor, R.J. (1980). Farmland habitat and Yellowhammer distribution in Britain. *Bird Study,* **27,** 155–62.

O'Connor, R.J. (1980). Population regulation in the Yellowhammer *Emberiza citrinella. Bird Census Work and Nature Conservation* (Ed. by H. Oelke), pp. 190–200. Proceedings of the VI International Conference on Bird Census Work. Dachverbandes deutscher Avifaunisten, Lengede.

O'Connor, R.J. (1981). Comparisons between migrant and non-migrant birds in Britain. *Animal Migration* (Ed. by D.J. Aidley), pp. 167–95. Cambridge University Press, Cambridge.

O'Connor, R.J. (1985). Behavioural regulation of bird populations: a review of habitat use in relation to migration and residency. *Behavioural Ecology: Ecological Consequences of Adaptive Behaviour* (Ed. by R.M. Sibly & R.H. Smith), pp. 105–42. Symposia of the British Ecological Society, 25. Blackwell Scientific Publications, Oxford.

O'Connor, R.J. & Fuller, R.J. (1985). Bird population responses to habitat. *Bird Census and Atlas Studies* (Ed. by K. Taylor, R.J. Fuller & P.C. Lack), pp. 197–212. British Trust for Ornithology, Tring.

O'Connor, R.J. & Marchant, J.H. (1981). *A Field Validation of Some Common Birds Census Techniques.* Nature Conservancy Council Chief Scientist Team Commissioned Research Report. NCC, Huntingdon.

O'Connor, R.J. & Shrubb, M. (1986a). *Farming and Birds.* Cambridge University Press, Cambridge.

O'Connor, R.J. & Shrubb, M. (1986b). Recent changes in bird populations in relation to farming practices in England and Wales. *Journal of the Royal Agricultural Society of England,* **147,** 132–41.

Parr, R. (1979). Sequential breeding by Golden Plovers. *British Birds,* **72,** 499–503.

Perrins, C.M. (1979). *British Tits.* Collins New Naturalist, London.

Pimm, S.L., Rosenzweig, M.L. & Mitchell, W. (1985). Competition and food selection: field tests of a theory. *Ecology,* **66,** 798–807.

Ricklefs, R.E. (1980). Geographical variation in clutch size among passerine birds: Ashmole's hypothesis. *Auk,* **97,** 38–49.

Rosenzweig, M.L. (1979). Optimal habitat selection in two-species competitive systems. *Fortschritte der Zoologie,* **25,** 283–93.

Rosenzweig, M.L. (1981). A theory of habitat selection. *Ecology, 62,* 327–33.

Rosenzweig, M.L. (1985). Some theoretical aspects of habitat selection. *Habitat Selection in Birds* (Ed. by M.L. Cody), pp. 517–40. Academic Press, Orlando.

Rotenberry, J.T. & Wiens, J.A. (1980a). Habitat structure, patchiness, and avian communities in North American steppe vegetation: a multivariate analysis. *Ecology, 61,* 1228–50.

Rotenberry, J.T. & Wiens, J.A. (1980b). Temporal variation in habitat structure and shrub-steppe bird dynamics. *Oecologia, 47,* 1–9.

Strong, D.R., Simberloff, D., Abele, L.G. & Thistle, A.B. (Eds) (1984). *Ecological Communities: Conceptual Issues and the Evidence.* Princeton University Press, Princeton, New Jersey.

Wiens, J.A. (1983). Avian community ecology: an iconoclastic view. *Perspectives in Ornithology* (Ed. by A.H. Brush & G.A. Clark), pp. 355–403. Cambridge University Press, Cambridge.

Wiens, J.A. (1984). On understanding a nonequilibrium world: myth and reality in community patterns and processes. *Ecological Communities: Conceptual Issues and the Evidence* (Ed. by D.R. Strong, D. Simberloff, L.G. Abele & A.B. Thistle), pp. 439–57. Princeton University Press, Princeton, New Jersey.

Wiens, J.A. & Rotenberry, J.T. (1981). Habitat associations and community structure of birds in shrubsteppe environments. *Ecological Monographs, 51,* 21–41.

Williamson, K. (1969). Habitat preferences of the Wren on English farmland. *Bird Study, 16,* 53–9.

Williamson, K. (1971). A bird census study of a Dorset dairy farm. *Bird Study, 18,* 80–96.

Winstanley, D., Spencer, R. & Williamson, K. (1974). Where have all the Whitethroats gone? *Bird Study, 21,* 1–14.

9. VARIATION IN DESERT RODENT GUILDS: PATTERNS, PROCESSES, AND SCALES

JAMES H. BROWN

Department of Ecology and Evolutionary Biology, University of Arizona, Tucson, Arizona 85721, USA

INTRODUCTION

An important ultimate goal of community ecology is to explain the organization of species assemblages. This means not only acounting for the number, identity, and ecological attributes of species that coexist within a region, but also understanding the variation in species composition over space and time. This is a formidable task, because literally millions of species of plants, animals, and microbes inhabit the earth, and hundreds of different kinds may coexist locally within a small patch of habitat. By definition, each of these species is unique; it has a unique spatial and temporal distribution which reflects its particular requirements for physical conditions and its interactions with other species. Empirical studies of the diversity of species and the complexity of their interrelationships are necessary to discover general patterns and processes that characterize the organization of communities.

In the present paper I use the seed-eating rodents that inhabit the arid region of south-western North America to show that guilds of coexisting species exhibit non-random patterns of organization, and that a combination of field experiments and non-manipulative observations can elucidate some of the processes that produce this structure.

BACKGROUND: GRANIVOROUS DESERT RODENTS

Like many others, the granivorous desert rodent guild is a rather arbitrarily-defined assemblage of species that share certain attributes and potentially interact with each other. It includes those representatives of the genera *Dipodomys*, *Microdipodops*, *Chaetodipus*, and *Perognathus* of the Family Heteromyidae, and the genera *Peromyscus* and *Reithrodontomys* of the Family Cricetidae that occur together in the arid habitats of south-western North America and feed to a large extent on the dry seeds produced primarily by desert annual plants (Brown, Reichman & Davidson 1979).

These granivorous desert rodents are by no means a discrete, well-defined group in time, space, or biological attributes. Their taxonomic and biogeographic affinities indicate a complex evolutionary history. Some forms have probably inhabited arid environments for at least 5 million years, as indicated by their fossil record, morphological and physiological specializations, and restricted geographic and habitat distributions (for examples, see appropriate chapters in Genoways & Brown 1987). For example, the bipedal kangaroo mice, genus *Microdipodops*, are endemic to the Great Basin Desert and restricted entirely to desert shrub habitats. In contrast, other rodents have invaded the desert so recently that they lack obvious specializations for arid environments. For example, the deer mouse, *Peromyscus maniculatus*, occurs over almost the entire North American continent in habitats that include grasslands, chaparral, deciduous and coniferous forests, and even salt marshes. Although the granivore guild is defined largely on the basis of a shared diet consisting largely of seeds, in this respect, too, the species are highly variable. Many are almost exclusively granivorous, but some heteromyids eat large quantities of green vegetation and some cricetids include many invertebrates and fleshy fruits in their diet.

TEMPORAL AND SPATIAL VARIATION IN SPECIES COMPOSITION

The observation that species composition varies in space and time is trivial, but the magnitude and the spatial and temporal scales of this variation warrant investigation. A recent study uses data from the literature to analyse spatial variation in desert rodent guilds throughout the south-western United States (J. Brown & M. Kurzius, unpublished data). The analysis is based on a total of twenty-nine species of granivorous rodents that were collected at 202 local sites, each consisting of a few hectares of relatively homogeneous desert habitat. The results show that the distribution of each species reflects its unique requirements and is largely independent of the occurrence of other species. Over its geographic range each species encounters many different species and combinations of species. The size of local guilds varies only from one to nine, but each species coexists with from two to twenty-five different species and with from one to seventy-six different combinations of species over its geographic range (Fig. 9.1). This kind of spatial variation in species associations can also be observed at very local spatial scales. For example, Kingsley (1981) sampled nineteen different combinations of eleven granivore species at twenty-three collecting sites within an area of

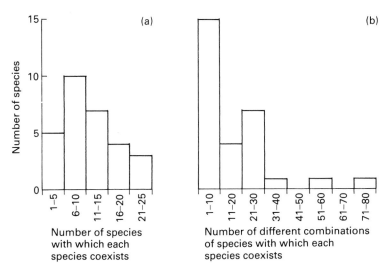

FIG. 9.1. Frequency distributions for, (a) the number of other rodent species, and (b) the number of different combinations of rodent species, with which each of twenty-nine rodent species coexist in 202 local patches of desert habitat in the south-western United States (Fig. 9.2). Note that most species occur together locally with many different species and combinations of species over their geographic range.

approximately 750 km^2 in the Death Valley region of southern Nevada and adjacent California.

Temporal variation in the composition of rodent assemblages is as dramatic as spatial variation. Perhaps the best documentation of post-Pleistocene changes in small mammal assemblages comes from more mesic habitats in the central United States. Graham (1986) and his co-workers have shown not only that fossil assemblages contain species that no longer occur in the region, but also that in some cases species that coexisted in the Pleistocene presently occupy geographic ranges separated by hundreds of kilometres.

A similar situation almost certainly occurred in south-western North America where palaeoclimatic and biogeographic evidence indicates that the present arid habitats have changed enormously during the last 2 million years. The Great Basin region provides an excellent example. During the glacial or pluvial periods of the Pleistocene, and as recently as 12000 years ago, what are now arid valleys covered with desert shrub habitats were filled with large lakes surrounded by extensive marshes. Some of the approximately fifteen species of desert rodents that now inhabit these valleys, including the two species of the endemic Great Basin genus *Microdipodops*, are almost certainly old residents that survived the pluvial periods in the small, isolated patches of arid habitats that persisted. Other species, such as the kangaroo rats *Dipodomys merriami* and *D. deserti*, are more recent

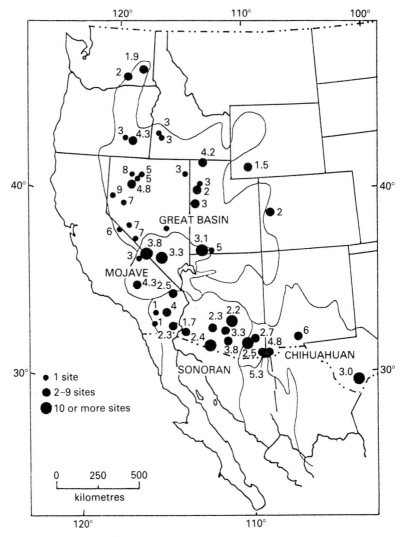

FIG. 9.2. A map showing the locations of the North American deserts and the 202 local sites for which J. Brown and M. Kurzius (unpublished) have analysed the co-occurrence of granivorous desert rodents. The numbers indicate the number of coexisting species per site (mean number of species is given for sites located too close together to be plotted individually). Note that with the exception of its northernmost part, the western Great Basin, which has been accessible to post-Pleistocene colonization, has local guilds that contain more species than those in the eastern Great Basin, which has long been isolated by biogeographic barriers.

colonists which survived the pluvial periods chiefly in larger refuges of desert habitat at lower latitudes and elevations in the present locations of the Mojave and Sonoran Deserts (shown in Fig. 9.2).

One interesting legacy of this biogeographic history is that several of the desert species, including the two kangaroo rats mentioned above, were able to colonize only the western part of the Great Basin where the north–south oriented valleys provided almost continuous corridors of suitable desert habitat (Reveal 1979). Dispersal barriers in the form of mountain ridges with rocky substrates and mesic vegetation prevented these species from colonizing the eastern part of the Great Basin. The substantial differences in the contemporary diversity and composition of desert rodent guilds between the eastern and western Great Basin (Fig. 9.2) can be attributed largely to this biogeographic history. Many contemporary assemblages are relatively recent, assembled from a pool of species that have been able to colonize or persist in the region since the end of the Pleistocene.

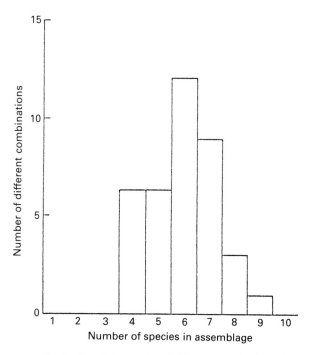

FIG 9.3. Frequency distribution of the number of different combinations of ten granivorous rodent species captured in ninety monthly samples over an 8-year period at a single local site of approximately 20 ha of desert shrub habitat in the Chihuahuan Desert of south-eastern Arizona. Note the large number of different combinations (37), indicative of short-term temporal variation in species composition.

Short-term changes in guild composition are hardly less dramatic than those that took place during the last several thousand years. Major changes have occurred in the combinations of species sampled on our experimental study site in south-eastern Arizona. This area has been sampled at monthly intervals since 1977 using an intensive, standardized programme of live-trapping at permanent grid stakes (see Brown & Munger 1985, for a description of the site and trapping regime). In ninety monthly samples during this 8-year period, we caught thirty-seven different combinations of ten granivorous rodent species (Fig. 9.3).

The apparent explanation for this enormous temporal and spatial variation is that these guilds are very dynamic systems. Each species is limited by a different combination of physical and biotic environmental variables. The variable composition of local guilds simply reflects the occurrence of combinations of variables that permit survival and reproduction of different species. There is no evidence of a small number of repeated species combinations which would suggest sets of coadapted species or alternate stable equilibria. Instead, these guilds seem rarely to be in any sort of equilibrium, because they are always responding to environmental changes that occur on a wide range of spatial and temporal scales.

PATTERNS OF SPECIES COMPOSITION

The wide temporal and spatial variation in species composition at all scales and the seemingly independent co-occurrence of species might suggest that rodent guilds are simply random assemblages. Nothing could be further from the truth. A set of strongly deterministic but subtle rules dictates the number and kinds of species that coexist. The following rules are empirical generalizations, which can be made by distinguishing patterns of guild composition from those expected on the basis of null models in which species are assumed to associate at random:

1   *The number of species that coexist in a local habitat is less than expected from random association of species whose geographic ranges encompass the site.* J.Brown & M. Kurzius (unpublished data) found that fifty-six different combinations of species occurred at seventy-nine local sites within the geographic ranges of fourteen species. The frequency distribution of these associations shows that they contained between one and seven species, substantially fewer than expected from the distribution of possible combinations of the fourteen species in the regional pool (Fig. 9.4). This pattern implies that local ecological interactions limit the number of coexisting species to a small fraction of the regional species pool.

2   *When physical habitat structure is held essentially constant in large-scale*

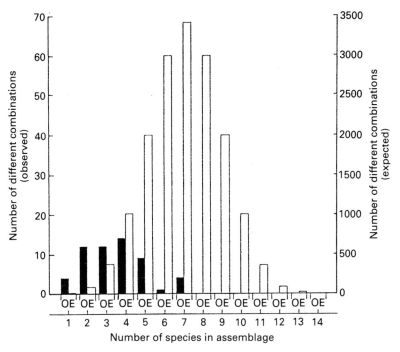

FIG. 9.4. The actual frequency distribution of fifty-six different local rodent guilds, each of which occurred within the geographic ranges of fourteen rodent species, as a function of the number of species per assemblage (shaded bars, left axis). This distribution is compared with the frequency distribution of the different possible combinations of fourteen species (unshaded bars, right axis). Note that the real assemblages contained significantly fewer species than expected from random combinations of the species in the regional pool ($P \ll 0.001$; from J. Brown & M. Kurzius unpublished data).

geographic and climatic gradients, the number of species, number of individuals, and biomass are all positively correlated with precipitation and primary productivity. Brown (1973, 1975) showed that this pattern holds along two different gradients, north–south in the Great Basin and Mojave Deserts, and east–west in the Chihuahuan and Sonoran Deserts, and in three different habitat types defined on the basis of soil and vegetation structure: sand dunes, sandy flatlands and rocky hillsides. This suggests that if the effects of soil and vegetation are controlled, the capacity of environments to support species depends largely on productivity and the availability of food resources. Local regions, such as the eastern Great Basin (see above), that have long been isolated by effective biogeographic barriers, provide important exceptions to this pattern and have fewer species than expected on the basis of climate and productivity (Brown 1973, 1975).

3 *Within a small geographic region, where climate is essentially constant, the number and identity of species varies predictably with physical habitat structure.* This pattern was first documented by Rosenzweig & Winakur (1969), and it has been reaffirmed by many subsequent studies (e.g. Rosenzweig 1973; Rosenzweig, Smigel & Kraft 1975; M'Closkey 1976; Hafner 1977; Price 1978; Thompson 1982a,b). Specific features of the soil and vegetation appear to be necessary for the occurrence of certain species, but in general, structural diversity of the habitat promotes the coexistence of species. This pattern implies that species differ in their requirements, and they are able to meet these requirements by utilizing different components of the habitat. It does not necessarily imply that resource partitioning has evolved to facilitate the coexistence of particular combinations of species. Note that this pattern and Pattern 2, above, are not inconsistent; they describe relationships between species and their environment on different scales.

4 *The species that coexist locally tend to be more different in body size than expected on the basis of chance.* The ratios of body mass of coexisting species are distributed more uniformly than random (Brown 1973; Bowers & Brown 1982; Hopf & Brown 1986). Species of similar size also tend to overlap less in their geographic ranges than expected on the basis of chance (Bowers & Brown 1982).

TABLE 9.1. Observed and expected (parentheses) frequencies of co-occurrence of rodents of different taxonomic and functional groups in sixty-one local two-species assemblages. Rodent species are divided into three groups: bipedal heteromyids (*Dipodomys* and *Microdipodops*), quadrupedal heteromyids (*Perognathus* and *Chaetodipus*), and quadrupedal cricetids (*Peromyscus* and *Reithrodontomys*). The observed frequency of co-occurrence of species among these groups is significantly different from that expected on the basis of random association ($\chi^2 = 31.42$; $P \ll 0.001$). Pairs of species in the same groups and also bipedal heteromyids and quadrupedal cricetids occur together less frequently than expected on the basis of chance. (From J. Brown & B.A. Harney unpublished data)

		Coexistence of:		
		Bipedal heteromyid	Quadrupedal heteromyid	Quadrupedal cricetid
with	Bipedal heteromyid	4 (7.38)	18 (13.30)	4 (9.33)
	Quadrupedal heteromyid		5 (23.94)	26 (16.81)
	Quadrupedal cricetid			4 (11.80)

5 *Coexisting species within local guilds tend to belong to different taxonomic and functional groups.* This can be shown most clearly by the composition of two-species combinations in the data set compiled by Brown and Kurzius (J. Brown & B.A. Harney unpublished data). When the rodents are divided into three taxonomic and functional groups (bipedal heteromyids, quadrupedal heteromyids, and quadrupedal cricetids), it is apparent that species in the same group occur together much less frequently than expected by chance (Table 9.1). This pattern, together with Pattern 4, above, suggests that coexistence is facilitated by differences in ecological requirements that are a consequence of evolutionary divergence.

6 *If additional variables are measured and/or controlled, it is often possible to document situation-specific patterns that describe additional aspects of the observed variation in the number and identity of coexisting species.* M'Closkey (1978) adduced 'assembly rules' that described and accounted for the frequencies with which different combinations of species occurred together in local patches of desert shrub habitat within a small region of southern Arizona. A more general example of such an assembly rule is the less frequent than expected coexistence of bipedal heteromyids and quadrupedal cricetids in local two-species guilds (Table 9.1). I hypothesize that such assemblages are unstable, because they are always susceptible to invasion by a quadrupedal heteromyid that would convert them into a three-species combination.

These empirical patterns suggest that guild organization reflects the operation of deterministic processes. This is not to imply that chance plays no role. At the present state of our knowledge, there are many relationships that appear random. Although some of these may ultimately turn out to be regular and predictable, others, especially those that are the result of sampling processes that involve small numbers, will probably always be viewed as stochastic phenomena. It is important to recognize these stochastic elements, but it is more profitable to frame and test rigorous hypotheses to account for the deterministic patterns.

CAPACITY RULES

A few years ago (Brown 1981) I suggested that the mechanistic processes that determine the diversity of species and the functional organization of communities can be characterized in terms of 'capacity rules' and 'allocation rules'. Capacity rules are the *extrinsic* patterns and processes that describe the relationships between physical and biotic variables and the capacity of the environment to support the group of organisms in question. Characteristics of the physical environment must ultimately account for the deterministic spatial and temporal variation in species composition. If it were not for differences in geology and climate, either at present or in the past, all communities should have similar numbers and kinds of species, except for a certain amount of apparently random variation. Capacity rules are not limited to the direct effects of physical factors, however. They also include the effects of all biotic interactions that are not confined within the assemblage, because the distribution and abundance of predators, prey, competitors, and mutualists largely reflect the ultimate influence of the physical environment.

The first three of the patterns described above reflect primarily the influence of extrinsic climatic and geological variables on the number and kinds of rodent species that occur at a site. The most important of these appear to be the effect of climate on productivity, the effect of geology on biogeographic barriers and soil type, and the combined influence of climate and soil on the structure and composition of the vegetation. If other factors are held relatively constant, the production and availability of seeds appears to be closely correlated with primary productivity, and food availability limits the rodent biomass that a site can support. Biogeographic barriers act on large spatial and temporal scales to affect the number and identity of species in the regional pool, thereby limiting the species that are available to colonize any site. Soil and vegetation influence the kinds of rodents that can occur in a habitat through the effects of habitat structure on burrowing, locomotion, foraging, and predator avoidance (Reichman & Price 1987; Brown & Harney 1987). In addition to these more direct and better documented effects, climate and geology undoubtedly have more indirect influenced by environmental temperature; Brown & Harney 1987).
and predators (e.g. ants and snakes, respectively, both of which are strongly influenced by environmental temperature; Brown and Harney, in press).

In some cases the evidence for the operation of these capacity rules is based primarily on correlations, but in others the proposed mechanisms have been supported by experimental results. For example, Brown (1973, 1975) amassed a variety of correlational evidence that the positive relationship between annual precipitation and rodent species diversity within large climatic gradients was owing to the effect of rainfall in increasing the production of seeds utilized as food by the rodents. Abramsky (1978) provided strong experimental support for this interpretation by documenting the local colonization of an additional kangaroo rat species in response to seed addition. Similarly, Rosenzweig & Winakur (1969) documented correlations between habitat structure and the occurrence of particular functional groups of rodents: bipedal kangaroo rats were associated with open habitats and friable soils, whereas quadrupedal pocket mice and cricetids were most abundant in brushy and rocky areas. When Rosenzweig (1973) performed 'habitat tailoring experiments' and removed brush from some areas and piled it up in others, he observed the predicted response: the quadrupedal species increased in the brushy areas, and the kangaroo rats shifted their activity to the cleared patches (see also Price 1978; Larsen 1986). Similarly, Thompson (1982b) altered local species composition through experimental modification of habitat structure: by adding small artificial shelters that apparently made the environment more suitable for certain species by reducing their exposure to predators.

Other capacity rules could potentially be investigated more rigorously by experimental manipulations, but many of these would be impractical or unethical. For example, it would be logistically difficult to exclude predators on a sufficiently large scale to analyse their effects in a realistic context (but see Kotler 1984 for alternative approaches). Introduction experiments to test hypotheses about biogeographic barriers are now illegal and unethical, but one such experiment, conducted in the 1930s, yielded fascinating results. Several individuals of the kangaroo rat, *Dipodomys ordii* from Oklahoma were introduced to a sand dune on the shore of Lake Erie in Ohio, approximately 1200 km east of their range. A substantial population became established and survived at least until the early 1950s (Bole & Moulthrop 1942; N. C. Negus & J. S. Findley personal communication). This introduction clearly demonstrates the role of biogeographic barriers in preventing species from colonizing otherwise suitable habitats.

ALLOCATION RULES

In addition to the extrinsic environmental relationships embodied in capacity rules, species composition also reflects the resolution of interactions among the species themselves. I have called these patterns and processes that are *intrinsic* to the community, 'allocation rules'. In cases such as the granivorous desert rodent guild, when the community is arbitrarily defined to include only a single trophic level, the important direct interactions should be primarily competitive.

The last three patterns of species composition documented above reflect primarily the operation of allocation rules based on interspecific competition. Given that local guilds are limited to a small subset of the species in the regional geographic pool (Pattern 1), what determines which of the possible combinations of species actually occur together? Although some of the patterns, such as the absence of certain taxonomic and functional groups from particular soil and vegetation types, almost certainly can be attributed to extrinsic limiting factors (see above), local competitive exclusion seems to be both necessary and sufficient to account for many of the remaining patterns of coexistence.

Patterns 4 and 5 indicate that differences among species promote coexistence, presumably because such variation allows different species to use different resources and thereby to avoid competitive exclusion. There is abundant evidence, much of it experimental, that the differences in morphology and taxonomy among species that frequently coexist are related to differences in resource utilization. Numerous studies document consistent differences in the microhabitats where the rodents forage which reduce competition for food (e.g. bipedal heteromyids tend to forage in the open, whereas quadrupedal heteromyids and circetids tend to remain under shrub canopies; Brown & Lieberman 1973; Rosenzweig 1973; Brown 1975; Lemen & Rosenzweig 1978; Price 1978; M'Closkey 1981; Bowers 1982; Price & Brown 1983; Kotler 1984; Larsen 1986). Some investigators have claimed that the pronounced variation in body size among coexisting species is related to foraging specializations based on the sizes and/or spatial dispersion of seeds, but this has proven to be controversial and difficult to resolve (e.g. Rosenzweig & Sterner 1970; Brown & Lieberman 1973; Smigel & Rosenzweig 1974; Mares & Williams 1977; Reichman & Oberstein 1977; Hutto 1978; Lemen 1978; Frye & Rosenzweig 1980; M'Closkey 1980; Price 1983; Harris 1984). Logically, however, there must be significant differences in food exploitation associated with differences in body size, because of the allometric scaling of energy requirements. Another consequence of asymmetries in body size is that aggressive interference probably plays a major role in resource allocation and coexistence (Frye 1983; M. A. Bowers personal communication).

The most direct evidence for interspecific competition is an increase in the abundance of one species after another member of the trophic guild has been experimentally excluded. Several-fold increases in population density of the remaining species in response to such removals have been documented in both short-term and long-term experiments (Munger & Brown 1981; Freeman & Lemen 1983; Brown & Munger 1985; Fig. 9.5). These experiments provide strong evidence that substantial competition occurs even among those species that regularly coexist. The most intense competition would be expected between closely-related species of similar morphology, physiology and behaviour, but these rarely occur together in local habitats (patterns 4 and 5, above). Thus, the experimental evidence supports the interpretation that the limited number of locally coexisting species can be attributed largely to competitive exclusion of species with similar requirements for limited resources.

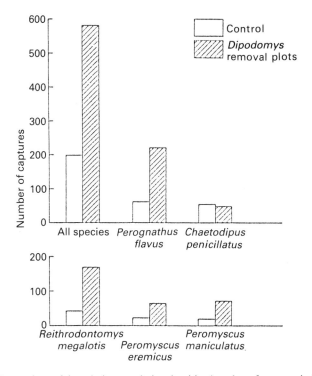

FIG. 9.5. Comparison of the relative population densities (number of captures in 8 years of standardized live-trap censuses) of five small, granivorous rodent species on four control plots where three kangaroo rat (*Dipodomys*) species were present (unshaded bars), compared to four experimental plots from which the kangaroo rats had been removed (shaded bars). All of the small rodents except for *Chaetodipus penicillatus* showed significant increases in density when relieved of competition from kangaroo rats. For details see Brown & Munger (1985).

The distinction between capacity and allocation rules emphasizes that once we define a community, its organization is determined by environmental variables and dynamic processes that are both intrinsic and extrinsic to that particular assemblage. This is not to say, however, that the intrinsic allocation rules should be viewed as independent of the extrinsic environmental capacities. The extrinsic variables ultimately determine the outcomes of the interactions within the assemblage. For example, in desert rodents there is increasing evidence, much of it experimental, to show how differences in physical habitat structure influence the outcome of competitive interactions, either directly by affecting exploitation of food or aggressive interactions or indirectly by altering the risk of predation (e.g. Rosenzweig 1973; Schroder & Rosenzweig 1975; Hoover, Whitford & Flavill 1977; Thompson 1982b; Kotler 1984).

GENERAL DISCUSSION

The foregoing summarizes both the empirical patterns of associations among species and the dynamic processes of interaction with the environment that characterize the organization of guilds of granivorous desert rodents. The question immediately arises, to what extent are these patterns and processes specific to desert rodents and to what extent can they be generalized to other assemblages?

Undoubtedly many features of the rodent assemblages emphasized above are system-specific, because they reflect unique attributes of these rodent taxa and their past and present environments. The effects of productivity and habitat structure on diversity and the importance of body size and other morphological differences in determining species composition are difficult to generalize to other communities. Qualitatively similar relationships have been found in other guilds, but there is no reason to expect particular habitat features or body size ratios to have similar consequences for organisms that interact with their environments in different ways. To cite just one example, Davidson (1977a,b) has studied the seed-eating ants that inhabit the same North American deserts as the rodents. The similarities and differences are striking. Locally coexisting ant species also exhibit regular differences in body size, but these result in resource partitioning on the basis of seed size that is much less ambiguous than in rodents. Ants and rodents show parallel increases in species diversity in an east–west gradient of increasing rainfall and productivity where temperature remains approximately constant, but ant diversity does not increase with rodent diversity in a north–south gradient where decreasing temperatures accompany increasing precipitation and productivity (Brown & Davidson 1977).

If the specific structure and dynamics cannot readily be generalized to other assemblages, is there another level at which we can identify more general patterns and processes of community organization? Some colleagues have tried to classify communities on the basis of whether their organization can be attributed primarily to the effects of the physical environment or to particular kinds of interspecific interactions, such as competition or predation. Unfortunately, I suspect that these classifications reflect the interests and biases of particular investigators rather than any useful generalities about the structure and function of complex ecological systems. The assembly and allocation rules derived for desert rodent assemblages suggest that such classifications are arbitrary and misleading. If we focus on physical factors, we can detect important effects of climate and geology; if we focus on biotic interactions within the guild, we can demonstrate an important influence of competition; if we focus on interactions with other trophic levels, there is increasing evidence that predation plays a major role. All of these processes interact over a wide range of spatial and temporal scales to determine guild organization.

There do, however, appear to be features of the rodent assemblages that are characteristic of virtually all communities. Many of these are also emphasized in other contributions to this volume. Below I list five attributes of structure and function of desert rodent guilds that appear to be very general and to warrant further investigation:

1 *Communities are open systems.* Ecosystem ecologists have long recognized that communities are open systems with respect to the flow of energy and materials. Community ecologists, however, have been slow to realize that communities are equally open with respect to their most important attribute: species composition. The arbitrary decisions that must be made to define membership in the granivorous desert rodent guild and to characterize the composition of assemblages in space and time are not unique to this system. Short of the entire biosphere, no natural community is really a discrete, closed system. Some insular habitats possess natural boundaries that limit exchange with other communities, but the important role of water birds in the ecology of both oceanic islands and lakes provides but one example of how open even these systems really are. Many continental and oceanic communities are so open that biotic as well as physical exchange across the arbitrarily-defined boundaries may be among the most important processes determining the organization of these systems. In this volume, O'Connor's analysis of migratory birds (Chapter 8) and Roughgarden and colleagues' studies of barnacles (Chapter 22) document extremely open communities, but Terborgh & van Schaik's observations of Neotropical primates (Chapter 10) suggest an almost closed assemblage of unchanging composition.

2 *Species composition varies in time and space.* This is a truism, but the magnitude of the variation suggests that traditional, highly-structured, equilibrial concepts of community organization are unwarranted. Assemblages probably are rarely in equilibrium, because they are open systems with respect to species composition and they are subject to continual, interacting perturbations on many spatial and temporal scales. Other examples of highly variable, nonequilibrial assemblages are described in the chapters on birds (O'Connor, Chapter 8), dung beetles (Doube, Chapter 12), microbial decomposers (Swift, Chapter 11), freshwater and marine plankton (Reynolds, Chapter 14; Harris, Chapter 15), and intertidal barnacles (Roughgarden, Gaines & Pacala, Chapter 22); tropical primates (Terborgh & van Schaik, Chapter 10) provide an apparent counter-example of an extremely stable assemblage. Extensive variation and continual perturbation does not necessarily imply that communities are random assemblages of unpredictable composition and behaviour. On the contrary, the variability in species composition and its deterministic relationship to variation in the extrinsic environment provide the basis for documenting the capacity and allocation rules that characterize community organization.

3   *The organization of any community is determined ultimately by characteristics of the physical environment.* Although some of the spatial and temporal variation in species composition can be regarded as essentially stochastic, most of this variation must be attributed to the deterministic effects of geology and climate. These physical factors exert their effects at a variety of spatial and temporal scales. At one extreme, large-scale conditions affect the composition of the species pool by influencing the probabilities of colonization, speciation, and extinction of species. At the other extreme, small-scale factors affect the composition of local assemblages by influencing the probabilities of migration, birth, and death of individuals. Similar relationships have been shown in this volume for birds (O'Connor, Chapter 8), dung beetles (Doube, Chapter 12), plankton (Reynolds, Chapter 14; Harris, Chapter 15), freshwater and marine benthos (Hildrew & Townsend, Chapter 16, Pearson & Rosenberg, Chapter 17), and intertidal barnacles (Roughgarden, Gaines & Pacala, Chapter 22).

4   *The organization of any community is determined proximally by both the extrinsic physical and biotic environment and intrinsic interactions among the species themselves.* Because communities are open systems, potentially they are subject to invasion of additional species that can tolerate the physical and biotic conditions. But, the numbers and kinds of species that actually live together on any spatial and temporal scale are also affected by their intrinsic interactions that either inhibit or facilitate coexistence. The outcome of these interactions is mediated ultimately by the physical environment and often more proximally by the indirect effects of other organisms. Other examples of relationships between intrinsic and extrinsic forces in the organization of plant and lizard assemblages are discussed by Fitter (Chapter 6) and Roughgarden, Gaines & Pacala (Chapter 22), respectively, in this volume.

5   *The composition of assemblages at all scales is influenced by processes that operate at other spatial and temporal scales.* The organization of local assemblages reflects in part processes that operate on evolutionary time scales and geographic spatial scales to determine the composition of the regional species pool, whereas the composition of the pool also reflects the cumulative effects of processes that operate on short time scales and local spatial scales to maintain populations of the species. This interaction

between scales has also been emphasized in chapters on plants (Grubb, Chapter 5), birds (O'Connor, Chapter 8), dung beetles (Doube, Chapter 12), plankton (Reynolds, Chapter 14; Harris, Chapter 15), and *Anolis* lizards and intertidal barnacles (Roughgarden, Gaines & Pacala, Chapter 22).

To investigate the implications of these general features of community organization will require a reversal of the trend toward specialization and reductionism that has dominated much of modern ecology. The approach that I have advocated suggests that the traditional distinctions—between autecology and synecology; among physiological, behavioural, population, and community ecology, and ecological and historical biogeography; between observational and experimental methods—are artificial and will be counterproductive to this endeavour. I am afraid that attempts to be reductionist, and to analyse the structure and function of communities solely or primarily in terms of the morphology, physiology, and behaviour of individual organisms or the dynamics of local species populations will continue to be plagued by the problems of generalizing the results to other systems that are subject to equally severe but different constraints. It remains to be seen whether the kinds of generalizations that I have made above can be developed further to provide a rigorous, quantitative basis for an alternative, more holistic and integrated approach to community organization.

ACKNOWLEDGMENTS

I am grateful to my wife, A. Kodric-Brown, to my students, to M. L. Rosenzweig and his students, and to numerous colleagues for their help in gathering the data and discussing the ideas presented above. B. A. Harney and M. Kurzius provided unpublished data and made valuable comments on the manuscript. The National Science Foundation has long supported my research, most recently with Grant BSR-8506729.

REFERENCES

Abramsky, Z. (1978). Small mammal community ecology: changes in species diversity in response to manipulated productivity. *Oecologia,* **34,** 113–23.

Bole, B.P. & Moulthrop, P.N. (1942). The Ohio recent mammal collection in the Cleveland Museum of Natural History. *Scientific Publications of the Cleveland Museum of Natural History,* **5,** 83–181.

Bowers, M.A. (1982). Foraging behavior of heteromyid rodents: field evidence of resource partitioning. *Journal of Mammalogy,* **63,** 361–67.

Bowers, M.A. & Brown, J.H. (1982). Body size and coexistence in desert rodents: chance or community structure? *Ecology,* **63,** 391–400.

Brown, J.H. (1973). Species diversity of seed-eating desert rodents in sand dune habitats. *Ecology*, **54**, 755–87.

Brown, J.H. (1975). Geographical ecology of desert rodents. *Ecology and Evolution of Communities* (Ed. by M.L. Cody & J.M. Diamond), pp. 314–41. Harvard University Press, Cambridge, Massachusetts.

Brown, J.H. (1981). Two decades of homage to Santa Rosalia: toward a general theory of diversity. *American Zoologist*, **21**, 877–88.

Brown, J.H & Davidson, D.W. (1977). Competition between seed-eating rodents and ants in desert ecosystems. *Science*, **196**, 880–2.

Brown, J.H. & Harney, B.A. (1987). Population and community ecology of heteromyid rodents in temperate habitats. *Biology of the Heteromyidae* (Ed. by H.H. Genoways & J.H. Brown). American Society of Mammalogists, Special Publication No. 10, in press.

Brown, J.H. & Lieberman, G.A. (1973). Resource utilization and coexistence of seed-eating desert rodents in sand dune habitats. *Ecology*, **54**, 788–97.

Brown, J.H. & Munger, J.C. (1985). Experimental manipulation of a desert rodent community: food addition and species removal. *Ecology*, **66**, 1545–63.

Brown, J.H., Reichman, O.J. & Davidson, D.W. (1979). Granivory in desert ecosystems. *Annual Review of Ecology and Systematics*, **10**, 201–27.

Davidson, D.W. (1977a). Species diversity and community organization in desert seed-eating ants. *Ecology*, **58**, 711–24.

Davidson, D.W. (1977b). Foraging ecology and community organization in desert seed-eating ants. *Ecology*, **58**, 725–37.

Freeman, P.W. & Lemen, C. (1983). Quantification of competition among coexisting heteromyids in the Southwest. *Southwestern Naturalist*, **28**, 41–6.

Frye, R.J. (1983). Experimental field evidence of interspecific aggression between two species of kangaroo rat (*Dipodomys*). *Oecologia*, **59**, 74–8.

Frye, R.J. & Rosenzweig, M.L. (1980). Clump size selection: a field test with two species of *Dipodomys*. *Oecologia*, **47**, 323–7.

Genoways, H.H. & Brown, J.H. (Eds) (1987). *Biology of the Heteromyidae.* American Society of Mammalogists, Special Publication No. 10, in press.

Graham, R.W. (1986). Response of mammalian communities to environmental changes during the Late Quaternary. *Community Ecology* (Ed. by J. Diamond & T.J. Case), pp. 300–13. Harper & Row, New York.

Hafner, M.S. (1977). Density and diversity in Mojave desert rodent and shrub communities. *Journal of Animal Ecology*, **46**, 925–38.

Harris, J.H. (1984). An experimental analysis of desert rodent foraging ecology. *Ecology*, **65**, 1579–84.

Hoover, K.D., Whitford, W.G. & Flavill, P. (1977). Factors influencing the distribution of two species of *Perognathus*. *Ecology*, **58**, 877–84.

Hopf, F.A. & Brown, J.H. (1986). The bull's-eye method for testing randomness in ecological communities. *Ecology*, **67**, 1139–55.

Hutto, R.L. (1978). A mechanism for resource allocation among sympatric heteromyid rodent species. *Oecologia*, **33**, 115–26.

Kingsley, K.J. (1981). *Mammals of the Grapevine Mountains, Death Valley National Monument.* Unpublished M.S. Thesis. University of Nevada, Las Vegas.

Kotler, B.P. (1984). Predation risk and the structure of desert rodent communities. *Ecology*, **65**, 689–701.

Larsen, E.C. (1986). Competitive release in microhabitat use among coexisting desert rodents in Great Basin sagebrush habitats. *Oecologia*, **69**, 231–7.

Lemen, C.A. (1978). Seed size selection in heteromyids. *Oecologia*, **35**, 13–19.

Lemen, C.A. & Rosenzweig, M.L. (1978). Microhabitat selection in two species of heteromyid rodents. *Oecologia*, **33**, 127–35.

M'Closkey, R.T. (1976). Community structure in sympatric rodents. *Ecology*, **57**, 728–39.

M'Closkey, R.T. (1978). Niche separation and assembly in four species of Sonoran Desert rodents. *American Naturalist*, **112**, 683–94.

M'Closkey, R.T. (1980). Spatial patterns in sizes of seeds collected by four species of heteromyid rodents. *Ecology*, **61**, 486–9.

M'Closkey, R.T. (1981). Microhabitat use in coexisting desert rodents—the role of population density. *Oecologia*, **50**, 310–15.

Mares, M.A. & Williams, D.F. (1977). Experimental support for food particle size resource allocation in heteromyid rodents. *Ecology*, **58**, 1186–90.

Munger, J.C. & Brown, J.H. (1981). Competition in desert rodents: an experiment with semipermeable exclosures. *Science*, **211**, 510–12.

Price, M.V. (1978). The role of microhabitat in structuring desert rodent communities. *Ecology*, **59**, 910–21.

Price, M.V. (1983). Laboratory studies of seed size and seed species selection by heteromyid rodents. *Oecologia*, **60**, 259–63.

Price, M.V. & Brown, J.H. (1983). Patterns of morphology and resource use in North American Desert rodent communities. *Great Basin Naturalist Memoirs*, **7**, 117–34.

Reichman, O.J. & Oberstein, D. (1977). Selection of seed distribution types by *Dipodomys merriami* and *Perognathus amplus*. *Ecology*, **58**, 636–43.

Reichman, O.J. & Price, M.J. (1987). Ecological aspects of heteromyid foraging. *Biology of the Hetromyidae* (Ed. by H.H. Genoways & J.H. Brown). American Society of Mammalogists, Special Publication No. 10, in press.

Reveal, J.N. (1979). Biogeography of the Intermountain Region: a speculative appraisal. *Mentzelia*, **4**, 1–92.

Rosenzweig, M.L. (1973). Habitat selection experiments with a pair of coexisting heteromyid rodent species. *Ecology*, **54**, 111–17.

Rosenzweig, M.L., Smigel, B. & Kraft, A. (1975). Patterns of food, space and diversity. *Rodents in Desert Environments. Monographiae Biologicae* (Ed. by I. Prakash & P. Gosh), pp. 241–68. Junk, The Hague.

Rosenzweig, M.L. & Sterner, P.W. (1970). Population ecology of desert rodent communities: body size and seed husking as bases for heteromyid coexistence. *Ecology*, **51**, 217–24.

Rosenzweig, M.L. & Winakur, J. (1969). Population ecology of desert rodent communities: habitats and environmental complexity. *Ecology*, **50**, 558–72.

Schroder, G.D. & Rosenzweig, M.L. (1975). Perturbation analysis of competition and overlap in habitat utilization between *Dipodomys ordii* and *Dipodomys merriami*. *Oecologia*, **19**, 9–28.

Smigel, B.W. & Rosenzweig, M.L. (1974). Seed selection in *Dipodomys merriami* and *Perognathus penicillatus*. *Ecology*, **55**, 329–39.

Thompson, S.D. (1982a). Microhabitat utilization and foraging behavior of bipedal and quadrupedal heteromyid rodents. *Ecology*, **63**, 1303–12.

Thompson, S.D. (1982b). Structure and species composition of desert heteromyid rodent species assemblages: effects of a simple habitat manipulation. *Ecology*, **63**, 1313–21.

10. CONVERGENCE VS. NONCONVERGENCE
IN PRIMATE COMMUNITIES

JOHN TERBORGH[1] and CAREL P. VAN SCHAIK[2]

[1]*Department of Biology, Princeton University, Princeton,*
New Jersey 08544, USA and [2]*Laboratory of Comparative Physiology,*
Jan van Galenstraat 40, 3572 LA Utrecht, the Netherlands

INTRODUCTION

It is almost an article of faith among biologists that species, communities or ecosystems which have evolved in isolation under similar physical conditions will show similar characteristics. This is a potentially testable proposition that has been given considerable attention, particularly by botanists. Without reviewing the details, suffice it to say that global vegetation provides very strong support for the convergence paradigm. Humid tropical climates, for example, everywhere support a type of vegetation that is universally described as rainforest (Richards 1952). Climates characterized by warm, dry summers and cool, moist winters support a shrubby, evergreen, fire-prone vegetation variously called chaparral, breccia or fynboss. Regardless of the names by which it is called, the vegetation that develops in so-called Mediterranean climates possesses numerous common features in such widely-scattered localities as South Africa, Chile, western Australia, California and Spain (Orians & Paine 1983; Walter 1984). Many additional examples are provided by Walter (1984).

Zoologists have not been so successful in their attempts to demonstrate convergence in animals. Serious efforts have been made to compare lizard assemblages in deserts around the world (Pianka 1975), the bird communities of tropical rainforests (Pearson 1977; Karr 1980), and the seed-eating rodent guilds of North American and South American shrub deserts (Mares 1976). In reviewing the results of these studies one is as impressed by the differences in the compared communities as by the similarities (Orians & Paine 1983; Terborgh & Robinson 1986). Why is this? The investigators have put forward various reasons: differing levels of resource productivity in the localities being compared, presence of distinct lineages of organisms at the initiation of geographical isolation, usurpation of ecological niches by unrelated organisms, etc.

While these arguments could be applied equally well to plants, botanists have had little occasion to evoke them. Non-convergence is far more often a dilemma encountered by zoologists, and for a good reason. Plants, as sessile photosynthetic machines are inescapably exposed to the physical environment, and are obliged to adapt above all to physiological stress. Animals, in contrast, can frequently avoid physiological stress by taking appropriate behavioural action. One might conclude, therefore, that the sizes, forms and appearances of animals should be selected by factors in addition to the physical environment.

One such factor is the available supply of food resources. Just as plant geographers have successfully compared vegetation at sites judged to experience similar climates, it might be possible to compare animal assemblages at sites assumed to offer similar suites of food resources. This idea is explored for the marine benthos by Pearson & Rosenberg, Chapter 17. Because rainforest vegetation is widely acknowledged to show convergence on a global scale, it seems reasonable to presume as a point of departure, that the resources available to animals would be similar in different rainforest regions. If so, one could anticipate that assemblages of rainforest animals in distant parts of the globe would show evidence of convergence. Because unrelated faunas possess distinct evolutionary histories, it would be naive to demand as one's criterion of convergence that every species in one assemblage have identifiable counterparts in another. Instead, it is more appropriate to ask about convergence in community level properties such as biomass, trophic structure, size distribution of species and species diversity (Terborgh & Robinson 1986). If rainforest ecosystems have converged beyond mere resemblance in their vegetative structure, one might, as a null hypothesis, expect similar concordance in the structure and organization of their animal communities.

To investigate this proposition, we shall focus on primates. In fact, in making intercontinental comparisons, one is restricted to primates because they are the only rainforest animals that have been widely censused. In spite of this arbitrary limitation, primates are nevertheless the most appropriate taxonomic common denominator because they are the preeminent group of arboreal consumers in the rainforests of South America (Terborgh 1983, 1986b), Africa (Emmons, Gautier-Hion & Dubost 1981) and South-East Asia (Leighton & Leighton 1983).

To begin the analysis we present comparative data on the structure of primate assemblages in four widely-separated rainforest regions. These data suggest that non-convergence rather than convergence is the rule. Next we shall look closely at one Neotropical locality in which the annual patterns of resource availability and of primate feeding behaviour are both well-known, making the point that the resources that sustain the community during periods of scarcity hold the key to understanding some adaptive features of the component species. Lastly, we examine the patterns of resource availability in rainforest localities in other parts of the world. Contrary to an *a priori* presumption of similarity, the resource regimes of these localities show important differences. These differences are interpreted as the principal causes of the observed non-convergences in the respective primate assemblages.

CONVERGENCE VS. NON-CONVERGENCE IN ASSEMBLAGES OF RAINFOREST PRIMATES

Four independently evolved primate assemblages

There are four major primate assemblages that inhabit the evergreen tropical forests of the world. These are found in the Neotropics, centred in the Amazon basin, the equatorial forest zone of Africa, the eastern rain-forest belt of Madagascar, and the Sunda Shelf region of south-east Asia. Each of these regions harbours eleven to sixteen rainforest genera and, with the exception of Madagascar, roughly forty-five species (Table 10.1). (One should keep in mind, however, that a sizeable fraction of the Malagasy fauna vanished after humans invaded the island about a thousand years ago.)

TABLE 10.1. Genera and species of primates in four rainforest regions[1]

	Total per realm		Rain forest habitats	
	Genera	Species	Genera	Species
Madagascar	12 (18[2])	28 (42[2])	12	14
Neotropics	16	52	16	47
Africa	16	54	12	42
Asia (Oriental)	12	51	11	43

[1] Taxonomy follows Jolly (1985).
[2] Includes recently extinct forms (Tattersall 1982).

With respect to phylogeny, these four assemblages are highly distinct. There is not one genus in common to any two of them, and with the exception of cercopithecids and pongids common to Africa and Asia, the remaining five or six possible pairwise comparisons contain no shared families at all. The lineages in the respective regions have been geographically isolated at least since the Oligocene (*c.* 30 million years ago), except in the Africa–Asia comparison where there was continuity until sometime in the Miocene (*c.* 15–20 million years ago). One can thus reasonably assert that the four assemblages represent four independent

Table 10.2. Biomass (and number of species) by trophic class in selected primate communities inhabiting tropical rainforest (kg km^{-2})

	Folivore–frugivore[1]	Frugivore–folivore	Frugivore	Frugivore–folivore–insectivore	Frugivore–insectivore	Gum-insects	Total biomass
Neotropics							
Cocha Cashu, Peru[2]		182 (2)	152 (2)		254 (6)	1 (1)	589
Barro Colorado Is., Panama[3]		416 (1)	4 (1)		25 (3)		445
Samiria, Peru[4]		217 (2)	9 (1)		119 (4)		345
Raleighvallen, Suriname[5]		94 (1)	84 (3)		73 (4)		251
La Macarena, Colombia[6]		112 (1)	82 (1)		36 (2)		230
Africa							
Kibale, Uganda[7]	2010 (2)		185 (2)		330 (1)		2525
Tai, Côte d'Ivoire[8]	558 (3)	42 (1)	167 (3)		35 (1)		802
Asia							
Kuala Lompat, Malaysia[9]	337 (2)	315 (2)		268 (2)	< 25 (1)		933
S. Kenyan, Malaysia[10]	161 (2)	400 (2)		> 36 (2)		597	
swamp forests, Malaysia[11]	117 (2)	84 (2)		425 (2)			626
non-riv. rainforests, Mal.[12]	381 (1)	251 (2)		7 (1)			639
Ketambe, Sumatra[13]	≥ 135 (1)	400 (3)		302 (2)			837
Kutai (Borneo), Indonesia[14]	82 (1+)	225 (3)	17 (1)				324
G. Mulu (Borneo), Malaysia[15]	112 (3)	35 (1)	85 (2)	232			

[1] Classes based on feeding times: Folivore–frugivore if > 50% leaves; frugivore–folivore if > 50% fruit; frugivore if < 10% leaves, etc.; species assigned to one of the insectivorous classes if > 10% insects. Nocturnal species often incompletely censused, but they contribute little to total biomass. Mean body weights employed in biomass calculations take average group composition into account.

[2] From Terborgh 1983.

[3] From Glanz 1982.

[4] From Freese *et al.* 1982.

[5] From Mittermeier & van Roosmalen 1981 (body weights adjusted).

[6] From Klein & Klein 1977a; midpoints of ranges given.

[7] From Struhsaker & Leland 1979. The data represent five common species; the community includes two additional uncommon species, *Cercopithecus l'hoesti* and *Pan troglodytes*, for which data are lacking.

[8] From Galat & Galat-Luong 1985, with *Pan* added from Bourliere 1985.

[9] From Raemaekers & Chivers 1980.

[10] From Marsh & Wilson 1981.

[11] From Marsh & Wilson 1981; mean of five swamp forest sites.

[12] From Marsh & Wilson 1981; mean of three non-riverine rainforest sites.

[13] After Rijksen (1978). Data on *Macaca nemestrina* from J.M.Y. Robertson, personal communication; data on *M. fascicularis* from van Schaik & van Noordwijk, unpublished.

[14] From Rodman 1978.

[15] From A. Mitchell, personal communication.

throws of the evolutionary dice. We can therefore assume that the characteristics shown by each of them are adaptations to local or regional environmental conditions rather than lingering reflections of descent from some ancient common ancestor.

Ecological characteristics of the four assemblages

The data presented in Table 10.2 are taken from the literature and represent the best-studied primate communities of their respective regions.

Biomass

Undisturbed rainforest primate communities vary over an order of magnitude in their standing biomasses (Table 10.2). Low levels are characteristic of the Neotropics, where most localities show values around 300 kg km^{-2}. An exceptionally fertile site on an alluvial plain near the base of the Andes (Cocha Cashu) sustains about twice this figure. In south-east Asia, the lowest values are found in Bornean localities (around 300 kg km^{-2}), but sites in West Malaysia and on Sumatra (volcanic soil) reach up to three times this value. Reliable estimates for Africa are scarce due to the nearly universal prevalence of hunting. Nevertheless, the available data suggest that African localities support the highest biomasses. No rainforest communities have been censused in Madagascar, but the biomass at a dry forest site (2720 kg km^{-2}: Morondava, Hladik 1979) is among the highest observed.

Most of the variation in these totals is accounted for by the folivore category. The Neotropical primate communities lack specialized folivores, while in the other three regions, highly folivorous species make up the bulk of the biomass. Nevertheless, the abundance of folivores tends to be greater in Africa than in south-east Asia. Later we shall consider possible reasons for these contrasts.

Trophic categorization

New World primate assemblages are as idiosyncratic in the over-representation of frugivore–insectivores as they are in the under-representation of specialized folivores (Table 10.2). Several genera of frugivore–insectivores (*Cebus, Saimiri, Aotus, Saguinus, Callimico*) include some of the most abundant and widespread Neotropical species. Their Old World counterparts (*Cercopithecus* and *Miopithecus* in Africa; *Macaca* in Asia) are not so diverse at the generic level and for the most part consume appreciable quantities of young foliage, something the New World species do not do. In the Old World fauna there are frugivore–insectivores among the nocturnal prosimians (*Galago, Perodicticus, Arctocebus, Nycticebus, Cheirogaleus, Microcebus*), but these animals contribute only a tiny fraction of the biomass of their respective communities.

In summary, we find that New World primate assemblages are peculiar in having few folivores (a fact that accounts for their reduced biomass) and unusually large numbers of frugivore–insectivores. African assemblages are rich and diverse with good representation in most trophic categories. The Malagasy assemblage, though relatively poorly known, is largely composed of diurnal folivores and small nocturnal forms of varied dietary proclivities (Tattersall 1982). Finally, the south-east Asian assemblages include many partly to strongly folivorous species but attain much lower biomasses than their African counterparts.

Size distributions of component species

Africa, Madagascar and the oriental region harbour several small, nocturnal prosimians that show insectivorous tendencies (Table 10.3). In this respect the three faunas are reasonably convergent. Again, the Neotropical realm represents the exception, as it contains only one nocturnal species, *Aotus*, a monkey with a frugivore–insectivore diet (Wright 1985). In contrast, the New World rainforests contain eighteen small, diurnal monkeys (mostly Callitrichidae) in a size class (< 1 kg) that includes only one diurnal Old World species. Even if one considers the Callitrichids to be the New World counterparts of the nocturnal prosimians of the Paleotropics, one still must answer the question of why they are diurnal.

TABLE 10.3. Size distributions of nocturnal (N) and diurnal (D) species in four rainforest regions

		Body size distribution (kg)								
		< 0.5	0.5–1	1–2	2–4	4–8	8–16	16–32	32–64	> 64
Madagascar	N	4	2		1					
	D		1	1	3	1	1			
Neotropics	N			1						
	D	7	11	7	9	10	2			
Africa	N	4		1						
	D			1	14	13	6		2	1
Asia (Oriental)	N	5		1						
	D				1	27	8		1	

As for the larger diurnal forms, the average size of the African and Asian species is greater than that of their Neotropical and Malagasy counterparts. Taking into account the fact that many of the recently extinct Malagasy lemurs were larger than those that survive, we note once again that the exception appears in the Neotropical fauna. There appear to be two reasons for this. First, Neotropical species are smaller than their Old World equivalents. For example, an adult female squirrel monkey (*Saimiri sciureus*) weighs between 700 and 900 g, whereas a female of the ecologically similar African talapoin (*Miopithecus talapoin*) weighs over 1 kg. Likewise, adult females of the four species of capuchins (*Cebus* spp.) weigh between 2 and 3 kg while those of the African guenons (*Cercopithecus* spp.) weigh between 2.5 and 5 kg, and those of the tropical macaques (*Macaca* spp.) weigh between 3.5 and 8 kg. A similar pattern is found in comparing female howler monkeys (*Alouatta* spp.: 5–6.5 kg) with African and Asian leaf monkeys (*Colobus* spp.: 6–9 kg; *Presbytis* spp.: 6–12 kg, respectively). Second, the Neotropics are deficient in large primates. This is probably related to the low representation of folivores noted above, since all species that weigh more than 8 kg, regardless of region, are partially to highly folivorous (Klein & Klein 1977b).

Terrestriality

No Neotropical primate habitually travels on the ground, whereas before the recent spate of extinctions several did in Madagascar, and a number still do in Africa and Asia. In the latter regions, most primate communities include at least one terrestrial species, usually a specialized frugivore. All but one of the terrestrially travelling species (the African *Allenopithecus*) weigh over 4 kg.

Diversity

The remarkable similarity in the regional (γ) diversities of the four rain forest realms (Table 10.1) belies strong contrasts in the diversities at individual sites (α-diversity). Geographically separated congeners contribute more to total diversity in the Oriental realm than elsewhere. The number of species found per site is predictably lower in the Orient (cf. Table II in Bourliere 1985). The diversity of 'ecospecies' (monotypic genera plus broadly overlapping species in other genera) is highest in the Congo Basin at 22, and lowest in the various subregions of the Orient. The whole Sunda Shelf harbours a mere thirteen ecospecies, hardly more than the single island of Madagascar.

TABLE 10.4. Non-convergent features of primate communities in the four main rainforest regions

Feature	Deviant region
Biomass	Low in Neotropics and parts of Oriental region
Folivory	Low in Neotropics
Nocturnality	Rare in Neotropics
Size distribution	Smaller in Neotropics
Terrestriality	Absent in Neotropics
Diversity	Reduced in Oriental region

In comparing the four rainforest regions, the primates of at least one (usually the Neotropical) have proven to deviate from the pattern established by the others (Table 10.4). In order to gain some insights into the possible adaptive causes underlying the observed patterns, we shall now examine the relationships between available food resources and the adaptive characteristics of one particularly well-studied community, that of Cocha Cashu in south-eastern Peru.

DRY SEASON FOOD RESOURCES AND THE COCHA CASHU PRIMATE COMMUNITY

Cocha Cashu is situated on the Amazonian lowland plain at the foot of the eastern Andes at 12°S. The climate is markedly seasonal, alternating each year between a rainy period of about 7 months' duration (November through May) and a dry period of 5 months (June through October). The Cocha Cashu field station is located beside an oxbow lake in tall, evergreen floodplain forest in the meander belt of the Manu River.

The primate community at this site is very well known, having been studied for a dozen years by a number of investigators (Terborgh 1983). Ten species are found in the immediate vicinity of the research station, and three more occur in nearby forests (Table 10.5). Eight species have been intensively investigated in studies lasting one to several years and the rest have received less attention.

In conjunction with behavioural and ecological studies of the animals themselves, we have measured the seasonal availability of the key resources consumed by primates. Foremost among these are plant reproductive parts, particularly fruit pulp. The seasonal character of the climate at Cocha Cashu results in a strongly cyclic pattern of fruiting in the surrounding forest (Fig. 10.1). Flowering is concentrated in the dry season and in the period of transition to the rainy season. Fruiting lags flowering by periods that vary

TABLE 10.5. Characteristics of primates of Cocha Cashu[1]

Species	Adult weight (kg)	Diet[2]	Position in forest[3]	Mean troop size	Population density[4]	Biomass (kg/km[2])
Ateles paniscus	8.0	F,L,U,Ne	C	Variable	25	175
Alouatta seniculus	8.0	F,L,U	C	6	30	180
Cebus apella	3.0	F,I,Nu,Ne,U	M,C,G	10	40	104
Cebus albifrons	2.8	F,I,Nu,Ne	M,C,G	15	35	84
Saimiri sciureus	0.9	F,I,Ne,Nu	M,U,C,G	35	60	48
Aotus trivirgatus	0.8	F,I,Ne	M,?	4	40	28
Callicebus moloch	0.8	F,L,U	M	3	24	17
Saguinus imperator	0.5	F,I,Ne	U,M,C	4	12	5
Saguinus fuscicollis	0.5	F,I,Ne,S	U,M,C	5	16	5
Cebuella pygmaea	0.1	S,I,Ne	U,M	5	5	1

[1]Adapted from Terborgh 1983.
[2]F = ripe fruit; I = insects; L = leaves; Ne = nectar; Nu = nuts; P = pith; S = sap; U = unripe fruit. Listed for each species in order of use.
[3]C = canopy; G = ground; M = midstory; U = understory. Listed in order of use.
[4]Number of individuals per km[2].

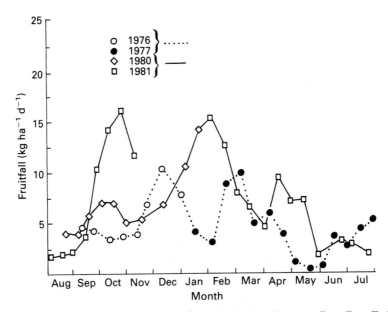

FIG. 10.1. Seasonal patterns of fruitfall at Cocha Cashu in south-eastern Peru. From Terborgh 1986a.

between species from one to several months. Overall fruiting activity reaches a maximum during the early and middle portions of the rainy season, and later declines to very low levels. The amount of fruit available from May through July (the bottom of the cycle, hereafter referred to as the 'dry' season) is estimated to be less than the amount required to sustain the frugivore biomass of the forest (Terborgh 1986a). The role of seasonality is discussed in more detail in Chapter 4.

The question of how frugivores survive the dry season has been analysed in depth in several previous publications (Terborgh 1983, 1986a, 1986b). Here, to illustrate the key points, we present data on the diets of five primate species in the wet and dry seasons (Table 10.6). These five species range in size from 0.5 kg (the tamarins: *Saguinus* spp.) to 3 kg (capuchins: *Cebus* spp.).

TABLE 10.6. Diets of five primate species (by per cent of feeding time) during the wet and dry seasons at Cocha Cashu [1]

Species	Fruit pulp Wet	Fruit pulp Dry	Palm nuts Wet	Palm nuts Dry	Nectar Wet	Nectar Dry	Other Wet	Other Dry
Brown capuchin (*Cebus apella*)	99	66		25		1	1	8
White-fronted capuchin (*Cebus albifrons*)	99	53		42		3	1	3
Squirrel monkey (*Saimiri sciureus*)	100	91[2]		9				
Emperor tamarin (*Saguinus imperator*)	97	41			1	52	2	7
Saddle-backed tamarin (*Saguinus fuscicollis*)	96	16		75	4	9		

[1]Adapted from Terborgh 1983.
[2]This value over-represents the actual importance of fruit in the dry season diet because it represents the percentage of plant material consumed. The total feeding time on plant matter is sharply reduced in the dry season.

During the wet season when fruit is superabundant, all five species feed on little else (except for arthropods which are taken on a year-round basis and which represent a relatively small proportion of the total caloric intake). The situation changes in the 3-month period from May through July when fruiting is at its annual nadir. All five species exhibit shifts in behaviour, as seen in altered ranging and activity patterns, increased or decreased tendencies to join mixed species associations and in the consumption of novel food resources (Terborgh 1983). The capuchins spend long hours laboriously opening and extracting the meat of *Astrocaryum* palm nuts, which require 140 kg of bite-force to break (Kiltie 1982). Squirrel monkeys roam over large areas to locate rare fruiting fig trees in which they feed to satiation. But during interludes when no figs are fruiting, the animals may go for a week or more without consuming any fruit at all. At these times their recourse is to hunt arthropods and other small prey in an effort that extends almost uninterruptedly from dawn to dusk. Finally, the two tamarin species switch to nectivory when their fruit supply runs out. During portions of July, nectar consumption accounts for > 90% of the total time spent feeding on plant materials (Terborgh 1983).

Observations made in the same forest on other frugivores (the remaining primate species plus procyonids, marsupials, peccaries, squirrels and other large rodents, etc.) greatly extend the results obtained with the five primates. Virtually the entire frugivore guild, which includes dozens of mammal and bird species ranging in size from 40 kg peccaries to 20 g birds, survive the annual period of scarcity by feeding on figs, palm nuts, nectar and small prey (Terborgh 1986a).

Consumption of foliage does not seem to represent another option in this forest. Howler monkeys eat leaves and so do the smaller titi monkeys, but even in the worst of times leaf consumption does not rise above 60% of the total intake of plant material in either species (Milton 1980; Wright 1985). Moreover, it should be noted that these two partial folivores contribute only a small portion of the community's total primate biomass. Non-primate folivores, i.e. sloths, are extremely rare at Cocha Cashu (Terborgh 1986a). The main component of the animal biomass is contained in frugivores that switch to resources other than leaves when fruit is scarce.

Adaptations for facultative use of alternative resources

There are obvious behavioural and/or morphological adaptations associated with feeding on each of the resources available to frugivores at Cocha Cashu during the period of fruit scarcity. Species that seek widely-scattered ripe crops of figs must possess exceptionally large home ranges, and if they are primates, must be able to obtain feeding positions under conditions of intense interspecific competition. Relatively large primates, such as spider and capuchin monkeys, can hold their own under these conditions; squirrel monkeys, although smaller, can compete by overwhelming larger adversaries through sheer force of numbers, while still smaller titi monkeys and tamarins are excluded entirely (Terborgh 1983; Wright 1985). Other frugivores, such as marsupials, procyonids and night monkeys avoid the aggression of large diurnal monkeys by using the same trees at night. All species that use figs must be able to switch to other resources during intervals when no figs are to be found.

Palm nuts sustain roughly a third of the animal biomass at Cocha Cashu during the early dry season, but they are exploited by a limited suite of species that possess appropriate adaptations: either an extraordinarily powerful bite (peccaries, capuchins) or gnawing ability (squirrels, other rodents and macaws) (Kiltie 1982; Terborgh 1986a). Although they are not the largest monkeys in the community, capuchins appear to be the only ones able to break hard nuts.

The use of nectar as an alternative resource carries other kinds of adaptive limitations. The nectars used by non-specialist birds and mammals are dilute and, even when available in copious amounts, provide a meagre energy budget (Terborgh & Stern 1987). No large animal could harvest enough to satisfy its metabolic needs. Facultative nectarivores are consequently small. These include marsupials, procyonids and most of the monkeys weighing < 1 kg, as well as numerous birds (Janson, Terborgh & Emmons 1981).

Body size also limits the use of arthropods and other small dispersed prey as an alternative food resource. The largest birds that are obligate arthropod feeders weigh around 200 g, while the only strictly predatory primates are the 60–80 g tarsiers. Squirrel monkeys can subsist for a time on prey. At 800–1000 g, they are much larger than any species that specializes on dispersed arthropods, though small enough to forestall starvation on what they can catch.

The purpose of this section has been to point out that nearly all the primates at Cocha Cashu, as well as many other animals, sustain themselves through the dry season on a very limited array of resources. Fruit pulp is the preferred food, but each species resorts to a fallback resource when no fruit is at hand. The evidence indicates that exploitation of alternative resources is efficiently accomplished only by species possessing appropriate adaptations (Terborgh 1986a). Virtually every species in the community can be seen to possess such adaptations. Is this true in other localities and of primate communities in general? We shall attempt to answer this important question in the remaining sections of the chapter.

FOOD RESOURCES AND THE STRUCTURE OF PRIMATE COMMUNITIES

We may now return to the issue of convergence vs. non-convergence in primate communities. We have found that the nature of the food resources available during cyclic episodes of scarcity can plausibly account for the trophic composition, size distribution and, less directly, the biomass of one primate community. Can we apply the same logic to others, or are the conclusions stated above peculiar to Cocha Cashu? We can begin to answer this question by attempting to draw inferences about the nature of resource regimes from the structure of the dependent consumer assemblages.

Predictions

Perhaps the most outstanding difference between New and Old World primate assemblages is the meagre representation, both relatively and absolutely, of folivores in the New World. At Cocha Cashu foliage does not constitute a major alternative resource for most primate species during times of fruit scarcity. Although data on the phenology of leafing from this locality are not available, it is our impression that leafing is at a minimum during the same May–July period when fruiting is also minimal. If true, this would explain the paucity of folivore–frugivores and frugivore–folivores. As species in these categories are under-represented in the Neotropics generally, one could predict that fruiting and leafing cycles in Neotropical forests will vary in parallel. An alternative explanation for the low biomass of folivores in the Neotropics, that plant defensive compounds are limiting, lacks any direct support at present. Such evidence as there is on this point suggests that levels of secondary compounds in leaves are related to soil fertility (McKey *et al.* 1978; Kinzey & Gentry 1979). If soils are the main controlling factor, the expectation would be of high intraregional differences in folivore densities, not of systematic interregional differences.

The small average size of New World primates could be a corollary of having to depend on nuts, nectar and arthropods when fruit is in short supply. And the low biomasses of New World communities, in turn, could be a reflection of the small sizes of the component species and the low abundance of their dry season resources relative to what might be available in the form of foliage were leafing activity more equitably distributed around the year.

In the Old World, we must try to explain the predominance of folivorous species in all the communities for which there are reliable census data. Following the line of reasoning developed above, we predict that leafing is seasonally less variable in the evergreen forests of the Old World, or if it is variable, that the periods of minimum availability of fruit and young foliage would be out-of-phase and hence offset each other. In comparing Africa and Asia, where the main differences are a reduced biomass and greater emphasis on folivory in Asia, we could expect a lower overall productivity in

Asia or a relatively lower minimum of fruit production in the annual cycle. In the case of Madagascar, where the primates display the strongest tendency toward folivory of all, we should expressly predict an unusual type of phenological cycle in which leafing is more or less continuous while fruiting is sharply seasonal. How well are these predictions supported by the data currently available?

Tests of the predictions

The prediction that leafing and fruiting cycles are expected to be in phase in many New World localities is confirmed in data from Barro Colorado Island, Panama. The availabilities of young foliage and fruit are positively correlated (Fig. 10.2a). As anticipated, foliage provides a poor to non-existent substitute for fruit when the latter is in short supply.

In the Old World where there are many more folivores, we expect leafing and fruiting cycles to be uncorrelated or even negatively correlated. This is borne out by the only three localities for which appropriate data exist: Makokou in Gabon (equatorial Africa), Ketambe, Sumatra (Indonesia) and

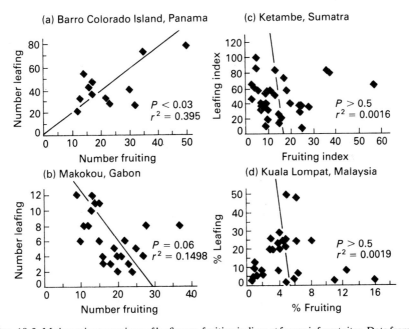

FIG. 10.2. Major axis regressions of leafing vs. fruiting indices at four rainforest sites. Data from published sources as follows: Barro Colorado Island, Panama: Leigh & Smythe 1978; Makokou, Gabon: Charles-Dominique 1977, Hladik 1973; Ketambe, Sumatra: Schaik 1986; Kuala Lompat, Malaysia: Raemaekers *et al.* 1980.

Kuala Lompat, Malaysia (Fig. 10.2b–d). Fruiting and leafing activity are uncorrelated or show a negative relationship at all three sites. Separate examination of the leafing and fruiting records for these localities reveals that the peaks of leafing and fruiting at Makokou are seasonally comple- mentary (Hladik 1973), while in the South-East Asian localities the lack of statistical correlation is due to highly variable patterns of fruit production superimposed on a more uniform background of leafing activity (Raemaekers, Aldrich-Blake & Payne 1980; Schaik & Noordwijk 1985). Under these circumstances, it is clear that resources are present to support specialized folivores on a year-round basis as well as frugivore–folivores that switch to leaf eating during periods of fruit scarcity.

We can now put together some more pieces of the puzzle. The occur- rence of year-round leafing activity at the African and Asian sites can account for the prevalence of folivores in their respective primate com- munities, but the matter of the body sizes of the component species remains to be considered. We concluded that the average size of New World primates is small because during periods of scarcity many of them switch to sparse and/or low energy resources—palm nuts, nectar and small prey. Published information on African and Asian rainforest primates gives no indications that any of these resources plays a major dietary role (Gautier- Hion 1980; MacKinnon & MacKinnon 1980). Palm nuts and nectar do not appear to feature at all, whereas small prey do contribute significantly to the diets of some of the smaller cercopithecines (16% of the stomach content of *Cercopithecus pogonias*; 35% of *Miopithecus talapoin*: Gautier-Hion 1980). Instead, young foliage appears to serve as the alternative resource for nearly all species. As leaves are more difficult to digest than fruit pulp, and require longer to pass through the gut, folivores tend to be of relatively large body size and to possess low metabolic rates (Clutton-Brock & Harvey 1977; Milton 1980; McNab 1986). The consumption of leaves as an alternative to fruit thus seems to account for the large size of African and Asian primates relative to their New World counterparts.

Asian primate communities contain a higher representation of folivores than African, and maintain lower average biomasses. This could result from a lower productivity of Asian forests, or from a more irregular fruit supply. A recent compilation of published data permits a pan-global comparison of litter-fall rates in lowland tropical rainforests (Table 10.7). Total production is highest in Africa, agreeing well with the high primate biomasses found in that region. Apart from the low value for South Asia (perhaps due to small sample size), leaf-fall is roughly similar in rainforests on all continents. However, fruit production in South-East Asia is only about half of that measured elsewhere. This finding is supported by an independent com-

TABLE 10.7. Litter production in tropical rainforests of various geographical regions[1] in t ha^{-1}

Region	Total small litter			Leaves			Reproductive parts		
	Mean	*n*	s.d.	Mean	*n*	s.d.	Mean	*n*	s.d.
Africa	11.13	11	2.08	7.03	8	0.65	0.76	5	0.33
Central America	9.84[2]	14	2.26	6.52[3]	6	0.71	0.80[4]	7	0.41
South America	7.97[5]	13	1.86	6.26	11	1.02	0.64[6]	14	0.32
South Asia	4.77	6	1.33	4.03	6	1.31	—		
South-East Asia	9.00[7]	23	2.29	6.48[8]	16	1.69	0.35	10	0.11
Papua/Australia	9.40	4	0.54	5.83	6	1.17	0.85	4	0.34
	$F(5,65) = 8.55$			$F(5,47) = 4.58$			$F(4,35) = 3.55$		
	$P < 0.001$[9]			$P < 0.01$[10]			$P < 0.05$[11]		

[1]Based on Proctor (1984), except where stated otherwise.
[2]Excluding outlier of 27.0 t ha^{-1}.
[3]Excluding outlier of 15.5 t ha^{-1}.
[4]Excluding outlier of 3.3 t ha^{-1}.
[5]Excluding outlier of 21.9 t ha^{-1}.
[6]Including three values from Cocha Cashu (Table 2.2 of Terborgh 1983), assuming 40% dry matter (cf. Smythe 1970).
[7]Excluding outlier of 23.3 t ha^{-1}.
[8]Including three values from Ketambe (low and high terraces and mountain slopes), taken from Schaik and Mirmanto (1985).
[9]South Asia significantly lower than all other (Tukey-Kramer method, at $P < 0.05$), South America significantly lower than Africa.
[10]South Asia significantly lower than all except Papua/Australia.
[11]South-East Asia significantly lower than Central America.

parison of fruit availability at Ketambe (Sumatra) and Makokou (Gabon). The monthly means for the number of fruit sources per 100 m of trail were 2.2 at Makokou and 1.4 at Ketambe (Gautier-Hion *et al.* 1985; Schaik 1986; Schaik & Noordwijk 1985).

The supply of fruit is also more irregular in south-east Asia than elsewhere. Asian forests are renowned for the irregular, mass-fruiting behaviour (masting) of trees in the Dipterocarpaceae (Janzen 1974; Ng 1981). A less well-known fact is that masting behaviour extends to families other than the Dipterocarpaceae, including many that have animal-dispersed fruits (Medway 1972; Leighton & Leighton 1983; Schaik & Noordwijk 1985). Consequently, in comparing fruiting phenology records from Gabon with those from Malaysia and Sumatra, we note that in the former there is high intra-annual but low inter-annual variability in fruit production while in the latter there is high variability at both the intra- and inter-annual levels (Charles-Dominique 1977; Putz 1979; MacKinnon & MacKinnon 1980; Gautier-Hion *et al.* 1985; Schaik & Noordwijk 1985). In

Gabon, fruit is in good supply for 8–9 months a year, while in the South-East Asian localities it is in short supply most of the time, except in occasional boom years. If this contrast between the African and Asian localities is typical, i.e. one that holds generally for the forests of both regions, it could easily account for the observed differences in biomass and development of folivory in the respective primate communities.

Finally, there is the question of why primate α-diversity is so much lower in South-East Asia than in the other regions. This is particularly remarkable as Sundaland was united during the glacial periods of the Pleistocene while the forests of Africa and South America were fragmented (Whitmore 1981). Even so, the large island of Borneo is depauperate in primate species although it is rich floristically. It is tempting to ascribe this to low fruit availability, though further corroborative evidence is sorely wanting.

CONCLUSIONS

In seeking evidence of convergence in the characteristics of primate assemblages representing the four major rainforest regions of the world (the Neotropics, Africa, Madagascar and South-East Asia), we more often found divergences in body size distributions, trophic relationships, biomasses and, to a lesser degree, local species diversities. The primates of the Neotropical region are especially distinctive in being of small size, and in eschewing foliage in their diets. In contrast, Old World species are larger, both in average and maximum size, and consume more foliage. Primate biomasses in African forests are particularly high, reaching levels of $> 2000 \text{ kg km}^{-2}$.

These aspects of non-convergence were investigated by examining the phenological patterns of resource production in forests representing the Neotropics, Africa and South-East Asia. At a locality in Amazonian Peru we found that the primates enjoyed a superabundance of fruit for much of the year, but experienced pronounced scarcity during the remaining months. During periods of scarcity, the animals switched to resources that were used relatively little at other times of year—figs, nectar and palm nuts, and some species greatly increased the time they spent searching for arthropods and other small prey. Significantly, most species did not resort to eating foliage during periods of fruit scarcity. Because only small animals can sustain themselves on nectar and small prey, these observations seem to account for the small body sizes and low biomasses of many New World primates.

Examination of phenological patterns at the few rainforest sites for which appropriate data are available revealed that leafing and fruiting were positively correlated at a Neotropical site (Barro Colorado Island, Panama), negatively correlated at an African site (Makokou, Gabon) and uncorrelated at two Asian sites (Kuala Lompat, Malaysia and Ketambe, Sumatra). The positive correlation between leafing and fruiting found in Panama indicates that scarcities of young foliage and fruit are coincident, a situation that could explain the low levels of folivory found among New World primates. In contrast, leafing reached seasonal peaks at times of low fruit production in Gabon, a situation that would allow primates the opportunity to concentrate on whichever of the two resources were most abundant. This could account for the mixed diets of most of the species in the Gabonese assemblage, and for the exceptionally high biomasses of forest primates in Africa. Lastly, the phenological data for the Asian forests show a strong year-to-year component in fruit production (masting) superimposed on a more continuous background of leaf production. The prolonged periods of fruit scarcity that characterize these forests are in accord with the well-developed folivory, low biomasses and low species diversity of their primate assemblages.

In sum, many of the non-convergences noted at the beginning of the chapter can potentially be explained by differences in climate-driven phenological patterns of resource production in the four regions. If confirmed, our interpretations could in the end provide a vindication of the convergence hypothesis. But, in caution, we must stress that many of the conclusions reached in this chapter are tentative because they are based on sometimes scanty data and rely on the strength of just one or two localities to represent whole regions of the world. Nevertheless, we are hopeful that our approach will eventually provide a powerful tool in advancing our understanding of the broad issue of convergence vs. non-convergence in animal communities.

As a postscript, we would like to draw the reader's attention to the fact that how far we can progress toward a fuller understanding of these matters may now be limited by time, the time over which unperturbed natural environments can still be studied. The situation is especially acute in West Africa, Madagascar and South-East Asia, where undisturbed tropical forests are unlikely to endure beyond the end of this century. That gives us maybe 15 years in which to decipher many remaining puzzles. Will that be enough?

ACKNOWLEDGMENTS

We wish to thank the editors and two anonymous reviewers for their sharp-eyed criticisms and for a number of constructive suggestions. We gratefully acknowledge financial support from the National Science Foundation (USA: Terborgh) and from the Netherlands Organization for the Advancement of Pure Research (Z.W.O.: Schaik).

REFERENCES

Bourlière, F. (1985). Primate communities: their structure and role in tropical ecosystems. *International Journal of Primatology,* **6**, 1–26.

Charles-Dominique, P. (1977). *Ecology and Behavior of Nocturnal Primates: Prosimians of Equatorial West Africa.* Columbia University Press, New York.

Clutton-Brock, T.H. & Harvey, P.H. (1977). Species differences in feeding and ranging behaviour in primates. *Primate Ecology: Studies of Feeding and Ranging Behaviour in Lemurs, Monkeys and Apes* (Ed. by T.H. Clutton-Brock), pp. 557–84. Academic Press, London.

Emmons, L.H., Gautier-Hion, A. & Dubost, G. (1981). Community structure of the frugivorous–folivorous forest mammals of Gabon. *Journal of Zoology (London),* **199**, 209–22.

Freese, C.H., Heltne, P.G., Castro, R.N. & Whitesides, G. (1982). Patterns and determinants of monkey densities in Peru and Bolivia, with notes on distributions. *International Journal of Primatology,* **3**, 53–90.

Galat, G. & Galat-Luong, A. (1985). La communauté de primates diurnes de la forêt de Tai, Côte d'Ivoire. *Revue d'Écologie (Terre et Vie),* **40**, 3–32.

Gautier-Hion, A. (1980). Seasonal variations of diet related to species and sex in a community of *Cercopithecus* monkeys. *Journal of Animal Ecology,* **9**, 237–69.

Gautier-Hion, A. Duplantier, J-M., Emmons, L., Feer, F., Heckestweiler, P., Moungazi, A. Quris, R. & Sourd, C. (1985). Coadaptation entre rythmes de fructification et frugivorie en forêt tropicale humide du Gabon: mythe ou réalité. *Revue d'Écologie (Terre et Vie),* **4(1)**, 405–29.

Glanz, W.E. (1982). The terrestrial mammal fauna of Barro Colorado Island: Censusses and long-term changes. *The Ecology of a Tropical Forest* (Ed. by E.G. Leigh, Jr., A.S. Rand & D.M. Windsor), pp. 455–68. Smithsonian Institution Press, Washington, DC.

Hladik, C.M. (1973). Alimentation et activité d'un groupe de chimpanzes reintroduits en forêt Gabonaise. *Revue d'Écologie Appliquée (Terre et la Vie),* **27**, 343–413.

Hladik, C.M. (1979). Diet and ecology of prosimians. *Study of Prosimian Behaviour* (Ed. by G.A. Doyle & R.D. Martin), pp. 307–57, Academic Press, London.

Janzen, D.H. (1974). Tropical blackwater rivers, animals, and mast fruiting by the Dipterocarpaceae. *Biotropica,* **6**, 69–103.

Janson, C.H., Terborgh, J. & Emmons, L.H. (1981). Non-flying animals as pollinating agents in the Amazonian forest. *Biotropica* suppl., 1–6.

Jolly, A. (1985). *The Evolution of Primate Behaviour,* 2nd edn. Macmillan, New York.

Karr, J.R. (1980). Geographical variation in the avifaunas of tropical forest undergrowth. *Auk,* **97**, 283–98.

Kiltie, R.A. (1982). Bite force as the basis for niche differentiation in rain forest peccaries (*Tayassu tajacu* and *T. peccari*). *Biotropica,* **14**, 188–95.

Kinzey, W.G. & Gentry, A.H. (1979). Habitat utilization in two species of *Callicebus. Primate Ecology: Problem Oriented Field Studies* (Ed. by R.W. Sussman), pp. 89–100. Wiley, New York.

Klein, L.L. & Klein, D.J. (1977a). Social and ecological contrasts between four taxa of neotropical primates. *Socioecology and Psychology of Primates* (Ed. by R. Tuttle), pp. 59–85. Mouton, The Hague.

Klein, L.L. & Klein, D.J. (1977b). Feeding behaviour of the Colombian Spider Monkey. *Primate Ecology: Feeding and Ranging Behaviour in Lemurs, Monkeys and Apes* (Ed. by T.H. Clutton-Brock), pp. 153–81. Academic Press, London.

Leigh, E.G. Jr. & Smythe, N. (1978). Leaf production, leaf consumption, and the regulation of folivory on Barro Colorado Island. *The Ecology of Arboreal Folivores* (Ed. by G.G. Montgomery), pp. 51–73, Smithsonian Institution Press, Washington, DC.

Leighton, M. & Leighton, D.R. (1983). Vertebrate responses to fruiting seasonality within a Bornean rain forest. *Tropical Rain Forest: Ecology and Management* (Ed. by S.L. Sutton, T.C. Whitmore & A.C. Chadwick), pp. 181–96. Special Publications of the British Ecological Society, 2. Blackwell Scientific Publications, Oxford.

MacKinnon, J.R. & MacKinnon, K.S. (1980). Niche differentiation in a primate community. *Malayan Forest Primates: Ten Years' Study in Tropical Rain Forest* (Ed. by D.J. Chivers), pp. 167–90. Plenum, New York.

Mares, M.A. (1976). Convergent evolution of desert rodents: multivariate analysis and zoogeographic implications. *Palaeobiology*, **2**, 39–63.

Marsh, C.W. & Wilson, W.L. (1981). *A Survey of Primates in Peninsular Malaysian Forests.* Universiti Kebangsaan Malaysia, Kuala Lumpur.

McKey, D., Waterman, P.G., Mbi, C.N., Gartlan, J.S. & Struhsaker, T.T. (1978). Phenolic content of vegetation in two African rain forests: ecological implications. *Science*, **202**, 61–3.

McNab, B.K. (1986). The influence of food habits on the energetics of eutherian mammals. *Ecological Monographs*, **56(1)**, 1–19.

Medway, L. (1972). Phenology of a tropical rainforest in Malaya. *Biological Journal of the Linnean Society*, **4**, 117–46.

Milton, K. (1980). *The Foraging Strategy of Howler Monkeys: A Study in Primate Economics.* Columbia University Press, New York.

Mittermeier, R.A. & Roosmalen, M.G.M. van. (1981). Preliminary observations on habitat utilization and diet in eight Surinam monkeys. *Folia Primitologica*, **36**, 1–39.

Ng, F.S.P. (1981). Vegetative and reproductive physiology of dipterocarps. *Malaysian Forester*, **44**, 197–221.

Orians, G.H. & Paine, R.T. (1983). Convergent evolution at the community level. *Coevolution* (Ed. by D.J. Futuyma & M. Slatkin), pp. 431–458. Sinauer Associates, Sunderland, Massachusetts.

Pearson, D.L. (1977). A pantropical comparison of bird community structure on six lowland forest sites. *Condor*, **79**, 232–44.

Pianka, E.R. (1975). Niche relations of desert lizards. *Ecology and Evolution of Communities* (Ed. by M.L. Cody & J.M. Diamond), pp. 292–314. Belknap, Cambridge, Massachusetts.

Proctor, J. (1984). Tropical rainforest litterfall II: The data set. *Tropical Rain-Forest: Ecology and Management Supplementary* (Ed. by A.C. Chadwick & S.L. Sutton), pp. 83–113. Proceedings of the Leeds Philosophical and Literary Society (Science Section), City Museum, Leeds.

Putz, F.E. (1979). Aseasonality in Malaysian tree phenology. *Malaysian Forester*, **42**, 1–24.

Raemaekers, J.J. & Chivers, D.J. (1980). Socio-ecology of Malayan forest primates. *Malayan Forest Primates: Ten Years' Study in Tropical Rain Forest* (Ed. by D.J. Chivers), pp. 279–316. Plenum, New York.

Raemaekers, J.J., Aldrich-Blake, F.P.G. & Payne, J.B. (1980). The Forest. *Malayan Forest Primates: Ten Years' Study in Tropical Rain Forest* (Ed. by D.J. Chivers), pp. 29–61. Plenum, New York.

Richards, P.W. (1952). *The Tropical Rainforest.* Cambridge University Press, Cambridge.

Rijksen, H.D. (1978). *A Field Study on Sumatran Orang Utans (Pongo pygmaeus abelii Lesson 1827).* Veenman en zonen, Wageningen.

Rodman, P.S. (1978). Diets, densities, and distributions of Bornean primates. *The Ecology of Arboreal Folivores* (Ed. by G.G. Montgomery), pp. 456–78. Smithsonian Institution Press, Washington, DC.

Schaik, C.P. van. (1986). Phenological changes in a Sumatran Rain-Forest. *Journal of Tropical Ecology*, **2**, 327–47.

Schaik, C.P. van & Mirmanto, E. (1985). Spatial variation in the structure and litterfall of a Sumatran rain forest. *Biotropica*, **17**, 196–205.

Schaik, C.P. van & Noordwijk, M.A. van. (1985). Interannual variability in fruit abundance and the reproductive seasonality in Sumatran long-tailed macaques (*Macaca fasicularis*). *Journal of Zoology* (London), **206**, 533–49.

Smythe, N. (1970). Relationships between fruiting seasons and seed dispersal methods in a Neotropical forest. *American Naturalist*, **104**, 25–35.

Struhsaker, T.T. & Leland, L. (1979). Socio-ecology of five sympatric monkey species in the Kibale Forest, Uganda. *Advances in the Study of Behaviour*, **9**, 159–227.

Tattersall, I. (1982). *The Primates of Madagascar.* Columbia University Press, New York.

Terborgh, J. (1983). *Five New World Primates: A Study in Comparative Ecology.* Princeton University Press, Princeton, New Jersey.

Terborgh, J. (1986a). Keystone plant resources in the tropical rainforest. *Conservation Biology: Science of Scarcity and Diversity* (Ed. by M. Soule). Sinauer Associates, Sunderland, Massachusetts.

Terborgh, J. (1986b). Community aspects of frugivory in tropical forests. *Frugivores and Seed Dispersal* (Ed. by A. Estrada & T.H. Fleming). Junk, The Hague.

Terborgh, J. & Stern, M. (1987). The precarious life of the saddle-backed tamarin. *American Scientist* (in press).

Terborgh, J. & Robinson, S. (1986). Guilds and their utility in community ecology. *Community Ecology: Pattern and Process* (Ed. by D.J. Anderson & J. Kikkawa). Blackwell Scientific Publications, Oxford.

Walter, H. (1984). *Vegetation of the Earth and Ecological Systems of the Geo-biosphere.* Springer-Verlag, New York.

Whitmore, T.C. (1981). Palaeoclimate and vegetation history. *Wallace's Line and Plate Tectonics* (Ed. by T.C. Whitmore), pp. 36–42. Clarendon Press, Oxford.

Wright, P.C. (1985). *The costs and benefits of nocturnality for* Aotus trivergatus (*the Night Monkey*). Ph.D. Thesis, City University of New York.

Microbial and Decomposer Assemblages

SCOTTISH AGRICULTURAL COLLEGE

AUCHINCRUIVE

LIBRARY

11. ORGANIZATION OF ASSEMBLAGES OF DECOMPOSER FUNGI IN SPACE AND TIME

M. J. SWIFT

*Department of Biological Sciences, University of Zimbabwe,
P.O. Box MP167 Harare, Zimbabwe*

INTRODUCTION

Although research into the functioning of the decomposer subsystem has greatly expanded in the last two decades, no consensus for a general model has emerged. It is therefore necessary to set this discussion of the organization of decomposer communities within a particular concept of how decomposition processes are regulated (Fig. 11.1). Three levels of factor are envisaged as regulating the processes of decomposition (dk) of any given resource; the physical environment (P) sets the initial limits to the rate of decay; this is 'fine-tuned' by the influence of the resource quality (Q); the composition and character of the decomposer community (O) can have a

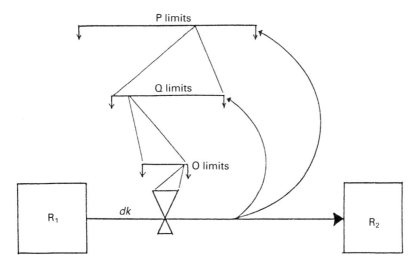

FIG. 11.1. A general descriptive model of the process of decomposition. An organic resource, R_1, is changed to state R_2 over time due to the action of decomposer organisms, O. The rate and pattern of decomposition (dk) is regulated by a hierarchy of factors, P the physical environment and Q, the resource quality acting through O. Limits to the rate of decay are set at each level, within which the factors at the succeeding level must operate. The changing state of the resource exerts feed-back effects which can change the setting of the regulatory factors (Swift & Heal 1986, after Swift, Heal & Anderson 1979).

further influence on the rate, but one determined and limited by the higher level factors (Swift & Heal 1986). The hierarchical structure is not rigid, however; lower levels may influence those above as shown by the feedback arrows. As decomposition proceeds, the changing state of the resource inevitably results in a shift in influence of both P and Q.

Odum & Biever (1984) have recently argued for the fundamental importance of resource quality as a determinant of food chain structure. In the following discussion the case is made that it is also a major determinant of the organization of decomposer communities.

THE DECOMPOSER COMMUNITY

Terrestrial decomposer communities are taxonomically diverse; prokaryotes, protists, fungi and a wide range of invertebrate phyla all being represented by high species numbers in most habitats (Wallwork 1976; Swift, Heal & Anderson 1979; Doube Chapter 12). The most useful trophic model, given the difficulties of applying conventional concepts of trophic structure, is that of Heal & MacLean (1975), which envisages cycles of processing of detritus, with transfer of material from microbe to animal at each pass. The model may be unravelled into a linear form but the positioning of individual groups or species of decomposers on such food chains remains uncertain. This suggests the possibility of more than one trophic system, which led Heal & Dighton (1985) to propose a sub-division of the decomposer community on the basis of size. Three sub-communities were suggested. The *microtrophic sub-community*, characterized by bacteria, protozoa and nematodes, lives largely in water films at the surface of roots or soil particles and utilizes either soluble organic compounds or food eroded from the surface of the particles (see Clarholm (1984) and Elliott *et al.* (1984) for reviews). In contrast the base of the *mesotrophic* and *macrotrophic sub-communities* lies in organic resources that are penetrated and exploited by fungi. The fungi are fed upon either by mesotrophic grazers, which consume hyphae at resource surfaces and in inter-particle space, or by macrotrophs which ingest detrital particles, usually after fungal colonization and proliferation. In these two systems bacteria tend to be characteristic of the later stages of decomposition and are often associated with senescent or dead fungal tissues and the products of animal comminution.

The present review is concerned with the organization of the fungal component of meso- and macrotrophic sub-communities. It is argued that the fungi, as primary colonizers of detritus and major agents of initial catabolism, act as determinants of the organization of the animal and prokaryote components of the community. These organisms in their turn affect the structure of fungal assemblages by their interactive effects and will therefore be treated as components of the biotic environment of the fungi.

ORGANIZATION IN SPACE

The community mosaic

The total census of a fungal assemblage in any habitat is difficult, if not impossible, because of the selective nature of all available techniques (Parkinson 1982; Swift 1982b). Nonetheless it is apparent that all ecosystems are rich in decomposer fungi (Hawksworth 1976; Swift 1976). Within a given vegetation type the fungi are partitioned between the different types and species of resource; the fungi of leaf litter are substantially (though not entirely) different to those of branches, roots or faecal pellets. This partitioning into a mosaic of different fungal species associations has been attributed to the selective influence of resource quality (Swift 1976). A distinction can be drawn between resource specific (RS) fungi (in which activity is restricted to particular types of resource) and resource non-specific (NS) species (which contribute to the decomposition of a variety of resources). Furthermore it may be possible to distinguish between selection operating at the resource level and that operating at the level of species of the same type. The limits and character of such specificity in the decomposer fungi have, however, not been well-defined and warrant further investigation.

Resource specificity

The acknowledged attributes of resource quality include varying levels of nutrient element content, the extent of lignification, the content of readily-metabolized carbon sources, and the spectrum of allelopathic and other modifier compounds (Swift, Heal & Anderson 1979). Resources of low quality (e.g. low nutrient concentrations) can be characterized as 'stressed' habitats in the sense used by Grime (1977, 1979) for higher plants, i.e. where stress is a condition that limits the productivity of an individual or a population below its genetic potential. The concept of adversity advocated by Southwood (1977) for animal populations is an equivalent term.

A number of authors have pictured the partitioning between different resource types as the operation of selection for stress tolerance (Pugh 1980; Cooke & Rayner 1984; Heal & Ineson 1984). This is constructive in that it leads to predictions concerning the characteristics of the occupant fungi. Classification of resources as stressed or non-stressed is, however, character specific as, although a resource may be stressed in relation to one factor, other factors may be initially non-limiting. Furthermore, some of the characters mentioned are relatively broad in influence (e.g. low nutrient content) but others are specific (e.g. the presence of particular allelopathic molecules). Thus, stress only provides a partial explanation of selection. For instance, wood-decay fungi can be described as stress-tolerant with respect to high C:N ratios, contents of lignin and refractory polysaccharides. This explains the dominance of certain Hymenomycetes which show tolerance or even preference for these factors. It offers no explanation, however, for the absence of others (of apparently similar capacities) or the presence of many Hyphomycetes and Zygomycetes which lack many of the presumptively adaptive characteristics of the Hymenomycetes. The key question may be not so much why a particular species is enabled to grow in a particular resource but rather what restricts the growth of other species. In determining the basis of apparent specificity derived from field studies it is therefore necessary to distinguish these factors from competitive and symbiotic aspects.

Community boundaries

A high degree of α-diversity characterizes fungal assemblages at the resource level. Swift (1976, 1984) proposed that this level of diversity was due in part to partitioning into component communities which he termed 'unit communities'.

The unit community was defined as a 'species assemblage which inhabits a volume of resource that is delimited in some clear and unequivocal way such that, whereas the species within the unit community may be expected to interact in some way, interactions between neighbouring unit communities will be minimized' (Swift 1984). I now propose to reconsider this concept in the light of recent developments in microbial ecology and to first consider the question of delimitation. Subdivision of assemblages into unit communities relates to two types of boundary; those due to the nature and pattern of distribution of resource units and those due to the limits of distribution of fungal mycelia.

Many fungi have the capacity to penetrate into organic resources, even of the hardest types such as wood. Others grow in the spaces between organic or mineral particles. The three-dimensional shape of the mycelium is thus determined both by its own internal regulatory mechanisms and by the confines of its immediate environment. Two strikingly different examples of this are shown in Fig. 11.2. In the one case (Fig. 11.2a) the shape and extent of the mycelium is determined by the extent of the resource and by the limits imposed by neighbouring mycelia; in the other (Fig. 11.2b) the mycelium, in the form of aggregated cords, proliferates through the interstices of the litter layer. Thompson & Rayner (1982b) demonstrated

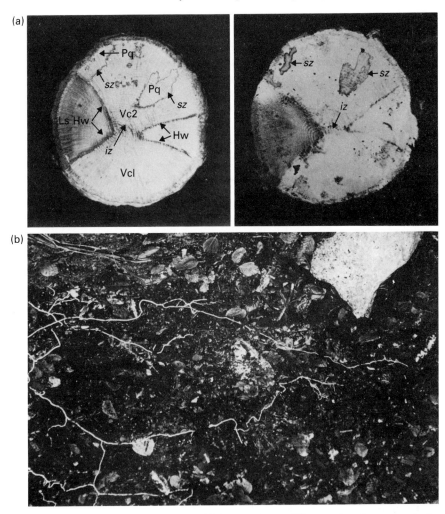

FIG. 11.2. Mycelia of wood-decay fungi. (a) Cross sections at different points along an attached branch of oak (*Quercus robur*) undergoing decomposition by two different individuals of *Vuelleminia comedens* (Vc1 and Vc2) and *Peniophora quercina* (Pq). The section on the left shows the appearance before incubation whilst that on the right shows outgrowth of mycelia after incubation. Ls = living sapwood with premature heartwood formed in wings (Hw) around it. Interactions between mycelia are clearly visible as *iz* (intra-specific zone lines) and *sz* (interspecific zone lines). (Courtesy of L. Boddy & A.D.M. Rayner.) (b) Part of a mycelial cord system of *Phanerochaete laevis* excavated from leaf litter. (From Thompson & Rayner 1982a with permission.)

by genetic analysis the existence of twenty-two mycelia of this latter type in about two hectares of woodland. The close proximity of samples of the same mycelium over fairly large areas suggested individual mycelia of considerable size, although there was no evidence to determine whether they were fragmented or continuous. In contrast the individual mycelia within a resource unit (Fig. 11.2a) may vary from the microscopic to over a metre in diameter. This discussion can be usefully furthered by considering two contrasting ecological groups, the wood decay fungi and the leaf litter fungi.

Wood decay fungi

Fallen wood is distributed in the form of discrete units—the individual branches and twigs—often with quite considerable distances between them. There is thus in these cases a clear discontinuity to the resource unit which may constitute a boundary to the distribution of occupant fungi. Insight into the spatial organization within such units has been gained during the last decade from studies by Rayner and his colleagues. When mycelia of Hymenomycetes (basidiomycetes) or ascomycetes encounter genetically distinct mycelia of the same or different species a reaction is commonly induced resulting in the formation of a 'zone line' of pigmented hyphae which is clearly visible to the naked eye (Fig. 11.2a). This reaction has been termed heterogenic somatic incompatibility (Rayner & Todd 1979, 1982; Rayner et al. 1984). It is possible, by the cutting of serial thick sections, to visualize the three-dimensional relationships of these mycelia within logs, and by microscopic examination and genetic analysis to determine their identity. Species associations within one unit of branch wood are commonly comprised of one to about five or six species of Hymenomycetes, each of which may be represented by a population of several individual dikaryotic mycelia. One or more species of *Xylareaceae* may also be associated with these basidiomycetes. Such associations have been detected in branches still attached to the tree (Boddy & Rayner 1981, 1983), after fall (Carruthers & Rayner 1979), and in stumps of both standing and fallen trees (Thompson & Boddy 1983).

Microfungi are also commonly isolated from wood dominated by Hymenomycetes. Rayner (1976) has shown the intriguing distribution of some of them, squeezed between the mycelia of neighbouring basidiomycetes, although others may intermingle with the basidiomycete mycelium. Boddy & Rayner (1984) have also reported on the fungi isolated from naturally abscissing twigs before and after fall as has Swift (1976) on those of small branches in the litter. These associations have a similar diversity but Hymenomycetes are less dominant and there is less apparent spatial demarkation between mycelia.

The unit community concept seems to hold up to scrutiny with respect to examples such as these. Deviations from this pattern are, however, evident among the wood decay fungi. Some species of wood-inhabiting Hymeno-mycetes are able to spread by means of hyphae or cords between individual units (e.g. *Phanerochaete* Fig. 11.2b). Individual mycelia of these fungi are thus not confined to single branches but may occupy several units as physiologically continuous mycelia. Resource non-specific (NS) fungi of this type may even be able to feed on other resources (e.g. leaf litter) in the intervening environment in contrast to RS types that are only able to obtain further nutriment once they have colonized a new branch.

Leaf litter fungi

Dead leaves, although particulate, commonly form a tightly-packed mat of continuous distribution. As Cooke & Rayner (1984) have pointed out, the mycelium of many leaf-decomposing Hymenomycetes (e.g. species of *Mycena, Marasmius* and *Collybia)* is formed between the leaves with decay taking place by erosion from the surface rather than by massive penetration. Mycelia of this kind are often perennial, and colonization of new resources after litter fall can occur both by spores and by hyphal growth from already established mycelia which may extend over a large area of the forest floor in a manner analogous to that of the cord-forming wood decay fungi (Swift 1982a; Frankland 1984).

This Hymenomycete assemblage is, however, associated with a large number of species of microfungi. These probably have less extensive growth, although there is regrettably little information on the distribution patterns of these mycelia as they are not easily susceptible to the methods used for the basidiomycetes. Shearer & Lane (1983) have, however, described some aspects of the spatial organization of unit communities of aquatic microfungi on individual leaves submerged in water. These fungi produce characteristic conidiospores from the leaf surface after incubation, thus enabling their location to be determined. In Fig. 11.3 more than a dozen fungi are shown to be cohabiting a single leaf. Whilst some species are spatially separated many clearly occur within the same quadrats. Nothing is known concerning the occurrence of genetically distinct individuals within this unit community, and hence of the number and size of mycelia. Nonetheless, there is a strong indication that the mycelia may in many cases have overlapping distributions rather than maintaining separate territories as in the wood decay Hymenomycetes. Chamier, Dixon & Archer (1984) have provided further evidence for mycelial intermingling in aquatic Hyphomycetes by utilizing even smaller leaf quadrats (2x2 mm).

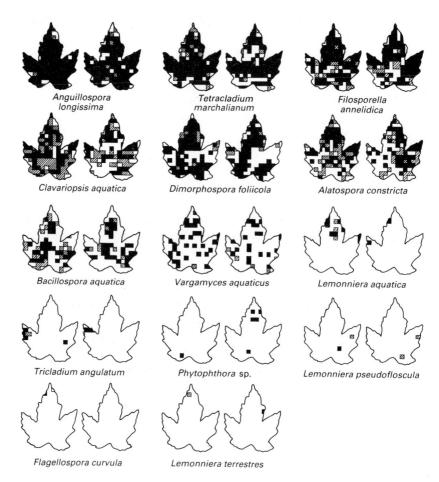

Fig. 11.3. Distribution of aquatic Hyphomycetes on a single leaf of silver maple. The map shows the presence of developing (lighter shading) or mature (heavier shading) conidia of a single species. Each square represents 6×6 mm. (From Shearer & Lane 1983 with permission.)

The picture that emerges of spatial organization in the leaf litter fungi is different to that of the wood. Some fungi show restricted mycelial growth confined by the boundaries of single resource units. This habit may be particularly characteristic of RS forms. Other fungi may show a limited extent of growth beyond the unit margin and share resource space with fungi in more than one unit. The extent of such growth for RS, but not for NS fungi will be more limited in mixed species litters than in monospecific ones. Basidiomycete mycelia may extend this unit-sharing phenomenon over a considerable distance in a manner which may be extensive horizontally but restricted vertically (Frankland 1984).

This leads to the conclusion that whereas the concept of a mosaic of clearly delimited unit communities can be clearly demonstrated for discrete resources such as branches, sharp boundaries between communities are less apparent in leaf litter. Nonetheless, there is evidence of interlocking and overlapping assemblages of varying size forming a mosaic on the forest floor. The existence of discontinuities of distribution at a variety of spatial scales does, however, seem to be a common feature of fungal assemblages.

Internal structure of unit communities

There is little precise information on levels of alpha-diversity at the resource unit scale but where this has been attempted, e.g. for branch wood (Swift 1976), submerged leaves (Chamier & Dixon 1982; Shearer & Lane 1983; Chamier, Dixon & Archer 1984), dung pats (Lussenhop *et al.* 1980), it is clear that a single unit may contain betwen one and at least eighteen species. The number of species may be related to the size of the resource unit (Sanders & Anderson 1979; Barlocher & Schweizer 1983; H.G. Wildman personal communication; Fig. 11.4). In studies on aquatic Hyphomycetes Sanders & Anderson (1979) and Chamier *et al.* (1984) reported a consistent association of six or seven species on most units, with an additional set of species which was distributed more randomly between units. Swift (1976)

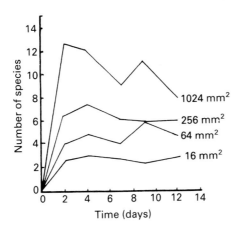

FIG.11.4. Colonization patterns of soil fungi over time. The numbers of fungi coexisting on squares of cellophane of different sizes are shown from the time of placement in the soil. (Unpublished work of Dr. H.G. Wildman with permission.)

also found considerable variation in species-composition of small branches on the forest floor.

This evidence for high species-packing into relatively small volumes of resource inevitably raises the question of how coexistence is maintained. This could be the consequence of strict niche partitioning. Alternatively it could be representative of a less deterministic situation in which a non-equilibrium and/or unstable state is maintained in which species, which may or may not be ecologically equivalent in terms of niche requirements, are enabled to coexist transiently. Any particular unit community may occupy a position on a spectrum of possibilities between these two extremes. The pros and cons of these two types of model will be discussed later. It is appropriate at this juncture, however, to review briefly some aspects of niche determination and competitive and other interactions between decomposer fungi.

Niche partitioning

Swift (1976) pictured the niche space available to decomposer fungi as being determined by two categories of factor: the physical environment and the range and character of chemical substrates (Fig. 11.5); further partitioning may be related to a variety of symbiotic associations between fungal species.

The need to clearly define functional groups among the decomposer organisms has recently been emphasized by Swift & Heal (1986). Early ideas suggested that clear distinctions in terms of enzymatic capability could be made taxonomically (Garrett 1951) but there is now increasing evidence that these characteristics are much more widely distributed than had been previously suspected (Thompstone & Dix 1985). The enzymatic spectrum is only one of the many characteristics that will determine the ecological success or failure of an organism in a given circumstance. Although there is a wealth of physiological information on decomposer fungi, very little progress has been made in defining spectra of adaptively significant features (but see Harper & Webster 1964 for an exception; see also the discussion below on selection models). For instance, Swift & Heal (1986) pointed out that we know little of the relative efficiency in decomposition by various combinations of species as compared with individuals. Many Hymenomycetes are quite capable of the total decomposition of a piece of wood and different species show varying capacities in this respect (e.g. in the rates and sequences of substrate utilization). What are the consequences when they act in concert as members of a community? What are the consequences of the addition of microfungal species to these basidiomycete assemblages?

The relationship between species-richness and unit size (Fig. 11.4) has some interesting implications for niche partitioning. One of the most obvious explanations for a greater species diversity on larger resources is the presence of a greater niche diversity. In the cases cited, however, there are unlikely to have been any differences between units of different size in the substrate niches because the materials are homogeneous in that respect; it is similarly difficult to perceive any widening of the range of physical environments in the regular-shaped units used. The most likely explanations seem to lie either in the increase in the number of biotic niches or in the alternative hypothesis of non-equilibrium dynamics.

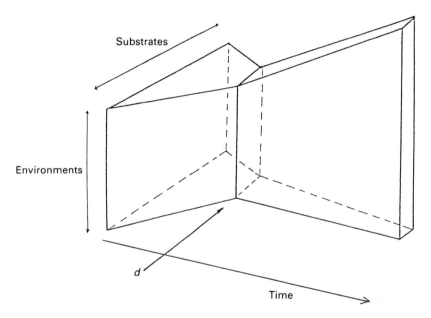

FIG.11.5. Niche partitioning in decomposer fungi. Niche space within a resource is shown as defined by the two axes of substrate and environmental diversity. The space changes with time as the character of the resource is altered by the processes of decomposition. The transverse arrow (*d*) indicates the point in time of intervention of microbivorous or detritivorous invertebrates. (From Swift 1976.)

A variety of symbiotic associations between species of decomposer fungi have been demonstrated or postulated (Cooke & Rayner 1984; Swift & Heal 1986). These range from strictly unidirectional benefits to mutualism. The observed diversity in unit communities may be due to a considerable extent to the existence of such relationships but there is little specific information on which to base such an assertion (Swift & Heal 1986).

Interspecific competition

Within any deterministic model of niche partitioning, coexistence is classically seen as the end product of competition. The topic of competition within fungal assemblages has been frequently reviewed (Clark 1965; Frankland 1981; Lockwood 1981; Wicklow 1981) but it is pertinent to restate some of the major features.

As competitive interactions occur between fungi either at the surface of, or actually within, resource units which they seek to occupy jointly, competition for resources must also necessarily involve competition for space. Arguments concerning distinctions between these types of competition are therefore irrelevant (Clarke 1965; Lockwood 1981). In a penetrating review of competitive and other interactions in the fungi, Rayner & Webber (1984) have emphasized the importance of distinguishing between the events of initial colonization of 'virgin' resources which, following Cooke & Rayner (1984), they term *primary resource capture*, and the competitive interactions occurring later in time which they call *combat*. Success in primary resource capture may depend on such factors as inoculum potential, the lag-time for germination of spores, the rate of hyphal extension, the ability to penetrate the resource, the range of extra-cellular enzyme capacity, and tolerance of allelopathic compounds or other features of initial resource stress. These adaptive features formed the basis of Garrett's (1956) concept of 'competitive saprophytic ability', although he did not distinguish between primary resource capture and combat.

The extent of intermycelial contact that occurs during initial colonization probably varies greatly and is determined primarily by the relationship between inoculum load and the surface area available for colonization. When mycelia come into contact following penetration of the resource then the phenomena associated with combat come into play. Here Rayner and Webber distinguish between *'defence'* and *'replacement'* (= *secondary resource capture*).

The collective term 'antagonism' has been used for mechanisms whereby some fungi are able to penetrate resource volume occupied by the mycelium of another species and replace it. These include both non-contact (e.g. antibiosis) and contact (e.g. hyphal interference) mechanisms. Although a good deal is known about the physiology of these interactions from pure culture studies, it has proved very difficult to establish their significance under natural conditions (Rayner & Todd 1979; Frankland 1981; Wicklow 1981). The ability to resist the action of a fungus of known antagonistic capability can also be readily demonstrated in culture but very little is known of the physiological basis of defence.

ORGANIZATION IN TIME

The most persistent concept in fungal ecology has been that of succession. Analysis of temporal patterns of the composition and abundance of fungi on different resources has been interpreted in terms of seral replacement determined by change in the chemical composition of the resource as decomposition proceeds. A number of predictions have been advanced distinguishing the patterns of decomposer successions from those of autotrophs (Swift 1982b; Campbell 1985). Whilst the weight of evidence is persuasive some doubt has been cast on the concept of directional succession in decomposer communities.

As Swift (1976) and Cooke & Rayner (1984) among others have pointed out, succession in the strictest terms must involve the replacement of the mycelium of one species by another at the same location, i.e. within the same resource unit. This has rarely been demonstrated unequivocally, largely due to deficiencies of both experimental design and method. In particular it has proved difficult to separate variations in community composition due to spatial effects from those attributable to temporal change. This problem can be illustrated by reference to a commonly-stated generalization: that the differences in community organization observed on a vertical scale in the soil and litter horizons represent changes due to time and the progress of decomposition. This is undoubtedly true to some extent; the contents of the fermentation and humus layers are the terminal stages of a leaf decomposition process initiated some time in the past in the litter layer. But these layers also consist of the residues of root decomposition. Furthermore, they clearly do not represent stages in the decomposition of the same units or set of units as those in the current layers above them. Differences between horizons may therefore reflect spatial differences (e.g. between assemblages from different resource units) as much as between times. Interpretations of successional patterns derived in other ways are equally subject to bias (see Swift 1976, 1982b; Parkinson 1982; Cooke & Rayner 1984 for discussion). Because of these reservations, interpretations of sequences of particular sets of species must be viewed with caution.

Arguments based on more general trends (e.g. changes in species-richness with time) may, however, be less subject to bias. In the earliest stages of decomposition the community is generally characterized by a relatively low species number which soon rises to a maximum (see Fig. 11.4). Pugh (1980) has designated these as the 'open' and 'closed' stages of community development respectively and raised the interesting question as to whether the 'closed' community is a stable climax or whether species replacement goes on whilst a near constant diversity is maintained.

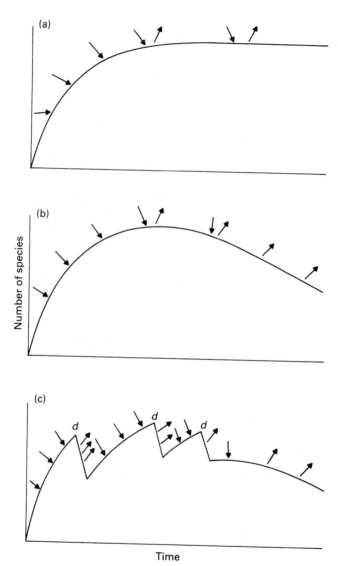

FIG.11.6. Changes in species diversity of island communities. Arrows indicate immigration or emigration of species. (a) Conventional islands; emigration and immigration balanced once community closure has been attained. (b) Decomposer resource islands showing a decrease in the available niche space; emigration exceeds immigration during the latter stages. (c) Resource islands subject to intermittent disturbances (*d*) and showing waves of emigration and immigration. (From Swift 1984.)

Decomposer resources show some obvious resemblances to islands. Classic island biogeography theory (e.g. MacArthur & Wilson 1967; Diamond & May 1981) predicts that species replacement is a continuous process with immigration and emigration remaining broadly balanced in the absence of major disturbing factors. This case is illustrated in Fig. 11.6a. There is no published evidence from which to test this for decomposer 'islands' although the unpublished results of Wildman (Fig. 11.4) were consistent with this interpretation.

Decomposer resources differ from conventional islands in one major respect however—they progressively diminish in 'size'. As decomposition proceeds the volume of resource available for colonization and occupation is eroded. For this reason Swift (1984) proposed a modification of the biogeographic model (Fig. 11.6b). This model presupposes a third phase of community development, one of decline in species-richness associated with diminishing niche space. This pattern is largely confirmed by observation where the species number of the terminal stages of decay is usually less than that of mid-decay (Swift 1976; Pugh 1980).

MODELS OF CHANGE IN SPACE AND TIME

Despite the problems of interpretation described above a number of authors have constructed models of change in fungal assemblages which provide a predictive framework for experimentation into the factors underlying the observed or supposed patterns. The models can be grouped into four categories. The first three of these—the nutritional model of Garrett (1951); the niche model of Swift (1976); and the selective models proposed by Pugh (1980) and Cooke & Rayner (1984)—are relatively deterministic in character. Their main features are summarized in Fig. 11.7. They can all be seen as attempts to explain the type of pattern described in Fig. 11.6b). The fourth model, that proposed by Swift (1984), is in contrast less deterministic and is based on the assumption that fungal assemblages are rarely at equilibrium.

The nutritional model

Garrett provided a great stimulus to the study of decomposer communities by proposing that fungal succession (Fig. 11.7a) could be interpreted in physiological terms based on the selective effect of a sequence of chemical changes induced in the resource by the activity of the occupant fungi (Garrett 1951, 1956, 1963). Subsequent study (Harper & Webster 1964; Hudson 1968; Swift 1976) has shown however that this pattern cannot be upheld.

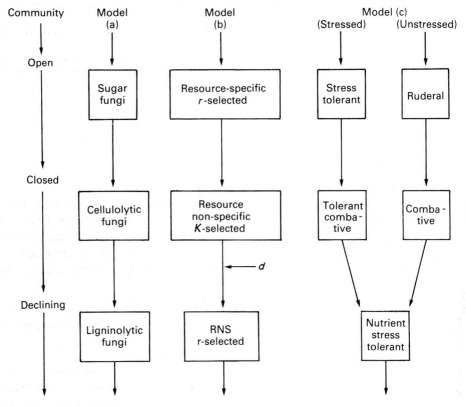

FIG. 11.7. Three models of change in community organization: (a) nutritional model after Garrett (1951); (b) niche partitioning model after Swift (1976); (c) selection model after Cooke & Rayner (1984).

The niche model

Swift's (1976) model of niche partitioning described earlier (Fig. 11.5) also depicts the availability and diversity of substrates diminishing with time and predicts a selective advantage for 'combative' (*sensu* Cooke & Rayner 1984) and resource non-specific species in the closed and early decline phases of community development. In contrast the open community stage is predicted to favour resource-specific species with opportunistic behaviour in terms of dispersal and colonization ability (Fig. 11.7b). It was also pointed out that the final stages of decomposition are commonly strongly influenced by the intervention of invertebrate animals in a process that was previously dominated by fungi. Mesofauna grazing directly on fungal hyphae, or macrofauna consuming resources inhabited by fungi, have been shown to reduce the biomass of resident fungi (Hanlon & Anderson 1980). Dependent on the intensity of the grazing, conditions may be created that favour recolonization by different fungal species or even a switch to a bacteria-dominated flora (Parkinson, Visser & Whittaker 1979; Lussenhop *et al.* 1980; Newell 1984a,b; Swift & Boddy 1984).

The intervention of invertebrates (d) is shown in the niche model by the break in the curves (Fig. 11.5). Whilst substrate diversity continues to decline after animal intervention, environmental diversity may actually increase due to the comminution of the resource by the feeding activity of the animals. The particulate structure so formed, being more environmentally varied than the massive resources of the earlier stages, was predicted to favour a third phase of organisms including many Mucorales and Fungi Imperfecti as well as bacteria. All these organisms were suggested to share an r-adapted strategy in comparison with the K-adapted combative species that precede them (Fig. 11.7b). Interventions of this kind can also be seen as examples of what Grime (1977, 1979) has defined as 'disturbances' (i.e. any factor that reduces biomass) and changes in the fungal assemblage as responses to disturbance.

Little evidence has been brought forward to confirm or refute the predictions of this model but a number of important modifications have been made as described in the following paragraphs.

Selection models

Many recent discussions of temporal changes have focused on grouping fungi into categories based on their adaptive characteristics and setting these in relation to varying conditions of their habitats. An initial proposal of this kind was that of Swift (1976) for differentiation of r and K strategies in relation to the niche model (Fig. 11.7b). The inappropriateness of this was pointed out by Wicklow (1980) and accepted by Swift (1982a) who therefore later (Swift 1984) proposed the use of more neutral terms to describe the different behaviours by fungi in response to the conditions described in that model. These were termed the 'window' and 'survival' strategies respectively.

Pugh (1980) adopted a different approach and advocated that fungi should be considered within the stress and disturbance matrix described by Grime (1977). On this basis a number of distinct strategies could be distinguished among the decomposer fungi, i.e. those relating to *combative* (*K*-related, characteristic of non-stressed and non-disturbed habitats), *ruderal* (*r*-related, characteristic of disturbed habitats) and *stress-tolerant*

(*A*-adversity-related, *sensu* Southwood 1977, characteristic of stressed habitats). These ideas have been discussed in detail by Heal & Ineson (1984), Cooke & Rayner (1984) and Rayner & Webber (1984). In the latter two papers predictive models have been developed for changes in the selective circumstances during community development (Fig. 11.7c). The resource is pictured as constituting a selective habitat definable within a stress by disturbance matrix. For instance, dung is an unstressed (in terms of nutrients) but disturbed habitat, and thus selects for ruderal species. Wood in contrast is a stressed resource and selects for stress-tolerant species. As decomposition proceeds so the selective features change; the model predicts a general progressive switch to combative species as the community closes but stress-tolerant species dominate in the terminal stages of decomposition as carbon sources are exhausted. The exact shift between strategies is variable according to the particular pattern of stress at each stage of decomposition and in response to the intervention of disturbance.

This model provides insight into the possible patterns of change in selective pressures with time and offers the opportunity of devising experiments to test its predictions. It also suggests some important modifications that need to be made to the stress by disturbance matrix as initially adopted by Pugh (1980) and others. Firstly, a distinction should be made between what may be termed '*intrinsic*' and '*extrinsic*' types of stress. The former includes those factors identified earlier as characteristic of a low resource quality. Stress which is imposed extrinsically is largely physical in nature (e.g. limiting levels of temperature or moisture availability) and generally varies independently of the type of resource. Yocom & Wicklow (1980) provided a good example of this when they showed how the composition and diversity of dung assemblages varied in relation to a gradient of external environmental stress.

Secondly, initial stress and change in stress need to be explicitly distinguished as selective features. During decomposition the intrinsic stress character of the resource invariably changes. For instance the C:N ratio narrows as C is dissipated and N conserved; thus a resource that was initially N-limited may no longer be so, implying an improvement in the stress condition. By virtue of the same processes, however, the availability of carbon sources may become limiting, implying an increase in stress. The simultaneous occurrence of increase and alleviation of stress during decomposition makes it difficult to predict what selective effects will be characteristic of a given stage of decay, except in relation to specifically defined circumstances. This not only suggests that the matrix should be extended, but also cautions against categorizing organisms into a small number of groups in general terms such as 'stress tolerant'.

This relates back to the caution urged by Wicklow (1980) and Swift (1982a) concerning *r* and *K* strategies. Most fungi have multi-stage life histories, each stage differing in its physiological characteristics. This enables an organism to respond to changing environmental circumstances. It also means that a single species may be able to adopt more than one ecological strategy. For instance, a Mucoraceous species in its initial growth may seem entirely ruderal (*r*-adapted) with rapid growth of hyphae to annex open resource space followed immediately by prolific asexual sporulation and dispersal. It may also be capable of adopting a much more *K*-type (defensive) strategy by the formation of resistant spores, either vegetatively (chlamydospores) or through sexual reproduction (zygospores). It should be emphasized that Cooke & Rayner (1984) did not make any taxonomic predictions in terms of their model, being well aware that ecological strategies should be seen as responses to habitat demands and not as genetically rigid characterizations of fungal species. It is nonetheless necessary to caution against this, for the distinction has not always been as clearly made. Grubb (1985, 1986) has argued against interpreting higher plant communities in terms of rigidly defined categories of ecological strategy; the same strictures apply to fungi.

Non-equilibrium model

All the above models assume a dynamic situation which nonetheless tends towards an equilibrium relationship between coexisting species. Above all it assumes a relatively consistent community composition as the most probable outcome of a particular set of circumstances.

An alternative is to assume that such a competitively-derived equilibrium is rarely or never attained. Arguments in favour of non-equilibrium models for community organization have gained favour in various areas of ecology. The model advanced by Shorrocks & Rosewell for instance (Chapter 2) has considerable relevance to fungal communities because of its applicability to island mosaics; Huston's (1979) 'dynamic competitive equilibrium' explanation for species diversity is similarly relevant. In such non-equilibrium models species may coexist that ostensibly should compete for the limited resources. Swift (1984) has suggested that such a situation might be maintained by the impact of frequent disturbances on fungal assemblages. Both the niche and selection models recognize the influence of disturbance; the main distinction is therefore quantitative; of what type, how intensive and how frequently does disturbance occur? No definitive answers can be given to these questions but a few pointers can be advanced.

Disturbance, by its very nature, is largely extrinsic in origin.

Disturbances may take many forms, such as grazing by microbivores, consumption by detritivores, fire, freezing, and other extreme and rapid changes in physical conditions such as desiccation or waterlogging. These events may occur at the initiation of decay and influence the pattern of colonization of a resource. More commonly, they intervene during the later stages of decomposition. Some disturbances (e.g. that of litter fall) occur invariably, although the exact timing in relation to community development is variable in relation to varying factors of the external environment. Many types of disturbance are indeed completely stochastic in relation to time in contrast to the generally progressive pattern of change that characterizes stress. This means that the ability of an organism to respond to disturbance may be quite different to its capacity to cope with a progressive stress of the same kind (e.g. temperature change) even though quantitatively they may attain similar levels.

Swift (1984) argued that disturbance of varying intensity was indeed a common feature of many fungal habitats. Fungi respond in differing ways to such disturbance. These habitats might therefore be characterized by waves of extinction/emigration followed by re-colonization as pictured in Fig. 11.6c). The diversity of such habitats need not be lower than those of equilibrium communities (Fig. 11.6b) as the dynamic nature of the system would enable species that would normally compete to coexist transiently, in the manner described above. This model would account for the apparent coexistence of ecologically equivalent species of fungi in some decomposer habitats (Swift 1976).

In so far as it is possible to perceive clear patterns in fungal sequences it seems evident that fungi with ruderal adaptations (e.g. Mucorales and Fungi Imperfecti) occur throughout the progress of decomposition and are by no means confined to the early stages. This is exactly as would be predicted for habitats in which disturbance plays an important role in determining structure.

CONCLUSION: THE HABITAT TEMPLET

Taken at any one time fungal assemblages present a picture of a highly organized structure in space. The major determinant of this organization is the nature and distribution of the resources they occupy. The structure can be seen as having a hierarchical nature, with the unit communities forming the base; these can be grouped to form the assemblages characteristic of the different species and types of resource; the fungal association characteristic of any ecosystem or vegetation type (Christensen 1981) is the sum of all the resource assemblages. This gives, however, a misleadingly static impression. The structure is constantly changing, the dominant agents of change being the animal members of the decomposer community whose distribution is less related to the distribution of resources than to environmental factors that operate at somewhat coarser scales in space and time (Anderson 1975, 1978). The fungal mosaic can thus be seen as operating within a larger mosaic determined by the meso- and macro- scale activities of the fauna.

There has been much discussion concerning the relative merits of deterministic and stochastic models of succession (Drury & Nisbet 1973; Connell & Slatyer 1977; Finegan 1984) and of the importance of competition in moulding community organization (Connell 1980, 1983; Roughgarden 1983). The outcomes of these debates are no more clear from analysis of fungal communities than for any other. What fungi may offer, however, are uniquely manipulable tools for testing the predictions of various models within the rigid protocols advocated by some authors. Moreover, there is little necessity to invoke a mutually exclusive outcome to the debate as far as fungi are concerned. It is highly probable that the models are differentially applicable to resources of different types. A massive and somewhat inaccessible resource such as the trunk of a tree has a stable internal environment, and is less susceptible to disturbance than other resources such as leaves, faecal pellets or soft fruit. Whilst developments in the latter resources may be largely at the mercy of external and generally stochastic events, those in the former may be determined by interspecific competition once the early stages of colonization have passed. The predictions for community structure and change for different resources are therefore different, allowing further opportunity for experimental testing.

The structure established by the model given in Fig. 11.1 sets P and Q as regulators of the activities of the decomposer community. Together with the influence of other organisms these factors define what Southwood (1977) has called 'the habitat templet' for decomposer fungi. Despite reservations expressed above against the too rigid application of this form of analysis, it nonetheless provides a framework for devising experiments to test the predictions of ecological theory. But perhaps the most telling point to emerge from a review of literature on fungal community organization is the danger of transgressing such axioms as the incompatibility of data gathered from investigations at different scales in time and space. Burges (1960) schooled microbial ecologists against this long ago; sadly the lesson still remains to be learned.

ACKNOWLEDGMENTS

I am most grateful for the comments of Drs Bruce Campbell, Jo Anderson and the editors on earlier versions of the manuscript and to Dr Howard Wildman for permission to use his unpublished material.

REFERENCES

Anderson, J.M. (1975). Succession, diversity and trophic relationships of some soil animals in decomposing leaf litter. *Journal of Animal Ecology*, **44**, 475–95.

Anderson, J.M. (1978). Inter- and intra-habitat relationships between woodland *Cryptostigmata* species diversity and the diversity of soil and litter microhabitats. *Oecologia*, **32**, 341–8.

Barlocher, F. & Schweizer, M. (1983). Effects of leaf size and decay rate on colonisation by acquatic hyphomycetes. *Oikos*, **41**, 205–10.

Boddy, L. & Rayner, A.D.M. (1981). Fungal communities and formation of heatwood wings in attached oak branches undergoing decay. *Annals of Botany*, **47**, 271–4

Boddy, L. & Rayner, A.D.M. (1983). Ecological roles of basidiomycetes forming decay communities in attached oak branches. *New Phytologist*, **93**, 77–88.

Boddy, L. & Rayner, A.D.M. (1984). Fungi inhabiting oak twigs before and at fall. *Transactions of the British Mycological Society*, **82**, 501–5.

Burges, A. (1960). Time and size as factors in ecology. *Journal of Ecology*, **48**, 273–85

Campbell, R. (1985). *Plant Microbiology*. Edward Arnold, London.

Carruthers, S.M. & Rayner, A.D.M. (1979). Fungal communities in decaying hardwood branches. *Transactions of the British Mycological Society*, **72**, 283–9.

Chamier, A.C. & Dixon, P.A. (1982). Pectinases in leaf degradation by aquatic hyphomycetes I: The field study. The colonisation-pattern of acquatic hyphomycetes on leaf packs in a Surrey stream. *Oecologia*, **52**, 109–15.

Chamier, A.C., Dixon, P. & Archer, S.A. (1984). The spatial distribution of fungi in decomposing alder leaves in a freshwater stream. *Oecologia*, **64**, 92–103.

Christensen, M. (1981). Species diversity and dominance in fungal communities. *The Fungal Community: Its Organisation and Role in the Ecosystem* (Ed. by D.T. Wicklaw & G.C. Carroll), pp. 201–32. Marcel Dekker, New York.

Clarholm, M. (1984). Heterotrophic, free-living protozoa: neglected microorganisms with an important task in regulating bacterial populations. *Current Perspectives in Microbial Ecology* (Ed. by M.J. King & C.A. Reddy), pp. 321–6. American Society for Microbiology, Washington.

Clark, F.E. (1965). The concept of competition in microbiology. *Ecology of Soil-borne Plant Pathogens* (Ed. by K.F. Baker & W.C. Snyder), pp. 339–47. University of California Press, Berkeley.

Connell, J.H. (1980). Diversity and the coevolution of competitors, or the ghost of competition past. *Oikos*, **35**, 131–8.

Connell, J.H. (1983). On the prevalence and relative importance of interspecific competition: evidence from field experiments. *American Naturalist*, **122**, 661–96.

Connell, J.H. & Slatyer, R.O. (1977). Mechanisms of succession in natural communities and their role in community stability and organisation. *American Naturalist*, **111**, 1119–44.

Cooke, R.C. & Rayner, A.D.M. (1984). *Ecology of Saprotrophic Fungi*. Longman, London.

Diamond, J.M. & May, R.M. (1981). Island biogeography and the design of native reserves. *Theoretical Ecology: Principles and Applications*, 2nd edn (Ed. by R.M. May), pp. 228–52. Blackwell Scientific Publiations, Oxford.

Drury, W.H. & Nisbet, I.C.T. (1973). Succession. *J. Arnold Arboretum,* **54,** 321–68.

Elliott, E.T., Coleman, D.C., Ingham, R.E. & Trofymow, J.A. (1984). Carbon and energy flow through microflora and microfauna in the soil subsystem of terrestrial ecosystems. *Current Perspectives in Microbial Ecology* (Ed. by M.J. King & C.A. Reddy), pp. 424–33, American Society for Microbiology, Washington.

Finegan, B. (1984). Forest succession. *Nature,* **312,** 109–14.

Frankland, J.C. (1981). Mechanisms in fungal successions. *The Fungal Community: Its Organisation and Role in the Ecosystem* (Ed. by D.T. Wicklaw & G.C. Carroll), pp. 403–26. Marcel Dekker, New York.

Frankland, J.C. (1984). Autecology and the mycelium of a woodland litter decomposer. *The Ecology and Physiology of the Fungal Mycelium.* (Ed. by D.H. Jennings & A.D.M. Rayner), pp. 241–60, Cambridge University Press, Cambridge.

Garrett, S.D. (1951). Ecological groups of soil fungi; a survey of substrate relationships. *New Phytologist,* **50,** 149–66.

Garrett, S.D. (1956). *Biology of the Root-Infecting Fungi.* Cambridge University Press, Cambridge.

Garrett, S.D. (1963). *Soil Fungi and Soil Fertility.* Pergamon, Oxford.

Grime, J.P. (1977). Evidence for the existence of three primary strategies in plants and its relevance to ecological and evolutionary theory. *American Naturalist,* **111,** 1169–94.

Grime, J.P. (1979). *Plant Strategies and Vegetation Processes.* Wiley, Chichester.

Grubb, P.J. (1985). Plant populations and vegetation in relation to habitat, disturbance and competition: problems of generalisation. *The Population Structure of Vegetation,* (Ed. by J. White), pp. 596–621. Junk, Dordrecht.

Grubb, P.J. (1987). Some generalizing ideas about colonization and succession in green plants and fungi. *Colonization, Succession and Stability.* (Ed. by A.J. Gray, M.J. Crawley & P.J. Edwards), pp. 81–102. Symposia of the British Ecological Society, 26. Blackwell Scientific Publications, Oxford.

Hanlon, R.D.G. & Anderson, J.M. (1980). The influence of macro-arthropod feeding activities on microflora in decomposing oak leaves. *Soil Biology and Biochemistry,* **12,** 255–61.

Harper, J.E. & Webster, J. (1964). An experimental analysis of the coprophilous fungal succession. *Transactions of the British Mycological Society,* **47,** 511–30.

Hawksworth, D.L. (1976). The natural history of Slapton Ley Native Reserve X. Fungi. *Field Studies,* **4,** 391–439.

Heal, O.W. & Dighton, J. (1985). Resource quality and trophic structure in the soil system. *Ecological Interactions in Soil* (Ed. by A.H. Fitter, D. Atkinson, D.J. Read & M.B. Usher), British Ecological Society Special Publication No. 4. Blackwell Scientific Publications, Oxford.

Heal, O.W. & Ineson, P. (1984). Carbon and energy flow in terrestrial ecosystems: relevance to microflora. *Current Perspectives in Microbial Ecology.* (Ed. by M.J. Klug & C.A. Reddy), pp. 394–404. American Society for Microbiology, Washington.

Heal, O.W. & MacLean, S.F. Jr. (1975). Comparative productivity in ecosystems—secondary productivity. *Unifying Concepts in Ecology.* (Ed. by W.H. van Dubben & R.H. Lowe-McConnelly), pp. 89–108. Junk, The Hague.

Hudson, H.J. (1968). The ecology of fungi on plant remains above the soil. *New Phytologist,* **67,** 837–74.

Huston, M. (1979). A general hypothesis of species diversity. *American Naturalist,* **113,** 81–101.

Lockwood, J.L. (1981). Exploitation competition. *The Fungal Community: Its Organisation and Role in the Ecosystem* (Ed. by D.T. Wicklow & G.C. Carroll), pp. 319–49. Marcel Dekker, New York.

Lussenhop, J., Kumar, R., Wicklow, D.T. & Lloyd, J.E. (1980). Insect effects on bacteria and fungi in cattle dung. *Oikos,* **34,** 54–8.

MacArthur, R.H. & Wilson, E.O. (1967). *The Theory of Island Biogeography.* Princeton University Press, Princeton, New Jersey.

Newell, K. (1984a). Interaction between two decomposer basidiomycetes and a collembolan under Sitka Spruce: distribution, abundance and selective grazing. *Soil Biology and Biochemistry*, **16**, 227–33.

Newell, K. (1984b). Interaction between two decomposer basidiomycetes and a collembolan under Sitka Spruce: grazing and its potential effects on fungal distribution and litter decomposition. *Soil Biology and Biochemistry*, **16**, 235–9.

Odum, E.P. & Biever, L.J. (1984). Resource quality, mutualism and energy partitioning in food chains. *American Naturalist*, **124**, 360–76.

Parkinson, D. (1982). Procedures for the isolation, cultivation and identification of fungi. *Experimental Microbial Ecology* (Ed. by R.G. Burns & J.H. Slater), pp. 22–30. Blackwell Scientific Publications, Oxford.

Parkinson, D., Visser, S. & Whittaker, J.B. (1979). Effects of collembolan grazing on fungal colonisation of leaf litter. *Soil Biology and Biochemistry*, **11**, 529–535.

Pugh, G.J.F. (1980). Strategies in fungal ecology. *Transactions of the British Mycological Society*, **75**, 1–14.

Rayner, A.D.M. (1976). Dematiaceous Hyphomycetes and narrow dark zones in decaying wood. *Transactions of the British Mycological Society*, **67**, 546–9.

Rayner, A.D.M., Coates, D., Ainsworth, A.M., Adams, T.J.H., Williams, A.N.D. & L. Todd, N.K. (1984). The biological consequences of the individualistic mycelium. *The Ecology and Physiology of the Fungal Mycelium* (Ed. by D.H. Jennings & A.D.M. Rayner). pp. 509–40. Cambridge University Press, Cambridge.

Rayner, A.D.M. & Todd, N.K. (1979). Population and community structure and dynamics of fungi in decaying wood. *Advances in Botanical Research*, **7**, 334–421.

Rayner, A.D.M. & Todd, N.K. (1982). Ecological genetics of basidiomycete populations in decaying wood. *Decomposer Basidiomycetes: Their Biology and Ecology* (Ed. by J.C. Frankland, J.N. Hedger & M.J. Swift), pp. 129–42. Cambridge University Press, Cambridge.

Rayner, A.D.M. & Webber, J.F. (1984). Interspecific mycelial interactions—an overview. *The Ecology and Physiology of the Fungal Mycelium* (Ed. by D.H. Jennings & A.D.M. Rayner), pp. 384–417. Cambridge University Press, Cambridge.

Roughgarden, J. (1983). Competition and theory in community ecology. *American Naturalist*, **122**, 583–601.

Sanders, P.F. & Anderson, J.M. (1979). Colonisation of wood blocks by aquatic Hyphomycetes. *Translations of the British Mycological Society*, **73**, 103–7.

Shearer, C.A. & Lane, L.C. (1983). Comparison of three techniques for the study of aquatic Hyphomycete communities. *Mycologia*, **75**, 498–503.

Southwood, T.R.E. (1977). Habitat, the templet for ecological strategies. *Journal of Animal Ecology*, **46**, 337–65.

Swift, M.J. (1976). Species diversity and the structure of microbial communities. *The Role of Terrestrial and Aquatic Organisms in Decomposition Processes*, Symposia of the British Ecological Society, 17 (Ed. by J.M. Anderson & A. MacFadyen), pp. 185–222. Blackwell Scientific Publications, Oxford.

Swift, M.J. (1982a). The basidiomycete role in forest ecosystems. *Decomposer Basidiomycetes: Their Biology and Ecology*, (Ed. by J.C. Frankland, J.N. Hedger & M.J. Swift), pp. 307–38. Cambridge University Press, Cambridge.

Swift, M.J. (1982b). Microbial succession during the decomposition of organic matter. *Experimental Microbial Ecology* (Ed. by R.G. Burns & J.H. Slater), pp. 164–77. Blackwell Scientific Publications, Oxford.

Swift, M.J. (1984). Microbial diversity and decomposer niches. *Current Perspectives in Microbial Ecology* (Ed. by M.J. King & C.A. Reddy), pp. 8–16. American Society for Microbiology, Washington, DC.

Swift, M.J. & Boddy, L. (1984). Animal–microbial interactions in wood decomposition. *Invertebrate–Microbial Interactions* (Ed. by J.M. Anderson, A.D.M. Rayner & D.W.H. Walton), pp. 89–132. British Mycological Society Symposium, 6. Cambridge University Press, Cambridge.

Swift, M.J. & Heal, O.W. (1986). Theoretical considerations of microbial succession and growth strategies: intellectual exercise or practical necessity? *Microbial Communities of Soil.* FEMS Symposium, 33 (Ed. by V. Jensen, A. Kjoller & L.H. Sorensen), pp. 115–31. Elsevier, London.

Swift, M.J., Heal, O.W. & Anderson, J.M. (1979). *Decomposition in terrestrial ecosystems.* Blackwell Scientific Publications, Oxford.

Thompson, W. & Boddy, L. (1983). Decomposition of suppressed oak trees in even-aged plantations II: Colonisation of tree roots by cord- and rhizomorph-producing basidiomycetes. *New Phytologist,* **93,** 277–91.

Thompson, W. & Rayner, A.D.M. (1982a). Structure and development of mycelial and systems of *Phanerochaete laevis* in soil. *Transactions of the British Mycological Society,* **78,** 193–200.

Thompson, W. & Rayner, A.D.M. (1982b). Spatial structure of a population of *Tricholomopsis platyphylla* in a woodland site. *New Phytologist,* **92,** 103–4.

Thompstone, A. & Dix, N.J. (1985). Cellulase activity in the Saprolegniaceae. *Transactions of the British Mycological Society,* **85,** 361–6.

Wallwork, J.A. (1976). *The Distribution and Diversity of Soil Fauna.* Academic Press, London.

Wicklow, D.T. (1980). Biogeography and conidial fungi. *Biology of Conidial Fungi, Vol 1* (Ed. by G.T. Cole & B. Kendrick), pp. 417–47. Academic Press, New York.

Wicklow, D.T. (1981). Interference competition and the organisation of fungal communities. *The Fungal Community: its Organisation and Role in the Ecosystem* (Ed. by D.T. Wicklow & G.C. Carroll), pp. 351–75. Marcel Dekker, New York.

Yocom, D.H. & Wicklow, D.T. (1980). Community differentiation along a dune succession: An experimental approach with coprophilous fungi. *Ecology* **61(4),** pp. 868–80.

12. SPATIAL AND TEMPORAL ORGANIZATION IN COMMUNITIES ASSOCIATED WITH DUNG PADS AND CARCASSES

BERNARD M. DOUBE*

*c/o Dung Beetle Research Unit, Private Bag X5, Lynn East,
Pretoria, 0039 South Africa*

INTRODUCTION

Almost all studies on dung and carrion communities note the large number of species and the coexistence of many of them apparently feeding on the same food at the same time and place. The trophic and competitive interactions in these communities have been studied in an extensive but piecemeal fashion and a number of mechanisms have been proposed to account for their structure. Because such systems are non-interactive (Caughley & Lawton 1981) i.e. the activities of community members have no influence on the supply of food, they are less complex than many other communities, e.g. plant–herbivore systems, and so may be more readily comprehensible. Here I establish a spatial and temporal framework designed to facilitate analysis of the processes that give structure to these communities.

Dung pads, carcasses and other items such as fruits and rotting logs are discrete, ephemeral patches or 'islands' of highly concentrated energy which are widespread throughout a variety of habitats over the surface of the landscape. Because they contain only heterotrophic organisms, these units exist only until the energy they contain has been consumed or dispersed. The consumers in each pad or carcass produce an array of recruits that constitutes the next generation and which disperses and colonizes newly formed 'islands'. Ecological processes during decomposition of these entities, therefore, are primarily concerned with the rate and mechanisms of energy utilization and dispersal and the changes observed during this process are referred to as heterotrophic succession (Connell & Slatyer 1977). These are determined by the physical, chemical and biological characteristics of the pad or carcass (size, structure, composition) and by its physical and biological environment.

*Present address: C.S.I.R.O. Division of Entomology, P.O. Box 1700, Canberra, ACT 2601, Australia.

Most published studies of dung and carrion communities fall into one of three categories. These are studies of (i) species composition and successional processes for guild(s) (defined as a group of species which exploits the same class of environmental resources in a similar way, Root 1967; Adams 1985) or entire communities (dung: Mohr 1943; Lawrence 1954; Poorbaugh 1966; Valiela 1974; Merritt 1976; Koskela & Hanski 1977; Hanski 1983, 1986a; corpses: Howden 1950; Bornemissza 1957; Reed 1958; Payne 1965; Coe 1978; Putman 1983; Braack 1984); (ii) the seasonal and habitat associations and behaviour of species from guilds within the community (dung: Hammer 1941; Lawrence 1954; Landin 1961; Rainio 1966; Bornemissza 1976; Nealis 1977; Koskela & Hanski 1977; Lumaret 1978; Cambefort 1982; Doube 1983; Hanski 1983; corpses: Walker 1957; Johnson 1975; Katakura & Fukuda 1975; Denno & Cothran 1975; Anderson 1982); (iii) the feeding and reproductive biology of coprophagous, necrophagous and predaceous species (Payne 1965; Halffter & Matthews 1966; Putman 1983). With a few exceptions (e.g. Payne 1965; Denno & Cothran 1976; Kneidel 1984a,b) these studies have been descriptive and without an experimental component. In addition, simple species associations have been used in the analysis of trophic relations (Harris & Oliver 1979; Summerlin *et al.* 1981; Doube, Macqueen & Huxham 1986). Thus while there are descriptive accounts of the patterns of community structure and some speculation about their causes, there have been few attempts (e.g. Hanski & Koskela 1977; Hanski 1986a, 1987), to integrate the information into current ecological theory or to test or modify these theories in the light of such information.

In the following discussion I am primarily concerned with communities found within, and derived from, dung masses produced by land-dwelling mammals, and with those from carcasses or carrion in terrestrial ecosystems. I examine the characteristics of these resource patches, the way in which their communites are sampled and described, the types of organisms that comprise these communities and their trophic, spatial and temporal patterns. The processes that determine such patterns are examined in detail. I do not discuss the ecological communities associated with dung accumulations, e.g. bat guano, in feed-lots or rhinoceros middens, because such systems have regular additions of dung and so differ markedly from individual pads and corpses which usually persist for only one generation of consumers.

CHARACTERISTICS OF DUNG PADS AND CARCASSES

The spatial and temporal patterns of distribution of dung and corpses impose major constraints upon the types of organisms that colonize and consume them and hence influence the nature of dung and carrion communities. Whether mammals are migratory, nomadic or sedentary is, *a priori*, a major determinant of these patterns, but the consequences of these constraints on dung or carrion communities have been examined only cursorily (McClain 1983), and deserve further attention.

Nomadic animals provide a highly irregular local supply of dung and carcasses (McClain 1983). Migratory mammals such as the bison of North America and the wildebeest of East Africa provide, in any one region, a regular but seasonal cycle in the availability of dung and carrion. However, the majority of large mammals today are either domesticated or constrained within game parks, and so the annual and daily supply of dung and its quality is relatively predictable (Crespo & Gonzalez 1983; Matthiessen & Hayles 1983; Macqueen, Wallace & Doube 1986). In contrast the temporal supply of corpses in natural environments in Africa, and probably elsewhere, will often be irregular and unpredictable because it is strongly dependent upon sporadic natural catastrophes such as drought cycles and disease epidemics and upon man's activity, e.g. mass culling, as well as upon natural mortality which often varies with the season of the year (Putman 1983; Braack 1984). Most published information is concerned with the dung and carrion communities associated with large sedentary animals or with carcasses of small vertebrates.

The structural complexity of carcasses and faecal deposits, and hence the number of potential feeding niches present, depends largely upon their size. For example, the corpses of large animals contain conglomerates of structural material such as skin, sinew, horn and bone which provide a wider range of niches for specialist decomposers than are present in small corpses (Reed 1958; Coe 1978; Braack 1984; Kneidel 1984a). Large carcasses also provide shelter for generalist species (e.g. collembolans, blattids, isopods) which also assist in the later stages of carcass decomposition (Cornaby 1974; Johnson 1975). At the other end of the scale, the carrion community associated with dead snails and slugs consists of a relatively small number of sarcophagid and calliphorid flies (Beaver 1977; Kneidel 1984a).

The dominant consumers of dung and carcasses also change with the size and structure of corpses. For example, most large corpses are consumed primarily by vertebrate scavengers (Putman 1983; Braack 1984) whereas small mammal and bird carcasses and dead insects are often consumed by necrophagous beetles and fly larvae (Springett 1968; Beaver 1977; Young 1984a; Hanski 1986b, 1987). Similarly, dung type needs to be specified because the dung of herbivores, omnivores and carnivores attracts different arrays of colonists (Rainio 1966; Endrody-Younga 1982).

However, in some instances, dung pads and carcasses have types of prey (e.g. fly larvae) and physical and chemical characteristics in common, which leads to the presence of similar guilds and species in the two types of community. For example, in South Africa bovine dung and the rumen contents of herbivore carcasses attract similar arrays of coprophagous dung beetles (Doube 1983; Braack 1984) and species which are predators of fly larvae (e.g. histerid and staphylinid beetles) are common to both dung and carrion communities (Braack 1984 unpublished data). Similarly, omnivore and carnivore dung and carrion all have a similar stink and are often colonized by the same or similar species of anthomyiid, muscid and sarcophagid fly and scarabaeid and silphid beetle (Rainio 1966; Howden & Nealis 1975; Peck & Forsyth 1982; Hanski 1983; Ridsdill-Smith, Weir & Peck 1983; Young 1984b). However, there are many other species which are largely specific to one or other resource (*ib. sit.*).

DESCRIPTION AND SAMPLING OF COMMUNITIES

The structure of dung and carrion communities is usually described in terms of the number and relative abundance of species within taxonomic groups (e.g. staphylinid beetles, Rainio 1966; Koskela 1972) or guilds (e.g. coprophagous flies or beetles, Papp 1971, 1976; Peck & Forsyth 1982). Because species within relatively large taxonomic groups usually have similar diets (e.g. most staphylinids in the subfamily Staphylininae are predators whereas those in the subfamily Oxytelinae are considered to be coprophagous, Koskela & Hanski 1977), they can be allocated to guilds without a detailed knowledge of the feeding biology of every species. Such a procedure is, of course, fallible (Beaver 1984) but allows preliminary analysis of the way in which these communities function (Merritt 1976). Because the mass of individuals within guilds can span three orders of magnitude, biomass is also used to assess the relative abundance and ranking order of species within assemblages (Koskela & Hanski 1977; Peck & Forsyth 1982; Ridsdill-Smith & Kirk 1985).

As discussed above, it is also necessary to specify the type of dung or carrion community being considered because the physical characteristics (size, physical and chemical composition) of these resource patches usually but not always (e.g. Kuusela & Hanski 1982), has a major influence on the type and relative abundance of colonists (Cornaby 1974; Denno & Cothran 1975; Kneidel 1984a,b).

The apparent relative abundance of species encountered depends upon the sampling techniques employed, of which two have been used most commonly. One is a baited trap, the form of which varies with the type of organism being sampled. Baited traps are used to assess the patterns of colonization and the seasonal or habitat associations of species but, because traps tend to be selective and do not allow emigration, caution is needed when interpreting patterns of relative abundance of species. The second method involves evaluation (often by destructive sampling) of the fauna associated with pads or carcasses at different stages during decomposition. This gives information on species-specific patterns of colonization, departure and mortality throughout the process of decomposition and the survivors constitute the local community.

Some communities are easier to study than others. For example, in communities associated with gastropod carrion it is relatively easy to gain a complete census and by setting up many replicates to assess variation within and between habitats and seasons (Beaver 1977; Kneidel 1984a). Such detailed data for large mammal carcasses is virtually unattainable. Furthermore, vertebrate carcasses are frequently removed or largely devoured by vertebrate scavengers (Putman 1983) and so the bulk of the information on the invertebrate component of the carrion community comes from studies using carcasses protected from such scavengers.

The number of species encountered also depends upon the intensity of sampling and a complete census of species present often requires a very substantial effort. However, a less intense sampling effort will indicate the presence and relative abundance of the more common species, and this data can be used to give some indication of the way in which the communities are structured, how they function, and to predict (using Preston's veil line, Southwood 1978) the total number of species present in the community.

COMMUNITY COMPOSITION

Dung and carrion communities are characterized by abundant aerobic and anaerobic (e.g. rumen bacteria) micro-organisms, a diverse arthropod fauna and an array of vertebrates (reviewed by Putman 1983; Beaver 1984; Braack 1984; Hollis, Knapp & Dawson 1985). Apart from micro-organisms, the majority of species and individuals belong to the orders Coleoptera, Diptera, Hymenoptera and Acarina. Other invertebrate groups present include Lepidoptera, Hemiptera, Dictyoptera, Collembola and Isoptera. The vertebrates include scavenging birds, mammals, reptiles and amphibians.

The trophic relations in dung pads and corpses are similar to those in other heterotrophic systems, e.g. rotting logs or leaf litter. There are primary consumers (e.g. fly larvae, dung beetles, tineid moths, general scavengers) which consume a range of raw materials, (e.g. dung bacteria, cellulose, flesh, horn), and secondary and tertiary consumers (e.g. predatory beetles) which prey upon the primary and secondary consumers. However, these categories are not mutually exclusive for there are species of primary consumers which are facultative predators (e.g. silphid, trogid and clerid beetles, calliphorid fly larvae, vultures (Braack 1984)) or cannibals (Braack 1984) and other species which act as primary and secondary consumers at different stages in the life-cycle (e.g. some hydrophilid beetles have coprophagous adults and predaceous larvae).

Within the primary consumers there are guilds whose members consume moist flesh (sarcophagid and calliphorid fly larvae), dry flesh (phorid flies), skin (dermestid beetles), horn (tineid moth larvae), dung juices (muscid fly larvae, staphylinids), or dung juices and dung fibre (scarabaeid dung beetles). There are also guilds of secondary consumers consisting of predators of immature flies (ants, mites, fly larvae, staphylinids, histerids, larval hydrophilids) and pupal parasitoids (microhymenoptera, aleocharine staphylinids) (Merritt 1976; Koskela & Hanski 1977; Halffter & Edmonds 1982; Putman 1983; Braack 1984; Doube 1986).

The relative abundance of species within guilds or taxonomic groups has been examined for numerous assemblages, e.g. dung scarabs (Nealis 1977; Walter 1978; Hanski 1980a,b, 1983; Peck & Forsyth 1982; Doube 1983), carrion beetles (Schubeck 1969; Katakura & Fukuda 1975; Anderson 1982), staphylinids, hydrophilids and histerids (Rainio 1966; Koskela 1972; Hanski 1980a), dung-breeding flies (Papp 1971, 1976) and carrion flies (Fuller 1934; Denno & Cothran 1976; Beaver 1977; Putman 1978). The rank-abundances of species within guilds within individual pads and carcasses and at the local and regional levels of organization usually follow the general pattern detailed by Gray (Chapter 3), namely a few abundant species, some common species and many rare species (Hanski & Koskela 1977; Peck & Forsyth 1982; Cambefort 1982; Doube 1983; Hanski 1983). However, the species that comprise the abundant and common component of an assemblage often change with habitat, season and year (see later).

The structure of an assemblage of species can also be described in terms of the frequency distributions of the log abundances of species (Southwood 1978). Few authors have done so for dung or carrion assemblages but Hanski (1981, 1986a) found a bimodal distribution of log abundances and suggested that the assemblage was composed of a mixture of abundant local 'core' species and scarce non-local 'satellite' species. The niches of the core species were more evenly spaced out than were those of the satellite species and there is some evidence that many core species have persisted in local assemblages over long periods of time (up to 40000 years; Hanski 1986a).

Because there are large numbers of species in most dung and carrion communities there are numerous potential interspecific interactions. However, the majority of these occur only infrequently, if at all, because most species are relatively rare, are restricted to particular habitats in specific seasons of the year and have specific dietary requirements. Furthermore, it has been suggested that communities may consist of a series of 'loosely coupled subsystems', with strong interactions within subsystems but weak interactions between them (Begon & Mortimer 1986). Such an arrangement may account, in part, for the failure of some introduced species to affect the abundance of dung breeding pestilent flies in Australia even though the introduced species became abundant and had the capacity to cause fly mortality, e.g. dung beetles and macrochelid mites (Wallace & Holm 1983; Doube 1986; Doube, Macqueen & Huxham 1986).

The dominant primary and secondary consumers in dung and carrion communities vary with latitude, and between continents. For example, aphodid beetles and flies are the dominant primary consumers of dung in the northern latitudes of Europe whereas the geotrupid and scarabaeid dung beetles become important in southern Europe and the scarabaeid dung beetles dominate in many tropical regions of the world (Hammer 1941; Lawrence 1954; Landin 1961; Halffter & Matthews 1966; Hanski 1983, 1986a; Ridsdill-Smith & Kirk 1985; Doube 1986). Similarly the histerids are largely restricted to tropical regions (Hanski 1986b). There are also latitudinal changes in species diversity in dung in some groups. For example, species diversity in dung beetles is greater nearer the tropics (Beaver 1984, unpublished data), but for other groups, e.g. carrion flies, latitudinal gradients in diversity are not evident (Hanski 1981).

Despite these latitudinal trends there are striking similarities between regions in the composition of dung and carrion communities. For example, the dung communities in Europe, Africa, America and Asia contain representatives of most of the following groups: dung beetles (tunnellers and ball rollers), aphodids, mites phoretic on dung beetles, coprophagous and predaceous staphylinids, histerid and hydrophilid beetles, flies with coprophagous or predaceous larvae, and hymenopterous and staphylinid pupal parasitoids (Rainio 1966; Bornemissza 1976; Merritt 1976; Hanski 1983; Putman 1983; Doube 1986).

The composition of the mammal fauna in many regions has been influenced by broad-scale and/or long-term processes such as biogeographic accidents (e.g. continental drift), fire, periodic natural catastrophes and man's activities (see reviews in Bourlière 1983) and this in turn must influence the composition of the dung and carrion communities. For example, in Australia kangaroos were the primary producers of herbivore dung prior to European settlement. Since then, kangaroos have been largely replaced by cattle, horses and sheep which graze the extensive pastures that have replaced the pristine bushland. The original Australian dung fauna was dominated by beetles adapted to a bush environment and marsupial pellets but ill-adapted to dealing with bovine dung in a grassland environment (Waterhouse 1974). Bovine dung was poorly colonized by indigenous dung beetles and provided an ideal breeding medium for coprophagous flies which, in many regions, became the dominant element in the dung pad fauna. Thus the activities of European man caused major changes in the composition of the dung community in grasslands. The introduction of dung beetles and predators by CSIRO in Australia is an attempt to further alter that balance and so reduce the abundance of dung breeding flies (Waterhouse 1974; Bornemissza 1976; Doube 1986).

SPATIAL PATTERNS: THE RELEVANCE OF SCALE

The mechanisms that determine the composition and relative abundance of species change with the spatial scale being considered. Five distinct levels of spatial organization are examined here.

Level 1: The patch or 'island' community

Individual pads or carcasses contain an invertebrate community of adult colonists and their progeny. Community members can interact with each other but are more or less temporarily isolated from members of all other such communities. They differ in several respects from the geographic islands considered by MacArthur & Wilson (1967) and from habitat 'islands' on mainlands and from perennial plants considered as 'islands' (Southwood & Kennedy 1983; Giller 1984). In particular each 'island' is rarely suitable for growth and reproduction for long enough to allow colonists to complete more than one generation and so they can never reach an equilibrium (Beaver 1977, 1984; Kneidel 1984a).

The composition of these 'island' communities at any one time is largely determined by factors that influence patterns of colonization, survival and emigration. Colonization patterns are determined by factors that include (i) the abundance and activity of the species in the region which change with season, prevalent weather conditions and the recent history of the populations in the area (Kingston 1977; Tyndale-Biscoe, Wallace & Walker 1981; Beaver 1984; Ridsdill-Smith & Kirk 1985); (ii) the time of day at which the 'island' becomes available to colonists (Koskela 1979) and the time for which it remains available; (iii) the location of the pad or carcass in relation to the source of colonists, the dispersal behaviour of the colonizing species and their ability to locate new habitats (Holter 1979; McClain 1983; Hanski 1987). Patterns of survival in, and emigration from, pads and carcasses are determined by the limitations of the physical environment (e.g. desiccation) and by biotic interactions (e.g. competition, predation and mutualism) within the pad or carcass during decomposition (Springett 1968; Denno & Cothran 1975, 1976; Braack 1984; Wilson, Knollenburg & Fudge 1984; Doube 1986). There is clear evidence from field and laboratory studies on dung beetles (Holter 1979; Ridsdill-Smith, Hall & Craig 1982), dung breeding flies (Wasti, Hosmer & Barney 1975; Sands & Hughes 1976; Doube & Moola 1987) and carrion breeding flies (Denno & Cothran 1976; Hanski 1986b) that intra- and interspecific competition occurs and can result in mortality of inferior competitors. Similarly, predators and parasites (Patterson & Rutz 1986) and mutualistic interactions (see later) can affect the patterns of survival within pads and carcasses.

Hence community composition at this level can be highly variable, depending upon local circumstances, many of which are the outcome of stochastic processes. Frequently the communities from adjacent 'islands' of a similar nature and size (e.g. dung pads, snail corpses) show substantial differences in species composition and their relative abundance even when the 'islands' occur in an apparently uniform habitat (Papp 1971, 1976; Beaver 1977; Holter 1979, 1982; Kneidel 1984a; Hanski 1987). When these 'islands' are small, each harbours only a small proportion of those species present in the region (Beaver 1984). In contrast, when the 'islands' are relatively large, e.g. impala carcasses, they appear to harbour a high proportion of the potential colonizing species (Braack 1984).

Level 2: The local community

This is the standing crop of organisms that results from an integration over time and space of the recruits from pads or carcasses occurring in an area of uniform habitat in which 'edge effects' (e.g. immigration from adjacent local communities) are minimal.

Community composition at the local level is largely determined by processes that influence the number and longevity of recruits from the pads or carcasses, namely (a) interactions of species with the physical environment; (b) pattern of colonization and utilization of dung pads and corpses; and (c) interactions between individuals associated with the pad or carcass. Information on these processes (discussed below) is relevant to the debate about the relative roles of interspecific competition, predation, aggregation, and weather and other forms of disturbance as determinants of the structure of natural local communities (Schoener 1974; Connell 1978; Begon & Mortimer 1986; Hanski 1986b, 1987; Shorrocks & Rosewell,Chapter 2).

(a) The physical environment is important, in part because specialization along environmental niche axes (Schoener 1974) is highly developed amongst the members of the dung and carrion communities. The adults of most species show a degree of specialization with regard to the soil type and vegetative cover in which they occur (Walker 1957; Koskela 1972; Katakura & Fukuda 1975; Howden & Nealis 1975; Nealis 1977; Anderson 1982; Doube 1983; Fay 1986) and some groups, e.g. perching dung beetles and carrion flies, show vertical stratification as well (Howden & Nealis 1978; Braack 1984; Young 1984b). At least for some species these boundaries are sharp despite the obvious mobility of the adults. For example, Peck & Forsyth (1982) found that dung beetles in an Equadorian forest could travel hundreds of metres per day but still showed habitat specificity. Similarly, Doube (1983) found that the numbers of the dung beetle *Sisyphus seminulum* decreased from about 500 per trap to twenty per trap over 20 m at a bush-grass boundary. Most species show diel and seasonal specialization (Schubeck 1971; Johnson 1975; Hanski 1980a; Doube 1983; see later) and a degree of preference for particular types of dung or carcasses (Fincher, Stewart & Davis 1970; Gordon & Cartwright 1974; Papp 1976; Endrody-Younga 1982; Kneidel 1984b). Thus the local community needs to be defined in terms of the season, dung or carcass type, soil type and vegetative cover.

Despite this high level of niche partitioning, there are some species with similar biological characteristics that occur together. For example, in South Africa there are three dung beetle species from the genus *Onthophagus* (*O. lanista, O. tersidorsis, O. quadrituber*) which are very similar in size, coloration and morphology, show similar patterns of seasonal and diel activity, have similar preferences for soil type and degree of vegetative cover and have coexisted for at least 4 years in regions where cattle are the dominant herbivore (unpublished data). Clearly such instances require further detailed study to determine how such species coexist, as has been attempted for European hydrophilid beetles (Hanski 1980a) and carrion flies (Hanski 1987).

Changes in the relative abundance of species can also be due to differential breeding success as a result of sporadic fluctuations in the suitability of the environment. For example, brood production by adults of the dung beetle *Onthophagus granulatus* and survival of the immatures are reduced during times of drought (Tyndale-Biscoe, Wallace & Walker 1981) and so there are major year to year changes in its abundance as a response to altered weather conditions. Similarly, seasonal changes in dung quality affect the reproductive success of this species and other species of dung beetle (e.g. *Onthophagus binodis, Onitis alexis, Euoniticellus intermedius*) and of dung breeding flies (e.g. *Haematobia irritans exigua, Musca vetustissima*) (Ridsdill-Smith 1986; Macqueen, Wallace & Doube 1986) but not all species are affected to the same degree (*ib. sit.*). Soil type (Fincher 1973) and soil moisture (Barkhouse & Ridsdill-Smith 1986) can influence the reproductive success of dung beetles and hence their relative abundance in different seasons, years and localities.

(b) The patterns of colonization and utilization of dung pads and corpses, especially by species which are effective competitors for the resource, can have a major influence on the reproductive success of colonists and hence upon community composition.

The degree of independent spatial aggregation in colonizing species (Atkinson & Shorrocks 1984; Hanski 1986a; Shorrocks & Rosewell, Chapter 2) and the patchy distribution and size of pads and carcasses (Kneidel 1984b; Hanski 1987) strongly influence patterns of colonization, density per patch (hence competition) and breeding success. Intraspecific aggregation within habitats is well established for insects living in patchy and ephemeral habitats (Holter 1979, 1982; Kneidel 1984a; Hanski 1986b, 1987; Shorrocks & Rosewell, Chapter 2) and there is now a body of evidence suggesting that the independent aggregated distribution of species over discrete patches of resource favours the coexistence of competing species by reducing the intensity of interspecific competition (*ib sit.*). Furthermore Kneidel (1984a) and Hanski (1987) have produced experimental evidence showing that reduced patchiness led to competitive exclusion of some species and reduced species diversity.

Priority effects are also important determinants of reproductive success. For example, species which arrive early in the successional process and preempt resources (e.g. ball-rolling dung beetles, silphid beetles, flies which produce larvae rather than eggs) clearly have a competitive advantage over other colonists (Springett 1968; Denno & Cothran 1975, 1976; Wilson 1983). However, each of these strategies has corresponding disadvantages (Beaver 1984), for example, ovoviparious flies have low fecundity compared with egg-laying species.

(c) Interactions between individuals associated with dung and corpses can affect their breeding success; these interactions include inter- and intraspecific competition for resources, parasitism, predation and mutualism.

Intra- and interspecific competition can result in stunting, mortality and reduced fecundity of individuals (Ridsdill-Smith, Hall & Craig 1982; Hanski 1986b; Doube & Moola 1987). The effect of this on the composition of the local community will depend upon the abundance of, and the degree of spatial and temporal overlap between, potential competitors. There are many instances, especially in carrion communities (Denno & Cothran 1976; Hanski 1986b, 1987) and some tropical dung communities (Cambefort 1982; Peck & Forsyth 1982; Hanski 1983 unpublished data), in which competition appears to be intense. In these communities there is a dominance hierarchy in the competitive ability of species which in part determines community structure by reducing the abundance of competitively inferior species (Denno & Cothran 1976; Braack 1984; Hanski 1987). However, competitive superiority may not be absolute and can change with ambient temperature (Wilson, Knollenburg & Fudge 1984, see later). Furthermore, many potential competitors are partly separated in time by diel and seasonal activity patterns (Fuller 1934; Denno & Cothran 1975) and in space by a variety of mechanisms (aggregation, habitat preference, etc., see earlier) and so may not occur together in numbers sufficient for interspecific competition to determine their relative abundance most of the time. In other communities, e.g. dung communities in Europe (Hammer 1941; Holter 1982) and America (Valiela 1974) competition is considered to be, at most, a minor determinant of community structure. Hence competition can be a proximate determinant of community structure, but its overall importance varies with the circumstances and few generalizations are possible.

Exploitation of the competitive superiority of dung beetles is the rationale behind the introduction of dung beetles to Australia for the biological control of dung-breeding flies (Waterhouse 1974; Doube 1986). Success in this venture will require temporal and spatial coincidence, at the 'island' and local levels, of dung-breeding flies and dung beetles whose activity results in substantial fly mortality.

There is a high diversity of species of predator and parasite in dung and carrion communities and clearly they must influence levels of mortality and the relative abundance of species at the 'island' and local levels of community organization (Braack 1984; Doube 1986). Nevertheless their relative importance as determinants of community structure has rarely been examined. The introduction of a parasitic wasp to Mauritius caused a drastic reduction in the abundance of the dung-breeding stable fly (Greathead & Monty 1982) and the extensive studies on predators and parasites of dung-breeding flies (see reviews in Patterson & Rutz 1986) are based upon the premise that such organisms can permanently alter the patterns of relative abundance within dung communities. However, the introduction of predatory beetles and mites, parasitic wasps and dung beetles to America and Australia appear to have had little influence on the abundance of pestilent dung-breeding *Haematobia* flies (Patterson & Rutz 1985) and Fuller (1934) considered that parasites and predators were of minor importance in determining the relative abundance of carrion flies in Australia.

The mutualistic interaction between carrion beetles and their phoretic mites which prey upon immature flies can be a major determinant of patterns or mortality and community structure in 'island' communities (Springett 1968; Wilson, Knollenburg & Fudge 1984) but its contribution to community structure at the local level has not been evaluated.

Level 3: Regional species pool

This results from the combination of a series of local communities (i.e. within-habitat, or α-diversity) and so the composition of the regional species pool is entirely dependent upon the relative abundance of each habitat type within a region (taken as a broad geographic area, e.g. the Natal lowveld) and the rate of change in species across habitats (β-diversity).

The way in which one can examine species assemblages at the first three levels of scale is illustrated by some data from Mkuze Game Reserve, South Africa. Dung-baited pitfall traps were placed in adjacent areas of grassveld and bushveld on four different soil types. The mean number of species per trap varied from twenty-eight to forty-seven, which was substantially less than the number of species recorded in the habitat in which the traps were set (Table 12.1). That the number of species recorded from most habitats (Table 12.1) was a complete census at that time is indicated by the failure to

TABLE 12.1. The effect of habitat-type on the diversity of dung beetle species trapped in Mkuze Game Reserve in December 1980. A total of 116 species was trapped

Soil type and vegetative cover (sample size in brackets)	Numbers of species		
	Mean±SD per trap	Total per habitat	Total per soil type
Deep Sand			
Grass (25,157)	47±7	86	94
Bush (25,080)	45±4	75	
Duplex (sand over clay)			
Grass (17,269)	40±8	81	97
Bush (31,074)	40±7	89	
Clay			
Grass (9,364)	28±9	65	78
Bush (18,114)	32±6	68	
Skeletal loam			
Grass (7,180)	40±7	75	81
Bush (6,535)	36±4	63	

detect additional species with additional sampling (Fig. 12.1). Thus the local assemblage at that time consisted of sixty-three to eighty-nine species per habitat but the regional pool consisted of seventy-eight to 116 species depending upon the habitat composition of the region being considered.

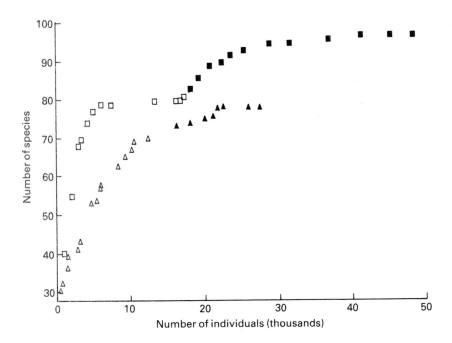

FIG. 12.1. The relationship between the number of individuals and the number of species present in a sample taken from an assemblage of dung beetles in Mkuze Game Reserve, Natal, South Africa. Each point represents the results of the addition of the data from one trap catch (number of individuals and number of new species) to the cumulative totals. The data for grassveld sites (open symbols □, △) and the sum of bushveld sites upon the cumulative total of grassveld sites (closed symbols ■, ▲) in two soil types (□■, sand over clay = duplex; △▲ deep sand) are presented. Dung baited pitfall traps were left exposed to colonists for 24 h.

Two further levels of spatial scale involve latitudinal and geographic species pools (level 4) within continents and global or intercontinental species pools (level 5) (γ-diversity). The species composition at these spatial scales appears to be associated with the carrying capacity of the environment, latitudinal gradients in species diversity (Hanski 1983; Beaver 1984) and long-term processes such as the rates of speciation, extinction and immigration (Anderson 1982).

TEMPORAL PATTERNS

The composition of the regional species pool as revealed by trapping is determined by the long-term processes just mentioned, by medium-term processes such as local extinctions caused by drought cycles, etc., and by short-term processes such as time of day, responses to weather and seasonal activity patterns. Here I examine three aspects of the temporal processes that affect the composition and relative abundance of species in dung and carrion communities at the 'island' and local levels of organization; namely (a) successional processes within individual pads or carcasses, (b) short-term changes in weather and resource supply, and (c) seasonal processes.

Succession: an island level process

The term 'succession' is used to denote the serial change in species composition in individual pads and corpses throughout the process of decomposition (Putman 1983). A series of two to eight successional phases or seres have been described for carrion communities on large carcasses (reviewed by Braack 1984) and the completion of each phase is often defined in terms of the departure of the members of a specific guild. However, small carcasses are often quickly and completely consumed by scavenging vertebrates, necrophagous beetles or flies and so the usual successional processes associated with large carcasses are abbreviated or do not occur.

Most species present in the early stages of succession are specialists in one habitat type (Mohr 1943; Beaver 1984). They usually develop rapidly and are present for only a short time. During succession there is a trend towards a greater proportion of generalist species able to live in a wide range of habitats containing dead organic matter (Mohr 1943; Elton 1966; Davidson 1979). Furthermore, the early successional stages are characterized by a high diversity of species which declines as decomposition proceeds.

Because most communities in pads and carcasses disappear after only one generation of colonists, the successional processes in these heterotrophic systems are markedly different from those observed in autotrophic systems, e.g. forests and aquatic systems. In the latter, succession involves a systematic change in species composition of a patch or region (see Harris, Chapter 15). This is usually caused by serial competitive replacements of species which, in a stable environment, may give rise to a climax community in equilibrium, although whether such conditions occur generally is in dispute (Connell & Slatyer 1977; Connell 1978).

In contrast, heterotrophic successional processes are determined largely by species-specific patterns of colonization, development, resource consumption, survival and departure (Johnson 1975; Braack 1984), and only occasionally by interspecific competition (but see Swift, Chapter 11 for fungal communities). For example, most dung- and carrion-breeding flies colonize the pad or corpse only during the first day or so although other groups, e.g. the predators of flies or consumers of skin and horn, arrive over a much longer period of time (reviewed by Putman 1983). Similarly, the fly larvae usually remain for up to a week or so whereas other groups, e.g. dermestids, occupy corpses for many weeks (Putman 1983). Competition between species within guilds is often intense (Denno & Cothran 1976) but this does little to alter the course of succession although it may determine the relative abundance of species within guilds (Fuller 1934; Denno & Cothran 1976; Beaver 1977).

There are, however, some instances in which interspecific competition does appear to determine the course of succession at the island level. For example, in England, Springett (1968), found that the necrophagous beetle *Nicrophorus investigator* carried the phoretic mite *Poecilochinus necrophori* which preyed upon immature carrion-breeding calliphorid flies. In the absence of these mites the fly larvae devoured the corpse and prevented breeding by the beetles whereas in the presence of the mite, the flies failed to survive and the mites and silphid beetles reproduced. Wilson (1983) was unable to repeat Springett's finding using the beetle *N. vespilliodes* but found evidence for mutualism between *P. necrophori* and an American silphid beetle. Thus the mutualistic relationship between a beetle and its phoretic mites can determine the course of succession by preventing competitive dominance.

Connell & Slatyer (1977) proposed three models to account for the mechanisms of succession and included some examples of heterotrophic succession in island communities in their facilitation model. However, as the mechanisms of heterotrophic succession in pads and carcasses rarely involve serial competitive replacement of species, climax communities are a logical impossibility. It is therefore more useful to consider autotrophic and heterotrophic succession as distinct processes.

Short-term processes at the local level

Sporadic changes in the pattern of supply of dung and carcasses can have a major impact on community structure. For example; in normal circumstances in both temperate and tropical regions, vertebrate scavengers are the dominant primary consumers of vertebrate carcasses, although the proportion of a carcass taken varies with the type of carcass, season and the habitat in which it occurs (Putman 1983; Braack 1984). However, there are abnormal circumstances in which the supply of food far exceeds the consumption capacity of these scavengers, e.g. mass mortality of elephants in Tsavo National Park (Coe 1978) or mass culling of buffalo in Kruger Park (Braack 1984). In these circumstances, *r*-selected species such as carrion flies become the dominant primary consumers.

Similarly, short-term fluctuation in weather within seasons is an important determinant of breeding success and of the instantaneous composition of the communities associated with dung pads and carcasses. For example, two sympatric species of *Nicrophorous* differ in their temperature thresholds for activity (Wilson *et al.* 1984) and so the relative activity and reproductive success of the two species changes with ambient temperature. Likewise, soil moisture influences the reproductive success of some species of dung beetles more than that of others (Barkhouse & Ridsdill-Smith 1986) and so short-term changes in soil moisture can favour some species over others.

Seasonal and annual patterns at the local level

There are clear seasonal and annual patterns in the activity of most species in dung and carrion communities and, for most species, the period of maximal activity occurs during the warmer, moist seasons of the year although some specialized groups, e.g. termites, are active in the dry season (dung: Lawrence 1954; Poorbaugh 1966; Rainio 1966; Ferrar & Watson 1970; Merritt & Anderson 1977; Coe 1977; Cambefort 1982; Ridsdill-Smith & Hall 1984; carrion: Putman 1978; Anderson 1982; Wilson, Knollenburg & Fudge 1984). For example, in the regions of Europe, America, Australia and Africa with a mediterranean-type or even-rainfall-type climate, the activity of most scarabaeine and geotrupine dung beetles is largely restricted to autumn and spring (Merritt & Anderson 1977; Tyndale-Biscoe, Wallace & Walker 1981; Ridsdill-Smith & Kirk 1985; Davis 1987), whereas in tropical regions dung beetles are most abundant during the warm, moist summer months (Walter 1978; Cambefort 1982). Seasonal activity patterns are strongly bimodal in regions where the monsoons provide a bimodal summer rainfall pattern (e.g. Kenya, Kingston 1977).

In regions in which the warmer, moist periods extend over many months of each year the level of activity tends to vary in response to fluctuating weather conditions. For example, in sub-tropical southern Africa, dung beetle activity is most intense immediately following rainfall but there is substantial activity throughout the wet season (Bernon 1981). Within these coarse-grained seasonal patterns there are fine-grained patterns in which the activities of members of guilds are often largely separated in time (Johnson 1975; Denno & Cothran 1975; Anderson 1982; Cambefort 1982). Such temporal partitioning has often been taken as evidence for character displacement (e.g. Anderson 1982) but the validity of such assumptions is questionable (Den Boer 1986). Whatever the evolutionary mechanisms for such temporal separation, these activity patterns are governed by direct responses to environmental conditions in r-selected species and by physiological mechanisms such as dispause in K-selected species (Edwards 1986).

One consequence of the serial activity of species within seasons is that the composition and structure of dung and carrion communities changes with the season of the year. Therefore, the question of stability or equilibrium within these communities (MacArthur & Wilson 1967; Schoener 1974; Weins 1977; Connell 1978; Atkinson & Shorrocks 1981, 1984; Kneidel 1984a) needs to be addressed over a series of years on a time scale that encompasses a number of generations of the dominant organisms and at the local level of spatial organization.

A lack of consistent community patterns over a series of years would indicate a non-equilibrium condition. However, consistent long-term patterns of community structure do not necessarily demonstrate equilibrium conditions because such patterns can be maintained away from such an equilibrium by periodic disturbance which prevents competitive displacement (Connell 1978; Sousa 1979).

The question of long-term regional stability in community structure has, with few exceptions (e.g. Hanski 1980a, 1986a), rarely been considered in dung and carrion communities, despite its obvious relevance to biological theory and to practical problems such as the biological control of dung- and carrion-breeding pests. In the remainder of this chapter I will present evidence which indicates that an assemblage of southern African dung beetles is a non-equilibrium system at the local level of spatial organization.

Pitfall traps baited with fresh cattle dung were placed at one locality in an extensive tract of grassveld on a clay-loam soil inside Hluhluwe Game Reserve. The traps were cleared after 24 hours and all the dung beetles present were collected and counted. Traps were set at regular intervals throughout each of 5 years (see Table 12.2). For each summer season (October to March inclusive) the abundance ranks in the 114 species present were calculated. The twenty most abundant species made up between 76 and 94% of all individuals trapped and the rank of most species varied widely between the years (Table 12.2). Even the most abundant and least variable species, *O. sugillatus*, made up between 11 and 41% of the total numbers trapped, depending on the year. Extensive trapping during the 1985/6 summer at five sites in an area of 200 km² of grassveld similar to that surrounding the trap sites in Hluhluwe Game Reserve, showed that the relative abundance of species was similar at all comparable sites (unpub-

TABLE 12.2. Evidence indicating that the dung beetle assemblage attracted to cattle dung in the grassveld of the Inzimane River flat in Hluhluwe Game Reserve, South Africa, was not in equilibrium during the period 1980–86. The total number of each species trapped during the summer seasons (October to March) was used to rank the species in order of abundance and the within-year ranks of the twenty most abundant species is given. A total of 90 000 individuals and 114 species were trapped. — indicates absence of that species

Overall rank	Species	Rank within years				
		1980/1	1982/3	1983/4	1984/5	1985/6
1	*Onthophagus sugillatus* Klug	4	2	1	1	1
2	*Sisyphus seminulum* (Thunburg)	3	32	2	4	2
3	*Onthophagus carbonarius* Klug	7	10	3	3	7
4	*Sisyphus ruber* Paschalidis	1	5	5	16	17
5	*Drepanocerus kirbyi* Kirby	9	8	4	8	3
6	*Euoniticellus intermedius* (Reiche)	10	1	7	20	4
7	*Liatongus militaris* (Laporte de Castelnau)	2	3	10	2	8
8	*Tineocellus spinipes* Roth	8	15	8	5	27
9	*Sisyphus spinipes* (Thunburg)	5	13	13	24	19
10	*Sarophorus costatus* (Fahraeus)	12	7	9	6	6
11	*Onthophagus vinctus* Erichson	29	28	6	7	25
12	*Onthophagus tersidorsis* d'Orbigny	11	12	11	14	18
13	*Onthophagus stigmosus* d'Orbigny	—	22	12	—	—
14	*Onthopagus gazella* (F.)	23	4	29	12	5
15	*Onthophagus pullus* Roth	—	24	14	15	12
16	*Sisyphus goryi** Thunberg	6	16	44	30	10
17	*Garreta nitens* (Olivier)	26	6	18	—	—
18	*Sisyphus mirabilis* Arrow	24	—	15	—	—
19	*Onthophagus aeruginosus* Roth	25	69	16	21	16
20	*Garreta unicolor* (Fahraeus)	19	21	17	41	—
% of total numbers		93.8	80.6	92.3	76.0	82.8
Total numbers		5824	8175	74102	1535	163
Number of sampling occasions		12	24	18	11	5

*Note: The taxonomy of some of the *Sisyphus* species is uncertain. The names used are those given by Paschaledis (1974).

lished data) and so it appears likely that the one site in the Reserve was representative of the local dung beetle assemblage.

Dung beetle faunas can be classified into groups of functionally analogous species according to the methods of use and disposal of dung. In the southern African dung beetle fauna I recognize seven such 'functional groups' (unpublished data) and the structure of dung beetle assemblages can be expressed in terms of the relative abundance of members of the groups. A functional analysis of the 1980/86 data (Table 12.3) suggests that the relative

TABLE 12.3. A functional analysis of the structure of the dung beetle assemblage attracted to cattle dung in the grassveld of the Inzimane River Flat in Hluhluwe Game Reserve, South Africa, during the summer season (October to March) during the period 1980–86. The dung burial behaviour of a number of the smaller beetles was not known and so kleptocoprids and members of functional group V could not always be distinguished and were combined for this analysis

	Functional group	Number of species	Per cent of total numbers within years				
			1980/81	1982/83	1983/84	1984/85	1985/86
I	Larger ball rollers	7	0.4	0.7	0.2	0.1	0.6
III	Fast burying tunnellers	15	0.2	1.2	0.2	0.3	0.6
VII	Endocoprids	2	0.2	0.3	0	0.6	0
V & VI	Smaller slow burying tunnellers and Kleptocoprids	28	18.4	19.4	55.4	50.9	37.6
II	Smaller ball rollers	19	52.5	17.2	20.6	9.1	23.6
IV	Larger slow burying tunnellers	43	28.3	61.2	23.7	39.4	37.6

numbers of individuals of each functional group varied substantially over the 5 years. Hence it appears that the way in which the assemblage of beetles use and dispose of dung changed over the five seasons and thus the assemblage was a non-equilibrium system at the functional level.

In conclusion, I suggest that future work on dung and carrion communities should pay particular attention to the interactions between short-term and seasonal fluctuations in the weather, and biotic factors (e.g. competition, predation) and spatial heterogeneity in the environment. These produce a mosaic of different levels of local favourability for each species which may well allow sporadic local extinction but regional persistence of species (Murdoch, Chesson & Chesson 1985). Interspecific competition, predation, weather-induced mortality and spatial aggregation of species and resources all play a part in determining community structure but their relative importance varies between situations. Evaluation of the roles of these factors should occur by experimental manipulation at both the island and local levels of spatial organization and special attention should be given to the degree of temporal and spatial co-occurrence of competitors, predators and prey.

ACKNOWLEDGMENTS

L. Barton-Browne, P. S. Giller, J. H. R. Gee and I. Hanski provided useful comments on the manuscript.

REFERENCES

Adams, J. (**1985**). The definition and interpretation of guild structure in ecological communities. *Journal of Animal Ecology*, **54**, 43–59.

Anderson, R.S. (**1982**). Resource partitioning in the carrion beetle (Coleoptera: Silphidae) fauna of southern Ontario: ecological and evolutionary considerations. *Canadian Journal of Zoology*, **60**, 1314–25.

Atkinson, W.D. & Shorrocks, B. (**1981**). Competition on a divided and ephemeral resource: a simulation model. *Journal of Animal Ecology*, **50**, 461–71.

Atkinson, W.D. & Shorrocks, B. (**1984**). Aggregation of larval diptera over discrete and ephemeral breeding sites: implications for coexistence. *American Naturalist*, **118**, 336–50.

Barkhouse, J. & Ridsdill-Smith, T.J. (**1986**). Effect of soil moisture on brood ball production by *Onthophagus binondis* Thunberg and *Euoniticellus intermedius* (Reiche) (Coleoptera: Scarabaeinae). *Journal of the Entomological Society of Australia*, **25**, 75–8.

Beaver, R.A. (**1977**). Non-equilibrium 'island' communities: Diptera breeding in dead snails. *Journal of Animal Ecology*, **46**, 783–98.

Beaver, R.A. (**1984**). Insect exploitation of ephemoral habitats. *South Pacific Journal of Natural Science*, **6**, 3–47.

Bernon, G. (**1981**). *Species abundance and the diversity of the coleopteran component of a South African cow dung community, and associated insect predators.* Unpublished Ph.D. Thesis, Bowling Green State University, Ohio, USA.

Begon, M. & Mortimer, M. (**1986**). *Population Ecology: A Unified Study of Animals and Plants*, 2nd edn. Blackwell Scientific Publications, Oxford.

Bornemissza, G.F. (**1957**). An analysis of arthropod succession in carrion and the effect of its decomposition on the soil fauna. *Australian Journal of Zoology*, **51**, 1–12.

Bornemissza, G.F. (**1976**). The Australian dung beetle project 1965–1975. *Australian Meat Research Committee Review*, **30**, 1–30.

Bourlière, A.F. (Ed.) (**1983**). Tropical Savannas. *Ecosystems of the World, Vol. 13*. Elsevier, Amsterdam.

Braack, L.E.O. (**1984**). *An ecological investigation of the insects associated with exposed carcasses in northern Kruger National Park: a study of populations and communities.* Unpublished Ph.D. Thesis, University of Natal, South Africa.

Cambefort, I. (**1982**). Les coleopteres scarabaeidae S. Str. de Lamto (Côte-D'Ivoire): Structure des perplements et rôle dans l'écosystème. *Annals of the Entomological Society of France*, **18**, 433–58.

Caughley, G. & Lawton, J.H. (**1981**). Plant herbivore systems. *Theoretical Ecology*, 2nd edn. (Ed. by R.M. May), pp. 132–66. Blackwell Scientific Publications, Oxford.

Coe, M. (**1977**). The role of termites in the removal of elephant dung in the Tsavo (East) National Park, Kenya. *East African Wildlife Journal*, **15**, 49–55.

Coe, M. (**1978**). The decomposition of elephant carcasses in the Tsavo (East) National Park, Kenya. *Journal of Arid Environments*, **1**, 71–86.

Connell, J.H. (**1978**). Diversity in tropical rainforests and coral reefs. *Science*, **199**, 1302–10.

Connell, J.H. & Slatyer, R.O. (**1977**). Mechanisms of succession in natural communities and their role in community stability and organization. *American Naturalist*, **111**, 1119–44.

Cornaby, B.W. (**1974**). Carrion reduction by animals in contrasting tropical habitats. *Biotropica*, **6**, 51–63.

Crespo, G. & Gonzalez, A. (**1983**). The quantity and distribution of faeces in a grassland and its influence on soil fertility. *Cuban Journal of Agricultural Science*, **17**, 1–10.

Davidson, S.J. (**1979**). Mesofaunal responses to cattle dung with particular reference to Collembola. *Pedobiologia*, **19**, 402–7.

Davis, A.L.V. (1987). Geographical distribution of dung beetles (Coleoptera: Scarabaeidae) and their seasonal activity in south-western Cape Province. *Journal of the Southern African Entomological Society* (in press).

Den Boer, P.J. (1986). The present status of the competitive exclusion principle. *Trends in Ecology and Evolution*, **1**, 25–8.

Denno, R.F. & Cothran, W.R. (1975). Niche relationships of a guild of necrophagous flies. *Annals of the Entomological Society of America*, **68**, 741–54.

Denno, R.F. & Cothran, W.R. (1976). Competitive interactions and ecological strategies of sarcophagid and calliphorid flies inhabiting rabbit carrion. *Annals of the Entomological Society of America*, **69**, 109–13.

Doube, B.M. (1983). The habitat preference of some bovine dung beetles (Coleoptera: Scarabaeidae) in Hluhluwe Game Reserve, South Africa. *Bulletin of Entomological Research*, **73**, 357–71.

Doube, B.M. (1986). Biological control of the buffalo fly in Australia: the potential of the Southern African dung fauna. *Miscellaneous Publications of the Entomological Society of America*, **61**, 16–34.

Doube, B.M., Macqueen, A. & Huxham, K.A. (1986). Aspects of the predatory activity of *Macrocheles peregrinus* (Acarina Macrochelidae) on two species of *Haematobia* fly (Diptera: Muscidae). *Miscellaneous Publications of the Entomological Society of America*, **61**, 132–41.

Doube, B.M. & Moola, F. (1987). Effects of intraspecific larval competition on the development of the African buffalo fly *Haematobia thirouxi potans*. *Entomologia Experimentalis et Applicata*, **43**, 145–52.

Edwards, P.B. (1986). Development and larval diapause in the southern African dung beetle *Onitis caffer* Boheman (Coleoptera: Scarabaeidae). *Bulletin of Entomological Research*, **76**, 109–17.

Elton, C.S. (1966). *The Pattern of Animal Communities*. Methuen, London.

Endrody-Younga, S. (1982). An annotated checklist of dung-associated beetles of the Savannah Ecosystem Project study area, Nylsvlei. *South African National Programmes Report, No. 59.*

Fay, H.A.C. (1986). Fauna induced mortality in *Haematobia thirouxi potans* (Bezzi) (Diptera: Muscidae) in buffalo dung in relation to soil and vegetation type. *Miscellaneous Publications of the Entomological Society of America*, **61**, 142–9.

Fay, H.A.C. & Doube, B.M. (1983). The effect of some coprophagous and predatory beetles on the survival of immature stages of the African buffalo fly, *Haematobia thirouxi potans*, in bovine dung. *Zeitschrift für angewandte Entomologie*, **95**, 460–6.

Ferrar, P. & Watson, J.A.L. (1970). Termites (Isoptera) associated with dung in Australia. *Journal of the Australian Entomological Society*, **9**, 100–2.

Fincher, G.T. (1973). Nidification and reproduction of *Phaneus* spp. in three textural classes of soil (Coleoptera: Scarabaeidae). *Coleopterist's Bulletin*, **27**, 33–7.

Fincher, G.T., Stewart, T.B. & Davis, R. (1970). Attraction of coprophagous beetles to faeces of various animals. *Journal of Parasitology*, **56**, 378–83.

Fuller, M.E. (1934). The insect inhabitants of carrion: a study in animal ecology. *Bulletin, Council for Scientific and Industrial Research, Commonwealth of Australia*, **82**, 5–62.

Giller, P.S. (1984). *Community Structure and the Niche*. Chapman & Hall, London.

Gordon, R.D. and Cartwright, O.L. (1974). Survey of food preferences of some North American *Canthonini* (Coleoptera: Scarabaeidae). *Entomological News*, **85**, 181–5.

Greathead, D.J. & Monty, J. (1982). Biological control of stable flies (*Stomoxys* spp.): Results from Mauritius in relation to fly control in dispersed breeding sites. *Biocontrol News Information*, **3**, 105–9.

Halffter, G. & Matthews, E.G. (1966). The natural history of dung beetles of the subfamily Scarabaeinae (Coleoptera: Scarabaeidae). *Folia Entomologica Mexicana*, **12–14**, 1–312.

Halffter, G. & Edmonds, W.D. (1982). *The Nesting Behaviour of Dung Beetles (Scarabaeinae).* Instituto de Ecologia, Mexico.

Hammer, O. (1941). Biological and ecological investigations on flies associated with pasturing cattle and their excrement. *Videnskabelige Meddeleser fra Dansk Naturhistorik Forening i Kjobenhaven,* **105,** 141–393.

Hanski, I. (1980a). Three coexisting species of *Sphaeridium* (Coleoptera: Hydrophilidae). *Annales Entomologica Fennici,* **46,** 39–48.

Hanski, I. (1980b). Patterns of beetle succession in droppings. *Annales Zoologica Fennici,* **17,** 17–25.

Hanski, I. (1980c). Spatial variation in the timing of the seasonal occurrence in coprophagous beetles. *Oikos,* **34,** 311–21.

Hanski, I. (1981). Carrion flies (Calliphoridae) in tropical rain forests in Sarawak, South-East Asia. *Sarawak Museum Journal,* **29,** 191–200.

Hanski, I. (1983). Distributional ecology and abundance of dung and carrion feeding beetles (Scarabaeidae) in tropical rain forests in Sarawak, Borneo. *Acta Zoologica Fennica,* **167,** 1–45.

Hanski, I. (1986a). Individual behaviour, population dynamics and community structure of *Aphodius* (Scarabaeidae) in Europe. *Acta Oecologia,* **7,** 171–87.

Hanski, I. (1986b). Nutritional ecology of dung- and carrion-feeding insects. *The Nutritional Ecology of Insects, Spiders and Mites* (Ed. by F. Slansky Jr. & J.R. Rodrigues), pp. 837–84. Wiley, New York.

Hanski, I. (1987). Colonization of ephemoral habitats. *Colonization, Succession and Stability.* (Ed. by A.J. Gray, M.J. Crawley & P.J. Edwards), pp. 155–86. Symposia of the British Ecological Society, 26. Blackwell Scientific Publications, Oxford.

Hanski, I. & Koskela, H. (1977). Niche relations among dung-inhabiting beetles. *Oecologia,* **28,** 203–31.

Harris, R.L. & Oliver, Lourdes, M. (1979). Predation of *Philonthus flavolimbatus* on the Horn Fly. *Environmental Entomology,* **8,** 259–60.

Heinrich, B. & Bartholomew, G.A. (1979). The ecology of the African dung beetle. *Scientific American,* **235,** 118–26.

Hollis, J.H., Knapp, F.W. & Dawson, K.A. (1985). Influence of bacteria within bovine faeces on the development of the face fly (Diptera: Muscidae). *Environmental Entomology,* **14,** 568–71.

Holter, P. (1979). Abundance and reproductive strategy of the dung beetle *Aphodius rufipes* (L) (Scarabaeidae). *Ecological Entomology,* **4,** 317–26.

Holter, P. (1982). Resource utilization and local coexistence in a guild of Scarabaeid dung beetles (*Aphodius* spp.). *Oikos,* **39,** 213–27.

Howden, A.T. (1950). *The succession of beetles on carrion.* Unpublished M.Sc. Thesis, University of North Carolina.

Howden, H.F. & Nealis, V.G. (1975). Effects of clearing in a tropical rain forest on the composition of the coprophagous scarab beetle fauna (Coleoptera). *Biotropica,* **7,** 77–83.

Howden, H.F. & Nealis, V.G. (1978). Observations on height of perching in some tropical dung beetles (Scarabaeidae). *Biotropica,* **10,** 43–6.

Johnson, M.D. (1975). Seasonal and microseral variations in the insect populations on carrion. *American Midland Naturalist,* **83,** 79–90.

Katakura, H. & Fukuda, H. (1975). Faunal makeup of ground and carrion beetles in Komiotoineppu, Hokkaido University Nagagawa Experimental Forest, northern Japan, with some notes on related problems. *Hokkaido Daigaku Nogakubu Enshurin Kenkyu Hokoku,* **32,** 75–92.

Kingston, T.J. (1977). *Natural manuring by elephants in Tsavo National Park, Kenya.* Unpublished D.Phil. Thesis, Oxford University.

Kneidel, K.A. (1984a). Competition and disturbance in communities of carrion breeding Diptera. *Journal of Animal Ecology,* **53,** 849–65.

Kneidel, K.A. (1984b). The influence of carcass taxon and size on species composition of carrion breeding Diptera. *The American Midland Naturalist,* **111,** 57–63.

Koskela, H. (1972). Habitat selection of dung inhabiting staphylinids (Coleoptera) in relation to age of the dung. *Annales Zoologica Fennici,* **9,** 156–71.

Koskela, H. (1979). Patterns of diel flight activity in dung inhabiting beetles: an ecological analysis. *Oikos,* **33,** 419–39.

Koskela, H. & Hanski, I. (1977). Structure and succession in a beetle community inhabiting cow dung. *Annales Zoologica Fennici,* **14,** 204–23.

Kuusela, S. & Hanski, I. (1982). The structure of carrion fly communities: the size and the type of carrion. *Holarctic Ecology,* **5,** 337–48.

Landin, B.O. (1961). Ecological studies on dung beetles. *Opuscula Entomologica,* Supplement, **19,** 1–228.

Lawrence, B.R. (1954). The larval inhabitants of cow pads. *Journal of Animal Ecology,* **23,** 234–60.

Lumaret, S.P. (1978). *Biographie et écologie des Scarabeides coprophages du sud de la France.* Doctoral Thesis, University of Montpellier.

MacArthur, R.H. & Wilson, E.O. (1967). *The Theory of Island Biogeography.* Princeton University Press, Princeton, New Jersey.

McClain, E. (1983). Effects of an arid environment on carrion-feeding insects. *Proceedings of the Fourth Entomological Congress of the Entomological Society of Southern Africa,* pp. 29–30.

Macqueen, A., Wallace, M.M.H. & Doube, B.M. (1986). Seasonal changes in favourability of cattle dung in Central Queensland for three species of dung inhabiting insect. *Journal of the Australian Entomological Society,* **25,** 23–9.

Matthiessen, J.N. & Hayles, L. (1983). Seasonal changes in characteristics of cattle dung as a resource for an insect in south western Australia. *Australian Journal of Ecology,* **8,** 9–16.

Merritt, R.W. (1976). A review of the food habits of the insect fauna inhabiting cattle droppings in North Central California. *The Pan-Pacific Entomologist,* **52,** 13–22.

Merritt, R.W. & Anderson, J.R. (1977). The effects of different pasture and rangeland ecosystems on the annual dynamics of insects in cattle droppings. *Hilgardia,* **45,** 31–71.

Mohr, C.O. (1943). Cattle droppings as ecological units. *Ecological Monographs,* **13,** 275–98.

Murdoch, W.M., Chesson, J. & Chesson, P.L. (1985). Biological Control in theory and practice. *American Naturalist,* **125,** 344–66.

Nealis, V.G. (1977). Habitat associations and community analysis of south Texas dung beetles (Coleoptera: Scarabaeinae). *Canadian Journal of Zoology,* **55,** 138–47.

Papp, L (1971). Ecological and production biological data on the significance of flies breeding in cattle droppings. *Acta Zoologica Academiae Scientiarum Hungaricae,* **17,** 91–105.

Papp, L. (1976). Ecological and zoogeographical data on flies developing in excrement droppings (Diptera). *Acta Zoologica Academiae Scientiarum Hungaricae,* **22,** 119–38.

Paschaledis, K.M. (1974). *The genus Sisyphus Lati. (Coleoptera, Scarabaeidae) in southern Africa.* Unpublished M.Sc. Thesis, Rhodes University.

Patterson, R.S. & Rutz, D.R. (Eds.) (1986). *Biological Control of Dung Breeding Flies.* Symposium, XVII International Congress of Entomology. *Miscellaneous Publications of the Entomological Society of America.*

Payne, J.A. (1965). A summer carrion study of the baby pig *Sus scrofa* Linnaeus. *Ecology,* **46,** 592–602.

Peck, S.B. & Forsyth, A. (1982). Composition, structure and competitive behaviour in a guild of Equadorian rain forest dung beetles (Coleoptera: Scarabaeidae). *Canadian Journal of Zoology,* **60,** 1624–34.

Poorbaugh, J.H. (1966). *Ecological studies on ecological communities utilizing undisturbed cattle droppings.* Unpublished Ph.D. Thesis, University of California.

Putman, R.J. (1978). The role of carrion-frequenting arthropods in the decay process. *Ecological Entomology,* **3,** 133–9.

Putman, R.J. (1983). *Carrion and Dung: The Decomposition of Animal Wastes.* Edward Arnold, London.

Rainio, M. (1966). Abundance and phenology of some coprophagous beetles in different kinds of dung. *Annales Zoologica Fennici,* **3,** 88–98.

Reed, H.B. (1958). A study of dog carcass communities in Tennessee, with special reference to the insects. *American Midland Naturalist,* **59,** 213–45.

Ridsdill-Smith, T.J. (1986). The effect of seasonal changes in cattle dung on egg production by two species of dung beetle (Coleoptera: Scarabaeidae) in south-western Australia. *Bulletin of Entomological Research,* **76,** 63–8.

Ridsdill-Smith, T.J., Hall, G.P. & Craig, G.F. (1982). Effect of population density on reproduction and dung dispersal by the dung beetle *Onthophagus binodus* in the laboratory. *Entomologia Experimentalis et Applicata,* **32,** 80–5.

Ridsdill-Smith, T.J. & Hall, G.P. (1984). Beetles and mites attracted to fresh cattle dung in south-western Australian pastures. *CSIRO Division of Entomology, Report No. 34.*

Ridsdill-Smith, T.J. & Kirk, A.A. (1985). Selecting dung beetles (Scarabaeinae) from Spain for bush fly control in south-western Australia. *Entomophaga,* **30,** 217–23.

Ridsdill-Smith, T.J., Weir, T.A. & Peck, S.B. (1983). Dung beetles (Scarabaeidnae and Aphodiinae) active in forest habitats in south-western Australia during winter. *Journal of the Australian Entomological Society,* **22,** 307–9.

Root, R.B. (1967). The niche exploitation pattern of the blue-grey gnat catcher. *Ecological Monographs,* **37,** 317–50.

Sands, P. & Hughes, R.D. (1976). A simulation of seasonal changes in the value of cattle dung as a food resource for an insect. *Agricultural Meteorology,* **17,** 161–83.

Schoener, T.W. (1974). Resource partitioning in ecological communities. *Science,* **185,** 27–39.

Schubeck, P.P. (1969). Ecological studies on carrion beetles in Hutcheson Memorial Forest. *Journal of New York Entomological Society,* **77,** 138–51.

Schubeck, P.P. (1971). Diel periodicities of certain carrion beetles. *Coleopterist's Bulletin,* **25,** 41–6.

Springett, B.P. (1968). Aspects of the relationship between burying beetles *Necophorus* spp. and the mite, *Poecilochirus necrophori* Vitz. *Journal of Animal Ecology,* **37,** 417–24.

Sousa, W.P. (1979). Disturbance in marine intertidal boulder fields: the non-equilibrium maintenance of species diversity. *Ecology,* **60,** 1225–39.

Southwood, T.R.E. (1978). *Ecological Methods.* Chapman & Hall, New York.

Southwood, T.R.E. & Kennedy, C. (1983). Trees as islands. *Oikos,* **41,** 359–71.

Summerlin, J.W., Bay, D.E. Harris, R.L. & Russell, D.J. (1981). Laboratory observations on the life cycle and habits of two species of Histeridae (Coleoptera): *Hister coenosus* and *H. incertus. Annals of the Entomological Society of America,* **74,** 316–19.

Tyndale-Biscoe, M., Wallace, M.M.H. & Walker, J.M. (1981). An ecological study of an Australian dung beetle, *Onthophagus granulatus* Boheman (Coleoptera: Scarabaeidae) using physiological age grading techniques. *Bulletin of Entomological Research,* **71,** 137–52.

Valiela, I. (1974). Composition, food webs and population limitation in dung arthropod communities during invasion and succession. *American Midland Naturalist,* **92,** 370–85.

Walker, T.W. Jr. (1957). Ecological studies of the arthropods associated with certain decaying material in four habitats. *Ecology,* **38,** 262–76.

Wallace, M.M.H. & Holm, E. (1983). Establishment and dispersal of the introduced predatory mite, *Macrocheles peregrinus* Krantz, in Australia. *Journal of the Australian Entomological Society,* **22,** 345–8.

Walter, P. (1978). *Recherches écologiques et biologiques sur les Scarabeides coprophages d'une savanne du Zaire.* Unpublished Doctoral Thesis, University of Montpellier.

Wasti, S.S., Hosmer, D.W. & Barney, W.E. (1975). Population density and larval competition in Diptera. I. Biological effects of intraspecific competition in three species of muscid fly. *Zeitschrift für angewandte Entomologie,* **79,** 96–103.

Waterhouse, D.F. (1974). The biological control of dung. *Scientific American,* **230,** 100–9.

Weins, J.A. (1977). On competition and variable environments. *American Scientist,* **65,** 592–7.

Wilson, D.S. (1983). The effect of population structure on the evolution of mutualism: a field test involving burying beetles and their phoretic mites. *American Naturalist,* **121,** 851–70.

Wilson, D.S., Knollenburg, W.G. & Fudge, J. (1984). Species packing and temperature dependent competition amongst burying beetles (Silphidae: *Nicrophorous*). *Ecological Entomology,* **9,** 209–16.

Young, O.P. (1984a). Utilization of dead insects on the soil surface in row crop situations. *Environmental Entomology,* **13,** 1346–51.

Young, O.P. (1984b). Perching of neotropical dung beetles on leaf surfaces: an example of behavioral thermoregulation? *Biotropica,* **16,** 324–7.

13. PATTERNS IN MICROBIAL AQUATIC COMMUNITIES

TOM FENCHEL

Marine Biological Laboratory, University of Copenhagen,
DK-3000 Helsingør, Denmark

INTRODUCTION

A consensus on the definition and delimitation of biotic communities has not been achieved. However, most ecologists consider a community to consist of a number of species populations that interact and are confined in time and space (see Southwood, Chapter 1). Although not often explicitly stated, it is not likely that different populations of species that differ considerably in size and in generation time will show strong or direct interactions (viz. prey–predator relationships or competition for common resources). This view accords with the theoretical results of May (1973) which suggest that communities are not likely to include a large number of strong interspecific interactions. It is also supported by the considerations of Levins (1979) showing that generation time (and hence size) is *per se* a niche component in a variable environment. Thus the scale of spatial and temporal hetero-geneity may separate different communities within a certain area. In nature, a physical space can, therefore, be thought of as harbouring a hierarchy of, albeit overlapping, communities, each representing a few trophic levels and characteristic scales of time and space. This view is at least implicitly accepted by many ecologists who may, for example, speak about bird or insect communities without the need to refer to coexisting populations of bacteria in any detail.

In this sense the concept of 'microbial communities' has a meaning. However, it must be acknowledged that 'microbes' (= unicellular organisms) range from 0.5 μm long bacteria to 5 mm long foraminifera and so even microbes may represent several levels in the 'community hierarchy' as discussed above. Since I will demonstrate properties of microbial communities by two isolated examples, this aspect will not be discussed further.

In many respects there are no fundamental differences between microbial communities and those formed by metazoa or vascular plants, except for scales of time and space. However, regarding prokaryote organisms, two properties, their inability to perform phagocytosis and their diversity in metabolic pathways, add qualitatively different aspects to the communities they form in nature. In addition to competition for common substrates, interactions between bacterial populations take place through one physiological type of bacterium utilizing metabolites of another type ('syntrophic interactions').

In order to get a feeling for microbial communities, it may be useful to list some figures for characteristic scales of micro-organisms. The generation time for any given species depends on environmental parameters such as temperature and, especially, the availability of food resources/substrates, but the potential range within which balanced growth is possible is limited. Many strains of bacteria can divide at intervals of less than 20 min under optimal conditions, but with low substrate concentrations generation times as long as several days may be achieved. Small protozoa may display generation times from around 3 h to about 24 h, while the largest protozoa have minimum generation times which exceed 24 h. Many bacteria and microalgae and most protozoa are motile; swimming or gliding bacteria move around at velocities of about 30 μm s^{-1}, swimming flagellates reach velocities of around 100 μm s^{-1} while ciliates approach 1 mm s^{-1}.

The spatial structure of microbial communities, especially those consisting of prokaryotes, is strongly influenced by the transport rate of solutes. In stagnant environments (sediments, microbial films on surfaces) molecular diffusion (typically around 10^{-5} cm^2 s^{-1} for low-molecular compounds in water) is important. If transport of dissolved substances also takes place through turbulent water movement the spatial scale of zonation patterns of microbial communities may expand considerably.

How small can spatially-defined microbial communities (homogeneous patches or zones characterized by certain species populations) become? Motile micro-organisms move by a sort of random walk with more or less straight 'runs' interrupted by 'tumbles', after which the cells start off with a run in a new, random direction. The motility of a microbial population can therefore be quantified by a diffusion coefficient, D (Berg 1983). Now consider a patch with a diameter, L, within which the cells grow with a rate, μ, while growth is zero elsewhere. The question is then how small can L become so that the patch still maintains a population in spite of the diffusional loss of cells to the surroundings. It can be shown (Okubo 1980) that this is given by the expression, $L = k \, (D/\mu)^{1/2}$, where k is a constant of the order of unity (the exact magnitude depends on the geometry and dimensionality of the patch). For a bacterium, suppose $\mu = 0.69$ h^{-1} (generation time $= 1$ h) and that $D = 6.8 \times 10^{-5}$ cm^2 s^{-1} (corresponding to a swimming velocity of 30 μm s^{-1}, a tumbling interval of 30 s and a 2-dimensional universe). The minimum patch diameter can then be calculated to be about 6 mm. For a ciliate population (with a much lower rate of division and a much higher swimming velocity) a similar estimate yields something of the order of 20 cm.

Any microbial naturalist (or even one looking at cultures which include some environmental heterogeneity) will know that spatial organization of microbial communities occurs at a much finer scale than the above considerations suggest. The discrepancy primarily illustrates the significance of behavioural (especially chemosensory) responses which the model above ignores because D is considered to be invariant. Thus micro-organisms are capable of orienting themselves in gradients of chemical species or light. They may do so by 'phobic responses' in which the cells react by tumbling upon entering adverse conditions (thus increasing the probability of returning to more benign regions) and by kineses, in which the cells permanently (that is, as long as the stimulus prevails) increase their motility (increasing swimming velocity or decreasing tumbling frequency) under adverse conditions and decrease their motility under favourable conditions. This mechanism will lead statistically to the aggregation of cells in favourable patches. 'Taxic' responses (in which the cells show an oriented response by sensing the direction of a stimulus or a gradient) are rare among micro-organisms since this requires more complex organelles which can analyse the environment. However, phototaxis is widespread among photosynthetic flagellates, and there exist ciliates and bacteria which use the direction of gravity or of magnetic field lines as a clue to the direction of oxygen gradients (Blakemore, Fraenkel & Kalmijn 1980; Fenchel & Finlay 1984; Lapidus & Levandowsky 1981). Such mechanisms together explain why microbial communities may show spatial zonation patterns of populations with a scale of less than a millimetre (e.g. Fig. 13.1d).

Microbial communities are ubiquitous in aquatic environments; they occur in the plankton of lakes and oceans, associated with surface films of water, on all solid surfaces such as rocks, plants, algae and animals, within animals and in or on sediments. Sieburth (1979) gives a broad (and beautifully illustrated) overview of the diversity of microbial communities in the sea. The scope of the present paper is only to give a brief description of two types of microbial aquatic communities: the prokaryote communities of anoxic environments and oxyclines, and the prey–predator relationship between bacteria and phagotrophic protozoa in marine plankton. The former example will emphasize especially the role of interspecific metabolite transfer and competition in explaining the spatial structure and diversity of a prokaryote community; the latter example will demonstrate the characteristic time-scale of events in microbial communities.

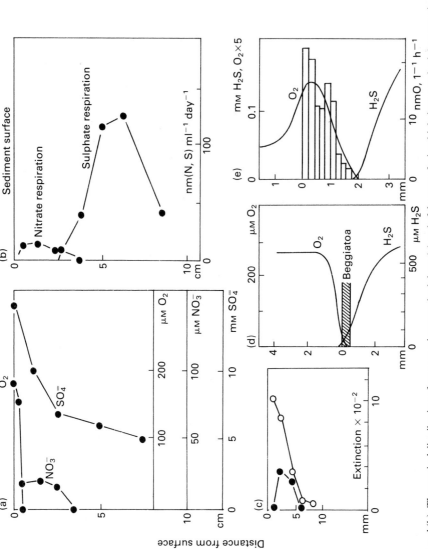

FIG. 13.1(a) and (b). The vertical distribution of oxygen, nitrate and sulphate (a) and of the rate of microbial nitrate and sulphate respiration (b) in an estuarine sediment (Sørensen, Jørgensen & Revsbech 1979). (c) The vertical distribution of bacteriochlorophyll a (solid circles) and chlorophyll a (open circles) measured as the extinction coefficient of ether extracts of sediment slices at 770 and at 665 nm respectively. The two curves reflect the distribution of purple sulphur bacteria and of eukaryotic algae + blue green bacteria, respectively (Fenchel & Straarup, 1971). (d) The vertical distribution of oxygen and hydrogen sulphide and of the white sulphur-bacterium, Beggiatoa, in the surface of a sediment. (e) Vertical distribution of oxygen and hydrogen sulphide and of oxygenic photosynthesis in an illuminated sediment (0 mm corresponds to the sediment surface). Photo-

PROKARYOTE COMMUNITIES OF ANOXIC ENVIRONMENTS

Metabolic processes

Anoxic habitats occur for different reasons (including such relatively exotic mechanisms as the emission of sulphide-containing water from hydro-thermal vents). The most widespread and important example, however, is provided by aquatic sediments which are always anoxic and chemically reducing below a certain depth. Many stagnant water bodies are also permanently or seasonally anoxic below some depth in the water column. The remains of planktonic and non-planktonic organisms sink to the bottom of the water column, and so aquatic sediments accumulate dead organic material. The aerobic (bacterial) degradation of this material soon exhausts the available oxygen because it can only be supplied from the surface via molecular diffusion, and so at some depth (from zero to several centimetres depending mainly on the load of organic material) sediments are completely anoxic.

Further degradation of organic material is carried out by fermenting bacteria which produce short chain fatty acids (mainly acetate, propionate and butyrate) and molecular hydrogen as metabolites. These compounds are in turn further mineralized by a variety of physiological types of bacteria that employ 'anaerobic respiration'; that is, they use inorganic, terminal electron acceptors other than oxygen in a respiratory process. Thus the 'denitrifying bacteria' use nitrate as an oxidant and produce mainly N_2 (and to some extent ammonia and nitrous oxide) as end-products. Sulphate-reducing bacteria employ sulphate for respiration and excrete sulphide (HS^- at the prevailing pH). Finally, methanogenic bacteria use CO_2 as a terminal electron acceptor for the oxidation of acetate or H_2 to produce CH_4. These processes all depend on the downward diffusion of the respective electron acceptors from the surface of the sediment. The energy yield (and hence the growth efficiency) of these processes differs according to the particular electron acceptor used in the respiratory process (nitrate being the most and carbon dioxide the least favourable one).

Spatial patterns

The outcome of competition for organic substrates between the different types of bacteria depends on the identity of the hydrogen acceptor which is both available and energetically most favourable; methanogenesis, for example, does not take place before sulphate has become depleted. Denitrifying bacteria are active immediately beneath the oxic zone. Since nitrate is usually not very abundant in seawater it is rapidly depleted and the process is of a relatively small quantitative importance. Sulphate, however, is extremely abundant in seawater (representing nearly fifty times as many oxidation-equivalents as does oxygen in atmospherically saturated seawater). Sulphate reduction is therefore the quantitatively dominating process and may account for more than 50% of the mineralization of dead organic material in the sea-bed. Sulphate is usually depleted only at considerable depths in the sediment, beneath which methanogenesis takes over. In freshwater sediments where sulphate is less abundant methanogenesis plays a correspondingly larger role. The one-way supply of electron acceptors in conjunction with competition for organic substrates explain the spatial zonation patterns of anaerobic bacteria in the sediment (Fenchel & Blackburn 1979; see also Fig. 13.1a,b).

The end-products of these anaerobic metabolic processes (that is methane, sulphide and ammonia, although the latter derives mainly from the mineralization of nitrogenous organic compounds rather than from the dissimilatory processes) diffuse upwards in the sediment following concentration gradients. In marine, sulphate-rich environments, methane is largely oxidized anaerobically by sulphate-reducers although some methane may escape from the sediment as gas bubbles. As sulphide reaches the oxic zone of the sediment it is utilized by 'chemolithotrophic' or 'white' sulphur bacteria. These organisms gain energy by oxidizing sulphide, first to elemental sulphur and then to sulphate, thus completing the microbial sulphur cycle. These bacteria are 'gradient organisms'; since sulphide oxidizes spontaneously in the presence of oxygen the organisms must situate themselves precisely in the transition zone between these two compounds (Fig. 13.1d). They are capable of tracing vertical migrations of this chemocline, which in shallow waters take place on a diurnal basis (see below). Ammonia is also utilized by chemolithotrophic bacteria, the 'nitrifying bacteria', among which some oxidize ammonia to nitrite and others oxidize nitrite to nitrate.

In sediments that are exposed to light, sulphide and sulphur may become oxidized anaerobically by a photosynthetic process carried out by 'purple' and 'green' sulphur-bacteria and also by some kinds of cyanobacteria. In this process sulphide is used as an electron donor in a photosynthetic process analogous to the way plants use water. The photosynthetic organisms are anaerobes and they thrive where the anoxic zone reaches the photic zone of sediments. The photic zone may extend about 3 mm beneath the surface and often beneath a zone of organisms with oxygenic photosynthesis (mainly cyanobacteria and unicellular eukaryotes), which explains the green-red banding often observed in the surface layers of estuarine sediments (Fig. 13.1c,e). Since the absorption spectra of bacterial chlorophylls differ from those of plant chlorophylls the bacteria are not 'shaded out'.

The activity of bacterial (anoxygenic) and oxygenic photosynthesis in the surface of sediments leads to diurnal vertical migrations of entire microbial and chemical zonation patterns. In sediments the vertical migrations extend perhaps 0.5 cm; in stratified lakes where the oxycline (and immediately beneath it, in the anoxic zone, a layer of photosynthetic bacteria) is found in the water column such migrations may extend for more than 1 metre.

Diversity

The concept of species diversity of prokaryote communities differs in one important respect from that of eukaryote communities. Among the latter, species are defined, at least in principle, as populations sharing a common gene pool, and are identified on the basis of more or less subtle morphological differences. Thereafter the ecologist may study the species discovered and defined by the taxonomist in order to find features (in terms of habitat preferences, resource utilization, or life-cycles) that explain species interactions and coexistence in nature. In the case of prokaryotes, species are largely defined by physiological properties which also define their ecological niches. Explanations of species diversity in prokaryote communities may therefore approach a tautology; bacterial species are discovered exactly because they differ in the way they utilize resources or tolerate environmental factors. To this must be added that the fundamental evolutionary concept of species as populations sharing a common gene pool does not apply to prokaryote organisms.

Extensive studies of bacterial physiology show that many strains utilize only a very restricted number of substrates and that different strains differ considerably with respect to their response to different substrate concentrations. An example is provided by the sulphate-reducing bacteria. The classical *Desulfovibrio* uses only lactate or hydrogen as a substrate, but more recently a form utilizing acetate (*Desulfotomaculum*) has been discovered. Another genus, *Desulforomonas*, also utilizes acetate as an electron donor, but it uses sulphur rather than sulphate as the electron acceptor (Pfennig & Biebl 1976; Widdel & Pfennig 1977). Photosynthetic bacteria differ in their requirements for sulphide concentrations and also in their tolerance of high concentrations of this substance. Different types of these bacteria also contain different types of bacteriochlorophylls, so that they show differential absorption of wavelengths of light within the red and near-infrared part of the spectrum (Veldkamp & Jannasch 1972; Pfennig 1975).

The use of continuous cultures has provided direct evidence of the resource niches of bacteria. Essentially this has been done by obtaining 'Monod-curves' (i.e. the growth response as a function of the concentration of a limiting substrate in a chemostat). Such studies have shown differential responses to different concentrations; certain strains are superior competitors at low substrate concentrations but are inferior competitors at higher concentrations (Veldkamp & Jannasch 1972).

The above discussion has demonstrated that resource and habitat niches are closely interconnected in prokaryote communities of chemical gradients. This applies to strains that utilize qualitatively different substrates, but it also applies to strains with differential competitive ability at different concentrations of identical substrates. This is because a concentration gradient of a substrate may harbour several spatially disjunct populations. This can be demonstrated on gels in the laboratory, using steady state diffusion gradients in one or two dimensions. In such gradients different strains may develop in a spatially ordered manner according to their optimum substrate concentrations or in response to other chemical factors (Caldwell, Lai & Tiedje 1973; Wimpenny 1981). This type of experiment represent a direct visualization of 'Hutchinsonian niches'.

Syntrophic interactions have led in some cases to the evolution of real mutualistic interactions, that is, when the component species can only grow in intimate physical proximity. Such cases have been discovered where cultures of bacteria previously believed to contain only one strain later proved to consist of two species. In most cases these seem to involve hydrogen transfer in which a fermenting bacterium depends on the presence of a hydrogen-scavenging species. Consortia consisting of a sulphate-reducing bacterium covered by cells belonging to photosynthetic sulphur bacteria are also known (Kuznetsov 1975; Wimpenny 1981).

Although this description of the prokaryote communities of sediments is brief and in some details superficial, it does illustrate some of their salient characteristics. Thus, properties like the diversity, spatial structure and population interactions can be understood on the basis of the physiological properties of the individual strains of bacteria, and of the physical and chemical properties of the environment. This understanding has been facilitated by growing the component organisms in pure and chemically defined cultures, and by using model microcosms in which environmental conditions (e.g. light, addition or removal of certain substrates) can be varied.

PELAGIC BACTERIA AND THEIR PROTOZOAN GRAZERS

The planktonic populations and their mutual interactions to be described below can hardly qualify as a complete 'community', but should really be described in the context of the producers of substrates for bacteria (phytoplanktonic algae) on the one hand and predators on the small protozoa (to a large extent other protozoa) on the other. However, looking at bacteria and their protozoan grazers in plankton communities in isolation will still give some impression of time scales of population events in microbial communities.

It has been recognized recently that suspended bacteria play a much larger role in marine (and in freshwater) plankton than hitherto believed. Bacteria typically occur in numbers of 0.5 to 2×10^6 cells per ml of seawater. Various methods (such as measurement of the rate at which bacteria incorporate tritium-labelled thymidine or the frequency of dividing cells) indicate that in summer the average generation time of the bacteria may be as short as 6–12 h. This rate of growth results in organic production that may constitute around 20% of the primary production of the entire system. It is based on the excretion of dissolved organic material from planktonic algae which have been shown to lose as much as 50% of their production as exudates (Azam *et al.* 1983; Riemann *et al.* 1984). Since bacterial numbers remain relatively constant over time it is likely that they are controlled by predation. The predominant grazers of these suspended bacteria have been shown to be small (3–10 μm long) heterotrophic flagellates. They represent a number of different types including choanoflagellates, chrysomonads, bicoecids and bodonids as the most important ones. Their occurrence in seawater is ubiquitous; their numbers fluctuate relatively more than do bacterial numbers, but on average there are about 10^3 cells per ml of seawater (Azam *et al.* 1983; Fenchel 1986).

Temporal patchiness

The microbial populations of the plankton show three types of temporal patchiness. Photosynthetic activity of the planktonic algae is confined to the daytime. This results in diurnal fluctuations in the excretion of dissolved organic material which constitutes the main energy and carbon source for the bacterial populations. The bacteria can respond to these fluctuations because of their short generation times; thus the mean volume of the bacteria increases around noon followed by an increase in the frequency of dividing cells and eventually, during the evening, the bacterial numbers reach a maximum (Riemann *et al.* 1984; Fig. 13.2a).

T. FENCHEL

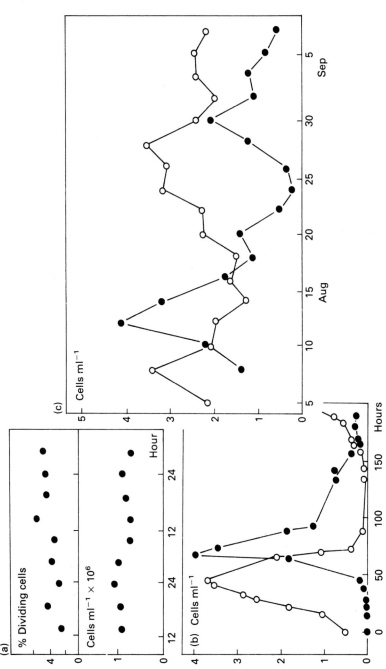

FIG. 13.2. (a) Percentage of cells undergoing division and numbers of bacteria in the water column in Limfjorden, Denmark during a 2-day period in September (Riemann *et al.* 1984). (b) Changes in numbers of bacteria (open circles × 10⁶) and phagotrophic flagellates (filled circles × 10³) during 200 h after collection and filtration through an 8 μm filter to remove larger organisms. If the flagellates are also removed, bacterial numbers increase to about 10⁷ ml⁻¹ within about 50 h and remain at this level (Andersen & Fenchel 1985). (c) Changes in the numbers of flagellates and bacteria in the water column of Limfjorden during a month. Symbols as in (b) (Fenchel 1986).

On a larger scale, the bacterial and protozoan populations show more or less regular and coupled fluctuations, the characteristics of which are determined by the predator–prey interaction. These are easily demonstrated in seawater samples which have been filtered to remove the large plankters which prey on the protozoa. Such samples will show regular population oscillations (Fig. 13.2b) until the substrate for the bacteria has been exhausted. The period of the oscillations can be accounted for on the basis of Lotka–Volterra type equations when parameters of growth and of functional and numerical responses of the flagellates are determined from pure cultures (Andersen & Fenchel 1985; Fenchel 1986). In nature such oscillations are more difficult to observe or to analyse in detail. This is in part due to the fact that plankton populations drift with the water currents. When sampling on a fixed location observed fluctuations may therefore represent spatial rather than temporal patchiness. Also the predation on the protozoa by other, larger protozoa and metazoan planktonic organisms may complicate the analysis. However, regular oscillations can sometimes be followed for several cycles (Fig. 13.2c). The period length of the oscillations is a function of water temperature which again determines the vital rates of the organisms; in the studied area the periods of the population oscillations ranged from 8–10 days during the warmest period and increased to about 20 days in early spring and late autumn. Peaks in the density of protozoa always lag about ¼ cycle behind the peak abundances of bacteria.

The third type of temporal pattern to be observed in the plankton microbial populations is a seasonal one. During winter, primary productivity is low and consequently bacterial and flagellate numbers remain relatively low and constant. The rapid increase in primary productivity in early spring leads to an initial increase in bacterial numbers, followed by an increase in the flagellate populations. This initial perturbation of the system triggers predator–prey oscillations which persist until productivity decreases in the beginning of autumn.

Diversity

The diversity of phagotrophic microplankton protozoa is still incompletely understood at the most basic level, that is, they are still poorly studied from a taxonomic point of view. Members of certain groups, notably the acanthocoeid choanoflagellates and certain chrysomonad species, carry siliceous spines or scales which render them easily recognizable as whole mounts in the transmission electron microscope, and a surprising number of such species have been described recently (see Fenchel 1986 for references). The species systematics of other groups is still very incomplete.

There is evidence that food particle size selectivity plays a role in niche-differentiation among these protozoa. The choanoflagellates capture food bacteria by sieving water through a tentacular filter which will retain particles down to a diameter of 0.3 μm, that is, the size of the smallest prokaryote cells. The helioflagellates also capture food particles with a tentacular sieve, but this is much coarser and will probably not retain particles smaller than 1–2 μm. Other types of flagellates capture suspended bacteria through direct interception ('raptorial feeding') and this mechanism will lead to a more graded size spectrum of ingested food particles. The helioflagellate, *Ciliophrys*, is not motile while feeding; it feeds on motile bacteria that adhere to pseudopodial tentacles radiating out from the cell, and in this respect resembles a heliozoan. Finally some of the microprotozoa of the plankton (such as bodonid flagellates) are associated with suspended, detrital particles and they are specialized to catch bacteria adhering to surfaces (Fenchel 1986).

An interesting aspect of the temporal patchiness of planktonic environments (and other microbial systems) is that it may allow for different time niches of the organisms. When starved, different types or strains of flagellates react in different ways. Some undergo one or two successive cell divisions resulting in two or four 'swarmer cells'. These are motile, but have a very low metabolic rate corresponding to the fact that macromolecular synthesis falls to a very low level. Due to the low metabolic rate these cells may survive without food for several days, but with prolonged starvation, the time-lag needed before cell divisions can take place following feeding increases since the synthetic apparatus has to be rebuilt. In this case the response to deprivation of food represents an evolutionary trade-off between long-term survival and the ability to resume growth once food is available again. Other forms or strains react to starvation by forming cysts. These are capable of a very long survival time, but the cost is that the metabolic machinery is nearly stopped and once food is available they are less competitive because it will take longer before cell divisions can take place. These different responses have so far only been studied in the laboratory in pure cultures (Fenchel 1982); however, it is likely that they reflect adaptations to different time scales of environmental changes and that such different life cycle characteristics will prove important for explaining the diversity of forms in microbial communities.

CONCLUSION

Microbial communities differ from those formed by larger organisms mainly with respect to the characteristic scales of time and space; these are determined by the rates of growth, by the motility of the individual organisms and by some physical properties of the environment such as molecular diffusivity.

From the viewpoint of community ecology these properties are useful. Simply observing microbial communities may add new perspectives to our ideas on community structure and function in general. Most important, however, is the fact that microbes lend themselves well to experimentation; the ecological niches of single species can be determined precisely in many cases. In addition, artificial or natural communities with different degrees of complexity can be manipulated and studied over a period of time which represents a large number of generations of the component organisms.

REFERENCES

Andersen, P. & Fenchel, T. (1985). Bacterivory by microheterotrophic flagellates in seawater samples. *Limnology and Oceanography,* **30,** 198–202.

Azam, F., Fenchel, T., Field, J.G., Gray, J.S., Meyer-Reil, L.A. & Thingstad, F. (1983). The ecological role of water-column microbes in the sea. *Marine Ecology Progress Series,* **10,** 257–63.

Berg, H.C. (1983). *Random Walks in Biology.* Princeton University Press, Princeton, New Jersey.

Blakemore, R.P., Fraenkel, R.B. & Kalmijn, Af. S. (1980). South-seeking magnetotactic bacteria in the Southern Hemisphere. *Nature (London),* **286,** 384–5.

Caldwell, D.E., Lai, S.H. & Tiedje, J.M. (1973). A two-dimensional steady-state diffusion gradient for ecological studies. *Bulletin of Ecological Research Communication* (Stockholm) **17,** 151–8.

Fenchel, T. (1982). Ecology of heterotrophic microflagellates. III. Adaptations to heterogenous environments. *Marine Ecology Progress Series,* **9,** 25–33.

Fenchel, T. (1986). The ecology of heterotrophic flagellates. *Advances in Microbial Ecology,* **9,** 57–97.

Fenchel, T. & Blackburn, T.H. (1979). *Bacteria and Mineral Cycling.* Academic Press, London.

Fenchel, T. & Finlay, B.J. (1984). Geotaxis in the ciliated protozoon Loxodes. *Journal of Experimental Biology,* **110,** 17–33.

Fenchel, T. & Straarup, B.J. (1971). Vertical distribution of photosynthetic pigments and the penetration of light in marine sediments. *Oikos,* **22,** 172–82.

Jørgensen, B.B. (1982). Ecology of the bacteria of the sulphur cycle with special reference to anoxic-oxic interface environments. *Philosophical Transactions of the Royal Society,* London, **B, 298,** 543–61.

Kuznetsov, S.I. (1975). Trends in the development of ecological microbiology. *Advances in Aquatic Microbiology,* **1,** 1–48.

Lapidus, R. & Levandowsky, M. (1981). Mathematical models of behavioral responses to sensory stimuli by protozoa. *Biochemistry and Physiology of Protozoa,* 2nd edn (Ed. by M. Levandowsky & S.H. Hutner), **4,** 235–60. Academic Press, New York.

Levins, R. (1979). Coexistence in a variable environment. *American Naturalist,* **114,** 765–83.

May, R.M. (1973). *Stability and Complexity in Model Ecosystems.* Princeton University Press, Princeton, New Jersey.

Okubo, A. (1980). *Diffusion and Ecological Problems: Mathematical Models.* Springer-Verlag, Berlin.

Pfennig, N. (1975). The phototrophic bacteria and their role in the sulfur cycle. *Plant and Soil,* **43,** 1–16.

Pfennig, N. & Biebl, H. (1976). *Desulforomonas acetoxidans* gen. nov. sp. nov., a new anaerobic sulfur-reducing, acetate-oxidizing bacterium. *Archives of Microbiology,* **110,** 3–12.

Riemann, B., Nielsen, P., Jeppesen, M., Marcussen, B. & Fuhrman, J.A. (1984). Diel changes in bacterial biomass and growth rates in coastal environments, determined by means of thymidine incorporation into DNA, frequency of dividing cells (FDC), and microradiography. *Marine Ecology Progress Series,* **17,** 227–35.

Sieburth, J. McN. (1979). *Sea Microbes.* Oxford University Press, New York.

Sørensen, J., Jørgensen, B.B. & Revsbech, N.P. (1979). A comparison of oxygen, nitrate and sulfate respiration in coastal marine sediments. *Microbial Ecology,* **5,** 105–15.

Veldkamp, H. & Jannasch, H.W. (1972). Mixed culture studies with the chemostat. *Journal of Applied and Chemical Biotechnology,* **22,** 105–23.

Widdel, F. & Pfennig, N. (1977). A new anaerobic, acetate-oxidizing, sulfate reducing bacterium *Desulfotomaculum* (emend.) *acetoxidans. Archives of Microbiology,* **112,** 119–22.

Wimpenny, J.W.T. (1981). Spatial order in microbial ecosystems. *Biological Reviews,* **56,** 295–342.

Aquatic Assemblages

14. COMMUNITY ORGANIZATION IN THE FRESHWATER PLANKTON

C. S. REYNOLDS

Freshwater Biological Association, The Ferry House, Ambleside, Cumbria LA22 OLP, UK

INTRODUCTION

Planktonic communities are adapted to live suspended in the open waters of lakes and rivers, coastal seas and the upper layers of the oceans. They comprise photoautotrophic plants, herbivores, carnivores and decomposers which assemble and recycle matter along analogous trophic pathways but, unlike terrestrial communities, the dominant life-forms are microscopic in size and move through three-dimensional planes rather than two.

There is a wide diversity of size, shape and phylogenetic representation among planktonic organisms. By implication, pelagic environments and the resource-bases on which the communities are built are highly diverse and/or variable. The present attempt to relate the organization of planktonic communities to the distributions of resources refers principally to lakes but analogous relationships may be found for the sea (Harris, Chapter 15). This review firstly examines the scales of variability in the principal resources and the ways in which their interactions shape the pelagic environment. The life-forms of phytoplankton that are selectively favoured at given spatial and temporal locations and the responses of zooplankton are evaluated in subsequent sections. Finally, it is argued that the internal organization of planktonic communities is frequently weak and far from equilibrium; rather their structure reflects the individualistic, mainly non-interactive responses of the component species to rapidly altering conditions in inherently unstable environments.

THE RESOURCE-BASE OF FRESHWATER PLANKTON

Light

Photoautotrophic primary production in open water is due largely to planktonic species of algae and cyanobacteria (phytoplankton). These small organisms sink slowly in a relatively dense and viscous medium which is itself

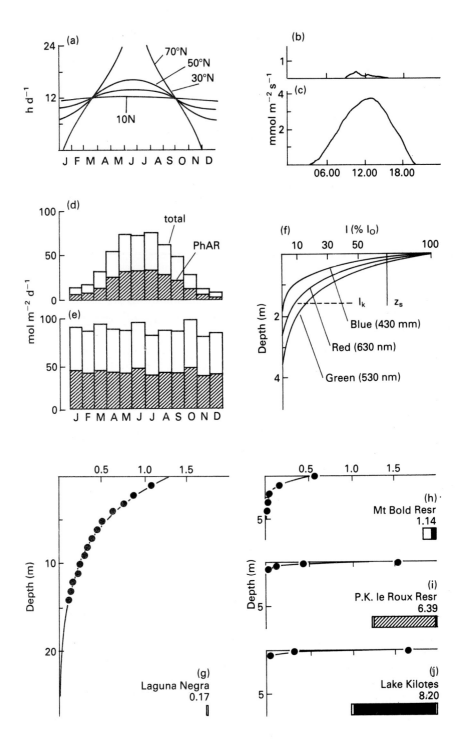

subject to continuous motion. With no need for non-productive supporting tissues, almost all vegetative cells contribute to net productivity. The time required for planktonic autotrophs to double their biomass is, potentially, hours or, at most, a few days (Hoogenhout & Amesz 1965). In natural systems, however, these rates are seldom achieved and the maximum autotrophic rate of increase of biomass per unit area of lake surface falls far short of the levels attained in terrestrial ecosystems (e.g. Margalef 1978).

The net productivity of planktonic systems is related principally to the availability of light. Annually recurrent variations in the duration and intensity of solar radiation are determined by latitude and modified by cloud structure (Fig. 14.1a–e). The photosynthetically-active radiation (PhAR, in mol photons m^{-2} s^{-1}) available to phytoplankton (wavelengths 360–

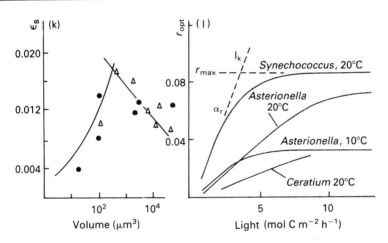

FIG. 14.1. Light and photosynthesis in planktonic systems: (a) seasonal variations in day-length at selected latitudes (data from List 1951); diel course of radiation input on (b) an overcast January day and (c) a cloudless June day at 53°N (redrawn from Talling 1971); mean monthly total solar radiation and PhAR (hatched) recorded (d) at 52°N (Brierley 1985) and (e) at 0° (Ganf 1974a); (f) attenuation with depth of selected spectral blocks (redrawn from Reynolds 1984a). Light penetration and (inset) absorbance components due to water (unshaded), non-living particulates (hatched) and algal material (solid) in (g) Laguna Negra, Chile (author's unpublished data), (h) Mount Bold Reservoir, Australia (Ganf 1980) (i) P.K. le Roux Reservoir, RSA (Allanson & Hart 1979) and (j) Lake Kilotes, Ethiopia (Talling *et al.* 1973); (k) the light absorption by phytoplankton cells (ϵ_s, m^2 (mg chl a)$^{-1}$) of differing shape (●sperical;△non-spherical) and size (redrawn from Reynolds 1984a); the straight line, Kirk's (1983) regression representing the diminishing efficiency of light interception with increasing algal size, does not respond to the diminution of area-specific chlorophyll content of smaller ($<300\mu^3$) cells to which data the curve is fitted; and (1) growth responses of selected algae (r_{opt}: units = mol C assimilated (mol cell C)$^{-1}$ h^{-1}) to light dose (data reviewed Reynolds 1984a).

750 nm) is further modified by surface reflectance (Weinberg 1976) and backscattering due to waves and foam-lines (Larkum & Barrett 1983). The fraction (generally < 0.9) penetrating the surface, I', is exponentially attenuated, through absorption and scattering, with increasing depth, as described empirically by the Beer–Lambert Law (Larkum & Barrett 1983). It should be noted that absorbance of PhAR is not uniform across the spectrum (Fig. 14.1f). Examples of attenuation profiles are given in Fig. 14.1g–j to illustrate the range of variation in separate components contributing to the absorbance. In Laguna Negra (Fig. 14.1g), the attenuation approximates to that of pure water. Elsewhere, yellowish-brown humic substances ('Gelbstoff') screen out blue light and, at high concentrations, contribute significantly to the overall attenuation (Fig. 14.1h) whilst high particulate loads (clay, Fig. 14.1i) and dense phytoplankton blooms (Fig. 14.1j) can each substantially reduce light penetration. The range of morphological adaptations to light interception by algae (Fig. 14.1k) are reviewed in Kirk (1983), Raven (1984) and Reynolds (1984a). Several additional mechanisms respond to low light-income (for reviews, see Falkowski 1980; Richardson, Beardall & Raven 1983).

The light energy absorbed drives the photoautotrophic growth, primarily through the photosynthetic fixation of carbon and its subsequent (light-independent) assimilation into new cell material. As with the photosynthetic behaviour of phytoplankton (Bannister 1974; Tilzer 1984), the rate of growth can be related to available light energy in terms of the onset of growth-saturating irradiance (I_k) and the efficiency of light-limited growth (α_r in Fig. 14.1l).

Morphological, physiological and biochemical differences among the planktonic algae contribute to considerable interspecific variation in I_k and α_r (Fig. 14.1l) which is presumably matched to the diversity of light environments in freshwaters. It is only in the upper layers, however, that algae can receive irradiance levels comparable with those experienced in terrestrial 'light' environments; much of the light gradient offers the equivalent of shade habitats. Most phytoplankton have evolved a facultative capacity for low-light adaptation, which provides a striking indication of the frequency with which light limitation may occur in aquatic environments.

Nutrients

The broad relationship between mean chlorophyll concentrations of temperate lakes and various measures of phosphorus availability (Fig. 14.2a) and areal loadings (Fig. 14.2b), together with the results of numerous bioassay experiments of differing scale (Schindler 1977; Schelske 1984), have established that phosphorus is, indeed, the main chemical resource limiting phytoplankton development. In many fresh waters, especially continental lakes at low latitudes (Owens & Esaias 1976), and in phosphorus-rich systems at higher ones (Reynolds 1979), nitrogen is the limiting resource. At times, silicon may be so depleted from solution as a result of biological uptake by diatoms so as to limit their further growth (Lund 1964; Paasche 1980). The photosynthetic depletion of dissolved carbon dioxide from lake water, the attendant pH rise and the predominance of bicarbonate or carbonate ions as carbon source also restrict the growth of certain species (Talling 1976, 1985).

These relationships conform to Liebig's Principle, as restated by Talling (1979), that crop yield is proportional to the amount of the nutrient present in the medium relative to the needs of the organism. However, precursor processes involving the absorption and assimilation of limiting nutrients will determine the rate at which the resource-limited yield (or carrying capacity, K) is approached. Under optimal conditions of supply, the resource-saturated rate of (exponential) population change is represented by

$$dN/dt = Ne^r \qquad (1)$$

(where N is the population and e^r is an expression of the maximum intrinsic rate of growth; r is the exponential growth constant). Under conditions of resource-limitation, the rate of change is diminished in accord with the classic Lotka–Volterra equation of population growth:

$$dN/dt = Ne^r \, (1 - N/K) \qquad (2)$$

The complex relationship between the growth rate of phytoplankton and the availability of given nutrients has been extensively investigated (for recent reviews see Rhee 1982; Tilman, Kilham & Kilham 1982; Turpin 1987). Generally, the relationship of nutrient-limited growth rate to the resource takes the form of a Monod curve (Figs. 14.2c,d).

Algae differ in their maximal rates of uptake and assimilation of nutrients. It is possible to differentiate species, on the basis of the ratio between maximal rates of nutrient uptake and growth, as being either (velocity-) adapted to capitalize upon favourable nutrient supplies or (storage-) adapted to fluctuating resources. Chronically low nutrient concentrations select (affinity-adapted) species with low half-saturation characters (Sommer 1984). Interspecific differences are of sufficient scale that, under a given steady state of nutrient limitation, one species is always likely to be able to maintain a faster rate of growth than its competitors. Moreover, the outcome is influenced by different limiting resources

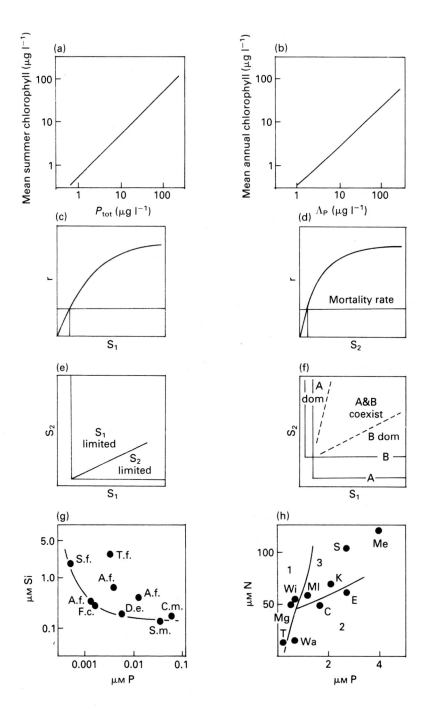

(Tilman & Kilham 1976), which can be represented by the graphical-mechanistic model of Tilman (1980). If allowance is made for mortality, a positive increase of alga A can be said to occur when the supply of resources exceeds the minimal requirements (Fig. 14.2e); depending on the ratio of available resources it may be said to be S_1- or S_2-limited. Superimposition of the corresponding representation for species B indicates the sectors where A and B will dominate the other and those where they may coexist for long periods (Fig. 14.2f).

The outcome of this kind of species interaction is clearly relevant to the structure of communities. On the basis of data reviewed in Tilman, Kilham & Kilham (1982), a declining Si:P ratio should select for particular diatom species at given points along the gradient (Fig. 14.2g) and, eventually, for green algae or cyanobacteria. Changes in dominance of natural plankton assemblages have been shown to conform to this hypothesis (Kilham 1978; Sommer 1983), provided the bias of inoculum size is not overwhelming (Reynolds 1986). Similarly, a declining gradient of N:P ratio favours the replacement of green algae by Cyanobacteria (Rhee 1978; Rhee & Gotham 1980). Certainly, there is evidence that characteristic differences in the assemblages dominating lakes are influenced by the relative loading rates of nitrogen and phosphorus (Fig. 14.2h), effects which have been imitated experimentally in enclosure experiments (Reynolds 1986). Moreover, just as there are several potential and simultaneously limiting resources, so there are many corresponding physiological niches (Tilman, Kilham & Kilham 1982) and simultaneously coexisting species (Petersen 1975). Even so, the number of species apparently coexisting in the plankton is recognized to be

---

FIG. 14.2. The role of limiting nutrients in planktonic systems: relationships between (a) mean summer chlorophyll concentration and epilimnetic total phosphorus concentration derived from closed systems by Lund & Reynolds (1982) and (b) mean annual chlorophyll concentration and the weighted phosphorus loading factor (Vollenweider 1976); idealized plots of the growth rate of an alga, A, against the supply of separate resources (c) S_1 and (d) S_2, from which graphical-mechanistic model (e) can be constructed and combined (f) with that of another algas B, to predict the outcome of direct competition (based on figures in Tilman, Kilham & Kilham 19a2); (g) competitive abilities of diatoms for silicate and phosphate (redrawn from Tilman, Kilham & Kilham 1982); A.f. = *Asterionella formosa;* C.m. = *Cyclotella meneghiniana;* D.e. = *Diatoma elongatum;* F.c. = *Fragilaria crotenensis;* S.f. = *Synedra filiformis;* S.m. = *Stephanodiscus minutus;* T.f. = *Tabellaria flocculosa*) and (h) the effect on phytoplankton assemblages in a selection of natural lakes of differing ratios of N to P (data from Kira *et al.* 1984): C, Crose Mere and Wi, Windermere, England; E, Esrum Sø, Denmark; K, Kasumigaura and S, Sagami-Ko, Japan; Me, Mendota, T, Tahoe and Wa, Washington, USA; Mg, Lago Magiorre, Italy and Ml, Malaren, Sweden; 1—lakes which are dominated by diatoms and chrysophytes; 2—lakes in which nitrogen-fixing cyanobacteria dominate for substantial parts of the year; 3—lakes dominated by *Microcystis* for long periods.

too large to be consistent with Hardin's (1960) principle of competitive exclusion. The eventual solution to this 'paradox of the plankton' (Hutchinson 1961) necessarily took account of additional factors operating in planktonic environments.

VERTICAL MIXING

The overriding importance of externally-driven hydraulic mixing on the structural organization of pelagic ecosystems has long been recognized (see Lund 1959; Margalef 1978; Legendre & Demers 1984). The energy sources are well-known, including Coriolis' Force (induced by the Earth's rotation), convection (caused by alternate heating and cooling at the air-water interface) and wind-stress (inducing surface waves, Langmuir circulations, deep return currents and a 'spectrum' of intermediate eddies). The broad scales of motion have been characterized, in general terms (Scott *et al.* 1969; Mortimer 1974; Bengtsson 1978; Csanady 1978) and in lakes of varying size (Boyce 1974; Thorpe 1977; Smith 1979. Quay *et al.* 1980; Lewis 1982), including some ice-covered examples (Colman & Armstrong 1983; Welch & Bergmann 1985). However, it is only relatively recently that attention has been focused towards the finer spatial and temporal scales of turbulent dissipation relevant to the lives of individual planktonic organisms (e.g. Abbot, Powell & Richerson 1982; Denman & Gargett 1983; Powell *et al.* 1984; Imberger 1985a).

The key variable is the vertical extent of mixing, which is related through the velocities of the major currents and eddies to the external driving energy, but is resisted by differences in water density, ρ, brought about by solar heating at the surface. Where the driving energy is insufficient to overcome the resistance, turbulence rapidly subsides and the temperature-induced density gradient stabilizes; the less dense (epilimnetic) water near the surface becomes separated from the denser (hypolimnetic) water below.

Once established, this thermal stratification effectively restricts active mixing to the epilimnion until surface cooling or severe storms depress and, eventually, overcome the intermediate (metalimnetic) density gradient.

Superimposed upon this seasonal variability of thermal stability (Figs. 14.3a–d) are diel cycles of warming and cooling, modified by day-to-day variations in solar imput and wind activity. During the course of consecutive days, the mixed-layer depth can fluctuate from a few millimetres to the full depth of the epilimnion (Figs. 14.3e,f). At another extreme, there are lakes which are permanently mixed (e.g. Volta Grande reservoir, Fig. 14.3d) or stratified (e.g. Lake Fryxell: Vincent 1981). Lakes therefore vary enormously in the stability and duration of stratification and, hence, degree of

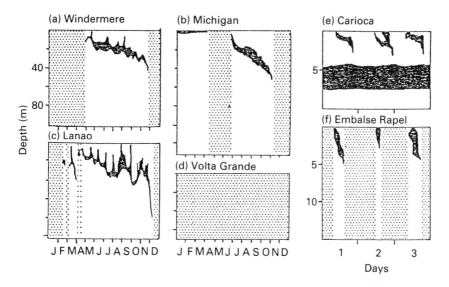

FIG. 14.3. Stratification and mixing in planktonic systems: seasonal metalimnia (silhouette) and periods of full mixing (stipple) in (a) Windermere, England, (b) Michigan, USA (from data in Kira *et al.* 1984), (c) Lake Lanao, Philippines (based on data in Lewis 1978) and (d) Volta Grande Reservoir, Brazil (author's unpublished data); diel stratification in (e) Lagoa Carioca, Brazil (data of Reynolds, Tundisi & Hino 1983) and (f) Embalse Rapel (Reynolds *et al.* 1986).

mixing according to their morphometry, exposure and geographical location (Lewis 1983; Patalas 1984).

Many of the direct consequences of fluid motion upon phytoplankton relate to the extent of their entrainment within turbulent eddies and to the relative delay in gravitational settling through the water column. The intrinsic settling velocities of even the dense, non-motile diatoms ($\rho = 1040$ to 1250 kg m^{-3}; Reynolds 1984a) are generally $\ll 50 \times 10^{-6}$m s^{-1}, whereas some of the larger dinoflagellates (Heaney & Furnass 1980) and colonial cyanobacteria (Humphries & Imberger 1982) regulate their own movements at velocities of 10^{-4} to 10^{-3}m s^{-1}. However, even light winds (< 3 m s^{-1}) generate vertical downwelling currents ($1-2 \times 10^{-3}$m s^{-1}; Arai 1984) sufficient to effectively entrain most plankton (George & Edwards 1976). The vertical extent of the entrainment approximates to the wind-mixed layer. Vertical advection times through the mixed layer (in the order of minutes to a few hours) are generally shorter than the times taken for algae to clear a quiescent layer of the same depth (hours to days), so they remain dispersed. (Depending upon the size and wind fetch of the water body, horizontal mixing times may be several orders of magnitude greater than vertical mixing times, allowing distinct horizontal patches to develop (George & Edwards 1976; Heaney 1976).

At the base of the mixed layer, where turbulent advection is minimal, it becomes possible for algae to maintain station within or to sink or swim through this layer at something approaching their still-water rates. Once disentrained however, motile or buoyant organisms can recover position within the turbulent flow. The extent of vertical mixing therefore has a profound influence on the spatial organization of the communities (Margalef 1978; Kemp & Mitsch 1979).

The rate and frequency of vertical mixing also impinges on the gross aspects of the environment experienced by the plankton. Vertical mixing promotes homogeneity in the distribution of solutes, often overriding the impact of localized inputs and uptake of dissolved gases and nutrients (George 1981). Nutrient uptake is concentrated in the epilimnion, often leading to local depletion. Sedimentation of particles, including algae, enriches the nutrient content of the hypolimnion but decomposition depletes the reserves of dissolved oxygen. The extent of this segregation depends upon the nutrient base (trophic level) of the lake and the relative volumes of the two layers (e.g. Jones 1976). Conversely, episodes of increased mixing return nutrients to the epilimnion and stimulate algal growth (Reynolds 1976; Stauffer & Armstrong 1984).

Vertical mixing also carries algae through the light gradient. Qualitative effects upon photosynthesis have been recognized for some time, especially with respect to the avoidance of near-surface photoinhibition (Harris 1978). The light received by single cells under certain idealized conditions is represented in Fig. 14.4 given vertical mixing times of around 30 minutes. The responses of cells circulating through very clear epilimnia (Fig. 14.4b) must be primarily at the biochemical level, through altered fluorescence, carbon metabolism, photorespiration and glycollate production (Harris 1980). Deeper mixing with respect to light penetration (Figs. 14.4c,d) also diminishes the time spent under conditions where net photosynthesis is possible. Such 'Lagrangian' views of the light received by mixed-layer algae can be integrated into realistic ('Eulerian') quantifications of their mean photosynthetic production (Falkowski & Wirick 1981; Farmer & Takahashi 1983) and light/shade adaptation (Falkowski 1983; Lewis, Cullen & Platt 1984).

Assuming algae accommodate to these high-frequency fluctuations in the light environment, population growth rates inevitably respond to reductions in the light 'dose' (irradiance × time: Gibson & Foy 1983) consequential upon mixing through optically deep water columns, (Fig.

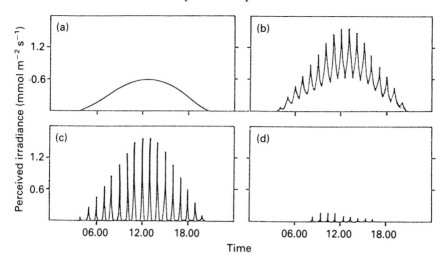

FIG. 14.4. Effect of full mixing on daily course of light received by phytoplankton cells (a) in a stable metalimnion, 5 m below the surface irradiated as in Fig. 14.1 (c) and vertical extinction coefficient, $\epsilon = 0.2\,m^{-1}$; (b) entrained in the epilimnion, mixed to 5 m depth; (c) entrained in a deeper and more turbid epilemnion (z_m ' 10 m; $\epsilon = 1.5\,m^{-1}0$; and (d) similarly entrained but under the winter insolation shown in Fig. 14.1(b). Epilimnetic mixing time of 30 minutes is assumed in each instance.

14.4c) which may be simultaneously subject to short day-length and weak insolation (Fig. 14.4d).

Abrupt variations in the depth of the mixed layer and integral light lead to altered rates of algal growth and loss, which, if sustained, lead to altered community composition. While the dynamic responses of individual species are apparently related to their morphological and physiological properties (Reynolds 1984b), the community responses tend to be time-lagged (Wall & Briand 1980; Trimbee & Harris 1983; Lhotsky 1985) pending the establishment of dominance by the species selected. The more frequent are the fluctuations in stability, then the poorer are the opportunities for either individual species to develop resource-limited populations (Robinson & Sandgren 1983; Smith 1985) or to interact with other species (Reynolds & Reynolds 1985) or for the community to approach some stable, equilibrium condition (Harris 1983; Powell & Richerson 1985). Only where the frequency of stability alterations is similar to or less than the generation times of algae (or instance, in warm-water lakes mixing and restratifying on diel cycles; see Ganf 1974b; Vincent, Neale & Richerson 1984) does the species composition accommodate, rather than respond, to the cycles of mixing. It is clear that planktonic algae collectively possess a suite of responses, ranging from the biochemical, through the physiological to the behavioural and the autoecological, that enables them to survive in the variable environments of lakes. These responses are summarized in Fig 14.5; the scale of the response is appropriate to the temporal scale of the fluctuation.

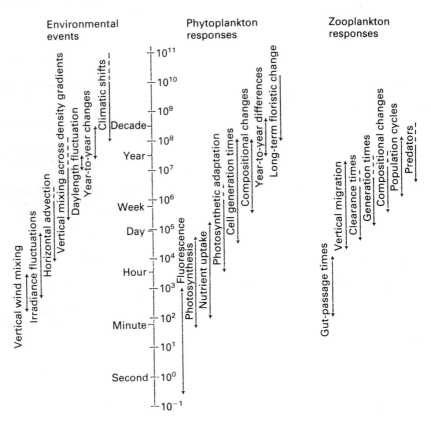

FIG. 14.5. Temporal scales of environmental variability in planktonic systems and the periods over which the various responses of phytoplankton and zooplankton are manifest. Based on Fig. 1 of Harris (1980) with additional data from Imberger (1985b) and author's unpublished information.

PHYTOPLANKTON ASSEMBLAGES

In addition to light and nutrients, many abiotic (temperature, trace elements, toxic substances) and biotic (including microbial heterotrophs,

pathogens, parasites, herbivorous feeders and—indirectly—their predators) factors influence the species composition of the phytoplankton. With
so many simultaneous and changing factor interactions, together with the
inherent stochasticity of key events (Talling 1951; Robinson & Sandgren
1983), it is scarcely surprising that the prediction of phytoplankton
dominance at the species level is still poor. However, by grouping species
with similar morphological and physiological properties, Reynolds (1984b)
showed that community responses to given sets of environmental conditions
could be predicted with a high level of probability. These associations are
listed in Fig. 14.6 together with summaries of their characteristics; several
idealized lake environments are represented in Fig. 14.7.

On the premise that (i) planktonic algae will grow wherever the complete
spectrum of their minimal environmental requirements is met and that (ii)
their optimal performance is achieved when all material requirements are
saturated, it is possible to conceive a natural limnetic system (Fig. 14.7a)
which would support near-maximal growth rates of most species. In this
ideal environment, the species sustaining the fastest rates of growth (r)
would be expected to dominate initially. Those which either maintained
slower rates of growth for longer or exploited the resources available (K) to
lower limits or were able to field a sufficiently large initial inoculum, might
eventually gain dominance. This succession would be accelerated by
differential rates of loss, for instance, arising through selective grazing of
small cells (see below) or sinking out of non-motile species.

The fastest-growing forms are generally small-celled, metabolically-
active species of Groups X and Y (Fig. 14.6). Shallow ponds and epilimnia
experiencing high nutrient loads, represented in Fig. 14.7a, provide good
examples of systems briefly supporting dominant populations of these
organisms; the superimposition of short hydraulic retention times
apparently extends the phase of their dominance, largely by excluding
slow-growing organisms, as in sewage lagoons (Uhlmann 1971) and in the
small, ground-water fed hollow, Montezuma's Well in Arizona (Boucher,
Blinn & Johnson 1984). Small algae may also proliferate near the surface of
larger fertile basins under conditions of extreme thermal stability, including
those obtaining under ice (Kalff & Welch 1974).

Where the nutrient-resource base is smaller, or is depleted by algal
uptake, the productive capacity is correspondingly reduced (Fig. 14.7b). In
extremes, only the hypolimnion might be capable of supplying nutrients and
the main productive base is removed to stable layers near the bottom of the
euphotic zone (Moll & Stoermer 1982), as shown in Fig. 14.7c. The
metalimnetic maxima of (group E) Chrysophytes (Pick, Nalewajko & Lean
1984) and, perhaps, of filamentous Cyanobacteria (group R) observed in

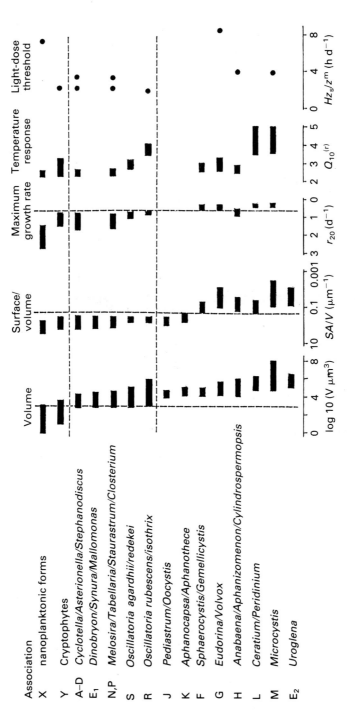

FIG. 14.6. Some properties (volume; surface area/volume ratio; r_{20} = max. growth rate at 20°; Q_{10} of growth rate 10–20°; the light-dose threshold) of phytoplankteon algae grouped into the ecological categories of Reynolds (1984b). The horizontal lines separate the small, high surface area/volume forms from larger, high-SA/V and larger low SA/V organisms (data mainly from Reynolds 1984a, 1986 and unpublished).

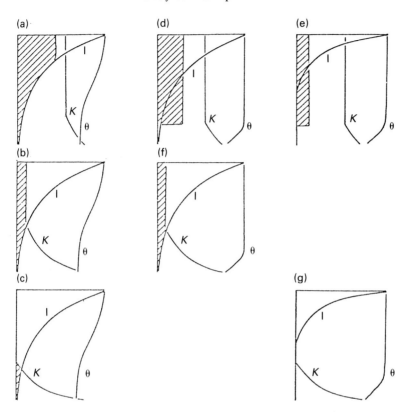

FIG. 14.7. Idealized profiles of the depth distribution of irradiance (I), limiting-nutrient concentration (K) and temperature $(\theta$, to indicate relative mixed depth) and a representation of their capacities to support net phytoplankton production (shaded areas); in (a) nutrient resources are saturating, the limit on productivity in the zone of light saturation being imposed by the intrinsic biochemical properties of the algae; (b) and (c) show the limitation imposed by low or depleted nutrients; (d) and (e) show increased depths of mixing relative to light penetration and the imposition of light limitation on productivity; (f) shows a combination of mild resource and light reduction but (g) simultaneous extremes of both factor groups cannot sustain growth.

many stratified nutrient-deficient lakes (Reynolds & Walsby 1975; Klemer 1976) exemplify this behaviour. In strongly segregated environments, selection favours algae capable of self-regulated movements, either up, to gain access to light, or down, to gain access to nutrients (Ganf & Oliver 1982; Raven & Richardson 1984). Motility, coupled with large unit size and storage- or affinity-adaptation of nutrient uptake, is shown variously by members of the E, G, H, L and M groupings (Reynolds 1984b). The relative efficiencies and nutrient limitations of these species are echoed in the well-known successional sequences of dominance (E/G → H → M → L) during the stratified phase and increasingly severe epilimnetic nutrient depletion (Reynolds 1976, 1980; Lewis 1978, 1986; Ashton 1985). Such sequences have also been imitated in enclosure experiments (Reynolds 1984a, 1986).

As populations develop, they become increasingly subject to light limitation, or there may be increased wind-mixing, such that the optical depth of the mixed column is increased. These conditions (Figs. 14.7d,e) are frequently generated at the onset of the autumn and spring-mixing phases in temperate lakes when stratification breaks down, as well as during seasonal episodes of increased wind activity and/or surface cooling in tropical lakes (Lewis 1983). In time, they usually come to be dominated by non-motile species (desmids and, especially, diatoms of groups A, B, C, D, N, P) or weakly motile species of *Cryptomonas* (Y) and *Oscillatoria* (R, S). The species selected have moderate to high ratios of surface area to volume, some storage-adaptation and, above all, a capacity for efficient harvesting of light energy at high frequencies and amplitudes of fluctuation (Reynolds 1984b).

These two distinct trends of increasing resource-stress (reading downward in Fig. 14.7) and of increasing hydraulic disturbance of the light field (rightward) are not independent: well-mixed, nutrient-deficient epilimnia (see Fig. 14.7f) probably typify many larger examples of the world's lakes, supporting plankton dominated by diatoms (A, B, N), chrysophytes (E) and colonial Chlorophyceae (F). However, extreme combinations of high hydraulic disturbance and low resources (Fig. 14.7g) offer untenable habitats for autotrophs (cf. Grime 1979). Moreover, these axes also interact with time because the relative importance of stress and disturbance alters with season (see Southwood Chapter 1). The range of possible responses on either axis is represented by the extremes of a species whose growth and biomass is very sensitive to factor variation and another whose growth is checked but whose biomass persists. Organisms that conserve elaborated biomass in this way include many of the larger, slower-growing species, whereas the more sensitive species are potentially able to recoup biomass losses by virtue of faster growth rates.

The distribution of Reynolds' (1984b) algal groupings in relation to increasing disturbance and increasing stress are represented in Fig. 14.8a. Those occupying the top left corner are less tolerant of extremes but are more responsive to favourable environments. Those extending towards the edges of the ranges are more likely to dominate in either disturbed or stressed environments. Superimposed upon the same outline (Fig. 14.8b) are shapes enclosing dominant phytoplankton associations represented in named lake systems. It is proposed that seasonal changes in dominance are

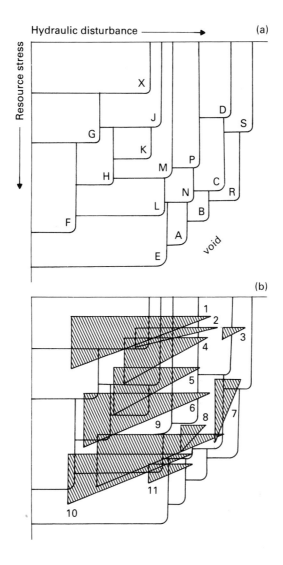

FIG. 14.8. Approximate ecological distributions of the species categories (Fig. 14.6) in relation to resource depletion and relative disturbance as represented in Fig. 14.7. In the lower panel, triangles are superimposed on the same outline to delimit the categories of species reported to dominate the named lakes in the literature: 1, North Shropshire pools, England (Reynolds 1973); 2, Montezuma's Well, USA (Boucher, Blinn & Johnson 1984); 3, Polder lakes, Netherlands (Berger 1975); 4, Norfolk Broads, England (Leah, Moss & Forrest 1980); 5, Hamilton Harbour, Canada (Harris & Piccinin 1980); 6, Crose Mere and, 10, Windermere, England (Reynolds 1980); 7, Lough Neagh, Northern Ireland (Gibson *et al.* 1971); 8, Volta Grande Reservoir and 9, Lagoa Carioca, Brazil (unpublished data of the author); 11, Millstättersee, Austria (Findenegg 1943).

community responses to alternating episodes of developing stress and increased frequency of hydraulic disturbance.

ZOOPLANKTON ASSEMBLAGES

Species composition and abundance in the zooplankton are influenced, on the one hand, by specific dietary preferences, feeding mechanisms and the availability of suitable foods and, on the other, by the intensity of predation. These factors are not independent and the interrelationships between them vary in time and space (see also Harris, Chapter 15).

Each of the major animal groups represented in the freshwater plankton—rhizopods, ciliates, rotifers and crustaceans—include species which are partly or wholly herbivorous. Some of these, especially among the rhizopods (Canter 1979) and rotifers (Pourriot 1977), are specialized feeders. In general, however, most common species of planktonic herbivores potentially compete for suspended particles—whole algae, bacteria and organic detritus—in overlapping size ranges (Gliwicz 1977, 1980). The lower and upper limiting sizes of particle that can be ingested are determined by the structure of the feeding apparatus, between 0.5 and 20 μm among rotifers (Pourriot 1977) and between c. 1 and 60 μm in the cladocera (Gliwicz 1980; Ganf & Shiel 1985). Calanoids supplement filter feeding by grasping and fragmenting larger algae (up to c. 50 μm) that they locate and select individually (Alcaraz, Paffenhofer & Strickler 1980). Filtering small particles and grasping large particles have been recently recognized as alternative feeding modes in *Bosmina* (DeMott & Kerfoot 1982; Bleiwas & Stokes 1985).

Species composition of zooplankton is supposed to be governed by the provisions of the well-known size-efficiency hypothesis (SEH) of Brooks & Dodson (1965; see also Hall *et al.* 1976). This recognizes (i) the selective advantage to larger animals that could feed more efficiently and on a wider size-range of foods; and (ii) that predation, especially by planktivorous fish, malacostracans and insect larvae tends to select against the larger cladocerans and calanoids, allowing small cladocerans and rotifers to dominate. The second part of this hypothesis has been tested in many experimental investigations but control by predators has been difficult to demonstrate other than in short-term bag experiments (Andersson *et al.* 1978), or in shallow ponds (Leah, Moss & Forrest 1980; Shapiro *et al.* 1982). The problem is one of scaling in that the generation times of herbivorous zooplankton are generally much shorter (weeks) than those of their main predators (months, years). A population of planktivorous fish, which was sufficiently large (some 600–900 kg fresh weight ha^{-1}, Gliwicz & Preis 1977; Koslow 1983) to suppress an increase in the planktonic herbivores would starve when prey was scarce. In reality, natural stocks of facultative planktivores (typically < 300 kg ha^{-1}) are geared to some lower, ambient level of food resources which nevertheless permits the zooplankton to increase under favourable conditions.

As with the phytoplankton, the size and anatomical complexity of animals influence both the maximal rates of population increase and their sensitivity to low temperature. Ferguson, Thompson & Reynolds (1982) demonstrated the generation times of small rotifers (*Keratella, Ascomorpha* spp.) in experimental enclosures, to be about 5 days at 15–20°, and of 9–10 days at *c.* 5°. *Daphnia* scarcely grew below 7–8° but an unpredated population was able to double its biomass in about 4.5 days and to recruit the next generation of individuals (nearly five times as many) within 13 days. This resulted in an 11- or 12-fold increase in the daily food requirement over about 2 weeks (Reynolds 1984a).

Such growth depends upon a ready supply of suitable foods. In his exhaustive study on the nutrition of *Daphnia pulex,* Lampert (1977a,b) derived regressions for the minimum maintenance and maximum assimilation requirements (as organic carbon) in terms of body size and weight. Given the individual filtration rates, which are also described empirically as a function of body size (Burns 1969), the range of food concentrations between the minimum for survival and the requirement to sustain maximal growth can be shown to be 0.08–0.5 μg C ml^{-1}. Equivalent concentrations of specific algal and bacterial cells are given in Reynolds (1984a).

The impact of grazing on the abundance (N) and rates of change (dN/dt) of phytoplankton may be approximated from the proportion (a) of water notionally swept clear of food particles by the filter-feeding population

$$a = B.F \tag{3}$$

where B is the density of grazers (l^{-1}) and F is the average individual filtration rate (ml d^{-1}). If there are i species (or ontogenetic stages) of grazers with differing filtration rates, the community filtration rate is given by the sum of the i products of densities and filtration rates (Reynolds 1986).

$$dN/dt = Ne^{r-a} \tag{4}$$

Increase remains positive so long as $r > a$. Although grazing regenerates some nutrients from digested algae (e.g. Lehman 1980), the finite resource base of the algae contributes to a decline in r, while increased feeding of an expanding grazer population raises the level of a, until eventually N decreases. As N falls to the equivalent of 0.08 μg C ml^{-1}, further growth of the animal population can no longer be sustained. Thus, there is little evidence of any correlation between the numbers of phytoplankton species reaching their peak populations and the biomass of unpredated planktonic herbivores (Fig. 14.9a). There is, however, a clear response of filter-feeders to concentrations of filterable foods exceeding 0.08 μg C ml^{-1} (Lampert 1977c; Reynolds 1984a); the depletion of filterable foods to < 0.5 μg C ml^{-1}, even when other, non-available foods are present, is soon followed by collapse of the filter-feeding biomass (Fig. 14.9b–d).

Evidence points to the resource side of the SEH being as important as its predation aspects, at least so far as the filter feeders are concerned (Gliwicz

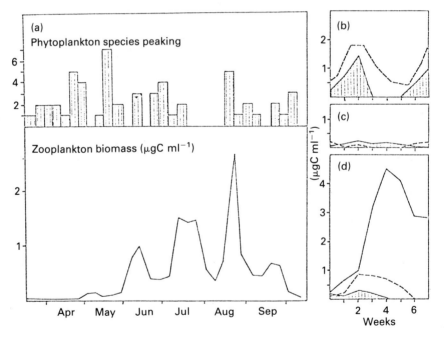

FIG. 14.9 (a) Season-long comparison between the numbers of phytoplankton species attaining maximal populations and the biomass of the filter-feeders in an experimental enclosure, in which phytoplankton was nutrient-saturated and herbivores were unconstrained by predation; original representation of data of Ferguson, Thompson & Reynolds (1982); (b)–(d) typical variations in the filter-feeding biomass (broken heavy line) in relation to fluctuations in the biomass of phytoplankton (fine, continuous line) and the size-fraction available to filter-feeders (hatched). Details extracted from other enclosure experiments, shown in Fig. 14.8. of Reynolds (1986).

1980). Indeed, fluctuations in their abundance probably represent lagged, tracking responses to the fluctuating availability of suitable foods (Harris 1985; see also Fig. 14.5) which itself is regulated by environmental constraints. At high concentrations of food particles, filter feeding is a better foraging strategy than selective feeding, yielding a high nutritive return for feeding effort (Lehman 1976). Moreover, while individual needs are satisfied by the concentration of available foods, there is little likelihood of competition between the responding species. Competition increases when food concentrations decline. Near the threshold levels, however, or where there is a high concentration of non-edible particles (Allanson & Hart 1979; Arruda, Marzolf & Faulk 1983), selective feeding (for instance, of calanoids) maintains a more productive food return for feeding effort (Reynolds 1984a). This offers a functional explanation for the frequent observation that the zooplankton in less productive lakes, as well as at later successional stages in more eutrophic waters, is typically dominated by calanoids and/or small cladocera, rather than by filter-feeding Daphniids and rotifers (Lampert 1977c; Gliwicz 1977, 1980; Hillbricht-Ilkowska, Spodniewska & Weglenska 1979; Bays & Crisman 1983; Byron, Fott & Goldman 1984).

THE ORGANIZATION OF PLANKTON COMMUNITIES

The foregoing sections have emphasized the small scales of organism size and generation times of plankton. These are considered to be essential adaptations to life in a fluid, three-dimensional environment that is susceptible to density-dependent fluctuations in resources and to frequent disturbance by external kinetic energy, affording relatively brief opportunities for their exploitation. Given environmental conditions of some constancy, community structure can follow an ecological succession of dominant species towards a climax characterized by the clear dominance of one species. To judge from the experiments of Tilman & Kilham (1976) and of Sommer (1983), in which mixed-species associations were raised in continuous culture at realistic temperatures and day–night cycles, achievement of the equilibrium state required some 30–45 days. Comparable examples drawn from nature (Fig. 14.8b), such as the spring dominance of diatoms or *Oscillatoria* in ice-free lakes, the spatial organization observed in very stable continental (Reynolds, Tundisi & Hino 1983) and ice-covered polar lakes (Vincent 1981) or the dominance of *Microcystis* in tropical systems with diel mixing (Ganf 1974b), each develop some sort of equilibrium only after substantial periods of unchanging ambient conditions.

Without necessarily entering the debate as to whether 'equilibrium' or 'non-equilibrium' concepts should be applied in evolutionary ecology (see, especially, Roughgarden 1983; Connell 1983; Hoffman 1985), it is a fact that the atmospheric changes that generate variations in the intensity of stratification and wind-mixing often occur on scales appreciably shorter than 30 days: the 4–11 days detected by Trimbee & Harris (1983) is a more realistic estimate for weather events in middle latitudes. Each event represents an environmental perturbation which may be sufficient to halt or reverse progress towards the equilibrium point or to divert towards another. The outcome is approximately equivalent to Connell's (1978) intermediate disturbance hypothesis. Disturbance plays a similar role in some freshwater benthic communities (see Hildrew & Townsend, Chapter 16).

Planktonic communities in freshwaters are, consequently, often far from being at equilibrium with the contemporaneous environmental conditions, their composition having been influenced by either spatially or temporally remote events. The view has been advanced that similar non-equilibrium conditions in the sea are consequential upon spatial heterogeneity ('patchiness') and local differentiation between species with (unaltered) equilibrium-based relations. Yet McGowan & Walker (1985) were unable to demonstrate significant change in the rank-order of species abundance on spatial scales up to 800 km, despite strong small-scale, seasonal and inter-year heterogeneity. In all but the largest lakes, it has to be argued that spatial patches rarely survive long enough for growth responses to become manifest (Reynolds 1984a). On the other hand, temporal variations in the scale of mixing and diffusivity destabilize any establishing equilibrium by invoking different interactions among physical and biological processes and selecting alternative species responses. Community responses are the inevitable consequence of the individualistic responses to change but they are both lagged and damped by the populations existing hitherto. This temporal disequilibrium must contribute to the 'paradoxical' (Hutchinson 1961) structure of planktonic communities.

Hydrodynamic properties of lakes and oceans are now widely recognized as the driving force in planktonic systems; physical, chemical and biological factors are but proximal agents through which the hydrodynamic variability is transmitted to the living organisms (Legendre & Demers 1984). Given the temporally patchy, frequently disturbed nature of the aquatic environment and ephemeral resources, the community responses will always be led by individualistic responses of the component species (Price 1984). There is, then, abundant evidence to support Margalef's (1978) contention that the internal organization of pelagic ecosystems is generally weak and that it is dominated by the input of external energy.

ACKNOWLEDGMENT

I am grateful to Jessie Maslen for the careful preparation of the typescript of this paper.

REFERENCES

Abbot, M.R., Powell, T.M. & Richerson, P.J. (1982). The relationship of environmental variability to the spatial patterns of phytoplankton biomass in Lake Tahoe. *Journal of Plankton Research*, **4**, 927–41.

Alcaraz, M., Paffenhofer, G.A. & Strickler, J.R. (1980). Catching the algae: a first account of visual observations on filter-feeding calanoids. *Evolution and Ecology of Zooplankton* (Ed. by W.C. Kerfoot), pp. 241–8. University Press of New England, Hanover, New Hampshire.

Allanson, B.R. & Hart, R.C. (1979). Limnology of P.K. le Roux Dam. *Reports, Rhodes University Institute for Freshwater Studies.* **11** (7), 1–3.

Andersson, G., Berggren, H., Cronberg, G. & Gelin, C. (1978). Effects of planktivorous and benthivorous fish on organisms and water chemistry in eutrophic lakes. *Hydrobiologia*, **59**, 9–15.

Arai, T. (1984). Measurement of vertical movement of lake water. *Verhandlungen der internationale Vereinigung für theoretische und angewandte Limnologie*, **22**, 108–11.

Arruda, J.A., Marzolf, G.R. & Faulk, R.T. (1983). The role of suspended sediments in the nutrition of zooplankton in turbid reservoirs. *Ecology*, **64**, 1225–35.

Ashton, P.J. (1985). Seasonality in Southern Hemisphere freshwater phytoplankton assemblages. *Hydrobiologia*, **125**, 179–90.

Bannister, T.T. (1974). Production equations in terms of chlorophyll concentration, quantum yield and upper limit to production. *Limnology and Oceanography*, **19**, 1–12.

Bays, J.S. & Crisman, T.C, (1983). Zooplankton and trophic state relationships in Florida lakes. *Canadian Journal of Fisheries and Aquatic Sciences*, **40**, 1813–19.

Bengtsson, L. (1978). Wind-induced circulation in lakes. *Nordic Hydrology*, **9**, 75–94.

Berger, C. (1975). Occurrence of *Oscillatoria agardhi* Gomont in some shallow eutrophic lakes. *Verhandlungen der internationale Vereinigung für theoretische und angewandte Limnologie*, **19**, 2687–97.

Bleiwas, A.H. & Stokes, P.M. (1985). Collection of large and small food particles by *Bosmina*. *Limnology and Oceanography*, **30**, 1090–2.

Boucher, P., Blinn, D.W. & Johnson, D.B. (1984). Phytoplankton ecology in an unusually stable environment (Montezuma Well, Arizona, USA). *Hydrobiologia*, **119**, 149–60.

Boyce, F.M. (1974). Some aspects of Great Lakes physics of importance to biological and chemical processes. *Journal of the Fisheries Research Board of Canada*, **31**, 689–730.

Brierley, S. (1985). *The Effects of Artificial Overturn on Algal Populations.* Research & Development Project Report RP 85–070, Severn-Trent Water Authority, Birmingham.

Brooks, J.L. & Dodson, S.J. (1965). Predation, body-size and composition of the plankton. *Science*, **150**, 28–35.

Burns, C.W. (1969). Relation between filtering rate, temperature and body size in four species of *Daphnia*. *Limnology and Oceanography*, **14**, 693–700

Byron, E.R., Fott, C.L. & Goldman, C.R. (1984). Copepod and Cladoceran success in an oligotrophic lake. *Journal of Plankton Research*, **6**, 45–65.

Canter, H.M. (1979). Fungal and protozoan parasites and their importance in the ecology of the phytoplankton. *Report of the Freshwater Biological Association*, **47**, 43–50.

Colman, J.A. & Armstrong, D.E. (1983). Horizontal diffusivity in a small, ice-covered lake. *Limnology and Oceanography*, **28**, 1020–6.

Connell, J.H. (1978). Diversity in tropical rain forests and coral reefs. *Science,* **199,** 1302–10.

Connell, J.H. (1983). On the prevalence and relative importance of interspecific competition: evidence from field experiments. *American Naturalist,* **122,** 661–96.

Csanady, G.T. (1978). Water circulation and dispersal mechanisms. *Lakes: Chemistry, Geology, Physics.* (Ed. by A. Lerman), pp. 21–64. Springer-Verlag, New York.

DeMott, W.R. & Kerfoot, W.C. (1982). Competition among cladocerans: nature of the interaction between *Bosmina* and *Daphnia. Ecology,* **63,** 1949–66.

Denman, K. & Gargett, A.E. (1983). Time and space scales of vertical mixing and advection of phytoplankton in the upper ocean. *Limnology and Oceanography,* **28,** 801–15.

Falkowski, P.G. (1980). Light–shade adaptation in marine phytoplankton. *Primary Productivity in the Sea* (Ed. by P.G. Falkowski), pp. 99–119. Plenum, New York.

Falkowski, P.G. (1983). Light-shade adaptation and vertical mixing of marine phytoplankton: a comparative field study. *Journal of Marine Research,* **41,** 215–37.

Falkowski, P.G. & Wirick, C.D. (1981). A simulation model of the effects of vertical mixing on primary productivity. *Marine Biology,* **65,** 69–75.

Farmer, D.M. & Takahashi, M. (1983). Effects of vertical mixing on photosynthetic responses. *Japanese Journal of Limnology,* **43,** 173–81.

Ferguson, A.J.D., Thompson, J.M. & Reynolds, C.S. (1982). Structure and dynamics of zooplankton communities maintained in closed systems, with special reference to the algal food supply. *Journal of Plankton Research,* **4,** 523–43.

Findenegg, I. (1943). Untersuchungen über die Ökologie und die Produktionsverhältnisse des Planktons im Kärnter Seengebiete. *Internationale Revue der gesamten Hydrobiologie,* **43,** 368–429.

Ganf, G.G. (1974a). Incident solar irradiance and underwater light penetration as factors controlling the chlorophyl a content of a shallow equatorial lake (Lake George, Uganda). *Journal of Ecology,* **62,** 593–609.

Ganf, G.G. (1974b). Diurnal mixing and the vertical distribution of phytoplankton in a shallow equatorial lake (Lake George, Uganda). *Journal of Ecology,* **62,** 611–29.

Ganf, G.G. (1980). Factors controlling the growth of phytoplankton in Mount Bold Reservoir, South Australia. *Technical Papers of the Australian Water Resources Council* No 48, 109 pp.

Ganf, G.G. & Oliver, R.L. (1982). Vertical separation of light and available nutrients as a factor causing replacement of green algae by blue-green algae in the plankton of a stratified lake. *Journal of Ecology,* **70,** 829–44.

Ganf, G.G. & Shiel, R.J. (1985). Feeding behaviour and limb morphology of the cladocerans with small intersetular distances. *Australian Journal of Marine and Freshwater Research,* **36,** 69–86.

George, D.G. (1981). The spatial distribution of nutrients in the South Basin of Windermere. *Freshwater Biology,* **11,** 405–24.

George, D.G. & Edwards, R.W. (1976). The effect of wind on the distribution of chlorophyll a and crustacean plankton in a shallow eutrophic reservoir. *Journal of Applied Ecology,* **13,** 667–90.

Gibson, C.E. & Foy, R.H. (1983). The photosynthesis and growth efficiency of a planktonic blue-green alga, *Oscillatoria redekei. British Phycological Journal,* **18,** 39–45.

Gibson, C.E., Wood, R.B., Dickson, E.L. & Jewson, D.M. (1971). The succession of phytoplankton in Lough Neagh, 1968–1970. *Mitteilungen der internationale Vereinigung für theoretische und angewandte Limnologie,* **19,** 146–60.

Gliwicz, Z.M. (1977). Food size selection and seasonal succession of filter-feeding zooplankton in an eutrophic lake. *Ekologia Polska Seria A.* **25,** 179–225.

Gliwicz, Z.M. (1980). Filtering rates, food size selection and feeding rates in cladocerans— another aspect of interspecific competition in filter-feeding zooplankton. *Evolution and*

Ecology of Zooplankton Communities (Ed. by W.C. Kerfoot) pp. 282–91. University Press of New England, Hanover, New Hampshire.

Gliwicz, Z.M. & Preis, A. (1977). Can planktivorous fish keep in check planktonic crustacean populations? A test of size-efficiency hypothesis. *Ekologia Polska Seria A,* **25,** 567–91.

Grime, J.P. (1979). *Plant Strategies and Vegetation Processes.* Wiley-Interscience, Chichester.

Hall, D.J., Threlkeld, S.T., Burns, C.W. & Crowley, P.H. (1976). The size-efficiency hypothesis and the structure of zooplankton communities. *Annual Review of Ecology and Systematics,* **7,** 177–208.

Hardin, G. (1960). The competitive exclusion principle. *Science,* **131,** 1292–7.

Harris, G.P. (1978). Photosynthesis, productivity and growth: the physiological ecology of phytoplankton. *Ergebnisse der Limnologie,* **10,** 1–163.

Harris, G.P. (1980). Temporal and spatial scales in phytoplankton ecology. Mechanisms, methods, models and management. *Canadian Journal of Fisheries and Aquatic Science.* **37,** 877–900.

Harris, G.P. (1983). Mixed-layer physics and phytoplankton populations: studies in equilibrium and non-equlibrium ecology. *Progress in Phycological Research, Vol. 2* (Ed. by F.E. Round & D.J. Chapman), pp. 1–52. Elsevier, Amsterdam.

Harris, G.P. (1985). The answer lies in the nesting behaviour. *Freshwater Biology,* **15,** 375–80.

Harris, G.P. & Piccinin, B.B. (1980). Physical variability and phytoplankton communities. IV. Temporal changes in the phytoplankton community of a physically variable lake. *Archiv für Hydrobiologie,* **89,** 447–73.

Heaney, S.I. (1976). Temporal and spatial distribution of the dinoflagellate, *Ceratium hirundinella* O.F. Müller within a small productive lake. *Freshwater Biology,* **6,** 531–42.

Heaney, S.I. & Furnass, T.I. (1980). Laboratory models of diel vertical migration in the dinoflagellate, *Ceratium hirundinella. Freshwater Biology,* **10,** 163–70.

Hillbricht-Ilkowska, A., Spodniewska, I. & Weglenska, T. (1979). Changes in the phytoplankton–zooplankton relationship connected with the eutrophication of lakes. *Symposia Biologica Hungariae,* **19,** 59–75.

Hoffman, A. (1985). Island biogeography and palaeobiology: in search for evolutionary equilibria. *Biological Reviews of the Cambridge Philosophical Society,* **60,** 455–71.

Hoogenhout, H. & Amesz, J. (1965). Growth rates of photosynthetic microorganisms in laboratory cultures. *Archiv für Mikrobiologie,* **50,** 10–25.

Humphries, S.E. & Imberger, J. (1982). *The influence of the internal structure and dynamics of Burrinjuck Reservoir on phytoplankton blooms.* Environmental Dynamics Report ED 82-023, University of Western Australia, Nedlands.

Hutchinson, G.E. (1961). The paradox of the plankton. *American Naturalist,* **95,** 137–46.

Imberger, J. (1985a). The diurnal mixed layer. *Limnology and Oceanography,* **30,** 737–70.

Imberger, J. (1985b). Thermal characteristics of standing waters: an illustration of dynamic processes. *Hydrobiologia,* **125,** 7–29.

Jones, J.G. (1976). The microbiology and decomposition of seston in open water and experimental enclosures in a productive lake. *Journal of Ecology,* **64,** 241–78.

Kalff, J. & Welch, H.E. (1974). Plankton production in Char Lake, a natural polar lake and in Meretta Lake, a polluted lake, Cornwallis Island, Northwest Territories. *Journal of the Fisheries Research Board of Canada,* **31,** 621–36.

Kemp, W.M. & Mitsch, W.J. (1979). Turbulence and phytoplankton diversity: a general model of the 'paradox of plankton'. *Ecological Modelling,* **7,** 201–22.

Kilham, S.S. (1978). Nutrient kinetics of freshwater planktonic algae using batch and semicontinuous methods. *Mitteilungen der internationale Vereinigung für theoretische und angewandte Limnologie,* **21,** 147–57.

Kira, T., Kurata, A., Nakajima, T. & Takahashi, M. (1984). *Data Book of World Lakes.* National Institute of Japan for Research Advancement, Otsu.

Kirk, J.T.O. (1983). *Light and Photosynthesis in Aquatic Ecosystems.* Cambridge University Press, Cambridge.

Klemer, A.R. (1976). The vertical distribution of *Oscillatoria agardhii* var. *isothrix. Archiv für Hydrobiologie*, **78,** 343–62.

Koslow, J.A. (1983). Zooplankton community structure in the North Sea and Northeast Atlantic: development and test of a biological model. *Canadian Journal of Fisheries and Aquatic Sciences.* **40,** 1912–24.

Lampert, W. (1977a). Studies on the carbon balance of *Daphnia pulex* De Geer as related to environmental conditions. II. The dependence of carbon assimilation on animal size, temperature, food concentration and diet species. *Archiv für Hydrobiologie (Supplementband)*, **48,** 310–35.

Lampert, W. (1977b). Ibid. III. Production and production efficiency. *Archiv für Hydrobiologie (Supplementband)*, **48,** 336–60.

Lampert, W. (1977c). Ibid. IV. Determination of the 'threshold' concentration as a factor controlling the abundance of zooplankton species. *Archiv für Hydrobiologie (Supplementband)*, **48,** 361–8.

Larkum, A.W.D. & Barrett, J. (1983). Light-harvesting processes in algae. *Advances in Botanical Research. Vol. 10* (Ed. by H.W. Woolhouse), pp. 1–219. Academic Press, London.

Leah, R.T., Moss, B. & Forrest, D.E. (1980). The role of predation in causing major changes in the limnology of a hyper-eutrophic lake. *Internationale Revue des gesamtem Hydrobiologie und Hydrographie*, **65,** 223–47.

Legendre, L. & Demers, S. (1984). Towards dynamic biological oceanography and limnology. *Canadian Journal of Fisheries and Aquatic Sciences*, **41,** 2–19.

Lehman, J.T. (1976). The filter-feeder as an optimal forager and the predicted shapes of feeding curves. *Limnology and Oceanography*, **21,** 301–16.

Lehman, J.T. (1980). Nutrient recycling as an interface between algae and grazers in freshwater communities. *Evolution and Ecology of Zooplankton Communities* (Ed. by W.C. Kerfoot), pp. 251–63. University Press of New England, Hanover, New Hampshire.

Lewis, M.R., Cullen, J.T. & Platt, T. (1984). Relationships between vertical mixing and photoadaptation of phytoplankton: similarity criteria. *Marine Ecology, Progress Series*, **15,** 141–9.

Lewis, W.M. (1978). Dynamics and succession of the phytoplankton in a tropical lake: Lake Lanao, Philippines. *Journal of Ecology*, **66,** 849–80.

Lewis, W.M. (1982). Vertical eddy diffusivities in a large tropical lake. *Limnology and Oceanography*, **27,** 161–3.

Lewis, W.M. (1983). A revised classification of lakes based on mixing. *Canadian Journal of Fisheries and Aquatic Sciences*, **40,** 1779–87.

Lewis, W.M. (1986). Phytoplankton succession in Lake Valencia, Venezuela. *Hydrobiologia*, **138,** 189–204.

Lhotsky, O. (1985). The time factor in the evaluation of algal communities. *Verhandlungen der internationale Vereinigung für theoretische und angewandte Limnologie*, **22,** 2285–7.

List, R.J. (1951). *Smithsonian Meteorological Tables.* Smithsonian Institution, Washington.

Lund, J.W.G. (1959). Buoyancy in relation to the ecology of the freshwater phytoplankton. *British Phycological Bulletin*, **1,** (7), 1–17.

Lund, J.W.G. (1964). Primary production and periodicity of phytoplankton. *Verhandlungen der internationale Vereinigung für theoretische und angewandte Limnologie*, **15,** 37–56.

Lund, J.W.G. & Reynolds, C.S. (1982). The development and operation of large limnetic enclosures in Blelham Tarn, English Lake District, and their contribution to phytoplankton ecology. *Progress in Phycological Research* Vol. 1. (Ed. by F.E. Round & D.J. Chapman), pp. 1–65. Elsevier, Amsterdam.

Margalef, R. (1978). Life-forms of phytoplankton as survival alternatives in an unstable environment. *Oceanologia Acta,* **1,** 493–509.

McGowan, J.A. & Walker, P.W. (1985). Dominance and diversity maintenance in an oceanic ecosystem. *Ecological Monographs,* **55,** 103–18.

Moll, R.A. & Stoermer, E.F. (1982). A hypothesis relating trophic status and subsurface chlorophyll maxima of lakes. *Archiv für Hydrobiologie,* **94,** 425–40.

Mortimer, C.H. (1974). Lake hydrodynamics. *Mitteilungen der internationale Vereinigung für theoretische und angewandte Limnologie,* **20,** 124–97.

Owens, O.v.H. & Esaias, W.E. (1976). Physiological responses of phytoplankton to major environmental factors. *Annual Review of Plant Physiology,* **27,** 461–83.

Paasche, E. (1980). Silicon. *The Physiological Ecology of Phytoplankton* (Ed. by I. Morris), pp. 259–84. Blackwell Scientific Publications, Oxford.

Patalas, K. (1984). Mid-summer mixing depths of lakes of different latitudes. *Verhandlungen der internationale Vereinigung für theoretische und angewandte Limnologie,* **22,** 97–102.

Petersen, R. (1975). The paradox of the plankton: an equilibrium hypothesis. *American Naturalist,* **109,** 35–49.

Pick, F.R., Nalewajko, C. & Lean, D.R.S. (1984). The origin of a metalimnetic chrysophyte peak. *Limnology and Oceanography,* **29,** 125–34.

Pourriot, R. (1977). Food and feeding habits of Rotifera. *Ergebnisse der Limnologie,* **8,** 243–60.

Powell, T., Kirkish, M.H., Neale, P.J. & Richerson, P.J. (1984). The diurnal cycle of stratification in Lake Titicaca: eddy diffusion. *Verhandlungen der internationale Vereinigung für theoretische und angewandte Limnologie,* **22,** 1237–43.

Powell, T. & Richerson, P.J. (1985). Temporal variation, spatial heterogeneity and competition for resources in plankton systems: a theoretical model. *American Naturalist,* **125,** 431–64.

Price, P.W. (1984). Alternative paradigms in community ecology. *A New Ecology: Novel Approaches to Interactive Systems* (Ed. by P.W. Price, C.N. Slobodchikoff & W.S. Gaud), pp. 353–83. Wiley-Interscience, New York.

Quay, P.D., Broecker, W.S., Hesslein, R.H. & Schindler, D.W. (1980). Vertical diffusion rates determined by tritium tracer experiments in the thermocline and hypolimnion of two lakes. *Limnology and Oceanography,* **25,** 201–18.

Raven, J.A. (1984). A cost-benefit analysis of photon absorption by photosynthetic cells. *The New Phytologist,* **98,** 593–625.

Raven, J.A. & Richardson, K. (1984). Dinophyte flagella: a cost-benefit analysis. *The New Phytologist,* **98,** 259–76.

Reynolds, C.S. (1973). Phytoplankton periodicity of some north Shropshire meres. *British Phycological Journal,* **8,** 301–20.

Reynolds, C.S. (1976). Succession and vertical distribution of phytoplankton in response to thermal stratification in a lowland lake, with special reference to nutrient availability. *Journal of Ecology,* **64,** 529–51.

Reynolds, C.S. (1979). The limnology of the eutrophic meres of the Shropshire–Cheshire Plain: a review. *Field Studies,* **5,** 93–173.

Reynolds, C.S. (1980). Phytoplankton assemblages and their periodicity in stratifying lake systems. *Holarctic Ecology,* **3,** 141–59.

Reynolds, C.S. (1984a). *The Ecology of Freshwater Phytoplankton.* Cambridge University Press, Cambridge.

Reynolds, C.S. (1984b). Phytoplankton periodicity: the interactions of form, function and environmental variability. *Freshwater Biology,* **14,** 111–42.

Reynolds, C.S. (1986). Experimental manipulations of the phytoplankton periodicity in large limnetic enclosures in Blelham Tarn, English Lake District. *Hydrobiologia,* **138,** 43–64.

Reynolds, C.S., Montecino, V., Graf, M.E. & Cabrera, S. (1986). Short-term dynamics of a

Melosira population in the plankton of an impoundment in central Chile. *Journal of Plankton Research*, **8**, 715–40.

Reynolds, C.S. & Reynolds, J.B. (1985). The atypical seasonality of phytoplankton in Crose Mere, 1972: an independent test of the hypothesis that variability in the physical environment regulates community dynamics and structure. *British Phycological Journal*, **20**, 227–42.

Reynolds, C.S., Tundisi, J.G. & Hino, K. (1983). Observations on a metalimnetic *Lyngbya* population in a stably stratified tropical lake (Lagoa Carioca, Eastern Brazil). *Archiv für Hydrobiologie*, **97**, 7–17.

Reynolds, C.S. & Walsby, A.E. (1975). Water blooms. *Biological Reviews of the Cambridge Philosophical Society*, **50**, 437–81.

Reynolds, C.S., Wiseman, S.W., Godfrey, B.M. & Butterwick, C. (1983). Some effects of artificial mixing on the dynamics of phytoplankton populations in large limnetic enclosures. *Journal of Plankton Research*. **5**, 203–34.

Rhee, C.-Y. (1978). Effects of N:P atomic ratios and nitrate limitation, on algal growth, cell composition and nitrate uptake. *Limnology and Oceanography*, **23**, 10–25.

Rhee, G.-Y. (1982).Effect of environmental factors and their interactions on phytoplankton growth. *Advances in Microbial Ecology, Vol. 6* (Ed. by K.C. Marshall) pp. 33–74. Plenum, London.

Rhee, G. -Y. & Gotham, I.J. (1980). Optimum N:P ratios and co-existence of planktonic algae. *Journal of Phycology*, **16**, 486–9.

Richardson, K., Beardall, J. & Raven, J.A. (1983). Adaptation of unicellular algae to irradiance: an analysis of strategies. *The New Phycologist*, **93**, 157–91.

Robinson, J.V. & Sandgren, C.D. (1983). The effect of temporal environmental heterogeneity on community structure: a replicated experimental study. *Oecologia (Berlin)*, **57**, 98–102.

Roughgarden, J. (1983). Competition and theory in community ecology. *The American Naturalist*, **122**, 583–601.

Schelske, C. (1984). *In situ* and natural phytoplankton assemblage bioassays. *Algae as Ecological Indicators* (Ed. by L.E. Shubert), pp. 15–47. Academic Press, Orlando.

Schindler, D.W. (1977). Evolution of phosphorus limitation in lakes. *Science*, **196**, 260–2.

Scott, J.T., Myer, G.E., Stewart, R. & Walther, E.G. (1969). On the mechanism of Langmuir circulations and their role in epilimnion mixing. *Limnology and Oceanography*, **14**, 493–503.

Shapiro, J., Forsberg, B., Lamarra, V., Lindmark, G., Lynch, M., Smeltzer, E. & Zoto, G. (1982). Experiments and experiences in biomanipulation—studies of biological ways to reduce algal abundance and eliminate blue-greens. *Ecological Research Series, U.S. Environmental Protection Agency, EPA 600/3-82-096.* 251 pp.

Smith, I.R. (1979). Hydraulic conditions in isothermal lakes. *Freshwater Biology*, **9**, 119–45.

Smith, I.R. (1985). The influence of events on population growth. *Annual Report of the Institute of Terrestrial Ecology, 1984*, **33–4.**

Sommer, U. (1983). Nutrient competition between phytoplankton species in multispecies Chemostat experiments. *Archiv für Hydrobiologie*, **96**, 399–416.

Sommer, U. (1984). The paradox of the plankton: fluctuations of phosphorus availability maintain diversity of phytoplankton in flow-through cultures. *Limnology and Oceanography*, **29**, 633–6.

Stauffer, R.E. & Armstrong, D.E. (1984). Lake mixing and its relationship to epilimnetic phosphorus in Shagawa Lake, Minnesota. *Canadian Journal of Fisheries and Aquatic Sciences*, **41**, 57–69.

Talling, J.F. (1951). The element of chance in pond populations. *The Naturalist, Hull.* October–December 1951, 157–70.

Talling, J.F. (1971). The underwater light climate as a controlling factor in the production ecology of freshwater phytoplankton. *Mitteilungen der internationale Vereinigung für theoretische und angewandte Limnologie*, **19**, 214–43.

Talling, J.F. (1976). The depletion of carbon dioxide from lake water by phytoplankton. *Journal of Ecology*, **64**, 79–121.

Talling, J.F. (1979). Factor interactions and implications for the prediction of lake metabolism. *Ergebnisse der Limnologie*, **13**, 96–109.

Talling, J.F. (1985). Inorganic carbon reserves of natural waters and eco-physiological consequences of their photosynthetic depletion: microalgae. *Inorganic Carbon Uptake by Aquatic Photosynthetic Organisms* (Ed. by W. J. Lucas & J.A. Berry), pp. 403–20. American Society of Plant Physiologists, Washington.

Talling, J.F., Wood, R.B., Prosser, M.V. & Baxter, R.M. (1973). The upper limit of photosynthetic productivity by phytoplankton: evidence from Ethiopian soda lakes. *Freshwater Biology*, **3**, 53–76.

Thorpe, S.A. (1977). Turbulence and mixing in a Scottish loch. *Proceedings of the Royal Society in London*. A, **286**, 125–81.

Tilman, D. (1980). Resources: a graphical-mechanistic approach to competition and predation. *American Naturalist*, **116**, 362–93.

Tilman, D. & Kilham, S.S. (1976). Phosphate and silicate growth and uptake kinetics of the diatoms *Asterionella formosa* and *Cyclotella meneghiniana* in batch and semi-continuous culture. *Journal of Phycology*, **12**, 375–83.

Tilman, D., Kilham, S.S. & Kilham, P. (1982). Phytoplankton community ecology: the role of limiting nutrients. *Annual Review of Ecology and Systematics*, **13**, 349–72.

Tilzer, M.M. (1984). The quantum yield as a fundamental parameter controlling vertical photosynthetic profiles of phytoplankton in Lake Constance. *Archiv für Hydrobiologie (Supplementband)*, **69**, 169–98.

Trimbee, A.M. & Harris, G.P. (1983). Use of time-series analysis to demonstrate advection rates of different variables in a small lake. *Journal of Plankton Research*, **5**, 819–33.

Turpin, D.H. (1987). The physiological basis of phytoplankton resource competition. *Growth and Survival Strategies of Freshwater Phytoplankton* (Ed. by C.D. Sandgren), Cambridge University Press, New York (in press).

Uhlmann, D. (1971). Influence of dilution, sinking and grazing rate on phytoplankton populations of hyperfertilized ponds and micro-ecosystems. *Mitteilungen der internationale Vereinigung für theoretische und angewandte Limnologie*, **19**, 100–24.

Vincent, W.F. (1981). Production strategies in antarctic inland waters: phytoplankton ecophysiology in a permanently ice-covered lake. *Ecology*, **62**, 1215–24.

Vincent, W.F., Neale, P.J. & Richerson, P.J. (1984). Photoinhibition: algal responses to bright light during diel stratification and mixing in a tropical alpine lake. *Journal of Phycology*, **20**, 201–11.

Vollenweider, R.A. (1976). Advances of defining critical loading levels for phosphorus in lake eutrophication. *Memorie dell Istituto Italiano di Idrobiologia*, **33**, 53–83.

Wall, D. & Briand, F. (1980). Spatial and temporal overlap in lake phytoplankton communities. *Archiv für Hydrobiologie*, **88**, 45–57.

Weinberg, S. (1976). Submarine daylight and ecology. *Marine Biology*, **37**, 291–304.

Welch, H.E. & Bergmann, M.A. (1985). Water circulation in small arctic lakes in winter. *Canadian Journal of Fisheries and Aquatic Sciences*, **42**, 506–20.

15. SPATIAL AND TEMPORAL ORGANIZATION IN MARINE PLANKTON COMMUNITIES

ROGER P. HARRIS

Marine Biological Association, The Laboratory, Citadel Hill, Plymouth PL1 2PB, UK

INTRODUCTION

Complex three-dimensional motions are the characteristic feature of the turbulent medium that forms the pelagic environment of the ocean. Spatial and temporal organization of communities in this environment is therefore extremely dynamic, reflecting a combination of the effects of physical processes such as turbulence and advection on species assemblages, with biological processes characteristic of individual species such as reproduction, death, and motile behaviour. It is in this mobile, unstructured environment that Hutchinson (1961) drew attention to the 'paradox of the plankton'. This emphasized the apparent contradiction between current theories of competitive exclusion and the co-occurrence of large numbers of similar species dependent on the same resources in an environment with little possibility of spatial segregation.

Despite further intense interest in this special aspect of plankton communities, the relative importance of the mechanisms that permit coexistence of species in the pelagic environment remains unclear. This has led some authors, for example Ghilarov (1984), to question the fundamental applicability of the concept of competitive exclusion to plankton communities. The series of investigations by McGowan and co-workers in the North Pacific represents the most complete study in the marine environment relevant to this problem of species-richness (Hayward & McGowan 1979; McGowan & Walker 1985). The coexistence of 175 species of copepod (McGowan & Walker 1985) in this relatively uniform water mass provides a powerful focus for discussion of the factors responsible for organizing communities in what would appear to be one of the least structured environments known. Most explanations of such coexistence imply that either competition is reduced through specialization and niche separation, or competition and resultant competitive exclusion are prevented by non-equilibrium conditions resulting from some disturbing factors such as resource-patchiness, seasonality or predation. It is the evaluation of these two explanations which is the common theme that runs through all recent studies of marine planktonic community structure.

In the face of such species diversity it has proved productive to regard size categories of phytoplankton and zooplankton as 'species' (Sheldon & Parsons 1967). This approach has been employed in important simulation models of the structure of plankton communities (Steele & Frost 1977) and in recent work on spatial and temporal organization of plankton communities of the continental shelf. Here the significance of physical processes in controlling biological productivity is particularly well understood, and physical discontinuities provide favourable regions for the study of community structure. Examples of general patterns of organization, and their controlling processes, within such communities will be considered in this paper with special emphasis on the spatial and temporal dimensions. In a restricted treatment of what is an extensive and complex subject, emphasis will be placed on examples of particularly relevant recent studies, and reference will be made, where appropriate, to comprehensive review articles.

CONTINENTAL SHELF COMMUNITIES

One of the most important of the recent approaches to the study of planktonic community structure has been to consider boundaries or gradients as being areas where pattern may be most easily detected, and the role of biotic interactions in structuring communities most readily evaluated. On a small vertical scale, the steep temperature gradient of the seasonal thermocline separates a warm, wind-mixed, surface layer from cooler, tidally-mixed, bottom waters (Pingree 1975). On a larger horizontal scale, the thermocline may outcrop at the surface, resulting in a frontal boundary between tidally-mixed and thermally-stratified waters, with a consequent horizontal discontinuity in properties of the water mass (Simpson & Hunter 1974). Several recent studies have emphasized the relationships between vertical mixing, light penetration, and supply of nutrients to explain spatial and temporal differences in biological productivity both at frontal boundaries and at the thermocline. These physical controls on phytoplankton growth have been considered in a number of recent reviews (Holligan 1981; Loder & Platt 1985, and are covered in some detail in the present volume by Reynolds). However, the general patterns of organization of these communities, their spatial and temporal dimensions, and the processes that may control patterns of species-composition within these physically defined environments have received less attention.

In many areas of the continental shelf, one of the clearest examples of the importance of physical mixing for plankton communities is provided by the onset of the thermal stratification in the early part of the growth season, resulting in the classical spring phytoplankton bloom (Pingree 1978). Phytoplankton growth causes a depletion of inorganic nutrients above the thermocline, a resultant decline in standing stock to the low levels characteristic of surface stratified waters during summer, and a successional change in species composition. Subsurface chlorophyll maximum layers associated with the temperature and nutrient gradients at the thermocline are characteristic of such stratified regions during the summer (Holligan 1978). These zones of high phytoplankton biomass are predicted from models relating light, nutrients and the physical structure of the water column (Jamart *et al.* 1977; Reynolds, Chapter 14) and are a major feature of the production biology of continental shelf ecosystems in summer.

Consideration of the effects of vertical stability of the water column on phytoplankton growth (Pingree *et al.* 1976) may be used to identify a succession of phytoplankton assemblages characteristic of the different phases of the development and breakdown of the seasonal thermocline (Holligan & Harbour 1977). The spring bloom is typically dominated by diatoms which grow rapidly in the high levels of inorganic nutrients and the increasingly favourable light regime in the surface layers. When the thermocline is well-established, the increased stability in the water-column, together with increased light energy and a supply of inorganic nutrients from below (Holligan *et al.* 1984b) results in the establishment of phytoplankton assemblages characteristically dominated by dinoflagellates within the subsurface chlorophyll maximum (Holligan & Harbour 1977). During summer thermal stratification, inorganic nutrient levels in surface waters are extremely low, and small flagellates dominate the phytoplankton, an assemblage which is characteristic of many oligotrophic oceanic regions. A similar, nutrient-dependent succession of phytoplankton is found in fresh water (see Reynolds, Chapter 14).

In summer the thermocline represents a well-developed physical discontinuity where planktonic organisms may experience fine-scale gradients of such physical and biological properties as temperature, predator abundance, and food quality and concentration; these result in phytoplankton communities with quite distinct size compositions (Fig. 15.1). Frontal boundaries also result in resource gradients in a horizontal dimension, and delimit communities characteristic of tidally mixed, frontal and stratified waters (Holligan 1981; Holligan *et al.* 1984a). In summer the

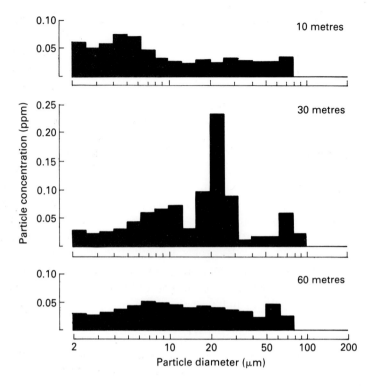

FIG. 15.1. Particle size distribution above (10 m), within (30 m) and below (60 m) the thermocline under stratified conditions in the western English Channel (46° 06′ N 06°30′ W) in late summer (6 September 1979).

phytoplankton community in mixed waters continues to be dominated by diatoms, whereas in frontal zones and within the thermocline dinoflagellates dominate. The general organization of each of these assemblages is quite different (Holligan *et al.* 1984a); the surface layers of stratified waters are dominated by heterotrophs, whereas within fronts phytoplankton are most important in terms of carbon biomass (Fig. 15.2), often with strong dominance by a single dinoflagellate species. Mixed water shows an intermediate distribution of biomass (Fig. 15.2).

Because vertical gradients of physical and biological variables are dominant in the sea, the structure of all plankton communities is best developed and most easily detected in the vertical axis. Hence there has been considerable recent interest in the fine-scale vertical distribution of organisms relative to physical structures such as the thermocline (Fiedler 1983; Herman 1983; Pugh & Boxshall 1984). Generally, concentrations of phytoplankton biomass at the thermocline (Fig. 15.3) are accompanied by

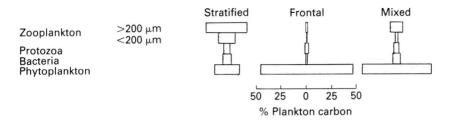

FIG. 15.2. Relative distribution of organic carbon in two size fractions of zooplankton, in bacteria, protozoa and phytoplankton in the surface 24 m of mixed, frontal and stratified waters of the English Channel in July 1981. (Adapted from Holligan *et al.* 1984a.)

aggregations of species from higher trophic levels (Ortner, Wiebe & Cox 1980; Herman, Sameoto & Longhurst 1981; Townsend, Cucci & Berman 1984). The biomass of autotrophs and heterotrophs makes the thermocline a site of intensified predation and competition. Longhurst (1985) has emphasized the importance of the stable physical structure of the thermocline in supporting zooplankton species diversity.

TEMPORAL PATTERN

Long-term patterns in the plankton communities of the continental shelf occur at frequencies much less than those of the seasonal cycle (Colebrook & Taylor 1984). In some cases where biological data series are available over many years, long-term, apparently cyclic, changes have been detected (Russell 1973). To what extent such patterns are a function of physical forcing due to climatic cycles remains uncertain (Southward, Butler & Pennycuick 1975) and similar uncertainties apply to the evaluation of the effects of water movements (Southward 1984). Such studies, however, emphasize the fact that a small change in an environmental variable such as temperature may result in a large biological change through its effect in altering the competitive advantage of one species over another. An example is seen in the pair of congeneric chaetognath predators *Sagitta elegans* and *Sagitta setosa* (Southward 1984).

Apart from such long-term patterns, and the seasonal cycle, the diel pattern is most characteristic of pelagic communities, and the most prominent example is the diel vertical migration of many groups of zooplankton and some motile phytoplankton (Kamykowski & Zentara 1977). This behaviour means that the structure of communities, for example above the thermocline, may vary significantly between day and night (Fig. 15.4). In addition, in many groups of planktonic organisms, for example copepods, vertical migration has an important ontogenetic component (Huntley &

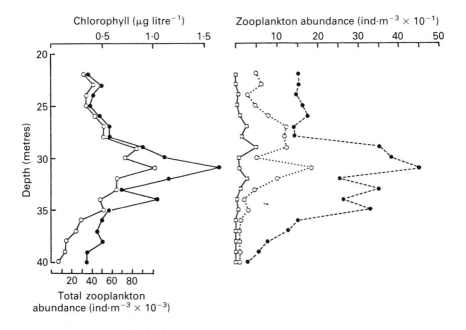

FIG. 15.3. Examples of fine-scale vertical profiles of zooplankton abundance relative to the subsurface chlorophyll maximum at a stratified station (46°06′N 06°30′W) in the western English Channel in August 1982. Chlorophyll concentration (—●—●—), abundance of total zoo-plankton greater than 80 μm (—○—○ —), and the copepods *Oithona similis* (... ● ... ● ...) *Oncaea subtilis* (—□—□—), and *Paracalanus parvus* (...○...○...).

Brooks 1982). The resultant temporal and spatial separations, through their influence on resource partitioning and predator avoidance, are the basis of the major theories proposed to explain the adaptive significance of such vertical migrations. These theories generally reflect either metabolic (McLaren 1963) and reproductive (McLaren 1974) advantages of migration in thermally stratified waters, or predator avoidance (Ohman, Frost & Cohen 1983).

Many aspects of general phytoplankton biology are also characterized by diel rhythms (Sournia 1981) as the division cycle is closely coupled with the light-dark cycle. For example, nutrient uptake may show diel patterns related to the photoperiod (Olson & Chisholm 1983) and may potentially influence competitive interactions. Similarly, many groups of grazing organisms, in particular copepods, exhibit diel feeding cycles (Baars & Oosterhuis 1984; Head, Wang & Conover 1984). Associated with such cyclic feeding activity, diel changes in ammonium excretion and hence the regenerated nitrogen available to phytoplankton have been observed in vertically migrating populations of *Calanus helgolandicus* by Harris & Malej (1986). The interactions between photoperiod, vertical migration of both zoo-plankton and phytoplankton, zooplankton feeding cycles, and the resulting diel changes in grazing mortality and nutrient availability for phytoplankton may have important effects on the organization of communities in surface stratified waters on the continental shelf.

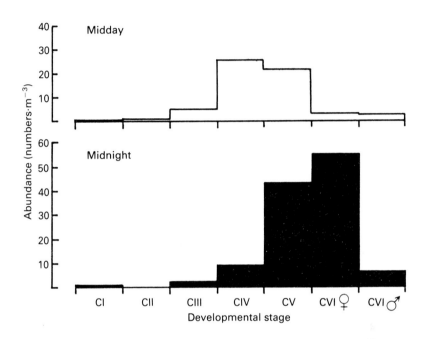

FIG. 15.4. Numbers of each copepodite stage of the copepod, *Calanus helgolandicus* observed above the thermocline (surface 20 m) at midday and midnight at a stratified station (46°06′N 06°30′W) in the western English Channel in August 1983. (Adapted from Harris & Malej 1986.)

SPATIAL HETEROGENEITY

Studies of patchiness or spatial heterogeneity have assumed particular significance in the pelagic zone, a variable and apparently unstructured environment. Spatial and temporal variability cause practical sampling problems, but also influence theoretical considerations concerning community structure (Haury, McGowan & Wiebe 1978; Shorrocks & Rosewell,

Chapter 2). Continuous sampling techniques have been used recently to detect the co-variation of properties such as temperature and chlorophyll on horizontal transects. Subsequent application of techniques such as spectral analysis (Denman 1976) has shown that the chlorophyll and temperature variance spectra are generally similar at wavelengths greater than 100 m, indicating that the observed phytoplankton pattern is essentially dominated by fluid motion. More recent developments of *in situ* particle-counting equipment (Herman & Dauphinee 1980) and multiple serial samples (Williams, Collins & Conway 1983) have started to extend our knowledge of such patterns to higher trophic levels (Mackas 1984).

At much smaller scales of both time and space, microscale patchiness has been the subject of considerable theoretical and experimental interest in relation to competitive interactions within the phytoplankton in nutrient-limited surface waters. For example, Goldman, McCarthy & Peavey (1979) have suggested that the apparently high growth rates observed in phyto-plankton communities at low ambient nutrient concentrations may be explained by re-cycling events occurring in micro-patches of nutrients, principally ammonia, excreted by zooplankton. However, it has been ques-tioned on theoretical grounds whether such small excretory pulses could be of significance for phytoplankton growth (Williams & Muir 1981). Diel changes in the flux of regenerated nitrogen due to zooplankton feeding cycles and vertical migration have been suggested by Doyle & Poore (1974) as a mechanism resulting in a phasing of cell division which would affect competition between phytoplankton species. Pulsed nitrogen regimes have been shown to affect the outcome of competition in continuous cultures of natural phytoplankton communities (Turpin & Harrison 1979) and evidence of diel changes in ammonium regeneration by zooplankton in surface-stratified waters has been reported recently by Harris & Malej (1986).

Interactions within the zooplankton under conditions of food limitation may similarly be influenced by microscale pattern within the phytoplankton. Mullin & Brooks (1976) showed that for *Calanus pacificus* phytoplankton concentrations were apparently not high enough to enable the copepods to balance their metabolic needs at a significant proportion of depths sampled in the water column. The importance of such patterns in terms of energy balance will depend on interactions between turbulence and phytoplankton growth and grazing mortality, and the vertical migratory behaviour of both plants and animals. When subjected to experimentally-induced patchiness, some species of copepod are better able to maintain their reproductive output during periods of starvation than others (Dagg 1977), suggesting that spatial heterogeneity within the phytoplankton might have an important influence on higher trophic levels in the food chain. Similarly, the developmental stages of an individual copepod species may vary considerably in their ability to withstand poor food conditions (Fig. 15.5). In such cases the young non-motile stages (Huntley & Brooks 1982), with lower resistance to starvation, should be more closely dependent on small scale spatial distribution of phytoplankton (Fig. 15.3) whereas adults with well-developed migratory ability can exploit the full range of phytoplankton concentrations in the water column. The interactions of rhythmic behaviour and spatial heterogeneity may have particular importance in modulating the effects of inter- and intra-specific competition.

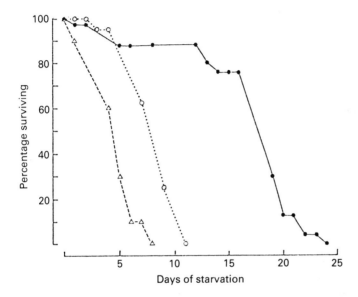

FIG. 15.5. Survival time of three developmental stages of the copepod *Calanus helgolandicus* under conditions of total starvation at 15°C. Adult females (—●—●—), copepodite I (. . . ○ . . . ○ . . .) and nauplius I (- - - △ - - - △ - - -).

COMPETITION AND PREDATION

The concept of nutrient limitation of phytoplankton growth is widely accepted and hence inter- and intraspecific competition are generally considered to be important in structuring phytoplankton assemblages (Maestrini & Bonin 1981a). Allelopathic relationships or interference competition between phytoplankton species may be of considerable importance (Maestrini & Bonin 1981b). Similar defences against grazing have been observed in a range of dinoflagellate species and associated monospecific blooms (red tides) in the field have been attributed to inhibition of zooplankton grazing (Fiedler 1982; Huntley 1982). In experiments such dinoflagellates may be rejected by the copepod *Calanus pacificus* when feeding on a mixture of similarly sized cells; this rejection is chemically mediated (Huntley, Sykes, Rohan & Marin, 1986). Hence chemical defence may confer a competitive advantage that may be an important factor in the formation and maintenance of monospecific dinoflagellate blooms in continental shelf communities. In these respects marine dinoflagellates appear to be ecological equivalents of blue-green algae in fresh water.

There has been considerable discussion as to whether food limitation is critical in structuring zooplankton assemblages and consequently whether inter- and intraspecific competition are significant. Evidence consistent with the food limitation hypothesis has been obtained for a range of continental shelf copepod species (Mayzaud & Poulet 1978) and the interactions between food limitation and, for example, diel vertical migration have been clearly demonstrated (Dagg 1985). On the other hand, food limitation does not appear to occur in other situations (Ohman 1985) in which predation is a more likely mechanism of population regulation. These conflicting views may be reconciled by the conclusion that copepods may sometimes experience food limitation and sometimes not, and that even among co-occurring species the incidence of limiting conditions varies both in space and time (Frost 1985).

Whereas there is little opportunity for phytoplankton to specialize along the food axis, there would appear to be considerable scope for niche separation in herbivorous zooplankton through size-selective feeding. However, in terms of the size range of phytoplankton consumed by co-occurring copepod species (Fig. 15.6) there is potential for interspecific competition for the same resource (Poulet 1978; Harris 1982) and hence grazing mortality on the same groups of phytoplankton. However, copepods do have well-developed food selection and post-capture rejection mechanisms (Huntley, Barthel & Star 1983; Price, Paffenhöfer & Strickler 1983), suggesting that mortality may be quite specifically related to certain cell types. Cell morphology (Gifford, Bohrer & Boyd 1981), supposed anti-predator devices such as spines, and differences in the biochemical composition and nutritive value of phytoplankton will also be important in influencing competitive interactions between grazers. Finally, it should be noted that many of these consumers of phytoplankton are omnivores and that omnivory is common in the plankton (Paffenhöfer & Knowles 1980; Poulet 1983). This fact, together with associated behaviour of switching

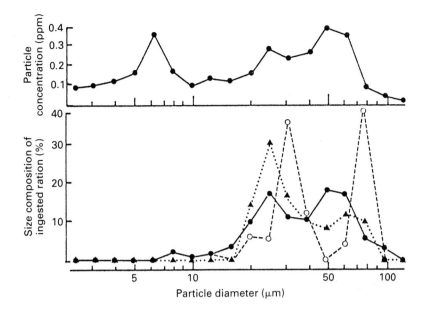

FIG. 15.6. Composition of the diet of adult females of three co-occurring copepod species feeding on natural phytoplankton in a diatom dominated community during a controlled ecosystem experiment. *Calanus (—●—●—)*, *Pseudocalanus (- - - ○ - - - ○ - - -)*, and *Paracalanus (. . . ▲ . . . ▲ . . .)* (Adapted from Harris *et al.* 1982).

from one food type to another (Landry 1981), introduces further complexity into an evaluation of the role of competition in structuring such assemblages.

The role of predation in the marine plankton is less certain than in freshwater communities where selective predation is recognized as being of central importance in structuring communities (Zaret 1975; Greene 1983; Reynolds, Chapter 14). Although advances have been made in the study of individual predatory interactions (Gophen & Harris 1981; Sullivan & Reeve 1982; Landry, Lehner-Fournier & Fagerness 1985), it is difficult to conclude from the existing evidence that the species composition of marine zooplankton communities is altered by selective predation. In comparison to freshwater environments, physical variability and patchiness are more pronounced influences on the marine plankton and may be particularly critical for predator–prey interactions. Marine planktivores are taxonomically more diverse than their freshwater counterparts, and also show a wider range of functional groups. In his functional classifications, Greene (1983, 1985) emphasized the dichotomy in freshwater predators between those that ambush their prey and those that are cruising raptors; in the sea there is an additional category of entanglers (ctenophores and medusae).

Detailed studies of predatory interactions within marine plankton have revealed a high degree of complexity, for example, cannibalism occurs in chaetognaths (Pearre 1982). Similarly, while most planktonic predators feed on organisms much smaller than themselves, there are species that are able to prey on organisms many times their own body size, for instance the copepod *Corycaeus anglicus,* one of the smallest predators in the marine plankton (Landry, Lehner-Fournier & Fagerness 1985). It has been further suggested that this same copepod species is not as actively preyed upon by ctenophores as are other copepods of similar size (Grice *et al.* 1980), which may explain its population increase during the rapid decline of calanoid copepods due to ctenophore predation in enclosed ecosystem experiments. The dominant predatory interaction in enclosure experiments has often been that between ctenophores and calanoid copepods (Harris *et al.* 1982). In enclosed systems it seems such voracious predators, which have the ability to withstand long periods of starvation, can make a dramatic impact on other components of the zooplankton community, whereas patchiness presumably provides a refuge for prey species in the physically hetero- geneous natural environment. However, even this predator–prey inter- action has further complexity as the copepods, the potential prey of the adult ctenophores, may under certain circumstances (Stanlaw, Reeve & Walter 1981) themselves inflict considerable damage on the minute and delicate larval stages of the predator.

EXPERIMENTAL MANIPULATION OF COMMUNITIES

One of the best methods for studying the biological interactions that produce community structure is the controlled field experiment (Connell 1975). In the sea the experimental manipulation of pelagic communities is extremely difficult; species cannot be added to or removed from natural systems in order to compare the response to such treatment with an adequate control. Natural assemblages of phytoplankton may be maintained and manipulated in quite small containers and competitive interactions investigated (Turpin & Harrison 1980), but the effects of grazing pressure cannot be included in such systems due to the low population densities of larger grazing organisms in the plankton. Such constraints have led to the development of large, enclosed ecosystems of up to 1335 m³ volume (Grice *et al.* 1980; Grice & Reeve 1982) to study the effects of experimental manipulation on plankton communities.

Enclosure experiments in general confirm the importance of nutrients, light and mixing for phytoplankton community structure. For example, populations dominated by microflagellates typically develop after the first 2 to 3 weeks if the enclosures are not stirred and become nutrient-limited at the surface. Eppley, Koeller & Wallace (1978) were able to replace the microflagellate assemblage with one dominated by large diatoms by stirring the enclosures. Davis (1982) convincingly demonstrated that experimental manipulation of nutrients, light and mixing in large enclosures would successfully produce diatom and flagellate dominated assemblages. In this experiment two communities were produced by such manipulation, one in which diatoms formed 75% of the total phytoplankton carbon and the other in which they formed only 11% in terms of biomass (Harris *et al.* 1982). Initial stabilization of the water column and experimental nutrient input produced an initial diatom bloom analagous to the spring bloom, which in turn resulted in a succession within the herbivorous zooplankton. The succession of phytoplankton, which in this case lasted for more than 100 days after initial stabilization, was probably due to the different growth rates of the groups (Davis 1982). The appearance of red-tide dinoflagellates later in the experiment was linked by Davis (1982) with field-studies such as those of Pingree, Holligan & Head (1977) in which the stability of the water column and a continual or periodic supply of nutrients was shown to be important for the development of red-tide conditions. It should be noted, however, that in this experiment the predation pressure on the phytoplankton was low, as the majority of herbivores were removed by intense predation by ctenophores (Harris 1982; Harris *et al.* 1982). Successional changes might therefore have been modified if selective mortality on the phytoplankton had been more intense.

Generally, a disadvantage of enclosure experiments is that vertical mixing may be reduced by as much as an order of magnitude compared to the open sea, resulting in sinking of non-motile phytoplankton. However, in a recent development of enclosure technology, Donaghay & Klos (1985) report the experimental induction of two-layered stratification in land-based enclosed ecosystems. As these authors point out, such systems have considerable promise for the analysis of the effects of mixing and spatial heterogeneity on the dynamics of coastal marine ecosystems.

In addition to successional events within the phytoplankton, controlled ecosystems have been used to provide important information on the effects of predation and spatial heterogeneity within plankton communities. For example, a continuing feature of these systems, the dominance of the community by a single group of predators such as ctenophores (Harris *et al.* 1982) has been considered to be a consequence of the lack of patchiness at the appropriate scale. The enclosure can be regarded as an isolated patch which is unaffected by infusions from adjacent patches (Reeve, Grice & Harris 1982). Such results lend support to the theory of contemporaneous disequilibrium (Richerson, Armstrong & Goldman 1970), which proposes that within a patch of water a species may be at a competitive advantage relative to others, but that such patches are disrupted frequently enough to prevent exclusive occupation of a niche by any one species. The interactions between ctenophores and copepods which developed in the enclosures were not characteristic of natural communities, but occurred because the developing ctenophore populations could not be advected out of the enclosed water body, or actively reproducing copepods be advected into it.

CONCLUSIONS

The broad features of species distributions and abundances in the marine plankton are undoubtedly controlled by physical factors. The examples of physical discontinuities represented by the seasonal thermocline and tidal fronts clearly demonstrate the physical control of production dynamics and temporal and spatial pattern within the phytoplankton (Loder & Platt 1985). The importance of nutrients, light and mixing in determining species composition of phytoplankton communities has been confirmed by experimental manipulations in enclosures (Davis 1982; Harris *et al.* 1982). In recent theoretical treatments of plankton community structure it has been suggested that the consequences of this physical control at the base of the food-web then propagate up through higher trophic levels as a result of size-selective grazing, these effects being also modified by selective predation (Steele & Frost 1977). This hypothesis has also been experimentally tested in enclosure experiments (Grice *et al.* 1980), but the results were inconclusive.

Although physical processes are clearly the major forcing function, biotic factors such as competition and predation are critical in controlling the pattern within plankton communities (McGowan & Walker 1985). However, the relative importance of these two structuring mechanisms and their interaction with physical variability remains unclear. The fact that many guilds of potentially competing species in the zooplankton show considerable dietary overlap (Poulet 1978; Harris 1982) and that omnivory is common (Poulet 1983) indicates that partitioning of food resources is not widespread. This in turn suggests specialized predation pressure and physical variability and structure as sources of pattern in these communities.

In the case of predation it seems that, in contrast to freshwater communities, there is little conclusive evidence for a dominant role of selective predation in shaping the structure of marine zooplankton communities, although the complexity of predatory interactions in the marine environment is very great.

Despite the uniform appearance of the pelagic environment, it is probable that it is physical structure and variability that provides the niche diversity required to explain the paradox of the plankton. This is also the conclusion reached for freshwater systems by Reynolds, Chapter 14. Longhurst (1985) has recently emphasized the importance of stable vertical structure, in particular the thermocline, in providing predictable environmental gradients and discontinuities, enabling species to exist in conditions ensuring competitive advantage over others. In contrast to these relatively predictable gradients, it is the physical variability in the euphotic zone above the thermocline that may enable the concept of contemporaneous disequilibrium (Richerson, Armstrong & Goldman 1970) to apply. However, as Longhurst (1985) has pointed out, the persistence time of such patchiness would seem to be more appropriate to the growth rates of phytoplankton than zooplankton.

As the diel time scale is so important in planktonic communities, and cyclical behaviour affects the partitioning of both food and habitat resources, an important area of interaction is that between cyclical behaviour and spatial heterogeneity. Tett & Edwards (1984) have suggested that species occupying the same physical space, but with slightly different biological characteristics, would produce pattern on different temporal and spatial scales matching different aspects of resource fluctuations in time and space. This idea is supported by work on the physiological responses of species to resource heterogeneity, for example that of Dagg (1977) for zooplankton and Turpin & Harrison (1979) for phytoplankton. Community interactions occurring on relatively small temporal and spatial scales are central to an understanding of processes structuring plankton communities, and will also be critical for the production of large-scale temporal and spatial changes in the plankton (Steele 1979).

The solution to the paradox of the plankton probably lies in a combination of factors rather than in one dominant process. Important among these interacting factors are physiological adaptation of species to physically mobile and heterogeneous conditions, rhythmic behaviour patterns such as vertical migration, competition between species at the same trophic level and biotic interactions between trophic levels, together with feedback such as nutrient regeneration, and the effects of persistent physical structure such as the thermocline. The elucidation of these interactions represents the

challenge for future investigations of what Dayton (1984) has categorized as
the world's most interesting, bewildering and difficult communities to study.

ACKNOWLEDGMENTS

Financial support for this work was received from the Ministry of
Agriculture, Fisheries and Food (UK).

REFERENCES

Baars, M. & Oosterhuis, S. (1984). Diurnal feeding rhythms in North Sea copepods measured
by gut fluorescence, digestive enzyme activity, and grazing of labelled food. *Netherlands
Journal of Sea Research*, **18**, 97–119.

Colebrook, J.M. & Taylor, A.H. (1984). Significant time scales of long-term variability in the
plankton and the environment. *Conseil Permanent International pour L'Exploration de la
Mer. Rapports et Procès Verbaux de Réunions*, **183**, 20–6.

Connell, J.H. (1975). Some mechanisms producing structure in natural communities: a model
and evidence from field experiments. *Ecology and Evolution of Communites* (Ed. by
M.L. Cody & J.M. Diamond), pp. 460–91. Harvard University Press, Cambridge,
Massachusetts.

Dagg, M. (1977). Some effects of patchy food environments on copepods. *Limnology and
Oceanography*, **22**, 99–107.

Dagg, M. (1985). The effects of food limitation on diel migratory behaviour in marine
zooplankton. *Archiv für Hydrobiologie, Beihefte, Ergebnisse der Limnologie*, **21**,
247–55.

Davis, C.O. (1982). The importance of understanding phytoplankton life strategies in the
design of enclosure experiments. *Marine Mesocosms: Biological and Chemical Research in
Enclosed Ecosystems* (Ed. by G.D. Grice & M.R. Reeve), pp. 323–32. Springer-Verlag,
New York.

Dayton, P.K. (1984). Processes structuring some marine communities: are they general?
Ecological Communities: Conceptual Issues and the Evidence (Ed. by D.R. Strong Jr., D.
Simberloff, L.G. Abele & A.B. Thistle), pp. 181–200. Princeton University Press,
Princeton, New Jersey.

Denman, K.L. (1976). Covariability of chlorophyll and temperature in the sea. *Deep-Sea
Research*, **23**, 539–50.

Donaghay, P.L. & Klos, E. (1985). Physical, chemical and biological responses to simulated
wind and tidal mixing in experimental marine ecosystems. *Marine Ecology Progress Series*,
26, 35–45.

Doyle, R.W. & Poore, R.V. (1974). Nurient competition and division synchrony in
phytoplankton. *Journal of Experimental Marine Biology and Ecology*, **14**, 201–10.

Eppley, R.W., Koeller, P. & Wallace, G.T. (1978). Stirring influences the phytoplankton
species composition within enclosed columns of coastal water. *Journal of Experimental
Marine Biology and Ecology*, **32**, 219–39.

Fiedler, P.C. (1982). Zooplankton avoidance and reduced grazing responses to *Gymnodinium
splendens* (Dinophyceae). *Limnology and Oceanography*, **27**, 961–5.

Fiedler, P.C. (1983). Fine-scale spatial patterns in the coastal epi-plankton off southern
California. *Journal of Plankton Research*, **5**, 865–79.

Frost, B.W. (1985). Food limitation of the planktonic marine copepods *Calanus pacificus* and *Pseudocalanus* sp. in a temperate fjord. *Archiv für Hydrobiologie, Beihefte, Ergebnisse der Limnologie,* **21,** 1–13.

Ghilarov, A.M. (1984). The paradox of the plankton reconsidered; or, why do species coexist? *Oikos,* **43,** 46–52.

Gifford, D., Bohrer, R.N. & Boyd, C.M. (1981). Spines on diatoms: Do copepods care? *Limnology and Oceanography,* **26,** 1057–61.

Goldman, J.C., McCarthy, J.J. & Peavey, D.C. (1979). Growth rate influence on the chemical composition of phytoplankton in oceanic waters. *Nature,* **279,** 210–15.

Gophen, M. & Harris, R.P. (1981). Visual predation by a marine cyclopoid copepod, *Corycaeus anglicus. Journal of the Marine Biological Association of the United Kingdom,* **64,** 391–9.

Greene, C.H. (1983). Selective predation in freshwater zooplankton communities. *Internationale Revue der gesamten Hydrobiologie und Hydrographie,* **68,** 297–315.

Greene, C.H. (1985). Planktivore functional groups and patterns of prey selection in pelagic communities. *Journal of Plankton Research,* **7,** 35–40.

Grice, G.D., Harris, R.P., Reeve, M.R., Heinbokel, J.F. & Davis, C.O. (1980). Large-scale enclosed water-column ecosystems: an overview of Foodweb 1, the final CEPEX experiment. *Journal of the Marine Biological Association of the United Kingdom,* **60,** 401–14.

Grice, G.D. & Reeve, M.R. (Eds.) **(1982).** *Marine Mesocosms: Biological and Chemical Research in Experimental Ecosystems.* Springer-Verlag, New York.

Harris, R.P. (1982). Comparison of the feeding behaviour of *Calanus* and *Pseudocalanus* in two experimentally manipulated enclosed ecosystems. *Journal of the Marine Biological Association of the United Kingdom,* **62,** 71–91.

Harris, R.P. & Malej, A. (1986). Diel patterns of ammonium excretion and grazing rhythms in *Calanus helgolandicus* in surface stratified waters. *Marine Ecology Progress Series,* **31,** 75–85.

Harris, R.P., Reeve, M.R., Grice, G.D., Evans, G.T., Gibson, V.R., Beers, J.R. & Sullivan, B.K. (1982). Trophic interactions and production processes in natural zooplankton communities in enclosed water columns. *Marine Mesocosms: Biological and Chemical Research in Enclosed Ecosystems* (Ed. by G.D. Grice & M.R. Reeve), pp. 353–87. Springer-Verlag, New York.

Haury, L.R., McGowan, J.A. & Wiebe, P.H. (1978). Patterns and processes in the time-space scales of plankton distributions. *Spatial Patterns in Plankton Communities* (Ed. by J.H. Steele), pp. 277–327. Plenum, New York.

Hayward, T.L. & McGowan, J.A. (1979). Pattern and structure in an oceanic zooplankton community. *American Zoologist,* **19,** 1045–55.

Head, E.J.H., Wang, R. & Conover, R.J. (1984). Comparison of diurnal feeding rhythms in *Temora longicornis* and *Centropages hamatus* with digestive enzyme activity. *Journal of Plankton Research,* **6,** 543–51.

Herman, A.W. (1983). Vertical distribution patterns of copepods, chlorophyll, and production in northeastern Baffin Bay. *Limnology and Oceanography,* **28,** 709–19.

Herman, A.W. & Dauphinee, T.M. (1980). Continuous and rapid profiling of zooplankton with an electronic counter mounted on a 'Batfish' vehicle. *Deep-Sea Research,* **27,** 79–96.

Herman, A.W., Sameoto, D.D. & Longhurst, A.R. (1981). Vertical and horizontal distribution patterns of copepods near the shelf-break south of Nova Scotia. *Canadian Journal of Fisheries and Aquatic Science,* **38,** 1065–76.

Holligan, P.M. (1978). Patchiness in subsurface phytoplankton populations on the northwest European continental shelf. *Spatial Pattern in Plankton Communities* (Ed. by J.H. Steele), pp. 221–38. Plenum, New York.

Holligan, P.M. (1981). Biological implications of fronts on the northwest European continental shelf. *Philosophical Transactions of the Royal Society* (A), **302,** 547–62.

Holligan, P.M. & Harbour, D.S. (1977). The vertical distribution and succession of phytoplankton in the western English Channel in 1975 and 1976. *Journal of the Marine Biological Association of the United Kingdom,* **57,** 1075–93.

Holligan, P.M., Harris, R.P., Newell, R.C., Harbour, D.S., Head, R.N., Linley, E.A.S., Lucas, M.I., Tranter, P.R.G. & Weekley, C.M. (1984a). Vertical distribution and partitioning of organic carbon in mixed, frontal and stratified waters of the English Channel. *Marine Ecology Progress Series,* **14,** 111–27.

Holligan, P.M., Williams, P.J. LeB., Purdie, D. & Harris, R.P. (1984b). The photosynthetic and respiratory activities, and nitrogen supply of summer plankton populations in stratified, frontal and mixed shelf waters. *Marine Ecology Progress Series,* **17,** 201–13.

Huntley, M.E. (1982). Yellow water in La Jolla Bay, California, July 1980. II. Suppression of zooplankton grazing. *Journal of Experimental Marine Biology and Ecology,* **63,** 81–91.

Huntley, M.E., Barthel, K.G., Star, J.L. (1983). Particle rejection by *Calanus pacificus:* discrimination between similarly sized particles. *Marine Biology,* **74,** 151–60.

Huntley, M. & Brooks, E.R. (1982). Effects of age and food availability on diel vertical migration of *Calanus pacificus. Marine Biology,* **71,** 23–31.

Huntley, M., Sykes, P., Rohan, S. & Marin, V. (1986). Chemically-mediated rejection of dinoflagellate prey by the copepods *Calanus pacificus* and *Paracalanus parvus:* mechanism, occurrence and significance. *Marine Ecology Progress Series,* **28,** 105–20.

Hutchinson, G.E. (1961). The paradox of the plankton. *American Naturalist,* **95,** 137–45.

Jamart, B.M., Winter, D.F., Banse, K., Anderson, G.C. & Lam, R.K. (1977). A theoretical study of phytoplankton growth and nutrient distribution in the Pacific Ocean off the northwestern U.S. Coast. *Deep-Sea Research,* **24,** 753–73.

Kamykowski, D. & Zentara, S.J. (1977). The diurnal vertical migration of motile phytoplankton through temperature gradients. *Limnology and Oceanography,* **22,** 148–51.

Landry, M.R. (1981). Switching between herbivory and carnivory by the planktonic marine copepod *Calanus pacificus. Marine Biology,* **65,** 77–119.

Landry, M.R., Lehner-Fournier, J.M. & Fagerness, V.M. (1985). Predatory feeding behaviour of the marine cyclopoid copepod *Corycaeus anglicus. Marine Biology,* **85,** 163–9.

Loder, J.W & Platt, T. (1985). Physical controls on phytoplankton production at tidal fronts. *Proceedings of the nineteenth European Marine Biology Symposium* (Ed. by P.E. Gibbs), pp. 3–21, Cambridge University Press, Cambridge.

Longhurst, A.R. (1985). Relationship between diversity and the vertical structure of the upper ocean. *Deep-Sea Research,* **32,** 1535–70.

McGowan, J.A. & Walker, P.W. (1985). Dominance and diversity maintenance in an oceanic ecosystem. *Ecological Monographs,* **55,** 103–18.

Mackas, D.L. (1984). Spatial autocorrelation of plankton community composition in a continental shelf ecosystem. *Limnology and Oceanography,* **29,** 451–71.

McLaren, I.A. (1963). Effects of temperature on growth of zooplankton and the adaptive value of vertical migration. *Journal of the Fisheries Research Board of Canada,* **20,** 685–727.

McLaren, I.A. (1974). Demographic strategy of vertical migration by a marine copepod. *American Naturalist,* **108,** 91–102.

Maestrini, S.Y. & Bonin, D.J. (1981a). Competition among phytoplankton based on inorganic macronutrients. *Physiological Bases of Phytoplankton Ecology* (Ed. by T. Platt), pp. 264–78. Canadian Bulletin of Fisheries and Aquatic Sciences, 210, Ottawa.

Maestrini, S.Y. & Bonin, D.J. (1981b). Allelopathic relationships between phytoplankton species. *Physiological Bases of Phytoplankton Ecology* (Ed. by T. Platt), pp. 323–38. Canadian Bulletin of Fisheries and Aquatic Sciences, 210, Ottawa.

Mayzaud, P. & Poulet, S.A. (1978). The importance of the time factor in the response of zooplankton to varying concentrations of naturally occurring particulate matter. *Limnology and Oceanography,* **23,** 1144–54.

Mullin, M.M. & Brooks, E.R. (1976). Some consequences of distributional heterogeneity of phytoplankton and zooplankton. *Limnology and Oceanography*, **21**, 784–96.

Ohman, M.D. (1985). Resource-satiated growth of the copepod *Pseudocalanus* sp. *Archiv für Hydrobiologie, Beihefte, Ergebnisse der Limnologie*, **21**, 15–32.

Ohman, M.D., Frost, B.W. & Cohen, E.B. (1983). Reverse diel vertical migration: an escape from invertebrate predators. *Science* **220**, 1404–7.

Olson, R.J. & Chisholm, S.W. (1983). Effects of photocycles and periodic ammonium supply on three marine phytoplankton species. I. Cell division patterns. *Journal of Phycology*, **19**, 522–8.

Ortner, P.B., Wiebe, P.H. & Cox, J.L. (1980). Relationships between oceanic epizooplankton distributions of the seasonal deep chlorophyll maximum in the Northwestern Atlantic Ocean. *Journal of Marine Research*, **38**, 507–31.

Paffenhöfer, G.- A. & Knowles, S.C. (1980). Omnivorousness in marine planktonic copepods. *Journal of Plankton Research*, **2**, 355–65.

Pearre, S.Jr. (1982). Feeding by chaetognatha: Aspects of inter- and intra-specific predation. *Marine Ecology Progress Series*, **7**, 33–45.

Pingree, R.D. (1975). The advance and retreat of the thermocline on the continental shelf. *Journal of the Marine Biological Association of the United Kingdom*, **55**, 965–74.

Pingree, R.D. (1978). Mixing and stabilization of phytoplankton distributions on the northwest European continental shelf. *Spatial Pattern in Plankton Communities* (Ed. by J.H. Steele), pp. 181–220. Plenum, New York.

Pingree, R.D., Holligan, P.M. & Head, R.N. (1977). Survival of dinoflagellate blooms in the western English Channel. *Nature*, **265**, 266–9.

Pingree, R.D., Holligan, P.M., Mardell, G.T. & Head, R.N. (1976). The influence of physical stability on spring, summer and autumn phytoplankton blooms in the Celtic Sea. *Journal of the Marine Biological Association of the United Kingdom*, **56**, 845–73.

Poulet, S.A. (1978). Comparison between five coexisting species of marine copepods feeding on naturally occurring particulate matter. *Limnology and Oceanography*, **23**, 1126–43.

Poulet, S.A. (1983). Factors controlling utilization of non-algal diets by particle grazing copepods. A review. *Oceanologica Acta*, **6**, 221–34.

Price, H.J., Paffenhöfer, G.-A. & Strickler, J.R. (1983). Modes of cell capture in calanoid copepods. *Limnology and Oceanography*, **28**, 116–23.

Pugh, P.R. & Boxshall, G.A. (1984). The small-scale distribution of plankton at a shelf station off the northwest African coast. *Continental Shelf Research*, **3**, 399–423.

Reeve, M.R., Grice, G.D. & Harris, R.P. (1982). The CEPEX approach and its implications for future studies in plankton ecology. *Marine Mesocosms: Biological and Chemical Research in Enclosed Ecosystems* (Ed. by G.D. Grice & M.R. Reeve), pp. 389–97. Springer-Verlag, New York.

Richerson, P., Armstrong, R. & Goldman, C.R. (1970). Contemporaneous disequilibrium, a new hypothesis to explain the"paradox of the plankton". *Proceedings of the National Academy of Sciences of the United States of America*, **67**, 1710–14.

Russell, F.S. (1973). A summary of the observations on the occurrence of the planktonic stages of fish off Plymouth 1924–1972. *Journal of the Marine Biological Association of the United Kingdom*, **53**, 347–55.

Sheldon, R.W. & Parsons, T.R. (1967). A continuous size spectrum for particulate matter in the sea. *Journal of the Fisheries Research Board of Canada*, **24**, 909–15.

Simpson, J.H. & Hunter, J.R. (1974). Fronts in the Irish Sea. *Nature*, **1250**, 404–6.

Sournia, A. (1981). Morphological bases of competition and succession. *Physiological bases of phytoplankton ecology* (Ed. by T. Platt), pp. 339–46. *Canadian Bulletin of Fisheries and Aquatic Sciences*, **210**, Ottawa.

Southward, A.J. (1984). Fluctuations in the "indicator chaetognaths" *Sagitta elegans* and *Sagitta setosa* in the western Channel. *Oceanologica Acta*, **7**, 229–39.

Southward, A.J., Butler, E.I. & Pennycuick, L. (1975). Recent cyclic changes in climate and in abundance of marine life. *Nature*, **153**, 714–17.

Stanlaw, K.A., Reeve, M.R. & Walter, M.A. (1981). Growth, food and vulnerability to damage of the early life history stages of the ctenophore *Mnemiopsis mccradyi*. *Limnology and Oceanography*, **26**, 224–34.

Steele, J.H. (1979). Interactions in marine ecosystems. *Population Dynamics* (Ed. by R.M. Anderson, B.D. Turner & L.R. Taylor), pp. 343–57. Symposia of the British Ecological Society. 20, Blackwell Scientific Publications, Oxford.

Steele, J.H. & Frost, B.W. (1977). The structure of plankton communities. *Philosophical Transactions of the Royal Society*, B, **280**, 485–535.

Sullivan, B.K. & Reeve, M.R. (1982). Comparison of estimates of the predatory impact of ctenophores by two independent techniques. *Marine Biology*, **68**, 61–5.

Tett, P. & Edwards, A. (1984). Mixing and plankton: an interdisciplinary theme in oceanography. *Oceanography and Marine Biology Annual Review*, **22**, 99–123.

Townsend, D.W., Cucci, T.L. & Berman, T. (1984). Subsurface chlorophyll maxima and vertical distribution of zooplankton in the Gulf of Maine. *Journal of Plankton Research*, **6**, 793–802.

Turpin, D.H. & Harrison, P.J. (1979). Limiting nutrient patchiness and its role in phytoplankton ecology. *Journal of Experimental Marine Biology and Ecology*, **39**, 151–66.

Turpin, D.H. & Harrison, P.J. (1980). Cell size manipulation in natural marine planktonic diatom communities. *Canadian Journal of Fisheries and Aquatic Sciences*, **37**, 1193–5.

Williams, P.J. LeB. & Muir, L.R. (1981). Diffusion as a constraint on the biological importance of microzones in the sea. *Ecohydrodynamics* (Ed. by J.C.J. Nihoul), pp. 209–18. Elsevier, Amsterdam.

Williams, R., Collins, N.R. & Conway, D.V.P. (1983). The double LHPR system, a high-speed micro- and macro-plankton sampler. *Deep-Sea Research*, **30**, 331–42.

Zaret, T.M. (1975). Strategies for existence of zooplankton prey in homogeneous environments. *Verhandlungen der Internationalen Vereinigung für theoretische und angewandte Limnologie*, **19**, 1484–9.

16. ORGANIZATION IN FRESHWATER BENTHIC COMMUNITIES

ALAN G. HILDREW[1] AND COLIN R. TOWNSEND[2]

[1]*School of Biological Sciences, Queen Mary College, University of London, Mile End Road, London E1 4NS, UK, and* [2]*School of Biological Sciences, University of East Anglia, Norwich NR4 7TJ, UK*

INTRODUCTION

Limnology has a long and respected history of contribution to the wider science of ecology, through the work of Hutchinson, Margalef, Rigler, Vollenweider and others. However, the study of the freshwater benthos (organisms living on the bed of the water body) has made less impact and, until recently (e.g. Barnes & Minshall 1983), has continued in relative isolation from the mainstream of ecology. Most studies were descriptive rather than experimental and related the distribution and abundance of benthic organisms to physico-chemical factors such as flow, substratum and oxygen concentration (see, for example, Hynes 1970 and Macan 1974).

In the last 10 years, benthic ecologists have increasingly turned away from this emphasis on physical factors and found evidence for the importance of biotic interactions in the ecology of benthic organisms (Townsend 1980; Barnes & Minshall 1983; Resh & Rosenberg 1984). This is ironic, because at the same time the wider ecological debate was moving away from a deterministic view that competition was overwhelmingly the most important process structuring communities (Cody & Diamond 1975), to a view giving non-equilibrial and stochastic factors more prominence (Strong *et al.* 1984). We therefore use this opportunity to assess the relative importance of deterministic and stochastic processes in structuring the freshwater benthos. Our treatment will emphasize the benthos of hard substrata and particularly that of rivers.

ORGANIZATION WITHIN HABITATS

Patterns in the use of space

Benthic organisms commonly show preferences for particular microhabitat features (Minshall 1984) and small-scale heterogeneity in factors such as substratum and flow allows these preferences to be expressed. Where

347

different microhabitat requirements are evident, particularly among closely-related species, ecologists have often focused on the possibility that spatial partitioning of a limited resource is occurring. For instance, pairs of coexisting species of net-spinning caddis larvae (Hydropsychidae) often partition net spinning sites in rivers, one species of each pair living beneath stones whilst the other is found on upper surfaces (Hildrew & Edington 1979; Tanida 1984).

However, in many studies of resource partitioning it is an unjustified assumption that in the absence of observed differences among coexisting species competitive exclusion would ensue. A more rigorous assessment of spatial partitioning involves experimental manipulations. Owl Creek, Montana, contains several shelter-building insects, the most abundant of which is a species of caddis, *Leucotrichia* (Trichoptera:Hydroptilidae). Larvae of the moth *Parargyractis* (Pyralidae) also occur, together with some species of tube-dwelling midge larvae (Chironomidae) and other, mobile grazers (McAuliffe 1984). *Leucotrichia* grazes algae on stone surfaces around its shelter and is highly territorial and aggressive, excluding other species by interference.

Although colonization of stones by *Leucotrichia* is itself inhibited by the presence of the silken retreats of *Parargyractis*, the caddis can ultimately establish a virtually complete competitive monopoly of space over the moth. *Leucotrichia* cases remain on stones after the adults have emerged. Larvae of the subsequent generation can reoccupy these cases; about 70% of remaining cases are taken over in this way. *Parargyractis* retreats, however, are less resilient and slough off the stones quite rapidly, freeing space that can be colonized by *Leucotrichia*. At least two kinds of disturbance account for the coexistence of other species with *Leucotrichia* in Owl Creek. Firstly, stones sometimes turn over during spates and, secondly, their upper surfaces may dry out at minimum discharge. Because building new cases takes longer than reoccupying old ones, it was reasoned that recently overturned stones would have relatively fewer *Leucotrichia* and more *Parargyractis* than stable stones: this expectation was confirmed experimentally (Fig. 16.1). When stone surfaces dry out in late summer, *Parargyractis* abandon their retreats but *Leucotrichia* are entombed in their cases and perish. In June, *Leucotrichia* were found only on surfaces below the previous August's low water. Above this level, the stone surface was dominated by the chironomid *Eukiefferiella*, a species with several generations per year that can best exploit the ephemeral resources of stone surfaces in shallow water. Densities of *Leucotrichia* and *Eukiefferiella* were negatively correlated and a vertical partitioning of space was occurring.

McAuliffe's clear demonstration that small scale, stochastic disturbance

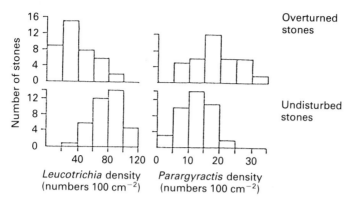

FIG. 16.1. McAuliffe (1984) selected eighty stones in Owl Creek. Forty were overturned in October 1980 and forty were undisturbed controls. This shows the frequency distribution of densities of *Leucotrichia* and *Parargyractis* on experimental and control stones in the subsequent August, after colonization.

can promote the coexistence of species, by preventing monopolization by a competitive dominant (see also Hemphill & Cooper 1983) closely parallels a study of seaweeds on a boulder beach in California (Sousa 1979). There are now many other examples from the freshwater benthos where interference competition for space has been demonstrated (Hildrew & Townsend 1980; Hart 1983; Wiley & Kohler 1984). The importance of the interaction of competition with processes ameliorating its effects has probably been underestimated.

Whilst experiments can test hypotheses unequivocally, this approach has been used in a tiny minority of cases. In this field, as in others, most studies have been descriptive and correlative and we have been too ready to accept competition as a structuring process if our data seem consistent with that view. More rigorous attempts to exclude other explanations, such as chance, for patterns in field data have not yet made much impact in the study of freshwater benthos. Recently, however, Tokeshi & Townsend (1987) subjected data on an assemblage of epiphytic Chironomidae, to a neutral model analysis. Nine species dominated and were found upon the apical leafy stems of the spiked water milfoil (*Myriophyllum spicatum*) in eastern England.

If competitive interactions play an important role in shaping the spatial organization of a community, then observed spatial overlap between species should be less than that expected by chance. Tokeshi & Townsend (1987) reasoned that spatial segregation along the short section of a suitable stem was unlikely. But segregation might be expected 'horizontally', among different stems or 'patches'. Thus, they chose single stems as the sampling unit and quantified spatial overlap between pairs of species in their regular samples. They then compared observed overlap with that generated by two models of random occurence.

In the first of these models, individuals of each species were randomly reassigned among the N habitat units (stems), and overlap between pairs of species was calculated. Two hundred replicates of this procedure were carried out for each monthly sample, and overlap values were compared with those from the real samples. A total of 140 real species pairs were examined from a year's sampling, and eighty (57.1%) had overlap values significantly smaller than expected from the neutral model, indicating a fairly widespread spatial separation. Is this, then, an assemblage structured by competition?

Many chironomid species have aggregated distributions and it is possible that low spatial overlap could result through species forming patches randomly and independently of one another, without involving interspecific competition. If competition is important, observed overlap should be smaller than that produced by a random dispersion of patches. When, in a second model, Tokeshi & Townsend (1987) randomly reassigned 'patches' (number of larvae of each species per stem), only four of the 140 species pairs showed overlap values smaller than expected by chance. Thus, the distribution of patches is generally random, and there is no evidence that spatial organization is determined competitively. Why should this conclusion be so radically different from that derived from studies of sessile animals on hard substrata, outlined previously (McAuliffe 1984)?

A stand of *Myriophyllum spicatum* stems is an ephemeral habitat of many discrete but qualitatively similar units which must be colonized afresh each growing season. Such habitats are difficult for one species to monopolize fully and there will always be, by chance, some stems available for others. Also, small differences between stems do not provide sufficient qualitative heterogeneity to form the basis for a systematic selection of habitat, i.e. resource partitioning. The habitat thus differs in some important respects from the relatively more constant and continuous, hard, stony habitat of Owl Creek.

One is struck by parallels with other communities of species colonizing divided and ephemeral resources, such as dung, carrion and fallen fruit (Doube, Chapter 12; Shorrocks & Rosewell, Chapter 2). Shorrocks & Rosewell show that regional coexistence between such species depends not on the strength of competition between them in the patch but rather their degree of independent aggregation between patches. Commonly observed values of independent aggregation among Diptera, particularly *Drosophila*, are sufficient to make competitive exclusion almost impossible.

Patterns in the use of food

It has often been suggested that exploitation of separate food resources is crucial for the coexistence of species, particularly predators. However, dietary differences have not always been found (see review by Hart 1983) and again, few attempts have been made to exclude other possible explanations.

By far the most complete case for the competitive division of food resources is that assembled by Prof. T.B. Reynoldson and co-workers for the four most common species of lake-dwelling triclads of the British Isles (Reynoldson 1983). Each of the three genera has a 'food refuge' although this specialization is by no means complete (Table 16.1). The two species of *Polycelis* have similar diets and appear to compete the most strongly. Indeed, when *Polycelis tenuis* gained access to a small lake in Anglesey, previously containing *P. nigra* only, the invader quickly almost completely replaced the resident species.

TABLE 16.1. The proportion of meals taken by the four common triclad species in British lakes on a number of prey taxa. Figures are the percentage of total meals per 500 triclads day^{-1}. Food refuges are shown in italics (Reynoldson 1983)

	Asellus	*Gammarus*	Oligochaeta	Gastropoda	Total meals
Polycelis tenuis	25	8	*57*	10	156
Polycelis nigra	8	7	*68*	17	155
Dugesia polychroa	16	4	22	*57*	176
Dendrocoelum lacteum	*63*	23	14	0	157

Patterns in time

There have been many reports of temporal separation among closely-related benthic animals, and particularly among northern-temperate stream insects (Hart 1983). For instance, the widely coexisting British caddis *Hydropsyche pellucidula* and *H. siltalai* have striking differences in their life-cycles (Hildrew & Edington 1979). Oswood (1976) also found staggered life-cycles among coexisting hydropsychids elsewhere.

Towns (1983) justifiably attacks the practice of interpreting observed differences in life histories 'on the assumption that they have been evolved as a tactic for ecological differentiation'. There was certainly no evidence of temporal segregation among six coexisting species of New Zealand Leptophlebiidae (Ephemeroptera) (Towns 1983), and several species had only weakly synchronized life cycles. However, the apparent temporal segregation revealed by so many other authors requires further examination. Greater rigour is needed in analyses and, again, Tokeshi (1986) has given a recent lead.

In this instance he assessed the temporal utilization of resources by each of his nine species of epiphytic chironomids by estimating their daily mean production. Overlap between pairs of resource utilization curves was then measured. In one neutral model, these curves were randomly rearranged through the year. In a second model, the procedure was identical except that resource utilization peaks in the model assemblages were constrained to fall between March and October (thus incorporating 'seasonality'). For both models observed overlap was far greater than would be expected by chance, and Tokeshi (1986) concluded that interspecific competition could not be an important factor in the temporal organization of this assemblage. In fact, all the species had resource utilization maxima between March and May, when diatoms were most abundant.

The general conclusion from Tokeshi's (1985, 1986) study of spatial and temporal patterns in an epiphytic chironomid assemblage is that competition is not an important structuring influence. Non-equilibrial processes also appear to structure some terrestrial guilds of phytophagous insects (Strong, Lawton & Southwood 1984), in which interspecific competition is rarely prominent (but see Southwood, Chapter 1; Claridge, Chapter 7). Strong (1984) has argued that natural enemies reduce the population of such insects below the level at which food shortage would result. Although Tokeshi (1985) does not believe predation was important for his epiphytic chironomids, it seems to be intense in some freshwater communities and we now turn our attention to its importance.

Predation and grazing as organizing processes

Predation and grazing have potentially similar consequences for community structure although there may be general differences between trophic levels (Hairston, Smith & Slobodkin 1960; Lawton & Strong 1981). Their influence can include changes in absolute and relative abundance of prey species and changes in species composition and diversity of the community. Community effects will be particularly apparent if the 'prey' removed is a 'strong interactor' (Paine 1980; Allan 1983). Predation can have important effects in other aquatic systems, notably the freshwater and marine plankton (Kerfoot 1980; Zaret 1980; Harris, Chapter 15; Reynolds, Chapter 14) and the marine intertidal (Paine 1980). How influential is it in freshwater benthic communities?

There are two conflicting lines of evidence. Indirect estimates of predator impact compare the calculated consumption rates of predators with estimates of standing crop or production of prey. In his review of stream studies, Allan (1983) concluded that such studies generally indicated a very heavy utilization of the invertebrate fauna by major predators (e.g. Benke 1978; Hildrew & Townsend 1982).

Direct experimental demonstrations of predator impact on the benthos have yielded equivocal results. When the numbers of invertebrate predators are increased the densities of individual prey species are often reduced but the results do not suggest a very strong or widespread 'keystone' effect (Peckarsky 1984; Thorp & Cochran 1984; Walde & Davies 1984). Results of manipulations involving fish have also often failed to show much effect on prey populations (Allan 1983). However, Flecker (1984) and Hemphill & Cooper (1984) reported positive results; the former showed that chironomid abundance in a stream was reduced in the presence of fish (other taxa were unaffected) and the latter that trout reduced the densities of some large insect predators in stream pools.

Turning now to grazers, benthic herbivores frequently depress the biomass of algae in streams (Gregory 1983) and may modify the extent of primary production (Lamberti & Resh 1983). Hart (1985) demonstrated a strong community effect by a herbivore on its algal 'prey'. The feeding territories of the sessile caddis *Leucotrichia* were covered with a 'lawn' of diatoms and the cyanobacterium *Schizothrix*. Outside the feeding areas a thick mat of filaments of the cyanobacterium *Microcoleus* was competitively dominant. Experimental removal of *Leucotrichia* larvae resulted in an almost complete coverage of their former territories by *Microcoleus* within 24 h. Hart (1985) showed that *Leucotrichia* was not strictly grazing on the *Microcoleus* but 'weeding' its territory and thus freeing it of this invasive species.

Power, Mathews & Stewart (1985) describe an equally striking example of strong interactions involving three trophic levels in small stream pools in the mid-western USA. A marked relationship exists between benthic algal standing crop, grazing minnows and predatory bass. In pools lacking bass algae are sparse because of grazing by minnows. Algae are abundant in pools with bass, because the minnows emigrate, hide or are eaten (Fig. 16.2).

A clue to the factor which may explain the variation in the prominence of obvious community effects via predation and grazing in the freshwater benthos is offered by Power, Mathews & Stewart (1985). The interaction between plants, herbivores and predators apparently breaks down during floods and may be absent altogether in streams with more erratic discharge.

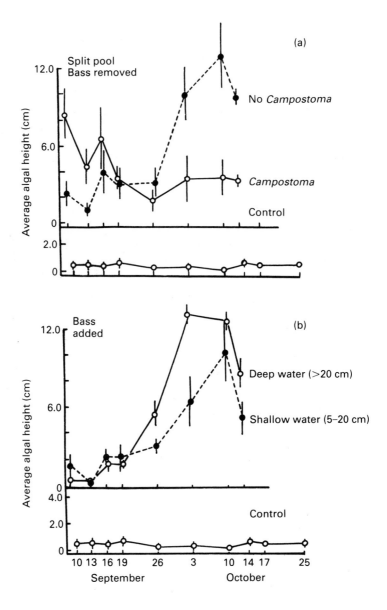

FIG. 16.2 Interaction between algae, herbivorous fish and predators: (a) a pool previously containing bass (*Micropterus salmoides*) was divided lengthwise and the bass and minnows removed. Grazing minnows (*Campostoma anomalum*) were added to one half and mean algal height (± 2 S.E.) subsequently assessed in both. The control pool contained minnows but no bass. (b) Bass were added to a pool previously containing minnows only, and algal height assessed in deep and shallow water. The control pool contained minnows and no bass (Power *et al.* 1985).

Perhaps, as others have suggested (Gregory 1983, Peckarsky 1984), in more physically disturbed systems prominent effects of species interactions will be masked or absent.

A rather different pattern has been described by Jeffries & Lawton (1985). They show that there is an orderly relationship between the numbers of predator and prey species among a miscellany of freshwater communities, with an average ratio of predators to prey of 0.36 (Fig. 16.3). A number of mechanisms could underlie such a pattern but Jeffries & Lawton (1984, 1985) propose 'apparent competition' among prey for enemy free space (see also Holt 1984). That is, a prey species is more likely to establish itself in a community if it does not resemble any existing prey species, because existing enemies will be less likely to recognize it as food. The consequences of competition for enemy free space are identical to those predicted for conventional interspecific competition. Predation would need to be pervasive and intense for this process to be a widespread phenomenon in freshwater communities.

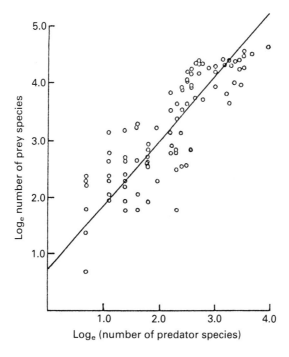

FIG. 16.3. The numbers of prey species plotted against the numbers of predator species for a miscellany of freshwater communities in Britain and North America (Jeffries & Lawton 1985). The line is the average of the two regressions with predator or prey species as the independent variable.

PATTERNS BETWEEN HABITATS

Longitudinal patterns in river communities

Differences in species composition between communities in different reaches of rivers have long been apparent and have been studied for over a century (Hawkes 1975). At first sight, the large scale, longitudinal change in river communities appears to be a particularly obvious example of a response to a continuous ecological gradient of conditions; and just as the arrangement of plant communities, for example, along such gradients has proved controversial (Austin 1985), so has the pattern of communities in rivers. The main argument is whether communities are gradually modified or whether there are abrupt zones. A current consensus might be that where conditions do change gradually then there are no sudden changes in community composition. However, there is some evidence that many rivers exhibit sharp changes in conditions, particularly in some features of water flow, and that this does then contribute to a zonation of biological communities (Statzner & Higler 1986).

In a thoughtful essay, Fisher (1983) points out that the unique feature of rivers is that communities upstream can affect those downstream via the transport of matter in the flowing water. He argues, therefore, that river communities represent a spatial analogue of site-specific succession through time. Early proponents of river zonation (Illies & Botasaneanu 1963) had not considered this possible functional linkage, and were effectively treating river communities as simple analogues of other communities along eco-logical gradients.

In their 'river continuum concept', Vannote *et al.* (1980) give prom-inence to the functional linkage of serially arranged communities. Community structure and function are conceived as varying gradually and continuously along rivers in response to a variety of hydrological, physical and biological variables. The 'river continuum concept' has been influential and widely quoted although some of its features have received justified criticism (Winterbourn, Rounick & Cowie 1981; Statzner & Higler 1985). We pursue just two aspects here.

Firstly, the argument is cast in teleological and clearly 'Clementsian' terms, and the collective attributes of ecological systems such as energy flow and nutrient cycling, are supposed to be 'adapted' to their physical environment. It is difficult to accept such an argument in the absence of a mechanism for evolution at the community level.

A second, positive, feature of the concept is the proposition that there are predictable downstream patterns in the trophic structure of animal communities in response to the changing pattern of energy inputs. The scheme assumes a downstream decrease in coarse particulate organic matter originating outside the stream (allochthonous) and a downstream increase in fine particulate allochthonous matter and also in primary production within the system. Vannote *et al.* (1980) thus expect a downstream decline in the relative importance of 'shredders' (species feeding upon autumn-shed tree leaves), and a consequent increase in 'collectors' (taking deposited or suspended fine particles, including shredder faeces). No doubt there are many exceptions to the proposed pattern and alternative explanations may exist (Winterbourn *et al.* 1981; Dudgeon 1984; Townsend & Hildrew 1984; Marchant *et al.* 1985; Bunn 1986). Nevertheless, the approach lends a focus to research and, most importantly, can generate testable hypotheses (Townsend & Hildrew 1984).

A completely different approach to classifying river communities uses statistical techniques originally developed for the study of vegetation (Gauch 1982). Three recent studies of British river invertebrates (Townsend, Hildrew & Francis 1983; Wright *et al.* 1984; Ormerod 1985) each ordinated sites on the basis of their species lists using detrended correspondence analysis (DECORANA, Hill & Gauch 1980). The survey by Wright *et al.* (1984) was based on small, qualitative collections of invertebrates at 268 sites throughout Britain, whereas Ormerod (1985) and Townsend *et al.* (1983) studied more restricted river systems. The results are in remarkable agreement.

In all three studies there were clear correlations between environmental variables and the first two axes of the ordinations. Axis 1 was highly correlated with substrate and alkalinity (Wright *et al.* 1984), with pH (Townsend *et al.* 1983) and hardness and pH (Ormerod 1985). Hardness, alkalinity and pH are closely related (Sutcliffe & Hildrew 1987) and thus this combined factor was prominent in all these studies. Axis 2 in each case seemed most closely related to factors changing from source to mouth of rivers (Ormerod—distance from source and discharge; Townsend *et al.*— summer temperature; Wright *et al.*—slope and stream link magnitude). These studies, and others not using multivariate statistical techniques, thus strongly suggest that the alkalinity of water makes a strong contribution to the variation in river benthos. Soft and acid waters are markedly impoverished both in abundance and diversity: the reasons for this are discussed elsewhere (Sutcliffe & Hildrew 1987). It is, finally, important to appreciate that the site groupings derived from multivariate statistical procedures do not necessarily correspond to natural 'zones' along rivers. Indeed, site groupings are usually established from collections from a number of streams and *not* from sites along a single waterway.

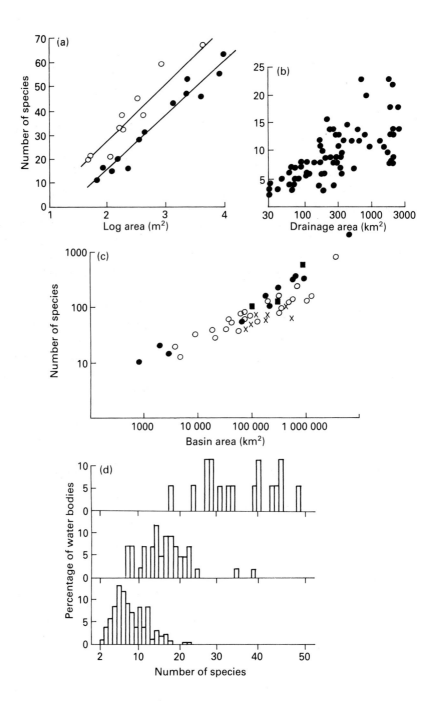

Habitat area and age

Species–area relationships have been found in very many groups of animals and plants, although their ecological basis is not clear (Williamson 1981). Patterns among water bodies are less well known, though inland waters are commonly quoted as habitat islands, analogous to islands of land in the sea (Giller 1984). We present four freshwater examples in Fig. 16.4, for aquatic vascular plants, mussels, fish and crustacea, which demonstrate the classic species–area relationship.

Lentic systems vary greatly in age. There are literally millions of temporary pools in the tropics and elsewhere with a lifespan of a few hours to a few months (e.g. Hildrew 1985). The oldest lakes, however, are millions of years old. Lake Baikal was probably formed in the middle Tertiary (Kozhov 1963), Lake Biwa (Japan) is about 5 million years old (Horie 1984), and the ancient African tectonic lakes may be 10–20 million years old (Beadle 1981). Some lacustrine systems may be even older, though probably not continuously in the same basin; for instance, there have been lakes in the Balkan peninsula, around the present Lake Ohrid, at least since the Oligocene (about 30 million years ago; Stankovic 1960).

The vast majority of still-water bodies, however, are geologically recent and even quite large lakes may not have been continuously suitable for animals and plants. A good example is Lake Chilwa in Malawi. This has a closed basin and exhibits remarkable variations in level and salinity (Kalk 1979); in 1960 and 1968 there were periods when the lake was completely dry. Studies of the benthos were made before and after 1968 and recolonization was shown to be rapid, with enormous populations of midge larvae (Chironomidae) having developed within weeks of refilling (McLachlan 1979).

The powers of dispersal of many freshwater animals are apparently very great. A recent compilation of data for the African lakes Albert (500000 years old) and Turkana (130000) and the recent man-made lakes Kariba (25) and Kainji (4) shows a remarkable insensitivity of species-richness to

---

FIG. 16.4. Species–area relationships for four groups of freshwater organisms: (a) aquatic vascular plants in ponds in two districts of Denmark where mean distance between ponds differed (○ ponds close; ● ponds far apart), (b) stream-dwelling mussels from southeastern Michigan, (c) fish species in various river systems (●South America; ○ Africa; ■Europe; x Asia), (d) frequency distributions of species-richness of pond-dwelling crustacea in Yorkshire, northern England (upper panel, large water bodies < 15000 m² (n = 17); middle panel, medium water bodies 5000–12000 m² (n = 41); lower panel, small water bodies < 5000 m² (n = 149),. After (a) Moller & Rordam (1985), (b) Welcomme (1979), (c) Strayer (1983), (d) Fryer (1985).

age (Green 1985). Rapid colonization has repeatedly been demonstrated for a variety of water bodies and presumably accounts for the cosmopolitan distribution of much of the fauna.

It is only in very large lakes, which are usually deep and ancient, that outbursts of speciation have occurred. This is particularly well known for the fish of African lakes (Fryer & Iles 1972) but has also occurred in the benthos, and outstandingly that of Lake Baikal (Kozhov 1963). For instance, this lake has about eighty species of endemic triclads, 240 species of gammarid Crustacea and eighty-four molluscs, of which more than fifty are endemic. There are groups in the lake, however, that have failed to radiate, including the isopod Crustacea, with only five species and a single genus.

Thus, the outstanding spatial feature of fresh water as a habitat is its severely limited overall extent and the insular, patchy nature of water bodies. In time they are usually rather ephemeral and often physically disturbed. Exceptions and variations on this theme yield the observed differences in community structure. However, these general characteristics seem to have contributed to an overall limitation in species-richness compared to terrestrial and marine communities.

Patterns among small water bodies

Temporary pools are the most numerous of all freshwater habitats (McLachlan 1985). They include ground pools on soil or rock but also numerous natural and artificial containers. Pockets of water sufficient to support an aquatic community have also been encountered on more than 1500 plant species (phytotelmata; Frank & Lounibos 1983), and Coleoptera and Diptera, including about 400 species of mosquito, dominate the fauna (Fish 1983).

Of all freshwater habitats, temporary pools appear to provide the most obvious parallel with other ephemeral and divided resources. There is a clear theoretical basis for the potential coexistence of species exploiting such habitats without resource partitioning (Shorrocks & Rosewell, Chapter 2). Further, the experimental work of the late R.P. Seifert (Seifert 1984) on communities in *Heliconia* inflorescences revealed little evidence of strong intra- or interspecific competition.

The situation may differ, however, for the midge larvae occupying temporary rock pools in Africa (McLachlan 1985). Two species, *Polypedilum vanderplanki* and *Chironomus imicola* were rarely found in the same pool, but partitioned them by area and depth. *Polypedilum vanderplanki* inhabited smaller, shallower pools; it is able to withstand desiccation as a larva (Hinton 1968) and can persist, in the dry sediment, between rainy seasons. *Chironomus imicola* needs at least 12 days from egg to adult (McLachlan 1983) and must emerge as an adult and disperse before the pool dries: but without further rain, small pools are short-lived. Larvae of *P. vanderplanki,* having persisted during the dry season, have an obvious size advantage at the onset of the rains. But why does *P. vanderplanki* not occur in the larger, *C. imicola* dominated pools? Experimental work suggests a competitive advantage for *C. imicola,* mediated in part via chemical 'conditioning' of the water (McLachlan 1985).

The important difference between these ground pools and other ephemeral and divided habitats, such as dung, carrion and, indeed, phytotelmata in relatively short-lived plants, is that the pools have considerable locational stability. To *P. vanderplanki* and other rainpool species (e.g. freshwater fairy shrimps, Hildrew 1985) the habitats are not temporary at all; they are permanent habitats which are temporarily favourable for activity. There is every likelihood that such species can monopolize habitats and persist for many generations. Such communities are more likely to be structured by species interactions than are those of other patchy habitats which are both ephemeral in time and unpredictable in space.

SYNTHESIS

Disturbance, productivity and diversity

Faced with a variety of contrasting results from many empirical studies can any systematic pattern be perceived in community structure?

Consider two dimensions characterizing the stony benthic habitat. These are firstly *disturbance,* which is any physical process removing residents from the surface, and secondly *productivity* at the stone surface, determined largely by water chemistry. Productivity here includes both autotrophic and heterotrophic elements and refers particularly to the rate of elaboration of the stone surface organic layer or 'epilithon' (Lock *et al.* 1984). This typically consists of a variety of algae, fungi and bacteria in a slimy matrix. Energy is drawn from photosynthesis and from dissolved organic molecules of diverse origin. Figure 16.5 indicates contours of species-diversity on a graph whose axes are frequency of disturbance and productivity. We use this simple scheme to organize discussion and identify trends.

The postulated relationship between disturbance and species-diversity is borrowed from Connell's (1975) intermediate disturbance hypothesis. Undisturbed surfaces, at least in reasonably productive waters, have surfaces dominated by sessile herbivores and filter-feeders. As disturbance

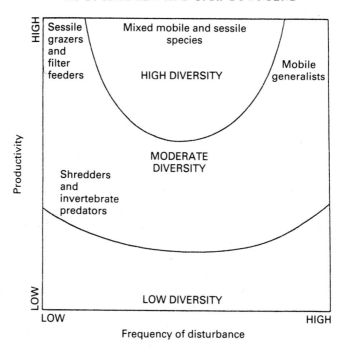

FIG. 16.5. A postulated relationship between frequency of disturbance and productivity and some structural features of the freshwater benthos of hard substrata.

increases, gaps become available and mobile species become more prominent. Sessile species are lost finally when disturbances occur so frequently that surface resources become indefensible and sessile life too hazardous. At very low productivity (equivalent here to nutrient-poor, often acidified waters), species diversity is low, presumably because few species can tolerate the conditions or find food (Sutcliffe & Hildrew 1987). In these situations and with a low frequency of disturbance, large, invertebrate predators are prominent, exclude vulnerable prey and may compete among themselves for prey (Hildrew & Townsend 1982). The appearance of fish as productivity increases may be important in removing invertebrate predation pressure (Townsend & Hildrew 1984). Where there is a source of terrestrial leaf litter, consumers of these coarse particles are prominent, particularly if the material persists through the year. For instance, frequent physical disturbance in spatey streams may remove the material before it can be consumed by animals and in hard, productive waters, microbial decompositon is rapid and the leaf litter quickly mineralized or dispersed. In other communities it has been suggested that medium primary productivity is associated with higher levels of diversity than lower or higher productivity (Grime 1979; Tilman 1982; Abramsky & Rosenzweig 1983; and see model proposed by Huston 1979): apart from artificially enriched situations it is not clear whether this pertains in the freshwater benthos of hard substrata.

Disturbance and succession

The effect of a disturbance on a climax, space-limited community is often to create openings which may be colonized by one or more of a group of opportunistic, early successional species. Later, these open areas regain the 'climax' biota when the most efficient competitors oust their neighbours. Early successional species can, however, persist in an area if sufficient new patches are made available by disturbance within the dispersal range of their propagules (Connell 1987). High diversity is maintained through a mosaic of successional stages being maintained in this way. Sousa (1984) suggests further that such a system could approach a regional steady state depending on the spatial scale of disturbances relative to the total landscape area.

Such processes underlie the postulated disturbance axis in Fig. 16.5. At one extreme, disturbance would be so frequent and widespread that 'later colonists' could not reliably persist and the overall community would consist of a mixture of resilient early colonists. At the other, newly disturbed patches would be infrequent and early colonists consequently vanishingly rare. But are there any empirical data on successional changes after disturbance of the freshwater benthos?

As Connell (1987) points out, after initial colonization patches may persist unchanged, change erratically or undergo a predictable interactive succession; and he gave marine examples of all three. Fisher (1983), in his review of stream succession, did not distinguish these possibilities and concluded that disturbed streams are colonized in a 'stochastic manner influenced by timing and severity of the disturbance and availability of nearby colonists'. Subsequent change could be due to further, stochastic colonization, but he could muster little evidence for or against predictable interactive succession (Fisher 1983).

Recent experiments, however, (Ladle *et al.* 1985; Pinder 1985), have yielded fascinating results. Two open-air, recirculating stream channels in south-west England have gravel beds like those of neighbouring streams and, presumably, draw colonists from such places. When the channels are first exposed to colonists, they can be considered as 'patches' recovering from disturbance. On day 1 of such an experiment, the two channels had virtually no macroinvertebrates but by day 37 they had accumulated thirty-three and thirty-five taxa, chironomid midges accounting for seventeen and twenty-one of them, respectively. However, on day 16 in both channels a single species of chironomid completely dominated the new communities with densities approaching 70000 and 30000 m^{-2}. It had virtually disappeared by day 35 (larval life took 16 days) when the total density of other chironomids was 110000 and 64000 m^{-2}, respectively. Remarkably, this early colonizer was a new species, *Orthocladius calvus* and this in an area where even the Chironomidae are relatively well-known. Further, previous experiments had seen the initial, rapid colonization by a variety of species not known locally or only recorded very rarely (Pinder 1985). In each case these disappeared rapidly to be replaced by a more diverse and characteristic community. In these channels, colonization was indeed stochastic and unpredictable but there was subsequently a predictable succession resulting in a similar community structure in each experiment. One is drawn to the conclusion that early colonists are maintained in this area by short-lived and patchy disturbances.

Disturbance and community persistence

There is little long-term data on freshwater benthos by which change or persistence can be judged. We have recently repeated our 1976 survey of stream communities in southern England (Townsend, Hildrew & Francis 1983) after a gap of 8 years, time for between about eight and thirty generations of most benthic species. Measures of persistence (the extent to which community composition was unchanged) at twenty-seven sites are given in Fig. 16.6 (Townsend, Hildrew, & Schofield, 1987). There has evidently been very substantial change, most obviously in the ranking of species-abundance. In terms of species composition, Jaccard's Index takes a modal value of between 0.4 and 0.5 (the index ranging from 0–1 for no to complete similarity) and a substantial proportion of the fifteen most abundant species in each community were still present in 1984 (modal value 11–12). Taken alone such measures cannot reveal the importance of stochastic or deterministic mechanisms; however, correlations between community persistence and environmental parameters revealed that those sites with lower temperatures, and a smaller annual range of temperature, possessed communities with a more persistent structure (Fig. 16.7a–c). When we analysed various taxa separately, we found that in some (Plecoptera, Chironomidae and non-dipterous insects), persistence was related to physicochemical variation among sites. The statistically significant relationships are given in Fig. 16.7(d–f) and show that the most persistent assemblages are found at upstream sites with a low discharge range or low mean pH.

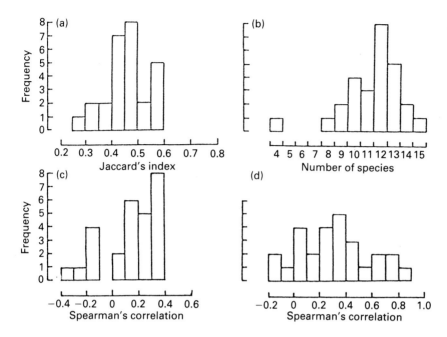

FIG. 16.6. Frequency distributions of four measures of community persistence in the macro-invertebrate communities at twenty-seven stream sites in southern England, sampled in 1976 and 1984: (a) Jaccard's Index (0–1, minimum to maximum persistence), (b) The number of the fifteen most abundant species in 1976 still present in 1984 (0–15), minimum to maximum persistence), (c) Coefficients of Spearman's rank correlation between species abundances in 1979 and 1984, calculated on all the species in each community(−1 to+1, minimum to maximum persistence), (d) Coefficients of Spearman's Rank correlation between species abundances in 1976 and 1984 calculated on only the fifteen most abundant in the 1976 survey (Townsend, Hildrew & Schofield 1987).

Perhaps then, among the sites in our study, those communities in cool, acid upstream sites with a low and stable discharge are the least disturbed and most persistent. Are such trends reflected in their food webs? Hildrew, Townsend & Hasham (1985) have recently constructed a food web for Broadstone stream, an acid, upstream site of low discharge. Connectance (number of interactions between species) was very high for the species-richness observed: very few of the forty food webs considered by Briand (1983) had higher values. Such features are theoretically associated with fragile and dynamically unstable food webs and were characteristic of communities from 'constant' environments in Briand's (1983) survey.

Finally, then, how are most benthic communities structured? We have tried to show the variation in the relative importance of deterministic and stochastic processes, and have suggested that perhaps this variation is systematic. We would not care to argue for the primacy of either. The

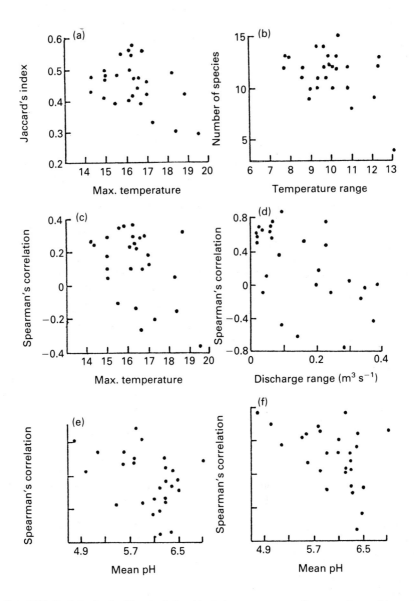

FIG. 16.7. Statistically significant relationships between measures of community persistence in twenty-seven stream invertebrate communities (see Fig. 16.6) and environmental variables: (a) Jaccard's Index for the whole community; (b) Number of fifteen most abundant species in 1976 still present in 1984; (c) Spearman's Rank correlation between abundances of all taxa in 1976 and 1984; (d) – (f) relate the Spearman's coefficients calculated for particular taxa, (d)Plecoptera, (e) Chironomidae, (f) non-dipterous insects.

answer to this final question is probably that most communities fall between the extremes. Sufficient controversy has already been generated in ecology by people working in contrasting but equally extreme systems. To make a final analogy with marine systems, the sea is both salt and water, but no one finds controversy in variations in the strength of brine.

REFERENCES

Abramsky, Z. & Rosenzweig, M.L. (1983). Tilman's predicted productivity–diversity relationship shown by desert rodents. *Nature (London)*, **309**, 150–1.

Allan, J.D. (1983). Predator–prey relationships in streams. *Stream Ecology: Application and Testing of General Ecological Theory* (Ed. by J.R. Barnes & G.W. Minshall), pp. 191–229. Plenum, New York.

Austin, M.P. (1985). Continuum concept, ordination methods, and niche theory. *Annual Review of Ecology and Systematics*, **16**, 39–62.

Barnes, J.R. & Minshall, G.W. (Eds) **(1983).** *Stream Ecology: Application and Testing of General Ecological Theory.* Plenum, New York.

Beadle, L.C. (1981). *The Inland Waters of Tropical Africa.* Longman, London.

Benke, A.C. (1978). Interactions among coexisting predators: a field experiment with dragonfly larvae. *Journal of Animal Ecology*, **47**, 335–50.

Briand, F. (1983). Environmental control of food web structure. *Ecology*, **64**, 253–63.

Bunn, S.E. (1986). Spatial and temporal variation in the macroinvertebrate fauna of streams of the northern jarrah forest, Western Australia: functional organisation. *Freshwater Biology*, **16**, 621–32.

Cody, H.L. & Diamond, J.M. (Eds) **(1975).** *The Ecology and Evolution of Communities.* Belknap Press, Cambridge, Massachusetts.

Connell, J.H. (1975). Diversity in tropical rain forests and coral reefs. *Science (N.Y.)* **199**, 1302–10.

Connell, J.H. (1987). Change and persistence in some marine communities. *Colonization, Succession and Stability.* Symposia of the British Ecological Society, 26 (Ed. by A.J. Gray, M.J. Crawley & P.J. Edwards), pp. 339–52. Blackwell Scientific Publications, Oxford.

Dudgeon, D.G. (1984). Longitudinal and temporal changes in functional organisation of macroinvertebrate communities in the Lam Tsuen River, Hong Kong. *Hydrobiologia*, **111**, 207–17.

Fish, D. (1983). Phytotelmata: flora and fauna. *Phytotelmata: Terrestrial Plants as Hosts for Aquatic Insect Communities* (Ed. by J.H. Frank & L.P. Lounibos), pp. 1–28. Plexus, Medford, New Jersey.

Fisher, S.G. (1983). Succession in streams. *Stream Ecology: Application and Testing of General Ecological Theory* (Ed. by J.R. Barnes & G.W. Minshall), pp. 7–28. Plenum, New York.

Flecker, A.S. (1984). The effect of predation and detritus on the structure of a stream insect community: a field test. *Oecologia (Berlin)*, **64**, 300–5.

Frank, J.H. & Lounibos, L.P. (1983). *Phytotelmata: Terrestrial Plants as Hosts for Aquatic Insect Communities.* Plexus, Medford, New Jersey.

Fryer, G. (1985). Crustacean diversity in relation to the size of water bodies: some facts and problems. *Freshwater Biology*, **15**, 347–61.

Fryer, G. & Iles, T.D. (1972). *The Cichlid Fishes of the Great Lakes of Africa.* Oliver & Boyd, Edinburgh.

Gauch, H.G. (1982). *Multivariate Analysis in Community Ecology.* Cambridge University Press, Cambridge.

Giller, P.S. (1984). *Community Structure and the Niche.* Chapman & Hall, London.
Green, J. (1985). Horizontal variations in associations of zooplankton in Lake Kariba. *Journal of Zoology (A),* **206,** 225–39.
Gregory, S.V. (1983). Plant–herbivore interactions in stream systems. *Stream Ecology: Application and Testing of General Ecological Theory* (Ed. by J.R. Barnes & G.W. Minshall), pp. 157–89. Plenum, New York.
Grime, J.P. (1979). *Plant Strategies and Vegetation Processes.* Wiley, Chichester.
Hairston, N.G., Smith, F.E. & Slobodkin, L.B. (1960). Community structure, population control, and competition. *American Naturalist,* **44,** 421–5.
Hart, D.D. (1983). The importance of competitive interactions within stream populations and communities. *Stream Ecology: Application and Testing of General Ecological Theory* (Ed. by J.R. Barnes & G.W. Minshall), pp. 99–136. Plenum, New York.
Hart, D.D. (1985). Grazing insects mediate algal interactions in a stream benthic community. *Oikos,* **44,** 40–6.
Hawkes, H.A. (1975). River zonation and classification. *River Ecology.* (Ed. by B.A. Whitton), pp. 312–74. Blackwell Scientific Publications, Oxford.
Hemphill, N. & Cooper, S.D. (1983). The effect of physical disturbance on the relative abundance of two filter-feeding insects in a small stream. *Oecologia,* **58,** 378–82.
Hemphill, N. & Cooper, S.D. (1984). Differences in the community structure of stream pools containing or lacking trout. *Verhandlungen der internationale Vereinigung für theoretische und angewandte Limnologie,* **22,** 1858–61.
Hildrew, A.G. (1985). A quantitative study of the life history of a fairy shrimp (Branchiopoda: Anostraca) in relation to the temporary nature of its habitat, a Kenyan rainpool. *Journal of Animal Ecology,* **54,** 99–110.
Hildrew, A.G. & Edington, J.M. (1979). Factors facilitating the coexistence of hydropsychid caddis larvae (Trichoptera) in the same river system. *Journal of Animal Ecology,* **48,** 557–76.
Hildrew, A.G. & Townsend, C.R. (1980). Aggregation, interference and foraging by larvae of *Plectrocnemia conspersa* (Trichoptera: Poly-centropodidae). *Animal Behaviour,* **28,** 553–60.
Hildrew, A.G. & Townsend, C.R. (1982). Predators and prey in a patchy environment: a freshwater study. *Journal of Animal Ecology,* **51,** 797–815.
Hildrew, A.G., Townsend, C.R. & Hasham, A. (1985). The predatory Chironomidae of an iron-rich stream: feeding ecology and food web structure. *Ecological Entomology,* **10,** 403–13.
Hill, M.O. & Gauch, H.G. (1980). Detrended Correspondence Analysis: an improved ordination technique. *Vegetatio,* **42,** 47–58.
Hinton. H.E. (1968). Reversible suspension of metabolism and the origin of life. *Proceedings of the Royal Society of London (B),* **171,** 43–57.
Holt, R.D. (1984). Spatial heterogeneity, indirect interactions and the coexistence of prey species. *America Naturalist,* **124,** 377–406.
Horie, S. (1984). Lake Biwa. *Monographiae Biologicae, Vol. 54,* Junk, Dordrecht.
Huston, M. (1979). A general hypothesis of species diversity. *American Naturalist,* **113,** 81–101.
Hynes, H.B.N. (1970). *The Ecology of Running Waters.* University of Liverpool Press, Liverpool.
Illies, J. & Botosaneanu, L. (1963). Probl'emes et methodes de la classification et de la zonation 'ecologique des eaux courantes consider'ee surtout du point de vue faunistique. *Mitteilungen der internationalen vereinigung für theoretische und angewandte Limnologie,* **12,** 1–57.
Jeffries, M.J. & Lawton, J.H. (1984). Enemy free space and the structure of ecological communities. *Biological Journal of the Linnean Society,* **23,** 269–86.
Jeffries, M.J. & Lawton, J.H. (1985). Predator–prey ratios in communities of freshwater invertebrates: the roles of enemy free space. *Freshwater Biology,* **15,** 105–12.

Kalk, M. (1979). Introduction: perspectives of research at Lake Chilwa. *Lake Chilwa* (Ed. by M. Kalk, A.J. McLachlan & C. Howard-Williams), pp. 3–16. Monographiae Biologicae, Vol. 35, Junk, The Hague.

Kerfoot, W.C. (Ed.) **(1980).** *Evolution and Ecology of Zooplankton Communities,* Special Symposium 3. American Society of Limnology and Oceanography. University Press of New England, Hanover, New Hampshire.

Kozhov, M. (1963). *Lake Baikal and its Life.* Monographiae Biologicae Vol.11. Junk, The Hague.

Ladle, M., Cooling, D.A., Welton, J.S. & Bass, J.A.B. (1985). Studies on Chironomidae in experimental recirculating stream systems. II. The growth, development and production of a spring generation of *Orthocladius (Euorthocladius) calvus* Pinder. *Freshwater Biology,* **15,** 243–55.

Lamberti, G.A. & Resh, V.H. (1983). Stream periphyton and insect herbivores: an experimental study of grazing by a caddisfly population. *Ecology,* **64,** 1124–35.

Lawton, J.H & Strong, D.R. (1981). Community patterns and competition in folivorous insects. *American Naturalist,* **118,** 317–38.

Lock, M.A., Wallace, R.R., Costerton, J.W., Ventullo, R.M. & Charlton, S.E. (1984). River epilithon: towards a structural functional model. *Oikos,* **42,** 10–22.

Macan, T.T. (1974). Freshwater Ecology (2nd edn). Longman, London.

Marchant, R., Metzeling L., Graesser, A. & Suter, P. (1985). The organization of macro-invertebrate communities in the major tributaries of the La Trobe River, Victoria, Australia. *Freshwater Biology,* **15,** 315–31.

McAuliffe, J.R. (1984). Competition for space, disturbance, and the structure of a benthic stream community *Ecology,* **65,** 894–908.

McLachlan, A.J. (1979). Decline and recovery of the benthic invertebrate communities. *Lake Chilwa* (Ed. by M. Kalk, A.J. McLachlan & C. Howard-Williams), pp. 143–60. Monographiae Biologicae, Vol. 35, Junk, the Hague.

McLachlan, A.J. (1983). Life-history tactics of rain-pool dwellers. *Journal of Animal Ecology,* **52,** 545–61.

McLachlan, A.J. (1985). What determines the species present in a rain-pool? *Oikos,* **45,** 1–7.

Minshall, G.W. (1984). Aquatic insect-substratum relationships. *The Ecology of Aquatic Insects* (Ed. by V.H. Resh & D.M. Rosenberg), pp. 358–400. Praeger, New York.

Moller, T.R. & Rordam, C.P. (1985). Species numbers of vascular plants in relation to area, isolation and age of ponds in Denmark. *Oikos,* **45,** 8–16.

Ormerod, S.J. (1985). *The distribution of macroinvertebrates in the upper catchment of the River Wye in relation to ionic composition.* Ph.D. thesis, University of Wales (U.W.I.S.T.).

Oswood, M.W. (1976). Comparative life histories of the Hydropsychidae (Trichoptera) in a mountain lake outlet. *American Midland Naturalist,* **96,** 493–7.

Paine, R.T. (1980). Food webs: linkage, interaction strength, and community structure. *Journal of Animal Ecology,* **49,** 667–85.

Peckarsky, B.L. (1984). Predator–prey interactions among aquatic insects. *The Ecology of Aquatic Insects* (Ed. by V.H. Resh & D.M. Rosenberg), pp. 196–254. Praeger, New York.

Pinder, L.C.V. (1985). Studies on Chironomidae in experimental recirculating stream systems. I. *Orthocladius (Euorthocladius) calvus* sp. nov. *Freshwater Biology,* **15,** 235–41.

Power, M.E., Mathews, W.J. & Stewart, A.J. (1985). Grazing minnows, piscivorous bass, and stream algae: dynamics of a strong interaction. *Ecology,* **66,** 1448–56.

Resh, V.H. & Rosenberg, D.M. (Eds.) **(1984).** *The Ecology of Aquatic Insects.* Praeger, New York.

Reynoldson, T.B. (1983). The population biology of Turbellaria with special reference to the freshwater triclads of the British Isles. *Advances in Ecological Research,* **13,** 236–316.

Seifert, R.P. (1984). Does competition structure communities? Field Studies on Neotropical *Heliconia* insect communities. *Ecological Communities: Conceptual Issues and the Evidence*

(Ed. by D.R. Strong, D. Simberloff, L.G. Abele & A.B. Thistle), pp. 54–63. Princeton University Press, Princeton, New Jersey.

Sousa, W.P. (1979). Disturbance in marine intertidal boulder fields: the non-equilibrium maintenance of species diversity. *Ecology* **60**, 1225–39.

Sousa, W.P. (1984). The role of disturbance in natural communities. *Annual Review of Ecology and Systematics*, **15**, 353–91.

Stankovic, S. (1960). *The Balkan Lake Ohrid and Its Living World*. Monographiae Biologicae, Vol. 9, Junk, The Hague.

Statzner, B. & Higler, B. (1985). Questions and comments on the river continuum concept. *Canadian Journal of Fisheries and Aquatic Sciences*, **42**, 1038–44.

Statzner, B. & Higler, B. (1986). Stream hydraulics as a major determinant of benthic invertebrate zonation patterns. *Freshwater Biology*, **16**, 127–39.

Strayer, D. (1983). The effects of surface geology and stream size on freshwater mussel (Bivalvia, Unionidae) distribution in south-eastern Michigan, USA. *Freshwater Biology*, **13**, 253–64.

Strong, D. (1984). Exorcising the ghost of competition past: phytophagous insects. *Ecological communities: Conceptual Issues and the Evidence* (Ed. by D.R. Strong, D. Simberloff, L.G. Abele & A.B. Thistle), pp. 28–41. Princeton University Press, Princeton, New Jersey.

Strong, D., Lawton, J.H. & Southwood, T.R.E. (1984). *Insects on Plants: Community Patterns and Mechanisms*. Blackwell Scientific Publications, Oxford.

Strong D.R., Simberloff, D., Abele, L.G. & Thistle, A.B. (Eds.) (1984). *Ecological Communities: Conceptual Issues and the Evidence*. Princeton University Press, Princeton, New Jersey.

Sutcliffe, D.W. & Hildrew, A.G. (1987). Invertebrate communities in acid streams. *Acid Toxicity and Aquatic Animals*. Seminar Series of the Society for Experimental Biology, Cambridge University Press, Cambridge, in press.

Tanida, K. (1984). Larval microlocation on stone faces of three *Hydropsyche* species (Insecta: Trichoptera) with a general consideration on the relation of systematic groupings to the ecological and geographical distribution among the Japanese *Hydropsyche* species. *Physiological Ecology, Japan*, **21**, 115–30.

Thorp, J.H. & Cochran, M.I. (1984). Regulation of freshwater community structure at multiple intensities of dragonfly predation. *Ecology* **65**, 1546–55.

Tilman, D. (1982). *Resource Competition and Community Structure*, Princeton University Press, Princeton, New Jersey.

Tokeshi, M. (1985). *The population and community ecology of chironomids in a small temperate stream*. Ph.D. thesis, University of East Anglia.

Tokeshi, M. (1986). Resource utilization, overlap and temporal community dynamics: a null model analysis of an epiphytic chironomid community. *Journal of Animal Ecology*, **55**, 491–506,

Tokeshi, M. & Townsend, C.R. (1987). Random patch formation and weak competition: coexistence in an epiphytic chironomid community. *Journal of Animal Ecology*, in press.

Towns, D.R. (1983). Life history patterns of six sympatric species of Leptophlebiidae (Ephemeroptera) in a New Zealand stream and the role of interspecific competition in their evolution. *Hydrobiologia*, **99**, 37–50.

Townsend, C.R. (1980). *The Ecology of Streams and Rivers*. The Institute of Biology's Studies in Biology, no. 122, Edward Arnold, London.

Townsend, C.R., Hildrew, A.G. & Francis, J. (1983). Community structure in some southern English streams: the influence of physicochemical factors. *Freshwater Biology*, **13**, 521–44.

Townsend, C.R. & Hildrew, A.G. (1984). Longitudinal pattern in detritivore communities of acid streams: a consideration of alternative hypotheses. *Internationale Vereinigung für theoretische und angewandte Limnologie*, **22**, 1953–8.

Townsend, C.R., Hildrew, A.G. & Schofield, K. (1987). Persistence of stream invertebrate communities in relation to environmental variability. *Journal of Animal Ecology,* **56,** 597–614.

Vannote, R.L., Minshall, G.W., Cummins, K.W., Sedell, J.R. & Cushing, C.E. (1980). The river continuum concept. *Canadian Journal of Fisheries and Aquatic Sciences,* **37,** 130–7.

Walde, S.J. & Davies, R.W. (1984). Invertebrate predation and lotic prey communities: evaluation of *in situ* enclosure exclosure experiments. *Ecology,* **65,** 1206–13.

Welcomme, R.L. (1979). *Fisheries Ecology of Flood Plain Rivers.* Longman, London.

Wiley, M. & Kohler, S. (1984). Behavioral adaptations of aquatic insects *The Ecology of Aquatic Insects* (Ed. by V.H. Resh & D.M. Rosenberg), pp. 101–33. Praeger, New York.

Williamson, M.H. (1981). *Island Populations.* Oxford University Press, Oxford.

Winterbourn, M.J., Rounick, J.S. & Cowie, B. (1981). Are New Zealand streams really different? *New Zealand Journal of Marine and Freshwater Research,* **15,** 321–8.

Wright, J.F., Moss, D., Armitage, P.D. & Furse, M.T. (1984). A preliminary classification of running water sites in Great Britain based on macro-invertebrate species and the prediction of community type using environmental data. *Freshwater Biology,* **14,** 221–56.

Zaret, T.M. (1980). *Predation and Freshwater Communities.* Yale University Press, New Haven, Connecticut.

17. FEAST AND FAMINE: STRUCTURING FACTORS IN MARINE BENTHIC COMMUNITIES

T. H. PEARSON[1] AND RUTGER ROSENBERG[2]

[1]*Scottish Marine Biological Association, The Dunstaffnage Marine Research Laboratory, P.O. Box 3, Oban, Argyll, PA34 4AD Scotland, UK and* [2]*University of Göteborg, Kristinebergs Marinbiologiska Station, S-45034 Fiskebäckskil, Sweden.*

INTRODUCTION

In a previous review (Pearson & Rosenberg 1978) we showed that, in general, benthic infaunal communities are organized structurally, numerically and functionally in relation to organic enrichment gradients. Several recent publications have confirmed the generality of these results (e.g. Bonsdorff, 1980; Hily 1984; Swartz *et al.* 1985). and indeed suggested that gradients of resource availability, although modified by interactions with other environmental factors, may underlie all marine faunal distributions (Valentine 1971; Rhoads & Boyer 1982; Smetacek 1985). However, recent general theorizing about the structuring factors underlying community organization has ignored the importance of food availability or at best relegated it to a subsidiary role in the pantheon of physical, chemical and biotic factors invoked to explain observed patterns of diversity (e.g. Cody & Diamond 1975; Connell & Slatyer 1977; Diamond & Case 1986). Here we will attempt to demonstrate the primary nature of food availability in constraining benthic community structure.

Studies of marine benthos arbitrarily divide the fauna into macro- (> 1000 μm), meio-($100-1000$ μm) and microfauna ($1-100$ μm), and further define organisms as infaunal (living within the sediment) or epifaunal (living at or on the sediment surface) (Parsons, Takahashi & Hargrave 1977). Most theoretical analyses of benthic community structure have concentrated on macro-infaunal data since this element of benthic populations is the most widely studied. In sub-tidal areas infaunal organisms are usually obtained quantitatively by remote sampling of soft sediments with grabs or corers and extraction of the animals from the substrate by sieving. Epifaunal populations on hard substrates are assessed by using diver-controlled or remote stereophotographic techniques, supplemented by dredging. Either technique necessarily samples only a segment of the total populations, thus their quantitative analysis and perceived community

373

organization are biased by such partial observations. Despite such qualifications it is thought that the general principles discussed here are widely applicable to all benthic organisms.

Structure is used here in a wide sense to include species composition, abundance and biomass. A continuous gradient of organic enrichment may be considered as the simplest case of variability. At one end of such a gradient inputs of degradable organic material exceed the assimilatory capacity of benthic communities, at the other end inputs are too low to support any organisms. Between these two extremes all benthic organisms and/or communities may be ordered in relation to their relative abilities to assimilate, survive and reproduce. In reality such idealized successions are modified in response to the variable effects of other environmental gradients and structuring factors that also influence and mould community composition. It is our intention here to order the relative importance of these gradients and factors in structuring benthic communities and to explore the resulting insights into community organization.

Since the primary food source for benthos originates, with a few localized exceptions, in euphotic surface waters, food availability necessarily attenuates with depth. Moreover, primary production is continuous in the tropics but becomes increasingly seasonal with increasing latitude. Thus both depth and latitude can be regarded as primary gradients affecting food availability. A third gradient, along which food availability attenuates, is declining current speed. Water movement driven by currents, tides, strong wind and other forces transports food particles in the water mass and causes resuspension of bottom sediments. This transport is of significant importance for the distribution of food to benthic animals and is here considered as a secondary factor modifying the two primary gradients. Species distributions may be seen as a response to the varying effects of these modified gradients. Such distributions are further affected, however, by another suite of factors, some of which are dependent on the primary gradients. These are, in general, physical factors contributing to the relative environmental harshness imposed on each species, e.g. sedimentary fluctuations in stability and in turbidity, salinity, oxygen, temperature and pressure. Other factors which exert effects independently of the primary gradients can be summarized as (a) stochastic events, e.g. vulcanicity, pollution, and (b) biotic interactions, e.g. competition, predation. These various factors influencing community distributions are further evaluated below before proceeding to a discussion of their comparative importance in the evolution of benthic community structure.

BENTHIC FOOD SOURCES

The food available to benthic organisms is derived from allochthonous or autochthonous sources in the surface waters. Allochthonous material, of terrestrial or freshwater origin, is most important in shallow sublittoral habitats and inputs from such sources attenuate rapidly with distance from the shore. Autochthonous production is confined to the euphotic upper layers and is greatest in areas of vigorous mixing where continuous nutrient renewal is occurring (Parsons, Takahashi & Hargrave 1977; Harris, Chapter 15). Such conditions are commonest in shallow inshore areas, thus carbon inputs from all sources are highest inshore and attenuate offshore as depth increases. Microbial biomass, which, in this context, may be taken as a function of the amount of labile carbon available, has been shown to decrease consistently with distance offshore over a range of environments (Sorokin 1978).

The availability of detrital material and the various pathways of its incorporation into pelagic and benthic food webs have been the subject of an exhaustive review by Conover (1978). In broad terms availability decreases with both increasing depth and latitude, and becomes more episodic at higher latitudes. The seasonality of inputs at higher latitudes is particularly important as a structuring influence on the benthos (Smetacek 1985). Episodic availability of food requires particular adaptive responses which impose specific constraints on the types of organisms able to exploit such resources. Thus the type of community which evolves as a response to irregular and/or episodic food availability is structured very differently from that which takes advantage of a continuously renewed food source. Smetacek (1985) also postulated that the higher the proportion of new to total production in a system ('new' production being dependent on nutrient input as opposed to nutrient regeneration within the water column) the greater the proportion of fixed energy diverted to the benthos. Indeed, a number of recent studies have suggested that in both inshore (Valderhaug & Gray 1984) and deep sea (Lampitt 1985) situations a large proportion of the spring bloom production may reach the benthos, and that this proportion may increase at higher latitude. Other minor but primary sources of food supply to the benthos should be noted, although their extent and generality are still the subject of considerable speculation, viz. the chemotrophic bacterial production from deep hydrothermic vents, which provides the base for the complex communities of the deep sea rifts (Jannasch 1984) and organics from sub-surface seeps (Spies & Davis 1979).

PRIMARY FACTOR 1—DEPTH

The organic carbon that can be exported from the total primary production in the oceans to the aphotic zone and/or stored in the sediments should be approximately equivalent to the 'new production', i.e. $3.4-4.7 \times 10^9$ tonnes C year^{-1} or 10% of total surface primary production (Eppley & Peterson 1979). Thus, globally, most of the algal biomass will be consumed in the photic zone, although locally there may be marked exceptions to this generalization, particularly in high latitudes (see above). In a summary of annual carbon deposition at various depths, Zeitschel (1980) concluded that 25–60% of the primary production sedimented in shallow areas and 1–10% in the deep ocean. The organic carbon content of such material in the deep sea was < 1% of the dry weight. In shallow water it was 5–10%. Riley (1951) estimated that about 10% of the primary production was utilized at depths greater than 200 m. It has also been demonstrated that the metabolism of organic matter in the sediment increases with increased surface primary production, but is inversely proportional to mixed water layer depth (depth of the thermocline during stratification) (Parsons, Takahashi & Hargrave 1977). Coupling between benthic metabolism and cycles of organic sedimentation has been shown in several shallow areas (Hargrave 1980; Graf *et al.* 1983). Seasonal vertical flux of organic matter from the surface has also been coupled with reproductive activity in the deep sea (Tyler, Gage & Pain 1983).

Benthic faunal abundance and biomass have been shown to decrease with increasing depth in many areas (e.g., Zenkevitch, 1963; Wigley & McIntyre 1964; Rowe & Menzel 1971; Marshall 1979; Rowe, Polloni & Haedrich 1982). Abundance and biomass seem in general to be comparatively high from close inshore to the edge of the continental shelf (Day, Field & Montgomery 1971). In some areas this could be due to regional nutrient inputs to this zone (Hanson *et al.* 1981). Between the shelf and the abyss is a transition zone at about 250 to 1000 m where macrofaunal structure changes markedly (Grassle, Sanders & Smith 1979; Rowe, Polloni & Haedrich 1982).

The inverse relation between depth and abundance-biomass seems to hold for both macrofauna and meiofauna (Wigley & McIntyre 1964; Pfannkuche, Theeg & Thiel 1983) and occurs at similar rates. Rowe & Menzel (1971) noted a tendency for reduced macrofaunal body size with increasing depth below the shelf break area and Shirayama (1983) showed this relationship for meiofauna in the upper 1 cm sediment layer.

Marshall (1979) has described the deep sea fauna as being composed of very diverse assemblages of tiny animals. Conspicuous animal groups are polychaetes, bivalves, peracarid crustaceans (especially tanaids), nematodes and foraminiferans. Characteristic features of the deep sea communities are thus low abundance, low biomass and high diversity (Hessler & Sanders 1967; Rowe, Polloni & Haedrich 1982). Rex (1983) has shown that the maximum diversity occurs at about 2000 to 3000 m and then decreases to the abyss (> 4000 m). The staple food for most organisms in the deep sea is most likely to be detritus, but dissolved organic carbon could be an important energy source for heterotrophic bacteria (Williams & Carlucci 1976). The metabolism of bacteria in the deep sea is probably 10–100 times less than in surface waters, mainly due to the high pressure (Marshall 1979). The low quantity and recalcitrant quality of available carbon and the low bacterial metabolism taken together suggest that food availability for the benthic infauna on the deep sea plain must be extremely limited.

Figure 17.1 summarizes diagrammatically the probable scale of the decline in benthic food availability as distance from the surface productive zone increases with increasing depth.

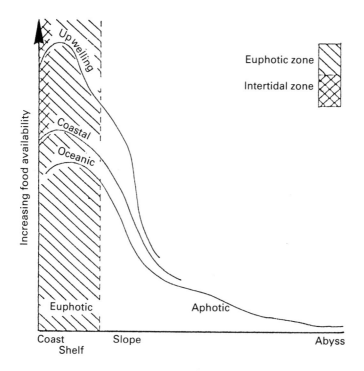

FIG. 17.1. Probable relative scale of the decline in benthic food availability as distance from the surface productive zone increases with increasing depth in three different marine systems (upwelling; coastal; oceanic).

PRIMARY FACTOR 2—LATITUDE

Total annual radiant energy incident at the earth's surface decreases with increasing latitude (see Reynolds, Chapter 14). In polar regions the light regime has a 365-day cycle, whereas the cycle in equatorial seas is 24 h (Petersen 1984). In Arctic and Antarctic waters light energy is sufficient for one summer phytoplankton bloom only (Heinrich 1962; Zenkevitch 1963). Because of upwelling, Antarctic waters are rich in nutrients which are not depleted during the bloom period. The annual primary production, although insufficiently known, seems to be similar but variable in Arctic and Antarctic coastal and shelf areas, i.e. 25–250 g C m^{-2} (Sakshaug & Holm-Hansen 1984). In the North Atlantic, phytoplankton biomass generally has two peaks, whereas in tropical waters the biomass does not show any seasonality (Heinrich 1962) (Fig. 17.2). Comparisons of primary production between regions are difficult to make because of the extent of both annual and local variations. It seems, however, that primary production and thus food availability for the benthos are episodic in polar regions and more

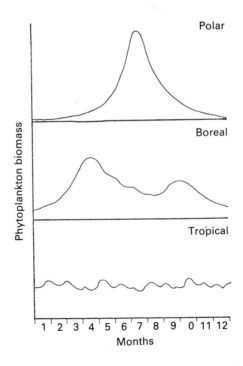

FIG. 17.2. Annual variation in total phytoplankton biomass at different latitudes (modified after Parsons *et al.* 1977, p. 30).

constant towards lower latitudes. In coral reefs primary production can be extremely high, but that energy is utilized mainly by the corals and other animals, including fish, associated with the reef and little is exported to surrounding sediments. The phytoplankton biomass in adjacent waters is low (Mann 1982) and food supply for benthic soft-bottom infauna is therefore poor.

The Antarctic continental shelf is narrow with a shelf/slope break at 400–800 m (Picken 1984). The central parts of the Arctic Ocean are deep, but surrounded by large areas of continental shelf. Benthic investigations of these areas are few. In the Barents Sea shelf area a macrobenthic biomass of 100–300 g m^{-2} has been recorded (Zenkevitch 1963), i.e. similar to other boreal and temperate areas at that depth. The production:biomass ratio was estimated at only 1:4–1:5 (Zenkevitch 1963), which is lower than at lower latitudes (e.g. Möller, Pihl & Rosenberg 1985). On the other hand Petersen (1984) has suggested that proportionately more energy flows through the benthic ecosystems of sub-arctic shelf seas than through the pelagos, in direct contrast to the relative flows in those compartments of tropical systems. Thus much of the spring bloom in subpolar areas sediments directly and is subsequently utilized by the benthos, whereas the continuous algal production in tropical areas is largely retained within the pelagic system.

Whilst several studies have demonstrated that species diversity increases with decreasing latitude (Thorson 1957; Sanders 1968; Wade 1972) others have failed to show this pattern (e.g. Longhurst 1957; Rosenberg 1975). Shallow (5–75 m) Antarctic waters can, for instance, accommodate high species diversity (Richardson & Hedgpeth 1977). Firm evidence is thus lacking to support the contention that there are basic latitudinal gradients in biomass, abundance or diversity of benthic organisms. More reliable and comparable benthic data from Arctic and tropical areas are required to match those available from boreal and temperate regions.

SECONDARY FACTOR — WATER MOVEMENT

In coastal areas geomorphological features control the direction and speed of water movements. Together these create the sedimentary conditions which influence both the structure and function of benthic communities (Möller, Pihl & Rosenberg 1985).

Turbulence is important in distributing particulate food (Fig. 17.3). Areas with high current speeds or strong wave action will keep food particles in suspension, thus favouring suspension feeding benthic animals. In coastal areas, material deposited on the bottom will periodically be resuspended by either wind driven or tidal forces. In sheltered areas particulate material

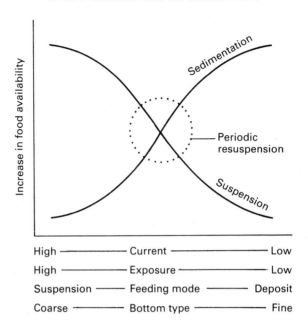

FIG. 17.3. Diagrammatic representation of variation in particulate material transport along a gradient of decreasing water movement related to consequent sedimentary and faunal gradients.

accumulates and, if currents are weak, stagnant conditions with anoxia can develop.

Estuaries function as traps for organic material. River-transported particles are precipitated in the mixing zone where fresh and saline waters meet. Coriolis forces may distribute such sedimenting organic material asymmetrically within estuaries (Rosenberg 1972).

Tidally-induced water transport affects vast areas of the oceans daily. In coastal areas it continuously redistributes sedimented material. Changes in tidal current velocity over short distances offshore in the southern North Sea have been shown to create a sharp transition zone between coarse and fine sediments. The sediments in that zone are enriched and harbour a benthic community with high biomass (Creutzberg et al. 1984). Pearson (1970) linked the changing distribution of trophic groups in the fauna of a sea loch system in Scotland to variations in tidally-induced water currents in the area.

In a number of studies in shallow, almost non-tidal areas on the Swedish west coast it has been shown that the interlinked factors of wave exposure, water movement and sediment structure affect the structure and function of both infaunal and mobile epibenthic faunal communities (Möller & Rosenberg 1983; Möller, Pihl & Rosenberg 1985; Pihl 1986). In exposed and semi-exposed sandy-silt areas, suspension feeders dominate the infauna. Their food, mainly phytoplankton, may be produced in water remote from their positions and transported to them by currents. This may elevate secondary production above the level expected from *in situ* primary production (Möller, Pihl & Rosenberg 1985) and contribute to the notably high productivity of coastal suspension feeders (Warwick 1979). Wildish (1977) has proposed a 'Trophic group mutual exclusion hypothesis' which postulates that current speed critically controls community composition, biomass and productivity through its effect on food supply and sedimentary composition; suspension feeders predominate in areas of strong water movement; deposit feeders in low flow areas.

In general then, it seems that water movement strongly influences the distribution of trophic modes in benthic organisms. Strong or moderate water movements in shallow waters favour suspension feeders, which may also be abundant on hard bottoms in eutrophic areas with low water exchange. Thus, high food availability favours suspension feeders. Conversely deposit feeders predominate in areas of low water exchange where sedimentation rates are high. There are, however, exceptions to this, e.g. in the Adriatic Sea, and in oligotrophic parts of the deep sea, where suspension feeders dominate in areas where both food availability and water movement are low (Sokolova 1972; Fedra *et al.* 1976). These points are further analysed below in the consideration of trophic structure in benthic communities.

TERTIARY MODIFYING FACTORS

The primary and secondary factors affecting food availability for benthic animals, briefly described above, are modified by several tertiary factors.

Dependent factors

These are physico-chemical variables that elicit an ecophysiological response from organisms. Evidence for the influences of such factors is extensive, e.g. temperature (Glemarec 1973; Möller & Rosenberg 1983; Rosenberg & Loo 1983); stratification of salinity (Jansson 1972; Rosenberg & Möller 1979) and oxygen (Rosenberg 1980; Rosenberg *et al.* 1983). Kinne (1978–83) provides an authoritative review.

Independent factors

Stochastic events

The intermittent effect of catastrophic environmental change caused by storm disturbance, vulcanicity, turbidity currents, etc. could influence community structure on a local scale by creating open space for recolonization and subsequent succession. The mosaic nature of many benthic community distributions has been attributed in part to such episodes caused by either physical or biotic factors. (Johnson 1970; Grassle & Sanders 1973; Probert 1984.) The former tend to act on a fairly wide scale and the latter only on a local scale. This is clearly seen in Roughgarden's discussion on barnacle distributions in intertidal systems (Chapter 22). The role of disturbance, particularly of sediment stability, in structuring benthic communities has been given detailed consideration by, among others, Thiery (1982) and Probert (1984).

Biotic interactions

Competition and/or predation are considered by some to be a strong influence on the evolution and maintenance of marine benthic communities. Theoretical and experimental treatments investigating the extent of such effects are numerous and often controversial. Branch (1984) provides an exhaustive account of competition effects, and predation effects have recently been reviewed by Hughes (1980a,b) for rocky shores and by Reise (1985) for tidal sediments. Commito & Ambrose (1985) have discussed infaunal predation in sedimentary systems. Contrary to earlier assumptions, recent evidence suggests that neither exploitation nor interference competition can account completely for intra-community differentation of population distributions, although many levels of both inter- and intraspecific interaction have been demonstrated. Similarly, although many instances of the control of population dynamics by predation have been described, the case for postulating predation as a major structuring force at the community level remains unproven for most marine habitats. It is of interest to note that some of the best documented examples of predatory control have been described from intertidal areas, e.g. Dayton (1975) and Menge (1976), where physical stresses may exert a disproportionate influence and food limitation is probably rare. Moreover the removal of dominants by predation leads directly to freeing of the major limiting resource in these two-dimensional habitats, i.e. space.

It is arguable that both competitive displacement (interference) and predation exert a greater impact on community dynamics in stressed than in unstressed environments. Thus areas subject to catastrophic disturbances on a relatively small spatial scale will present a mosaic of successional patches in various stages of development (*sensu* Johnson 1970; Thistle 1981). Such mosaics present almost continuous opportunities for resource (space and food) competition between colonizers, in addition to an excellent grazing ground for epibenthic predators. Conversely, more stable areas will be inhabited by a suite of fully established, older (generally larger), individuals less vulnerable to predation and accommodated to the presence of their neighours.

In general, biotic interactions must be considered as important modifiers of community structure acting independently within the structural framework of environmental gradients outlined above. With this suggestion we approach below the hierarchical concepts of community successional dynamics outlined for example by Wildish (1977) and Zajac & Whitlach (1985).

DEVELOPMENT OF TROPHIC GROUPS AND PREDICTED FUNCTIONAL ORGANIZATION OF COMMUNITIES

Identification of the major environmental gradients influencing benthic communities allows the analysis of those faunal adaptations that are a specific response to a particular gradient. The most general adaptations are those made to exploit a particular position on the gradient of food availability and the most fundamental is feeding mode. The feeding patterns of benthic organisms have been used frequently as a basis for distinguishing ecological zones (e.g. Hunt 1925; Turpajeva 1957; Sokolova 1972; Neyman 1979), but both the definition of trophic categories and their applicability within and across taxa are controversial (e.g. Dauer, Maybury & Ewing 1981), and are in need of more detailed examination. The most widely accepted classification recognizes four major trophic groups, namely suspension feeders, deposit feeders, carnivores and herbivores. Such simplicity conceals major trophic options and each of these categories can be subdivided, depending ultimately on individual choice, into as many sets as there are species in the sample. Some order was created out of this apparent chaos, in the case of polychaetes, by Fauchald & Jumars (1979) who defined twenty-two different guilds on the basis of relative motility, type of food taken and feeding habits, and suggested that the trophic composition of the polychaete fraction of the fauna of particular environments is partly predictable. Thus, among other hypotheses, they suggest that sessile guilds will diminish in nutrient-poor environments, since below a certain threshold of food availability motility is necessary to acquire sufficient food. In turbulent areas sessile guilds will also be at a disadvantage because of the clogging of feeding mechanisms and the risk of burial.

A less complex classification of macrofaunal trophic types related to a range of environmental characteristics has recently been proposed by Wildish (1985). This considers two categories of deposit feeders and four of suspension feeders. This scheme can be extended to include herbivores and carnivores (including epifaunal carnivores where information is available) and the hypotheses of Fauchald & Jumars regarding the divisions of polychaete feeding guilds. Thus five principal trophic types can be distinguished, i.e. herbivores, suspension feeders, surface deposit feeders, burrowing deposit feeders and carnivores. Each of these types can potentially belong to one of three categories of mobility, i.e. mobile, semi-mobile (capable of moving between feeding episodes, but sessile while feeding) and sessile, and one of four groups based on food gathering techniques, i.e. use of jaws, use of tentacles, use of entangling mechanisms and use of other techniques not covered by the first three categories. This last category is the least definitive and the most contentious, since a wide range of food-gathering techniques have evolved in benthic organisms which generally defy easy comparative definition. Some trophic categories can be placed with some precision along one or more environmental gradients. Thus the various categories of suspension feeding are closely related to current speed. Others are more generally spread along the gradients, e.g. the distributions of many wide spectrum carnivores are more closely related to that of their potential prey than to other variables. Nevertheless some broad generalizations linking distributions to particular gradients can be made. This can be demonstrated by an analysis of the marine benthic communities described by Jones (1950), who listed species characteristic of different types of sediment in boreal areas. If each listed species is assigned to an appropriate functional group and the percentage of species occurring in each group in each area is compared, then a distribution such as that shown in Fig. 17.4 is obtained. This demonstrates that along a gradient of increasing depth, as both current speed and turbulence decline and hard substrates grade into increasingly fine sediments, the predominant trophic mode changes.

Passive suspension feeders partition feeding space vertically above the sediment/water interface. Such organisms use hard substrate as an anchor for a variety of complex filtering mechanisms. As current speed declines with increasing depth and sedimentary substrata begin to predominate,

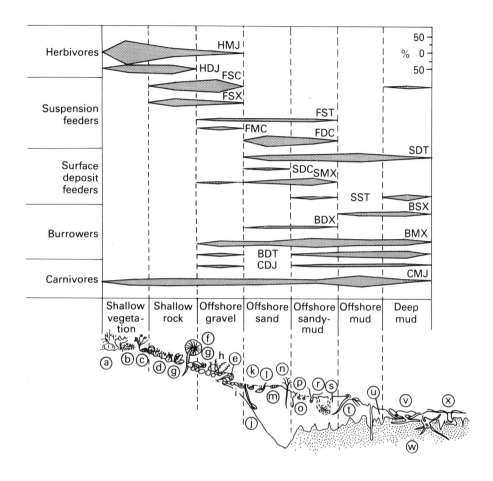

FIG. 17.4. Distributions of functional groups in boreal coastal communities (compiled from species listed by Jones 1950) along a gradient of decreasing food availability and water movement with increasing depth and sedimentation. Key to functional groups (Modified from Fauchald & Jumars 1979): Feeding type: H, herbivore; F, suspension feeder; S, surface deposit feeder; B, burrowing deposit feeder; C, carnivore. Degree of mobility: M, mobile, D, semi-mobile; S, sessile. Feeding habit: J, jawed; C, ciliary mechanisms; T, tentaculate; X, other types. Width of line representing each functional group along the gradient is proportional to its percentage contribution to the community at any particular point. A diagrammatic representation is included of the feeding position of taxa representative of the various groups relative to the sediment water interface along such gradients. Key to Taxa: (a) Macroalgae; (b) Echinoids, e.g. *Echinus* (HMJ); (c) Limpets, e.g. *Patella* (HDJ); (d) Barnacles, e.g. *Balanus* (FSX); (e) (f) Sepulids, Sabellids (FST); (g) Epifaunal bivalves, e.g. *Mytilus* (FSC); (h) Brittle stars, e.g. *Ophiothrix* (FMC); (i) *Venus* (FSX); (j) *Mya* (FSC); (k) *Cardium* (FDC); (l) *Tellina* (FDT); (m) *Turritella* (SMX); (n) *Lanice* (SST); (o) *Abra* (SDC);(p) *Spio* (SST); (r) *Amphiura* (FDT); (s) *Echinocardium* (BMX); (t) *Ampharete* (SST); (u) *Maldane* (BSX); (v) *Glycera* (CDJ); (w) *Thyasira* (BDX); (x) *Amphiura* (SDT).

active suspension feeders, which use various types of pumping or straining mechanisms to capture particles, become more numerous. These may combine suspension feeding from the water column with surface deposit feeding and may be mobile or semi-mobile to take advantage of the often variable current speeds in shallow coastal areas (e.g. ophiurid brittle stars, venerid and tellinid bivalves). These in turn are gradually replaced by obligate surface deposit feeders and sub-surface deposit feeders (burrowers) which are generally semi-mobile or sessile and partition food and space resources through a variety of tube building and burrowing habits. Such organisms are necessarily restricted to soft substrata and the predominance of subsurface deposit feeders increases with increasing depth and decreasing particle size. The degree of mobility in both burrowing and surface deposit feeding organisms decreases progressively with increasing depth. This reflects the development of larger and more complex food gathering mechanisms, prompted by an increasing scarcity of food, which often requires a larger maximum adult size. At some point, often at depths corresponding to the shelf/slope break, this trend is reversed when increasing scarcity of food re-emphasizes the advantages of mobility in allowing access to sparsely scattered resources (Fedra *et al.* 1976; Jumars & Fauchald 1977).

Herbivores are confined to the euphotic zone, although a few may be distributed a little deeper, consuming undegraded phytoplankton and macroalgal deposits. Infaunal carnivores tend to be fairly evenly spread along the gradients, generally comprising (numerically) between 10 and 20% of the total community (cf. Pearson 1970). Their greatest diversities coincide with the largest densities of their principal infaunal prey, i.e. the surface and sub-surface deposit feeders. In contrast, epifaunal predators with great mobility, e.g. crustaceans, are most abundant at the upper (shallower) end of the gradient, whereas slow-moving or territorial epifaunal predators have their greatest concentrations lower down the gradient (Pearson & Feder 1987). It thus follows that at any one point along the gradient of food availability the species present will exhibit a mix of trophic options whose constitution is to some extent predictable. The exact composition at a particular time and place will be dependent on the influence of the tertiary modifying factors outlined above.

This necessarily oversimplified picture must be further qualified by the recognition that the assignment of a species to a particular trophic category is often a fairly arbitrary decision. Many species are capable of switching feeding modes to suit changing circumstances. Most notably those feeding in the benthic boundary layer are often able to exploit both suspended and deposited particles (Pearson 1970). Nevertheless the broad categories outlined above do impart some recognizable structure to communities. Such communities will exhibit a functional similarity in areas subjected to common environmental conditions. Thus taking two extremes: in shallow current-swept tropical areas where a well-mixed water column ensures a high and continuous primary productivity, the communities will be composed predominantly of turbulent and laminar flow filter feeders, i.e. coral reef areas and high energy beaches with dense populations of lamellibranchs; conversely, in deep basin areas beneath a strongly stratified and extensive water column, the communities will be dominated by sub-surface deposit/absorption feeders, cf. abyssal populations. Between such extremes lie the various combinations which should constitute the observed groupings analogous to the parallel bottom communities suggested by Thorson (1957).

DISCUSSION

A general argument has been advanced that food availability is the most fundamental variable underlying the structure of marine benthic communities. The factors which influence food availability have been outlined briefly and some evidence for their relative importance has been discussed. Biomass, production and diversity have been shown to change in a relatively predictable manner along a gradient of declining food availability. Such changes may be summarized through a few broad generalizations which in turn lead to some general hypotheses as to how benthic communities are structured. In areas of high food availability communities are dominated by species which maximize turnover, i.e. rapid growth and high reproductive rate (r strategists). In areas of low but predictable food availability species diversity reaches a maximum since a complex range of feeding strategies has evolved necessitating considerable investment in somatic growth. Consequently communities in such areas are characterized by low growth rates and hence low turnover. Unpredictable and low food availability will also favour slow-growing organisms but also a maximization of individual biomas to allow persistence through long starvation periods. Very low and spasmodic food availability will favour mobility rather than biomass and lead to communities dominated by epifaunal organisms. Environmental perturbation will alter the relative availability of food to the various species in any given community in such a way as to favour one or other of the above tendencies resulting in a change in species composition.

Table 17.1 places the interacting environmental factors in a simplified hierarchy which not only summarizes the general theme but allows some

TABLE 17.1. Conceptual scheme of factors influencing community distributions with a projected evolutionary time scale for their impact and their probable structural effect

	Primary and secondary factors	General dependent modifying factors	Independent modifying factors
Factors influencing food availability	Depth Latitude Current speed	Environment harshness including Sedimentary instability and turbidity Salinity Oxygen Temperature } fluctuations Pressure	Stochastic events i.e. vulcanicity pollution Biotic interactions Competition Predation
Evolutionary time scale	Long term (years $\times 10^6$)	Medium term (years $\times 10^4$)	Short term (years)
Structural scale of impact on community	Megastructure, i.e. distribution of trophic and mobility groupings.	Macrostructure, i.e. successional phenomena and life history characteristics etc.	Microstructure, i.e. inter- and intraspecific interactions creating patchiness.

notional scaling of their evolutionary impact. Stochastic events and biotic interactions will have the most immediate and short-term effects on community structure (acting within a year or a period of years). Environmental harshness, which includes all aspects of seasonally-induced variability in resources will exert a more lasting influence on species composition (perhaps on a millennial scale). However, underlying both such groups of modifying factors are the gradients in food availability induced by variations in latitude, depth and current speed which, over geological times, will have been responsible for the evolution of the major communities.

Table 17.1 also proposes a possible scale of impact of the factors at each level on the structure of the community. We conceive that the mega-structure, i.e. the distribution of major functional groups within the community, is determined by the effect on food availability of changing depth, latitude and current speed. The macro-structure, principally the divisions within functional groups which define successional phenomena and selection of life history characteristics, is determined primarily by the general modifying factors which define the degree of environmental harshness. Independent modifying factors, principally biotic interactions, define the microstructure of the community, i.e. the small scale patchiness and degree of local variability. This conjectural scheme suggests therefore that the general features of marine benthic community structure are the product of habitat variability as it influences food availability, acting on the evolution of functional attributes.

Such a scheme is a logical extension of earlier attempts to perceive community development as proceeding along a unidimensional gradient of environmental harshness (e.g. Sanders 1968; Slobodkin & Sanders 1969). Such ideas have evolved into a range of theories emphasizing the multivariate nature of environmental and biotic influences. These have tended to follow either the biotic control explanations, for example of Connell & Slatyer (1977) and Hughes (1980a), or emphasized the role of environmental constraints, see Thistle (1981), Thiery (1982), Probert (1984). We suggest here that the influence of biotic interactions is more immediate than that of environmental effects and more likely to satisfactorily explain within habitat patchiness than regional and biogeographical differences in species associations. It is not surprising that, in our opinion, the most seminal recent work on the evolution of marine communities is found in the geological literature, e.g. the compilation by Tevesz & McCall 1983). Palaeoecology has developed with a full awareness of the extensive geological time scales underlying both the evolution of communities and the climatic changes which can be involved most plausibly as causative factors. Such a conceptual framework rarely influences ecologists concerned with contemporary communities, which no doubt explains, to some extent, their concentration on short-term biotic interactions as causative factors in community development. However, the comparative study of contemporary communities is a powerful tool in the interpretation of fossil community structure and vice versa (see Chapters 18 and 19). This has been elegantly illustrated by Larson & Rhoads (1983) who explain the development of Paleozoic infaunal communities by reference to the short-term succession of a modern community following disturbance (Fig. 17.5). These authors see the increasing trophic diversity and spatial complexity of infaunal communities from the Ordovician through to the Cretaceous as a response to the presence of previously unexploited food resources at depth in the sediments. Except in exceptional circumstances modern benthic communities appear to exploit all available resources. Documented exceptions are all at the extreme ends of the gradients described above, i.e. in shallow, highly disturbed or highly seasonal areas where unpredictably

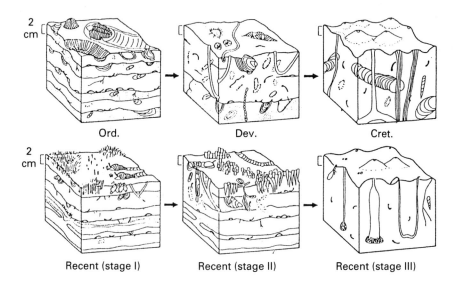

FIG. 17.5. Comparison of the evolution of community structure in ancient shelf environments with successional changes in a modern shelf community following disturbance. The depth of burrowing, functional composition and sedimentary fabric all change progressively as the infaunal habit develops from the Ordovician through the Devonian to the Cretaceous. The lower diagrams show successive stages in the development of a modern community on muddy substrates, progressing from initial colonization by surface deposit feeders to an equilibrium stage (111) dominated by deep deposit feeders. In terms of animal–sediment interactions recent community succession appears to recapitulate community evolution (after Larson & Rhoads 1983).

harsh conditions or temporary enrichment events may provide pools of food or space which are subsequently opened for successional exploitation whenever the relaxation of conditions allows recolonization. The course of such successions is usually predictable and results in the development of characteristic communities constrained by food limitation (Pearson & Rosenberg 1978; Rhoads & Boyer 1982).

REFERENCES

Bonsdorff, E. (1980). Macrozoobenthic recolonization of a dredged brackish water bay in SW Finland. *Ophelia, suppl.1,* 145–55.

Branch, G.M. (1984). Competition between marine organisms: ecological and evolutionary implications. *Oceanography and Marine Biology Annual Review,* **22,** 429–593.

Cody, M.L. & Diamond, J.M. (Eds) **(1975).** *Ecology and Evolution of Communities.* Harvard University Press, Cambridge, Massachusetts.

Commito, J.A. & Ambrose, W.G. (1985). Predatory infauna and trophic complexity in soft-bottom communities. *Proceedings of the Nineteenth European Marine Biology Symposium, Plymouth, Devon, UK 16–21 September 1984.* (Ed. P.G. Gibbs), pp. 323–34. Cambridge University Press, Cambridge.

Connell, J.H. & Slatyer, R.A. (1977). Mechanisms of succession in natural communities and their role in community stability and organization. *American Naturalist,* **111,** 1119–44.

Conover, R.J. (1978). Transformation of organic matter. *Marine Ecology IV, Dynamics* (Ed. by O. Kinne). pp. 221–499. Wiley, Chichester.

Creutzberg, F., Wapenaar, P., Duineveld, G. & Lopez, N.L. (1984). Distribution and density of the benthic fauna in the southern North Sea in relation to bottom characteristics and hydrographic conditions. *Rapports et Procès-verbaux des Réunions,* **183,** 101–10.

Dauer, D.M., Maybury, C.A. & Ewing, R.A. (1981). Feeding behaviour and general ecology of several spionid polychaetes from the Chesapeake Bay. *Journal of Experimental Marine Biology and Ecology,* **54,** 21–38.

Day, J.H., Field, J.G. & Montgomery, M.P. (1971). The use of numerical methods to determine the distribution of the benthic fauna across the continental shelf of North Carolina. *Journal of Animal Ecology* **40,** 93–125.

Dayton, P.K. (1975). Experimental evaluation of ecological dominance in a rocky intertidal algal community. *Ecological Monographs,* **45,** 137–59.

Diamond, J.M. & Case, T.J. (Eds) **(1986).** *Community Ecology.* Harper & Row, New York.

Eppley, R.W. & Peterson, B.J. (1979). Particulate organic matter flux and planktonic new production in the deep ocean. *Nature (London)* **282,** 677–80.

Fauchald, K. & Jumars, P.A. (1979). The diet of worms: a study of polychaete feeding guilds. *Oceanography and Marine Biology Annual Review,* **17,** 193–284.

Fedra, K., Olscher, E.M., Scherubel, C., Stachowitsch, I. & Wurzian, R.S. (1976). On the ecology of a North Adriatic benthic community: distribution standing crop and composition of the macrobenthos. *Marine Biology,* **38,** 129–45.

Glemarec, M. (1973). The benthic communities of the European north Atlantic continental shelf. *Oceanography and Marine Biology Annual Review,* **11,** 263–89.

Graf, G., Schulz, R., Peinert, R. & Meyer-Reil, L.-A. (1983). Benthic response to sedimentation events during autumn to spring at a shallow-water station in the Western Kiel Bight. I. Analysis of processes on a community level. *Marine Biology,* **77,** 235–46.

Grassle, J.F. & Sanders, H.L. (1973). Life histories and the role of disturbance. *Deep-Sea Research,* **20,** 643–59.

Grassle, J.F., Sanders, H.L. & Smith, W.K. (1979). Faunal changes with depth in the deep sea benthos. *Ambio Special Report*, **6**, 47–50.

Hanson, R.B., Tenore, K.R., Bishop, S., Chamberlain, C., Pamatmat, M.M. & Tietjen, J. (1981). Benthic enrichment in the Georgia Bight related to Gulf Stream intrusions and estuarine outwelling. *Journal of Marine Research*, **39**, 417–41.

Hargrave, B.T. (1980). Factors affecting the flux of organic matter to sediments in a marine bay. *Marine Benthic Dynamics* (Ed. by K.R. Tenore & B.C. Coull), pp. 243–63. University of South Carolina Press, Columbia, South Carolina.

Heinrich, A.K. (1962). The life histories of plankton animals and seasonal cycles of plankton communities in the oceans. *Journal du Conseil pour Exploration de la Mer*, **27**, 15–24.

Hessler, R.R. & Sanders, H.L. (1967). Faunal diversity in the deep-sea, *Deep-Sea Research*, **14**, 65–78.

Hily, C. (1984). *Variabilité de la macrofauna benthique dans les milieux hypertrophiques de la rade de Brest.* Université Bretagne Occidentale. Brest. Thèse Docteur ès-Sciences.

Hughes, R.N. (1980a). Predation and community structure. *The Shore Environment, Vol. 2: Ecosystems.* Systematics Association Special Vol. No. 17(b), (Ed. by J.H. Price, D.E.G. Irvine & W.F. Farnham), pp. 699–728. Academic Press, New York.

Hughes, R.N. (1980b). Optimal foraging theory in the marine environment. *Oceanography and Marine Biology Annual Review*, **18**, 423–81.

Hunt, O.D. (1925). The food of the bottom fauna of the Plymouth fishing grounds. *Journal of the Marine Biological Association of the United Kingdom*, **13**, 560–99.

Jannasch, H.W. (1984). Chemosynthesis: the nutritional basis of life at deep-sea vents. *Oceanus*, **27**, 73–8.

Jansson, B.O. (1972). Ecosystem approach to the Baltic problem. *Bulletins from the Ecological Research Committee NFR*, **16**, 1–82.

Johnson, R.G. (1970). Variations in diversity within marine benthic communities. *American Naturalist*, **104**, 285–300.

Jones, N.S. (1950). Marine bottom commmunities. *Biological Review*, **25**, 283–313.

Jumars, P.A. & Fauchald, K. (1977). Between community contrasts in successful polychaete feeding strategies. *Ecology of Marine Benthos* (Ed. by B.C. Coull), pp. 1–20. University of South Carolina Press, Columbia, South Carolina.

Kinne, O. (Ed.) (1978-83). *Marine Ecology*, Vols. I–V. Wiley, Chichester.

Lampitt, R.S. (1985). Evidence for the seasonal deposition of detritus to the deep-sea floor and its subsequent resuspension. *Deep-Sea Research*, **32**, 885–97.

Larson, D.W. & Rhoads, D.C. (1983). The evolution of infaunal communities and sedimentary fabrics. *Biotic Interactions in Recent and Fossil Benthic Communities* (Ed. by M.J.S. Tevesz & P.L. McCall), pp. 627–47. Plenum, New York.

Longhurst, A.R. (1957). Density of marine benthic communities off West Africa. *Nature (London)*, **179**, 542–3.

Mann, K.H. (1982). *Ecology of Coastal Waters. A Systems Approach.* Studies in Ecology, Vol. 8. Blackwell Scientific Publications, Oxford.

Marshall, N.B. (1979). *Developments in Deep-sea biology.* Blandford Press, Poole, Dorset.

Menge, B.A. (1976). Organization of the New England rocky intertidal community: role of predation, competition and environmental heterogeneity. *Ecological Monographs*, **46**, 355–93.

Möller, P., Pihl, L. & Rosenberg, R. (1985). Benthic faunal energy flow and biological interaction in some shallow marine soft bottom habitats. *Marine Ecology Progress Series*, **27**, 109–21.

Möller, P. & Rosenberg, R. (1983). Recruitment, abundance and production of *Mya arenaria* and *Cardium edule* in marine shallow waters, western Sweden. *Ophelia*, **22**, 33–55.

Neyman, A.A. (1979). Soviet investigations of the benthos of the shelves of the marginal seas. *Marine Production Mechanisms* (Ed. by M.V. Dunbar), pp. 269–84. Cambridge University Press, Cambridge.

Parsons, T.R., Takahashi, M. & Hargrave, B. (1977). *Biological Oceanographic Processes* (2nd edn). Pergamon Press, Oxford.

Pearson, T.H. (1970). The benthic ecology of Loch Linnhe and Loch Eil, a sea-loch system on the west coast of Scotland. I. The physical environment and the distribution of the macrobenthic fauna. *Journal of Experimental Marine Biology and Ecology*, **5**, 1–34.

Pearson, T.H. & Feder, H. (1987). The benthic ecology of Loch Linnhe and Loch Eil, a sea-loch system on the west coast of Scotland. V. Biology of the dominant soft bottom epifauna and its interaction with the infauna. *Journal of Experimental Marine Biology and Ecology*, (in press).

Pearson, T.H. & Rosenberg, R. (1978). Macrobenthic succession in relation to organic enrichment and pollution of the marine environment. *Oceanography and Marine Biology Annual Review*, **16**, 229–311.

Petersen, G.H. (1984). Energy-flow budgets in aquatic ecosystems and the conflict between biology and geophysics about earth-axis tilt. *Climatic Changes on a Yearly to Millennial Basis.* (Ed. by N.A. Morner & W. Karlen), pp. 621–33. D. Reidel, Massachusetts.

Pfannkuche, O., Theeg, R. & Thiel, G. (1983). Benthos activity, abundance and biomas under an area of low upwelling off Morocco, Northwest Africa. *"Meteor" Forschung - Ergebnisse, Reihe D*, **36**, 85–96.

Picken, G.B. (1984). Benthic research in Antarctica: past, present and future. *Marine Biology of Polar Regions and Effects of Stress on Marine Organisms* (Ed. by J.S. Gray & M.E. Christiansen), pp. 167–84. Wiley, Chichester.

Pihl, L. (1986). Exposure, vegetation and sediment as primary factors for mobile epibenthic faunal community structure and production in shallow marine soft bottom areas. *Netherland Journal of Sea Research*, **20**, 75–83.

Probert, P.K. (1984). Disturbance, sediment stability and trophic structure of soft-bottom communities. *Journal of Marine Research*, **42**, 893–921.

Reise, K. (1985). Predator control in marine tidal sediments. *Proceedings of the Nineteenth European Marine Biology Symposium, Plymouth, Devon, UK 16–21 September 1984* (Ed. by P.C. Gibbs), pp. 311–22. Cambridge University Press, Cambridge.

Rex, M.A. (1983). Geographic patterns of species diversity in the deep sea benthos. *Deep-Sea Biology* (Ed. by G.T. Rowe), pp. 453–72. Wiley, Chichester.

Rhoads, D.C. & Boyer, L.E. (1982). The effects of marine benthos on physical properties of sediments: a successional perspective. *Animal-Sediment Interactions* (Ed. by M.J.S. Tevesz & P.L. McCall), pp. 3–52. Plenum, New York.

Richardson, M.D. & Hedgpeth, J.W. (1977). Antarctic soft-bottom, macrobenthic community adaptations to a cold, stable, highly productive, glacially affected environment. *Adaptations Within Antarctic Ecosystems* (Ed. by G.A. Llano), pp. 181–96. Gulf, Houston, Texas.

Riley, G.A. (1951). Oxygen, phosphate and nitrate in the Atlantic Ocean. *Bulletin Bingham Oceanographic Collections*, **13**, 1–126.

Rosenberg, R. (1972). Benthic faunal recovery in a Swedish fjord following the closure of a sulphite pulp mill. *Oikos*, **23**, 92–108.

Rosenberg, R. (1975). Stressed tropical benthic faunal communities off Miami, Florida. *Ophelia*, **14**, 93–112.

Rosenberg, R. (1980). Effect of oxygen deficiency on benthic macrofauna in fjords. *Fjord Oceanography* (Ed. by H.J. Freeland, D.M. Farmer & C.D. Levings), pp. 499–517. Plenum, New York.

Rosenberg, R. & Loo, L.O. (1983). Energy-flow in a *Mytilus edulis* culture in western Sweden. *Aquaculture,* **35,** 151–61.

Rosenberg, R. & Möller, P. (1979). Salinity stratified benthic macrofaunal communities and long-term monitoring along the Swedish west coast. *Journal of Experimental Marine Biology and Ecology,* **37,** 175–208.

Rosenberg, R., Arntz, W.E., de Flores, E.C., Flores, L.A., Carbajal, G., Finger, I. & Tarazona, J. (1983). Benthos biomass and oxygen deficiency in the upwelling system off Peru. *Journal of Marine Research,* **41,** 263–79.

Rowe, G.T. & Menzel, D.W. (1971). Quantitative benthic samples from the deep Gulf of Mexico with some comments on the measurement of deep-sea biomass. *Bulletin of Marine Sciences,* **21,** 556–66.

Rowe, G.T., Polloni, P.T. & Haedrich, R.L. (1982). The deep-sea macrobenthos on the continental margin of the northwest Atlantic Ocean. *Deep-Sea Research,* **29,** 257–78.

Sakshaug, E. & Holm-Hansen, O. (1984). Factors governing pelagic production in polar oceans. *Marine Phytoplankton and Productivity.* Lecture Notes on Coastal and Estuarine Studies, 8. (Ed. by O. Holm-Hansen, L. Bolis & R. Gilles), pp. 1–18. Springer-Verlag, New York.

Sanders, H.L. (1968). Marine benthic diversity: a comparative study. *American Naturalist,* **102,** 243–82.

Shirayama, Y. (1983). Size structure of deep-sea meio- and macrobenthos in the western Pacific. *International Revue der gesamten Hydrobiologie* **68,** 799–810.

Slobodkin, L.B. & Sanders, H.L. (1969). On the contribution of environmental predictability to species diversity. *Brookhaven Symposia in Biology,* **22,** 82–93.

Smetacek, m. (1985). The supply of food to the benthos. *The Flow of Energy and Materials in Marine Ecosystems.* (Ed. by M. Farham), pp. 517–48. Plenum, New York.

Sokolova, M.N. (1972). Trophic structure of deep-sea macrobenthos. *Marine Biology,* **16,** 1–12.

Sorokin, Y. I. (1978). Decomposition of anoxic matter and nutrient regeneration. *Marine Ecology, IV, Dynamics.* (Ed. by O. Kinne), pp. 501–616. Wiley, Chichester.

Spies, R.B. & Davis, P.H. (1979). The infaunal benthos of a natural oil seep in the Santa Barbara Channel. *Marine Biology* **50,** 227–37.

Swartz, R.C., Schults, D.W., Ditsworth, G.R., DeBeh, W.A. & Cole, F.A. (1985). Sediment toxicity, contamination, and macrobenthic communities near a large sewage outfall. *Validation and Predictability of Laboratory Methods for Assessing the Fate and Effects of Contaminants in Aquatic Ecosystems, ASTM STP 865* (Ed. by T.P. Boyle), pp. 152–75. American Society for Testing and Materials, Philadelphia.

Tevesz, M.J.S. & McCall, P. (Eds) (1983). *Biotic Interactions in Recent and Fossil Benthic Communities,* Plenum, New York.

Thiery, R.G. (1982). Environmental instability and community diversity. *Biological Reviews,* **57,** 671–710.

Thistle, D. (1981). Natural physical disturbance and communities of marine soft bottoms. *Marine Ecology, Progress Series,* **6,** 223–8.

Thorson, G. (1957). Bottom communities (sublittoral and shallow shelf). *Geological Society of America, Memoir* **67,** 461–534.

Turpajeva, E.P. (1957). Food interrelationships of dominant species in marine benthic biocoenoses. *Transactions of the Institute of Oceanology.* XX. *Marine Biology* (Ed. by B.N. Nitchin). Translated by the American Institute of Biological Sciences, Washington D.C., 1959.

Tyler, P.A., Gage, J.D. & Pain, S.L. (1983). Reproductive variability in deep sea echinoderms and molluscs from the Rochall Trough. *Oceanologica Acta, Volume Spécial,* 191–5.

Valderhaug, V.A. & Gray, J.S. (1984). Stable macrofauna community structure despite fluctuating food supply in subtidal soft sediments of Oslofjord, Norway. *Marine Biology,* **82,** 307–22.

Valentine, J.W. (1971). Resource supply and species diversity patterns. *Lethaia,* **4,** 51–61.

Wade, B.A. (1972). A description of a highly diverse soft-bottom community in Kingston Harbour, Jamaica. *Marine Biology,* **13,** 57–69.

Warwick, R.M. (1979). Population dynamics and secondary production of benthos. *Marine benthic dynamics* (Ed. K.R. Tenore & B.C. Coull), pp. 1–24. University of South Carolina Press, Columbia.

Wigley, R.L. & McIntyre, A.D. (1964). Some quantitative comparisons of offshore meio-benthos and macrobenthos south of Martha's Vineyard. *Limnology and Oceanography,* **9,** 485–93.

Wildish, D.J. (1977). Factors controlling marine and estuarine sublittoral macrofauna. *Helgolander wissenschaftliche Meeresuntersuchungen,* **30,** 445–54.

Wildish, D.J (1985). Geographical distribution of macrofauna on sublittoral sediments of continental shelves: a modified trophic ratio concept. *Proceedings of the Nineteenth European Marine Biology Symposium, Plymouth, Devon, UK, 16–21 September 1984* (Ed. by P.E. Gibbs), pp. 335–46. Cambridge University Press, Cambridge.

Williams, P.M. & Carlucci, A.F. (1976). Bacterial utilisation of organic matter in the deep sea. *Nature (London)* **262,** 810–11.

Zajac, R.N. & Whitlach, R.B. (1985). A hierarchical approach to modelling soft-bottom successional dynamics. *Proceedings of the Nineteenth European Marine Biology Symposium, Plymouth, Devon, UK 16–21 September 1984.* (Ed. by P.E. Gibbs), pp. 265–76. Cambridge University Press, Cambridge.

Zeitschel, B. (1980). Sediment-water interactions in nutrient dynamics. *Marine Benthic Dynamics* (Ed. by K.R. Tenore & B.C. Coull), pp. 195–218. University of South Carolina Press, Columbia, South Carolina.

Zenkevitch, L. (1963). *Biology of the Seas of the U.S.S.R.* Allan & Unwin, London.

III

LESSONS FROM THE PAST

18. FACTORS CONTROLLING THE ORGANIZATION AND EVOLUTION OF ANCIENT PLANT COMMUNITIES

M. E. COLLINSON[1] AND A. C. SCOTT[2]

[1]*Department of Biology, King's College, Kensington Campus, University of London, Campden Hill Road, London W8 7AH, UK and* [2]*Department of Geology, Royal Holloway and Bedford New College, University of London, Egham Hill, Egham, Surrey TW20 0EX, UK*

INTRODUCTION

In contrast to their modern counterparts ancient plant communities must be inferred from plant parts entombed in sediment. Reconstructions of ancient plant communities have frequently been little more than paintings, using nearest living relatives or, preferably, reconstructed fossil plants in an artistically pleasing layout. Current palaeobotanical literature, using increasingly rigorous method, contains relatively few windows through which more reliable reconstructions of ancient plant communities can be seen. We restrict our information to those sources where the problems of community reconstruction are minimized and have been carefully evaluated. Where evidence is equivocal we indicate this.

In this chapter we review, at a very general level, some of the major events in the evolution of plant communities on land prior to the Quaternary (Fig. 18.1). The fossil record provides the only evidence of the pattern of community change on the larger temporal and spatial time scales; see definitions of Birks (1986) and Delcourt, Delcourt & Webb (1983). Much of the overall pattern we describe falls in the mega-scale category ($> 10^6$ years, $> 10^{12}$ m^2). This is interspersed with rarer and scattered insights into the smaller macro- and micro-scales. Time resolution in older strata is insufficient to quantify precisely the time involved, and spatial resolution is also complicated by transport and sorting phenomena prior to sedimentation. Therefore we are usually dealing with vegetation types, formations or regions and global terrestrial vegetation rather than with single stands.

RECONSTRUCTING ANCIENT PLANT COMMUNITIES

Recently, quantitative, bed by bed analysis of plant fossil distribution has

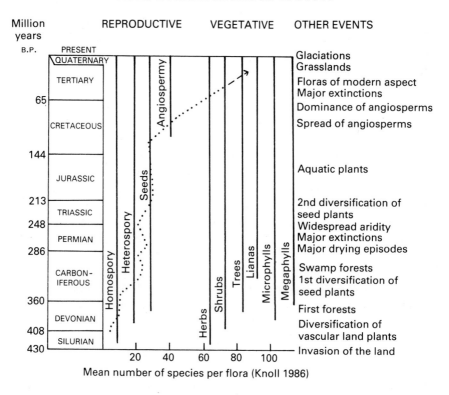

FIG. 18.1. Timing of the major innovations of reproductive biology and vegetative organization that are discussed in the text. Other major events that influenced ancient plant community organization and vegetation pattern are also included. Geological time scale follows Harland *et al.* (1982). The dotted line indicates community diversity, assessed on the basis of mean numbers of species in 391 fossil plant assemblages from former 'tropical to subtropical' lowlands (redrawn from Knoll 1986, Fig. 7.1). Continuation of this curve (dashed line) must be estimated in the absence of adequate fossil floras from comparable conditions (see Knoll 1986).

superseded previous qualitative assessments. Quadrat analyses of exposed layers of sediment enable both percentage plant cover and numbers of individual plant organs or fragments to be measured. The latter information may also be obtained by dissaggregation of sediments, which permits a volumetric analysis of plant fossil content (Spicer & Hill 1979; Scott & Collinson 1983a,b; Lapasha & Miller 1984). Quantitative data may also be obtained from permineralized peats (Phillips, Peppers & DiMichele 1985).

Advanced technology including transmitted, reflected, polarized and fluorescent light microscopy, X-ray and infra-red light (Schaarschmidt 1985) along with scanning and transmission electron microscopy (Mapes & Rothwell 1984; Friis 1985a,b; Walker & Walker 1985) is applied to fossils. The aspects of functional biology thus revealed help to elucidate the role of the plants in the community.

Laboratory experimentation and field investigation of the processes influencing plant representation in modern sediments (taphonomy) have enhanced our understanding of the relationship between fossil assemblages and ancient communities (Spicer 1980, 1981; Scheihing & Pfefferkorn 1984; Ferguson 1985, Rex 1986). Most plant fossil assemblages are composed of plant parts, organs or fragments which have been removed from their context in the original communities. Parent plant biology (e.g. deciduous habit, form and production of propagules) is clearly as important as physical and chemical processes (e.g. decomposition, abrasion, hydrodynamic sorting) in controlling likely representation in the fossil record (Scheihing & Pfefferkorn 1984). Herbaceous and low-growing forms within canopy vegetation, and even tall trees more than 50 metres away from a potential environment of deposition, may not be represented in sediments there as potential macrofossils (Ferguson 1985). An original plant community may well be represented as segregated components in several different types of sedimentary environments. It is therefore vital to include all sediment types and sources of plant material when examining a package of rocks with a view to reconstructing the ancient plant communities around their sites of deposition. A further complication is the likelihood that, within one deposit, plant material may have accumulated over a period ranging from a few hours (e.g. during a flood) to several months in 'normal' circumstances (Spicer 1980; Scheihing & Pfefferkorn 1984). Frequently, resolution of sediment sampling is not sufficient to distinguish different moments in depositional history. According to the time involved in deposition and the transport which has occurred, any one plant-bearing horizon may include elements from one or more, originally distinct, plant communities.

A thorough and detailed sedimentological analysis is a prerequisite for interpretation of the depositional history of plant fossil assemblages. Physical evidence, such as that from oxygen isotope studies (e.g. Collinson 1983a) and fossil soil profiles (e.g. Retallack 1983, 1985) may help to substantiate reconstructions. This data, combined with evidence from other associated biota, aspects of plant biology and taphonomic studies, all contribute to well-founded hypotheses of the original plant communities from which the plant fossil assemblages have been derived. Such reconstructions of ancient plant communities form the basis for any investigation of community evolution.

VEGETATIVE ORGANIZATION;
EFFECTS ON NICHE AND RESOURCE PARTITIONING

Papers in Chaloner & Lawson (1985) deal in detail with plant invasion of the land from the late Silurian (or possibly earlier) into the early Devonian. This topic is beyond the scope of the present paper and includes few instances where plant communities have been reconstructed. Acquisition of characters of ecophysiological significance is considered by Raven (1985).

Early Devonian plant communities were dominated by small, herbaceous plants, leafless or with microphyllous leaves and with limited branching. They tended to occupy damp habitats, often in monotypic stands (Edwards 1980). They provided humid and shaded microhabitats for the establishment of microarthropod communities (Rolfe 1985).

Subsequently, through the Devonian and early Carboniferous, niche diversity increased greatly due to two major innovations: megaphyllous leaves (coupled with increased branching) and the arborescent strategy (Chaloner & Sheerin 1979, Gensel 1984; Knoll *et al.* 1984). Shading of the ground by plants with megaphyllous leaves would have resulted in niche partitioning leading to specialization, perhaps with some plants adapting to less direct light. Competition for light may well have contributed to the selective pressure resulting in evolution of the tree habit. Arborescence first becomes common in the late Devonian in at least two groups of plants; the progymnosperms and the lycopsids (Chaloner & Sheerin 1979). This presence of trees and hence, presumably, forests (Beck 1964) by the late Devonian had several consequences. It produced a tiered vegetation with forest floor and canopy ecosystems (Scott 1980, 1984). The warm, shaded and humid forest floor may have provided an appropriate environment for the evolution of terrestrial amphibians (Thomson 1980; Scott 1984; Webb Chapter 20). Herbaceous ferns and fern-like plants have provided ground cover ever since vascular plants colonized the land (Thomas 1985). Such communities, beneath canopy vegetation, may be recognized through the fossil record (Lapasha & Miller 1984; Pelzer 1984) and persist to the present day (Thomas 1985). Bryophytes probably also contributed to these communities (Krassilov 1973) but their fossil record is generally poor.

Arborescence also provided niches for lianas, first recorded during the Carboniferous. These are mainly pteridosperms, species of the genera *Medullosa, Callistophyton, Neuropteris* and *Mariopteris* (see Scott & Rex 1985) and also some ferns (Scott & Galtier 1985). Increased branching, megaphyllous leaves and arborescence provided an increased area and elevation for production and liberation of propagules (see pp. 404–6). Such elevation of food resources may have contributed to selective pressures resulting in the evolution of flight in insects and of the arboreal habit in vertebrates.

The presence of the diverse forest communities in the Carboniferous (Wnuk & Pfefferkorn 1984; DiMichele & Phillips 1985; Phillips, Peppers & DiMichele 1985) might imply that niche availability and diversity by then was very similar to that of today. However, consideration of the distinctive vegetative organization and reproductive biology of the dominant members of the ancient communities argues against such a naive conclusion. Arborescent lycopsids for example, had determinate growth, produced only a simple branching canopy and many remained unbranched for much of their life (DiMichele & Phillips 1985). Thus, these forests and those domi-nated by tree ferns like *Psaronius* (Phillips, Peppers & DiMichele 1985), must have provided very different niches, both at canopy and ground level, to a forest dominated by much branched conifers or angiosperms. Further-more, these lycopsid forests were often of very low diversity, in itself a restriction on niche availability. In addition it must be noted that the forest communities were not subjected to the effects of vertebrate herbivory until at least the late Carboniferous (Milner 1980).

Certain familiar habitats occupied by diverse plants communities today were apparently scarcely exploited at this time. For example, there is no evidence for diverse communities of freshwater vascular plants prior to the Mesozoic, although certain Carboniferous plants probably occupied marginal habitats in standing water (Collinson 1987; Collinson & Scott 1987). Other habitats were occupied by widely different groups of plants. Marginal, tidally influenced, coastal habitats have been occupied, for example, by cordaites in the Carboniferous (Phillips, Peppers & DiMichele 1985), conifers in the Mesozoic (Upchurch & Doyle 1981; Francis 1984) and flowering plants more recently, e.g. *Nipa* palm and rhizophoraceous man-groves in the Tertiary (Collinson 1983a).

Comparatively little is known about the below-ground portion of ancient plant communities except in rare cases of permineralized peats (DiMichele & Phillips 1985). Compressed rooting systems tend to be difficult to detect and to assign to source plants unless remaining in organic connection. This is unfortunate given the possible significance of below-ground processes in structuring plant communities (see Fitter, Chapter 6). The scant evidence indicates that elaboration of rooting systems proceeded alongside that of the aerial portion. Some lower Devonian plants possessed only rhizoids whereas others had simple dichotomous rooting systems or rhizophores (Edwards 1980; Rayner 1983, 1984). Variations may have been related to features of environmental stress such as water availability and sediment stability. These early communities must have had relatively restricted anchoring and absorptive powers and formed only a shallow root mat with poor powers of soil stabilization (Banks 1985). Our very limited knowledge of pteridosperm and progymnosperm rooting systems is summarized by Matten, Tanner & Lacey (1984). Most evidence is derived from isolated pieces of secondary wood which do at least indicate the presence of extensive, woody, branched rooting systems in the late Devonian, contemporaneous with the early trees. It is also important to note that some ancient anchoring systems differ considerably from any known modern analogues with consequent influence on the plant communities (see Collinson & Scott 1987 for further discussion).

Mycorrhizal associations, within or around rooting systems, have an extremely scant fossil record and are difficult to recognize with certainty (Banks 1985). Available evidence (see also Berch & Warner 1985; Stubblefield, Taylor & Beck 1985) suggests their presence from the early Lower Devonian communities onwards.

Recycling, via decomposition in the litter layers is a very important part of the functioning of communities. Inevitably, decomposition being the antithesis of fossilization, our knowledge in this area is limited and analyses, like those described by Swift (Chapter 11) for contemporary fungal assemblages, are not possible. Rolfe (1985) considered that the terrestrial decomposer niche was occupied by higher fungi and saprophagous micro-arthropods by the late Silurian. According to Stubblefield, Taylor & Beck (1985) fungal wood-decay mechanisms have existed with little change since their first appearance in the late Devonian, contemporaneous with woody tissues. Collinson & Scott (1987) comment briefly on aspects of decomposition in peats.

REPRODUCTIVE BIOLOGY; EFFECTS ON LIFE HISTORY STRATEGY, DISPERSAL & COLONIZATION

Prior to angiosperm dominance

All early vascular plants were homosporous, colonizing via spore dispersal, and many subsequently spreading clonally to form monotypic stands (Chaloner & Sheerin 1981, Knoll *et al.* 1984). It has often been considered that this primitive reproductive biology was one which favoured the colonizing strategy in a pioneer species (Chaloner & Sheerin 1981). Dispersal of spores would have been mainly by wind, but increasing evidence from the Devonian and Carboniferous indicates that spore-eating arthropods may also have acted as minor dispersal agents (Scott, Chaloner & Paterson 1985).

By the end of the Lower Devonian we have the first evidence of hetero-spory (Chaloner & Sheerin 1981). The late Devonian and Carboniferous represent the acme of heterosporous plants with this reproductive strategy exhibited by several distinct groups of plants—progymnosperms, lycopsids and sphenopsids (and later by ferns). Arborescent heterosporous lycopsids show a wide range of sporophyll morphologies which, it has been argued, evolved to aid wind (Thomas 1985) or water (DiMichele & Phillips 1985) dispersal. The heterosporous strategy does confer some competitive advantage on the endosporous, protected, self-supporting prothallus (com-pared with the homosporous strategy) amongst vegetation cover or in slightly drier circumstances. Such plants dominated the Upper Carbon-iferous lowland tropical swamps of Euramerica (DiMichele & Phillips 1985). Their heterosporous strategy, determinate growth and limited secondary conducting tissues made them well-adapted for a wetland life. Several subcommunities may be recognized, reflecting environment variations.

An embryo nourished by the parent sporophyte, hence mature at dispersal, with protection and with a future food reserve (i.e. a seed), has a further competitive advantage in establishing under vegetation cover or in dry and disturbed conditions, however. The evolution of the megaphyllous leaf and arborescence may therefore have provided some of the selective pressures favouring seed production.

Seed plants are first recorded from the late Devonian (Chaloner & Sheerin 1981). Many early seeds were small, typical of the pioneer strategy today, and occur in floodplain sediments consistent with this interpretation (Scott & Rex 1987). However, as early as the Lower Carboniferous large seeds such as *Salpingostoma* (Gordon 1941) were present. In the late Carboniferous, there were diverse large seed, mainly of pteridosperms (Combourieu & Galtier 1985) which may be regarded as typical of the *k*-selected strategy, whose parent plants dwelt in stable communities such as swamps.

In drier habitats seed plants were already diverse in the Lower Carboniferous (Scott & Rex 1985, 1987) and came to dominate many communities, especially uplands, by the end of the Carboniferous (Scott 1980). The earliest conifers (early Upper Carboniferous) may have been either upland or dry lowland plants (Scott & Chaloner 1983). There is now considerable evidence that, during the later Carboniferous and early Permian, these plants spread to dominate lowland communities, probably as a response to major climatic change leading to increased aridity (Frederiksen 1972; Scott 1980).

Seeds provide food for foraging fauna (see Brown, Chapter 9) in discreet packages of high nutritive value. These are produced at various levels in a tiered ecosystem and also accumulate amongst forest floor litter. Specializations of jaws, limbs etc. may be needed to exploit this resource, although, as noted above, vertebrates did not exploit this until at least the late Carboniferous (Milner 1980) i.e. more or less contemporaneous with seed dominance in plant communities. Evidence of some past plant–animal interactions is given by Scott & Taylor (1983).

The shift of the moment of dissemination from pre to post-fertilization and from pre to post-embryo maturity permits greater dispersal distance, increasing the likelihood of outbreeding with resultant variability of offspring. These, and other factors mentioned above, combine to increase the geographic area and variety of environment open for colonization. Not only did the seed strategy open up new niches but it also increased community diversity.

Knoll (1986) analysed diversity within plant communities through time using an estimate of change in species-richness of fossil floras (i.e. fossil plant assemblages from time-bracketed blocks of sediment). Numbers of species per flora (Fig. 18.1) reveal three phases of diversity increase: the early Devonian, the Lower Carboniferous and the Lower Cretaceous onwards. These may be considered to reflect, with some lag in reaction, the evolution of vascular plants on land, the evolution of the seed strategy and the evolution of the angiosperm strategy (see below). They thus demonstrate the vital influence of evolutionary innovations of reproductive biology on the structure of ancient plant communities.

Angiosperm-dominated communities

The most recent innovation in plant reproductive biology was the evolution of the angiosperms by the Lower Cretaceous and possibly earlier (Crane 1985). In the angiosperms, presence of a pollen-receptive surface (stigma) on a structure (the carpel) more or less enclosing the egg, permits control over pollination and hence fertilization. The carpel and double seed integument provide for a huge variety of post-fertilization protective mechanisms (including dormancy) and attractant fruit and seed walls. Co-evolution of faithful animal pollinators and a variety of dispersal agents has resulted in extensive outbreeding with resultant enhanced variability. Furthermore, reduced gametophytes and double fertilization (producing a food reserve post-fertilization) all combine to speed up the life cycle and reduce the required energy input and establishment time prior to progeny production. Combined, these factors have resulted in the success and diversity of the angiosperms today (Regal 1977; Stebbins 1974, 1976, 1981; Crepet 1983, 1985).

Diversity within angiosperm-dominated communities has increased since their origin (Knoll 1986). This diversity, combined with other biological attributes, has resulted in the transformation of a variety of habitats. Present knowledge suggests that angiosperms provided the first herbaceous seed plants. (Those pteridosperms and gymnosperms which have been adequately reconstructed exhibited secondary growth and were lianas, shrubs or trees (e.g. Stewart 1983). Herbaceous ground cover was thus available for dry and disturbed areas and newly-exposed soils, colonization being from easily-dispersed seeds or from those dormant in the soil seed bank. In effect, the 'weeds' of today were born. (see also comments on disturbed habitats). Other habitats where spore-bearing plants had dominated and angiosperms effected major changes include the freshwater habitat (Collinson 1987) and, although evidence is largely by inference from nearest living relatives, the epiphytic habitat (Knoll 1986). Grasslands perhaps represent the ultimate in herbaceous cover and total transformation of a previously fern-dominated habitat (see below) with associated effects on the animal assemblages (see Webb, Chapter 20).

The earliest recognizable angiosperm pollen is found in the Lower Cretaceous about 116 million years ago in proportions of less than 1% of pollen and spore assemblages (Walker & Walker 1985). The earliest records are more or less contemporaneous in various parts of the world. Later phases of diversification seem to spread from lower to higher latitudes (Hickey & Doyle 1977; Hughes 1978; Crane 1985; Muller 1985). Angiosperm pollen reached up to 50% of pollen and spore assemblages in some early Upper Cretaceous examples, suggesting that competitive replacement within communities may well have been complete (Muller 1985).

Authors disagree concerning the geographical and altitudinal situation, the habitat and habit of early angiosperms (Takhtajan 1969; Axelrod 1970; Stebbins 1974; Doyle 1978; Retallack & Dilcher 1981a,b.). There is, however, general agreement that early angiosperms were pioneer strategists, invading communities dominated by ferns, conifers and other gymnosperms and pteridosperms. Evidence from seed size (Tiffney 1985) supports this hypothesis. At least in some of the earlier stages, though not known for the earliest, some angiosperms were living in unstable environments such as periodically inundated coastlines (Retallack & Dilcher 1981a) and the stream margin sites of a meandering river system (Hickey & Doyle 1977). In the latter case more stable, backswamp environments were dominated by cycadeoids and conifers. However, until more details are known about Lower Cretaceous plant communities (Lapasha & Miller 1984; Pelzer 1984) it is impossible to be more precise about which elements of established communities were displaced during the early phases of angiosperm evolution (see papers in Friis, Chaloner & Crane 1987).

By the late Cretaceous, pollen (Muller 1985) and macrofossil data including leaves (Hickey & Doyle 1977), fruits and seeds (Tiffney 1985) and flowers (Crane & Dilcher 1985; Crepet 1985; Dilcher & Crane 1985; Friis 1985a) show that angiosperms were diverse and dominated plant communities. Pollen and macrofossil evidence (Herngreen & Chlonova 1981; Batten 1984) also show distinct floral provinces by this stage. A further phase of diversification occurred in the early Tertiary (Muller 1985).

Subsequent fossil assemblages are usually assessed in terms of the proportions of nearest living relatives from various modern climatic zones, geographic areas or vegetation formations. Thus the proportions of tropical versus temperate; Arcto-Tertiary/Palaeotropical (or mastixioid); local/exotic; mixed mesophytic forest/tropical rain forest; etc., have been analysed (Wolfe 1975; Mai & Walther 1978; Gregor 1982; Collinson 1983a; Van der Burgh 1983, 1987; Friis 1985b). These analyses reveal some overall pattern such as a general decrease in the proportion of elements with tropical living relatives in assemblages from Europe from the early Tertiary to the present. The mere existence of such combinations of groupings now segregated today, emphasizes that these past communities are not directly comparable to any modern examples (Collinson 1983a). The former geoflora concept (see Wolfe 1975), in which plant groupings were said to have persisted unchanged for long periods of time, is now untenable.

There are cases where dominant elements in a community have remained similar, e.g. the *Typha/Acrostichum* association (Collinson 1983b) and the taxodiaceous swamp, a common feature of late Cretaceous and Tertiary wetlands (e.g. Teichmüller 1958; Wing 1980, 1984; Gregor 1982 and references cited; Cross & Taggart 1983; Krumbiegel, Rüffle & Haubold 1983; Van der Burgh 1983, 1987; Basinger 1984; Smiley and Rember 1985). However, close inspection reveals various differences between the ancient communities and their modern analogues. Usually the combination of elements in the ancient community no longer persists in any modern community. In the case of the taxodiaceous swamp the dominant conifer may be closest to modern *Metasequoia* (Basinger 1984), *Glyptostrobus* (Wing 1980, 1984) or *Taxodium* (Gregor 1982), or it may show a combination of features of several genera (Fowler, Edwards & Brett 1973). Often fossils assigned to two or three genera in the family occur together, e.g. *Taxodium* and *Glyptostrobus* (Krumbiegel, Rüffle & Haubold 1983). Smiley & Rember (1985) document a *Taxodium/Nyssa* swamp forest surrounded by slope forests with *Metasequoia*, thus combining features of modern eastern USA and south-east Asian communities. An additional complication is provided by the presence of slope forest conifers with western North American affinities.

Evidence is accumulating that some ancient angiospermous plants have retained broadly comparable reproductive biology since the Palaeocene or even latest Cretaceous (Crane & Stockey 1985, D. L. Dilcher, personal communication) whereas others differ from their nearest living relatives (Crane 1981). Amongst the more usual fossil records of isolated organs, many are known which differ from those of their nearest living relatives, whereas others are more or less indistinguishable from living examples (Collinson 1983a,b; Friis 1985b).

The precise significance of these differences in plant biology, and the effects of combinations within communities of species whose living relatives are now segregated, cannot be reliably assessed in the absence of many more well-founded reconstructions of Upper Cretaceous and Tertiary communities. However, it is clear that any implication that angiosperm dominated communities have remained stable for long periods is, at the very least, a gross oversimplification.

In rare cases, where a sequence of fossil plant assemblages covers a period of time in a small area unaffected by local geographic variation, fluctuations in local plant communities may be assessed (Collinson & Hooker 1987). Interactions with local mammalian faunas may also be evaluated (Collinson & Hooker 1987). Other evidence of interactions within angiosperm-dominated communities may be derived from vertebrate gut contents (e.g. references cited in Koenigswald & Michaelis 1984); by inference from floral structure to pollinators (e.g. Crepet 1985) and from a variety of other data (papers in Friis, Chaloner & Crane 1987).

Assessments of global vegetation may also be made (e.g. Wolfe 1980, 1985). Some of the most significant of these involve recognition of the past extent of forest growth. Broad-leaved evergreen forest extended to about 70–75° North during the early Eocene, and coniferous forests once blanketed the entire Arctic region (Wolfe 1980, 1985; Axelrod 1984; Creber & Chaloner 1984, 1985). Far from being speculative, these deductions are supported by evidence of *in situ* tree stumps. Possible explanations have included variations in continental positions, atmospheric composition and the axis of the earth's rotation, along with adaptations in the form of the physiological responses of the biota.

During the Cretaceous and Tertiary, after angiosperms came to dominate world floras, coniferous forests consistently dominated at higher latitudes (Wolfe 1985). This may be considered in one of two ways; as part of gymnosperm restriction from a previous dominance in world vegetation, or as the adaptation of conifers, in preference to angiosperms, for high latitude growth. Such biological features as tracheidal secondary xylem, rooting depth, xeromorphy, tree shape etc., may be responsible for this (Axelrod 1984, Creber & Chaloner 1984, 1985). Certainly, these facets of forest vegetation distribution are one of the most striking examples, from within a general pattern, of major fluctuations in the extent of vegetation zones which characterized the Tertiary (Wolfe 1985). These changes would certainly have had significant effects on the animal assemblages (for effects on vertebrates see Webb, Chapter 20).

Another major feature of the evolution of angiosperm-dominated communities occurred at the Cretaceous–Tertiary (K/T) boundary. This K/T event certainly involved extinction of previously dominant angiosperm forms and an overall 'modernization' of the flora (Hickey 1981; Tschudy et al. 1984; Muller 1985), although these authors differ as to the gradual or sudden nature of this change. Tschudy et al. (1984) document an increase in fern spores versus angiosperm pollen in sediment at several sites, coincident with other events at this time. This is colloquially referred to as the 'fern-spike' and considered to support the catastrophist explanation for K/T boundary events (see Officer & Drake 1985 for a recent summary). In view of our observations (see 'Disturbed habitats') concerning the association of ferns (and other homosporous plants) and disturbed volcanic terrains, it seems reasonable to accept that some major disruption of terrestrial plant communities did occur at this time. This presumably created new, open niches, into which angiosperms were able to rapidly diversify. The pollen record certainly shows rapid diversification in the early Tertiary (Muller 1985). Since submission of this manuscript these comments have been confirmed and expanded upon by several recent papers. These are summarized by Collinson (1986).

Details of the origin of Quaternary and Recent communities require individual consideration for any given area and are beyond the scope of this paper. It should, however, be clear from the foregoing that these represent merely one time slice of a continuum probably shifted onto its present course by a combination of K/T events and the later southern and northern hemisphere glaciations (beginning in the latest Oligocene and late Miocene respectively).

Grasslands had such a marked effect on faunal evolution (see Chapter 20) and subsequent exploitation by man that we feel it is worth considering their evolution here, even though their fossil record is very poor. Grass pollen is first securely encountered in the fossil record in the Palaeocene, according to Muller (1985), although in very small amounts. Scattered references to pre-Oligocene grass macrofossils are reviewed by Thomasson

(1980) and generally considered doubtful or insufficiently well-supported taxonomically. Well-substantiated grass macrofossils, related to plants forming grasslands today, are first recorded from the Upper Miocene of North America (references cited by Thomasson 1983). These are determined on the basis of epidermal characters of anthoecia. Whereas fruit and seed floras likely to yield such well-preserved fossils are rare through the Tertiary in the USA, it is pertinent to note that in Europe, where they are abundant and have been extensively studied, no earlier examples have been found. According to Wolfe (1985), grasslands in central North America can be no older that late Miocene, and according to Axelrod (1985) were probable not extensive until the Mio/Pliocene transition. A study using pollen, leaves and their sedimentological context (Leopold & Wright 1985) indicates that former inferences concerning Miocene savanna in the Pacific north-west of the USA (based on grazing mammals) should be revised. Extensive grasslands probably did not develop there until the Pliocene at the earliest.

Savanna vegetation generally seems to be of Miocene or younger origin based on increasing abundance of putative grass pollen (Wolfe 1985). However, most records are from low latitudes and the overall situation in higher latitudes is much less clear. Solbrig (1976) has suggested the existence of savanna-woodland in South America during the Eocene. However, the grass macrofossils of this area are very few amongst a large leaf flora of forest/woodland aspect (Menendez 1972). Furthermore, they fall within the doubtful category of Thomasson (1980).

Retallack (1983) has suggested on the basis of fossil soil profiles that savanna-woodland occurred during the early Oligocene in the Badlands National Park, south Dakota, USA. However, in his data, the first soil profile specifically associated with savanna today is in the mid-late Oligocene. Comparisons must also be complicated by the general absence of soil profiles from modern herbaceous vegetation in which grasses are absent.

Disturbed habitats

Various phenomena familiar today as disrupting habitats may be recognized in the fossil record, e.g. fire and volcanic activity. The first wildfires (recognized by fossil charcoal) are contemporaneous with the first forests in the late Devonian (Cope & Chaloner 1985). Thus, some climax communities may have been fire-maintained from the Devonian onwards much as they are today, e.g. the Florida pine flatwoods and other areas (Wade, Ewel & Hofstetter 1980).

Evidence from early Carboniferous floras shows that free-sporing plants (e.g. lycopsids and ferns) are often intimately associated with both lava and ash land surfaces (Scott & Galtier 1985; Rex & Scott 1987; Scott & Rex 1987). Some were probably creeping rhizomatous plants, which may have persisted through, or indeed been usefully spread by, volcanic eruptions (Scott & Galtier 1985). Some gymnosperms (i.e. seed bearers) as colonizers of Lower Carboniferous ash deposits are also known (Rothwell & Scott 1985). In a Tertiary volcanogenic sequence *Equisetum* is abundant along with several ferns and an assemblage of dicotyledons (Manchester 1981).

Free sporing ferns and sphenopsids are found today as early colonizers (via spores or rhizome fragments), followed by 'weedy' angiosperms, of volcanic ashes in several recent volcanoes, e.g. Mt. St. Helens, Krakatoa and El Chichon (Richards 1952, pp. 271–7; Flenley 1979; references cited by Leahy, Spoon & Retallack 1985; Moral & Clampitt 1985; Spicer *et al.* 1985). This pattern has thus changed little over millions of years in so far as the free-sporing plant role is concerned (although the taxa involved are quite different). The role of the seed plant was, however, revolutionized after the evolution of the angiospermous 'weed' (see p. 405). Cross & Taggart (1983) provide an elegant example of disruption to local vegetation succession caused partly by volcanic activity in the Miocene of the USA. Smiley & Rember (1985) record the influence of lake damming due to volcanic activity on local Miocene plant communities. Events at the *K/T* boundary (see p. 408) certainly indicate disturbance although whether this was volcanic or extraterrestrial is a matter of great dispute.

CONCLUSIONS

The major contribution of the studies outlined above is to provide an expanded time dimension to ecological work. Modern communities merely represent one small time slice. They should be considered less as finely-tuned entities in dynamic equilibrium and more as sampling points within an evolving continuum. We have demonstrated unequivocally that major community changes have occurred in the past and that succession of these communities, on various time scales, may be observed in the fossil record.

The fossil evidence does reveal communities whose dominant elements or general aspect have long been established, e.g. swamps dominated by taxodiaceous conifers; high latitude coniferous forest; ferns and other homosporous plants as early colonizers of volcanic terrain. However, even in these cases, the individual taxa differ from their nearest living relatives in a variety of ways and the entire communities are quite unlike any existing today. There are many examples where past communities have no close modern analogues, e.g. swamps with arborescent lycopsids. No pre-

Quaternary plant community should be considered exactly like any one existing today.

Innovations of reproductive and vegetative organization are a typical feature of the fossil record (Fig. 18.1). Turnover of past plant communities is largely related to these biological innovations, sometimes selected for by physical factors such as climatic change or volcanic activity, as well as other biological variables such as herbivory and co-evolution of pollinators and dispersers. Many of these changes took several to many millions of years before they altered the overall appearance of the landscape.

Three major evolutionary innovations of reproductive biology (free-sporing plants on land; the seed strategy; the angiospermous strategy) are reflected in colonization potential, with invasion and replacement of existing communities and occupation of previously scarcely or unexploited sites. Life history strategies (e.g. r and k-selectivity) passed through several cycles in response. In each case a 'new wave' of opportunists entered previously unexploited habitats, creating new communities. This was followed, as adaptation proceeded, by invasion of previously established communities with replacement of dominants by new k strategists. Such cycles may have also typified the earliest non-vascular land plant communities (see Gray 1985). Increases in community diversity (Knoll 1986) also reflect innovations of reproductive biology, especially in the case of the angiosperms where many communities were transformed.

Major innovations of vegetative organization, in particular the megaphyllous leaf and arborescence, were early events in the evolution of plant communities. They resulted in shaded microhabitats, tiered communities with forest floor and canopy layers, microhabitats for lianas and the occurrence of fire. They may have influenced the changes in reproductive biology in the case of evolution of the seed.

Man exerts a profound influence over community structure. It may no longer be entirely frivolous to look to the past community changes for analogues to potential current man-induced events. The likening of the postulated nuclear winter to events at the K/T boundary (e.g. Wolbach, Lewis & Anders 1985) and the comments by Delcourt, Delcourt & Webb (1983) on the role of past analogues in evaluating biotic response to CO_2-induced warming, are examples of this. Raven (1986) compares current extinction in tropical communities with events at the K/T boundary. Current forest destruction is several magnitudes faster than any of the major changes we have documented. This 'biological innovation' is without parallel in the past, as arborescent plants have been a major component of earth's vegetation since their evolution some 375 million years ago.

Evolutionary changes, at organismal through to ecosystem level, result from ecological pressures. This brief outline has examined the large-scale palaeoecological patterns in plant communities which represent an accumulation of smaller-scale ecological events. We hope that this will contribute to an increasing evolutionary perspective in ecological analyses.

ACKNOWLEDGMENTS

We thank W. G. Chaloner, F.R.S., J. J. Hooker and P. D. Moore for helpful discussion, the editors and an anonymous reviewer for valuable comments on an earlier draft of this paper and various conference participants for posing unanswerable questions. Collaborative work with J. Galtier and G. Rex described herein was made possible by grants from NATO (RG 361/83) and CNRS which are gratefully acknowledged. This work was undertaken by MEC whilst in receipt of a Royal Society 1983 University Research Fellowship which is gratefully acknowledged.

REFERENCES

Axelrod, D.I. (1970). Mesozoic paleogeography and early angiosperm history. *Botanical Review,* **36,** 277–319.

Axelrod, D.I. (1984). An interpretation of Cretaceous and Tertiary biota in polar regions. *Palaeogeography, Palaeoclimatology, Palaeoecology,* **45,** 105–47.

Axelrod, D.I. (1985). Rise of the grassland biome, Central North America. *Botanical Review,* **51,** 163–201.

Banks, H.P. (1985). Early land plants. *Philosophical Transactions of the Royal Society, B,* **309,** 197–200.

Basinger, J.F. (1984). Seed cones of *Metasequoia milleri* from the Middle Eocene of southern British Columbia. *Canadian Journal of Botany,* **62,** 281–9.

Batten, D.J. (1984). Palynology, climate and the development of Late Cretaceous floral provinces in the Northern Hemisphere: a review. *Fossils and Climate.* (Ed. by P. Brenchley), pp. 127–64. Wiley, New York.

Beck, C.B. (1964). Predominance of *Archaeopteris* in Upper Devonian flora of western Catskills and adjacent Pennsylvania. *Botanical Gazette,* **125,** 125–8.

Berch, S.M. & Warner, B.G. (1985). Fossil vesicular arbuscular mycorrhizal fungi: Two *Glomus* species (Endogonaceae, Zygomycetes) from late Quaternary deposits in Ontario, Canada. *Review of Palaeobotany and Palynology,* **45,** 229–37.

Birks, H.J.B. (1986). Late-Quaternary biotic changes in terrestrial and lacustrine environments, with particular reference to north-west Europe. *Handbook of Holocene Palaeoecology and Palaeohydrology* (Ed. by B.E. Berglund), pp. 3–65. Wiley, Chichester.

Chaloner, W.G. & Lawson, J.D. (Eds) (1985). Evolution and environment in the late Silurian and early Devonian. *Philosophical Transactions of the Royal Society, B,* **309,** 1–342.

Chaloner, V.G. & Sheerin, A. (1979). Devonian Macrofloras. *Special Papers in Palaeontology,* **23,** 145–61.

Chaloner, V.G. & Sheerin, A. (1981). The evolution of reproductive strategies in early land plants. *Evolution Today.* (Ed. by G.G.E. Scudder & J.L. Reveal), pp. 93–100.

Proceedings of the Second International Congress of Systematic and Evolutionary Biology. Carnegie-Mellon University, Pittsburg.

Collinson, M.E. (1983a). *Fossil Plants of the London Clay.* Palaeontological Association Field Guides to Fossils, 1. Palaeontological Association, London.

Collinson, M.E. (1983b). Palaeofloristic assemblages and palaeoecology of the Lower Oligocene Bembridge Marls, Hamstead Ledge, Isle of Wight. *Botanical Journal of the Linean Society,* **86,** 177–225.

Collinson, M.E. (1986). Catastrophic vegetation changes. *Nature (London),* **324,** 112.

Collinson, M.E. (1987). Freshwater macrophytes in palaeolimnology. *Palaeogeography, Palaeoclimatology, Palaeoecology,* **60,** (in press).

Collinson, M.E. & Hooker, J.J. (1987). Vegetational and mamalian faunal changes in the early Tertiary of Southern England. *Angiosperm Origins and Biological Consequences.* (Ed. by E.M. Friis, W.G. Chaloner & P.R. Crane), pp. 259–304. Cambridge University Press, Cambridge.

Collinson, M.E. & Scott, A.C. (1987). Implications of vegetational change through the geological record on models for coal-forming environments. *Coal and Coal-Bearing Strata: Recent Advances* (Ed. by A.C. Scott), pp. 67–85. Geological Society of London Special Issue, Blackwell Scientific Publications, Oxford,

Combourieu, N. & Galtier, J. (1985). Nouvelles observations sur *Polpterospermum, Polylophospermum, Colpospermum* et *Codonospermum* ovules de Pteridospermales du Carbonifère Superieur Français. *Palaeontographica, B,* **196,** 1–29.

Cope, M.J. & Chaloner, W.G. (1985). Wildfire: an interaction of biological and physical processes. *Geological factors and the Evolution of Plants* (Ed. by B.H. Tiffney), pp. 257–77. Yale University Press, CT.

Crane, P.R. (1981). Betulaceous leaves and fruits from the British Upper Palaeocene. *Botanical Journal of the Linnean Society,* **83,** 103–36.

Crane, P.R. (1985). Phylogenetic analysis of seed plants and the origin of angiosperms. *Annals of the Missouri Botanical Garden.* **72,** 716–93.

Crane, P.R. & Dilcher, D.L. (1985). *Lesqueria:* an early angiosperm fruiting axis from the mid-Cretaceous. *Annals of the Missouri Botanical Garden,* **71,** 385–402.

Crane, P.R. & Stockey, R.A. (1985). Growth and reproductive biology of *Joffrea speirsii* gen.et sp.nov., a *Cercidiphyllum*-like plant from the Late Paleocene of Alberta, Canada. *Canadian Journal of Botany,* **63,** 340–64.

Creber, G.T. & Chaloner, W.G. (1984). Influence of environmental factors on the wood structure of living and fossil trees. *Botanical Review,* **50,** 357–448.

Creber, G.T. & Chaloner, W.G. (1985). Tree growth in the Mesozoic and early Tertiary and the reconstruction of palaeoclimates. *Palaeogeography, Palaeoclimatology, Palaeoecology,* **52,** 35–60.

Crepet, W.L. (1983). The role of insect pollination in the evolution of the angiosperms. *Pollination Biology* (Ed. by L. Real), pp. 29–50. Academic Press, Orlando, Florida.

Crepet, W.L. (1985). Advanced (constant) insect pollination mechanisms: pattern of evolution and implications vis-a-vis angiosperm diversity. *Annals of the Missouri Botanical Garden,* **71,** 607–30.

Cross, A.T. & Taggart, R.E. (1983). Causes of short-term sequential changes in fossil plant assemblages: some considerations based on a Miocene flora of the northwest United States. *Annals of the Missouri Botanical Garden,* **69,** 676–734.

Delcourt, H.R., Delcourt, P.A. & Webb, T. III. (1983). Dynamic plant ecology: the spectrum of vegetational change in space and time. *Quaternary Science Reviews,* **1,** 153–75.

Dilcher, D.L. & Crane, P.R. (1985). *Archaeanthus:* an early angiosperm from the Cenomanian of the western interior of North America. *Annals of the Missouri Botanical Garden,* **71,** 351–83.

DiMichele, W.A. & Phillips, T.L. (1985). Aborescent lycopod reproduction and paleoecology in a coal-swamp environment of late Middle Pennsylvanian age (Herrin Coal, Illinois, U.S.A.). *Review of Palaeobotany and Palynology,* **44,** 1 – 26.

Doyle, J.A. (1978). Origin of the Angiosperms. *Annual Review of Ecology and Systematics,* **9,** 365 – 92.

Edwards, D. (1980). Early land floras. *The Terrestrial Environment and Origin of Land Vertebrates* (Ed. by A.L. Panchen), pp. 55 – 85. Academic Press, New York.

Ferguson, D.K. (1985). The origin of leaf-assemblages— new light on an old problem. *Review of Palaeobotany and Palynology,* **46,** 117 – 88.

Flenley, J.R. (1979). *The Equatorial Rain Forest: A Geological History.* Butterworths, London.

Fowler, K., Edwards, N. & Brett, D.W. (1973). *In situ* coniferous (Taxodiaceous) tree remains in the Upper Eocene of southern England. *Palaeontology,* **16,** 205 – 17.

Francis, J.E. (1984). The seasonal environment of the Purbeck (Upper Jurassic) fossil forest. *Palaeogeography, Palaeoclimatology, Palaeoecology,* **48,** 285 – 307.

Frederiksen, N.O. (1972). The rise of the Mesophytic flora. *Geoscience & Man,* **4,** 17 – 28.

Friis, E.M. (1985a). Preliminary report of Upper Cretaceous angiosperm reproductive organs from Sweden and their level of organization. *Annals of the Missouri Botanical Garden,* **71,** 403 – 18.

Friis, E.M. (1985b). Angiosperm fruits and seeds from the Middle Miocene of Jutland, Denmark. *Det Kongelige Danske Videnskaberne Selskab Biologiske Skrifter,* **24,** 1 – 165.

Friis, E.M., Chaloner, W.G. & Crane, P.R. (Eds) **(1987).** *Angiosperm Origins and the Biological Consequences.* Cambridge University Press, Cambridge.

Gensel, P.G. (1984). A new lower Devonian plant and the early evolution of leaves. *Nature (London).* **309,** 785 – 7.

Gordon, W.T. (1941). On *Salpingostoma dasu:* a new Carboniferous seed from East Lothian. *Transactions of the Royal Society of Edinburgh,* **60,** 427 – 44.

Gray, J. (1985). The microfossil record of early landplants: advances in the understanding of early terrestrialization. *Philosophical Transactions of the Royal Society,* **309,** 167 – 95.

Gregor, H-J. (1982). *Die jungtertiären Floren Süddeutschlands.* Ferdinand Enke Verlag, Stuttgart.

Harland, W.B., Cox, A. V., Llewellyn, P.G., Pickton, C.A.G., Smith, A.G. & Walters, R. (1982). *A Geologic Time Scale.* Cambridge University Press, Cambridge.

Herngreen, G.F.W. & Chlonova, A.F. (1981). Cretaceous microfloral provinces. *Pollen et Spores,* **23,** 441 – 555.

Hickey, L.J. (1981). Land plant evidence compatible with gradual, not catastrophic change at the end of the Cretaceous. *Nature (London).* **292,** 529 – 31.

Hickey, L.J. & Doyle, J.A. (1977). Early Cretaceous fossil evidence for angiosperm evolution. *Botanical review,* **43,** 3 – 104.

Hughes, N.F. (1978). *Palaeobiology of Angiosperm Origins.* Cambridge University Press, Cambridge.

Knoll, A.H. (1986). Patterns of change in plant communities through time. *Community Ecology* (Ed. by J. Diamond & T.J. Case), pp. 126 – 48. Harper & Row, New York.

Knoll, A.H., Niklas, K.J., Gensel, P.G. & Tiffney, B.H. (1984). Character diversification and patterns of evolution in early vascular plants. *Paleobiology,* **10,** 34 – 47.

Koenigswald, W.V. & Michaelis, W. (1984). Fossillagerstätte Messel—Literaturübersicht der Forschungsergebnisse aus den Jahren 1980 – 1983. *Geologisches Jahrbuch Hessen,* **112,** 5 – 26.

Krassilov, V. (1973). Mesozoic bryophytes from the Bureja Basin, far East of the U.S.S.R. *Palaeontographica, B* **143,** 95 – 105.

Krumbiegel, G., Rüffle, L. & Haubold, H. (1983). *Das Eozäne Geiseltal.* A. Ziemsen Verlag, Wittenberg Lutherstadt.

Lapasha, C.A. & Miller, C.N. (1984). Flora of the early Cretaceous Kootenai Formation in

Montana, paleoecology. *Palaeontographica*, B, **194**, 109–30.

Leahy, G.D., Spoon, M.D. & Retallack, G.J. (1985). Linking impacts and plant extinctions. *Nature (London)*, **318**, 318.

Leopold, E.B. & Wright, V.C. (1985). Pollen profiles of the Plio-Pleistocene Transition in the Snake River Plain, Idaho. *Late Cenozoic History of the Pacific Northwest* (Ed. by C.J. Smiley), pp. 323–48. Pacific Division of the American Association for the Advancement of Science, San Francisco, California.

Mai, D.H. & Walther, H. (1978). Die Floren der Haselbacher Serie im Weisselster-Becken (Bezirk Leipzig, D.D.R.). *Abhandlungen der Staatliches Museum für Mineralogie und Geologie zu Dresden*, **28**, 1–101.

Manchester, S.R. (1981). Fossil plants of the Eocene Clarno Nut Beds. *Oregon Geology*, **43**, 75–81.

Mapes, G. & Rothwell, G.W. (1984). Permineralized ovulate cones of *Lebachia* from the Late Palaeozoic limestones of Kansas. *Palaeontology*, **27**, 69–94.

Matten, L.C., Tanner, W.C. & Lacey, W.S. (1984). Additions to the silicified upper Devonian/lower Carboniferous flora from Ballyheigue, Ireland. *Review of Palaeobotany and Palynology*, **43**, 303–20.

Menendez, C. (1972). Paleofloras de la Patagonia. *La Region de los Boqies Andino Patagonicos* (Ed. by M.J. Dimitri), **10**, 129–84.

Milner, A.R. (1980). The tetrapod assemblage from Nyrany, Czechoslovakia. *The Terrestrial Environment and the Origin of Land Vertebrates* (Ed. by A.L. Panchen), pp. 439–96. Systematics Association Special Volume, 15. Academic Press, New York.

Moral, R., del., & Clampitt, C.A. (1985). Growth of native plant species on recent volcanic substrates from Mount St. Helens. *American Midland Naturalist*, **114**, 374–83.

Muller, J. (1985). Significance of fossil pollen for angiosperm history. *Annals of the Missouri Botanical Garden*, **71**, 419–43.

Officer, C.B. & Drake, C.L. (1985). Terminal Cretaceous environmental events. *Science*, **227**, 1161–7.

Pelzer, G. (1984). Cross section through a fluvial environment in the Wealden of Northwest Germany. *Third Symposium on Mesozoic Terrestrial Ecosystems, Short Papers.* (Ed. by W.-E. Reif and F. Westphal), pp. 181–6. ATTEMPTO Verlag, Tübingen.

Phillips, T.L., Peppers, R.A. & DiMichele, W.A. (1985). Stratigraphic and interregional changes in Pennsylvanian coal-swamp vegetation: Environmental inferences. *International Journal of Coal Geology*, **5**, 43–109.

Raven, J.A. (1985). Comparative physiology of plant and arthropod land adaptation. *Philosophical Transactions of the Royal Society*, **309**, 273–88.

Raven, P.H. (1986). The urgency of tropical conservation. *The Nature Conservancy News*, **36**, 7–11.

Rayner, R.J. (1983). New observations on *Sawdonia ornata* from Scotland. *Transactions of the Royal Society of Edinburgh: Earth Sciences*, **74**, 79–93.

Rayner, R.J. (1984). New finds of *Drepanophycus spinaeformis* Göppert from the Lower Devonian of Scotland. *Transactions of the Royal Society of Edinburgh: Earth Sciences*, **75**, 353–63.

Regal, P.J. (1977). Ecology and the evolution of flowering plant dominance. *Science*, **196**, 622–29.

Retallack, G.J. (1983). Late Eocene and Oligocene paleosols from Badlands National Park, South Dakota. *Special Papers of the Geological Society of America*, **193**, 1–82.

Retallack, G.J. (1985). Fossil soils as grounds for interpreting the advent of large plants and animals on land. *Philosophical Transactions of the Royal Society*, **309**, 105–42.

Retallack, G.J. & Dilcher, D.L. (1981a). A coastal hypothesis for the dispersal and rise to dominance of flowering plants. *Paleobotany, Paleoecology and Evolution, Vol. 2.* (Ed. by K.J. Niklas), pp. 27–77. Praeger, New York.

Retallack, G.J. & Dilcher, D.L. (1981b). Early angiosperm reproduction: *Prisca reynoldsii*, gen.et sp.nov. from mid-Cretaceous coastal deposits in Kansas, U.S.A. *Palaeontographica, B*, **179**, 103–37.

Rex, G. (1986). Further experimental modelling of the formation of plant compression fossils. *Lethaia*, **19**, 143–60.

Rex, G. & Scott, A.C. (1987). The sedimentology, palaeoecology and preservation of the Lower Carboniferous plant deposits at Pettycur, Fife, Scotland. *Geological Magazine*, **124**, 43–66.

Richards, P.W. (1952). *The Tropical Rain Forest, an Ecological Study*. Cambridge University Press, Cambridge.

Rolfe, W.D.I. (1985). Early terrestrial arthropods: a fragmentary record. *Philosophical Transactions of the Royal Society*, **309**, 207–18.

Rothwell, G.W. & Scott, A.C. (1985). Ecology of Lower Carboniferous plant remains from Oxroad Bay, East Lothian, Scotland. *American Journal of Botany*, **72**, 899.

Schaarschmidt, F. (1985). Flowers from the Eocene oil-shale of Messel: A preliminary report. *Annals of the Missouri Botanical Gardens*, **71**, 599–606.

Scheihing, M.H. & Pfefferkorn, H.W. (1984). The taphonomy of land plants in the Orinoco delta: a model for the interpretation of plant parts in clastic sediments of late Carboniferous age of Euramerica. *Review of Palaeobotany and Palynology*, **41**, 205–40.

Scott, A.C. (1980). The ecology of some Upper Palaeozoic floras. *The Terrestrial Environment and Origin of Land Vertebrates*. (Ed. by A.L. Panchen), pp. 87–115. Systematics Association Symposium, 15, Academic Press, New York.

Scott, A.C. (1984). The early history of life on land. *Journal of Biological Education*, **18**, 207–19.

Scott, A.C. & Chaloner, W.G. (1983). The earliest fossil conifer from the Westphalian B of Yorkshire. *Proceedings of the Royal Society of London, B*, **220**, 163–82.

Scott, A.C., Chaloner, W.G. & Paterson, S. (1985). Evidence of pteridophyte-arthropod interactions in the fossil record. *Proceedings of the Royal Society of Edinburgh, B*, **86**, 133–40.

Scott, A.C. & Collinson, M.E. (1983a). Investigating fossil plant beds I. The origin of fossil plants and their sediments. *Geology Teaching*, **7**, 114–22.

Scott, A.C. & Collinson, M.E. (1983b). Investigating fossil plant beds II. Methods for palaeoenvironmental analysis and modelling and suggestions for experimental work. *Geology Teaching* **8**, 12–26.

Scott, A.C. & Galtier, J. (1985). The distribution and ecology of early ferns. *Proceedings of the Royal Society of Edinburgh, B*, **86**, 141–9.

Scott, A.C. & Rex, G. (1985). The formation and significance of Carboniferous coal balls. *Philosphical Transactions of the Royal Society of London, B*, **311**, 123–37.

Scott, A.C. & Rex, G. (1987). The accumulation and preservation of Dinantian plants from Scotland and its borders. *European Dinantian Environments*. (Ed. by J. Miller, A.E. Adams & V.P. Wright), 329–44. Geological Journal Special Volume No. 11. Wiley, New York.

Scott, A.C. & Taylor, T.N. (1983). Plant–animal interactions during the Upper Carboniferous. *Botanical Review*, **49**, 259–307.

Smiley, C.J. & Rember, W.C. (1985). Composition of the Miocene Clarkia Flora. *Late Cenozoic History of the Pacific Northwest*. (Ed. by C.J. Smiley), pp. 95–112. Pacific Division of the American Association for the Advancement of Science, San Francisco, California.

Solbrig, O.T. (1976). The origin and floristic affinities of the South American temperate desert and semidesert regions. *Evolution of Desert Biota* (Ed. by D.S. Goodall), pp. 7–51. University of Texas Press, Austin & London.

Spicer, R.A. (1980). The importance of depositional sorting to the biostratigraphy of plant megafossils. *Biostratigraphy of Fossil Plants.* (Ed. by D.L. Dilcher & T.N. Taylor), pp. 171–83. Dowden, Hutchinson & Ross, Stroudsburg, Pennsylvania.

Spicer, R.A. (1981). The sorting and deposition of allochthonous plant material in a modern environment at Silwood Lake, Silwood Park, Berkshire, England. *United States Geological Survey Professional Paper, No. 1143,* 77 pp.

Spicer, R.A., Burnham, R.J., Grant, P. & Glicken, H. (1985). *Pityogramma calomelanos,* the primary, post-eruption colonizer of Volcan chichonal, chiapas, Mexico. *American Fern Journal* **75,** 1–5.

Spicer, R.A. & Hill, C.R. (1979). Principal components and correspondence analysis of quantitative data from a Jurassic plant bed. *Review of Palaeobotany and Palynology,* **28,** 273–99.

Stebbins, G.L. (1974). *Flowering Plants, Evolution above the Species Level.* Belknap Press of the Harvard University Press, Cambridge, Massachusetts.

Stebbins, G.L. (1976). Seeds, seedlings and the origin of angiosperms. *Origin and Early Evolution of the Angiosperms* (Ed. by C.B. Beck), pp. 300–11. Columbia University Press, New York.

Stebbins, G.L. (1981). Why are there so many species of angiosperms? *Bioscience,* **31,** 573–6.

Stewart, W.N. (1983). *Paleobotany and the Evolution of Plants.* Cambridge University Press, Cambridge.

Stubblefield, S.P., Taylor, T.N. & Beck, C.B. (1985). Studies of Paleozoic fungi. V. Wood-decaying fungi in Callixylon newberryi from the Upper Devonian. *American Journal of Botany,* **72,** 1765–74.

Takhtajan, A. (1969). *Flowering Plants, Origin and Dispersal.* Oliver & Boyd, Edinburgh.

Teichmüller, M. (1958). Rekonstruktion verschiedener Moortypen des Hauptflözes der niederrheinischen Braunkohle. *Fortschritte in der Geologie von Rheinland und Westfalen,* **2,** 599–612.

Thomas, B.A. (1985). Pteridophyte success and past biota—a palaeobotanist's approach. *Proceedings of the Royal Society of Edinburgh B,* **86,** 423–30.

Thomasson, J. (1980). Paleoagrostology: A historical review. *Iowa State Journal of Research,* **54,** 301–17.

Thomasson, J. (1983). *Carex graceii* sp.n., *Cyperocarpus eliasii* sp.n., *Cyperocarpus pulcherrima* sp.n. (cyperaceae) from the Miocene of Nebraska. *American Journal of Botany,* **70,** 435–49.

Thomson, K.S. (1980). The ecology of Devonian lobe-finned fishes. *The Terrestrial Environment and the Origin of Land Vertebrates* (Ed. by A.L. Panchen), pp. 187–222. Academic Press, New York.

Tiffney, B.H. (1985). Seed size, dispersal syndromes, and the rise of the angiosperms: evidence and hypothesis. *Annals of the Missouri Botanical Garden,* **71,** 551–76.

Tschudy, R.H., Pillmore, C.L., Orth, C.J., Gilmore, J.S. & Knight, J.D. (1984). Disruption of the terrestrial plant ecosystem at the Cretaceous–Tertiary boundary, Western Interior. *Science,* **225,** 1030–2.

Upchurch, G.R. & Doyle, J.A. (1981). Paleoecology of the conifers *Frenelopsis* and *Pseudofrenelopsis* (Cheirolepidiaceae) from the Cretaceous Potomac Group of Maryland and Virginia. *Geobotany II* (Ed. by R.C. Romans), pp. 167–202. Plenum, New York.

Van der Burgh, J. (1983). Allochthonous seed and fruit floras from the Pliocene of the lower Rhine Basin. *Review of Palaeobotany and Palynology,* **40,** 33–90.

Van der Burgh, J. (1987). Some local floras from the Neogene of the lower Rhenish Basin. *Tertiary Research Special papers,* (in press).

Wade, D., Ewel, J. & Hofstetter, R. (1980). *Fire in South Florida Ecosystems.* U.S. Department of Agriculture Forest Service General Technical Report SE-17, Southeastern Forest Experiment Station, Asheville, North Carolina.

Walker, J.W. & Walker, A.G. (1985). Ultrastructure of Lower Cretaceous angiosperm pollen and the origin and early evolution of flowering plants. *Annals of the Missouri Botanical Garden*, **71**, 464–521.

Wing, S.L. (1980). Fossil floras and plant-bearing beds of the central Bighorn Basin *University of Michigan Papers in Paleontology*, **24**, 119–25.

Wing, S.L. (1984). Relation of paleovegetation to geometry and cyclicity of some fluvial carbonaceous deposits. *Journal of Sedimentary Petrology*, **54**, 52–66.

Wnuk, C. & Pfefferkorn, H.W. (1984). The life habits and paleoecology of middle Pennsylvanian medullosan pteridosperms based on *in situ* assemblages from the Bernice Basin (Sullivan County, Pennsylvania, U.S.A.). *Review of Palaeobotany and Palynology*, **41**, 329–51.

Wolbach, W.S., Lewis, R.S. & Anders, E. (1985). Cretaceous extinctions: evidence for wildfires and search for meteoritic material. *Science*, **230**, 167–70.

Wolfe, J.A. (1975). Some aspects of plant geography of the northern hemisphere during the late Cretaceous and Tertiary. *Annals of the Missouri Botanical garden*, **62**, 264–79.

Wolfe, J.A. (1980). Tertiary climates and floristic relationships at high latitudes in the northern hemisphere. *Palaeogeography, Palaeoclimatology, Palaeoecology*, **30**, 313–23.

Wolfe, J.A. (1985). Distribution of major vegetational types during the Tertiary. *The Carbon Cycle and Atmospheric CO_2; Natural Variations Archaen to Present*, pp. 357–75. Geophysical Monograph 32, American Geophysical Union.

19. THE RESPONSE OF LATE QUATERNARY INSECT COMMUNITIES TO SUDDEN CLIMATIC CHANGES

G. R. COOPE

Department of Geological Sciences, University of Birmingham, P.O. Box 363, Birmingham B15 2TT, UK

INTRODUCTION

The special contribution that palaeontology makes to biological science is a unique appreciation of the history of life set in a context of geologic time. Classically, palaeontologists deal with fossils that are often many millions of years old which may represent the ancestral stages of living organisms or else totally extinct forms with no present-day close relative. Both of these groups of fossils present problems to the palaeoecologist since, for the most part, the environmental requirements of a fossil organism must be inferred indirectly from what is known of the ecology of its nearest modern relatives and its associates. The rocks in which the fossils are preserved also supply environmental information, particularly when the fossils have not undergone much post-mortem transportation. As a general rule, the more ancient the fossils, the more precarious become our environmental inferences. Conversely, palaeontologists who are concerned with the more recent of geological periods can be more confident of the ecological implications of their fossils in both the significance of the species themselves and in their associations. The fossils from these geologically recent periods are most relevant to our understanding of present-day communities.

My object here is to discuss some aspects of the palaeontology of the Quaternary period that spans the last 2 million years and which, by definition, thus extends up to the present day. Fossils from this period include the immediate precursors of modern species and their communities and they provide insights into the rates of evolution, biogeographic changes and species associations that led up to the communities of the present day.

There is, however, an additional factor that complicates the recent history of our fauna and flora. Throughout almost the whole of the Quaternary there have been numerous large-scale climatic fluctuations causing a succession of glacial/interglacial cycles. In western Europe these climatic oscillations have been particularly severe and their impact on the biota has been drastic. During episodes of extreme cold, large areas to the

421

south of the ice sheets were reduced to polar desert, but during the shorter temperate interludes the climate of England was warm enough occasionally to support warm, temperate vegetation, and even the hippopotamus thrived in central England. The recentness, frequency and severity of these climatic changes may well have left their mark on the modern ecosystems.

I am concerned here with fossil insects from the last few tens of thousands of years, a period that included at least one major glacial event when ice sheets extended as far south as the English Midlands, and with the faunal and floral recovery that has taken place during the present interglacial.

QUATERNARY ENTOMOLOGY

Amongst the most common Quaternary fossil invertebrates that can be recovered from terrestrial and freshwater deposits, are the well-preserved exoskeletons of insects, particularly coleoptera (beetles). They may be extracted from almost any waterlogged sediment, together with remains of a wide variety of other arthropods, molluscs, small vertebrates and all manner of plant debris.

It is especially fortunate that entomologists dealing with present day insects use exoskeletal characters almost exclusively in their taxonomy. Thus neontologist and palaeontologist can use the same criteria to recognize their species and there is no problem of differing species concepts between the two sister disciplines; we can use the same nomenclature in both. It is because the remains of coleoptera are so beautifully preserved, and therefore most easily compared with modern comparative material, that I am going to restrict my attention here to Quaternary fossil beetles. It must be emphasized, however, that there is nothing particularly unusual about them from an ecological or evolutionary point of view; they merely have robust enough skeletons to have left a fine fossil record. They should thus be looked upon as a model for the understanding of the recent history of other groups of terrestrial organisms that have a less adequate fossil record.

Because of the structural complexity and superb preservation of the fossil coleoptera, even down to the intimacies of their male genitalia, identification to the species level is often possible. However, the fact that the fossils are usually found as disarticulated skeletal elements presents special problems in their identification. The Quaternary entomologist may have little opportunity to follow through the keys to identification devised for use with modern material, because the fossils may well lack the crucial characters. Fortunately, numerous alternative diagnostic features may be utilized when the fossils are compared directly with well-identified modern specimens.   One of the most interesting results of the resurgence of interest in Quaternary entomology has been the discovery that there is virtually no evidence of any morphological change in insect species during this period. In the last three decades, during which we have investigated some hundreds of insect-bearing sites involving the recognition of almost 2000 species, we have rarely found distinctive pieces that suggest that we may be dealing with an extinct species or with specimens that suggest that there has been any significant evolution during the last few hundred thousand years. This was a rather unexpected outcome because earlier workers in this field had often credited the fossils with new names which implied that they were either the precursors of living species, but not conspecific with them, or else totally extinct forms (see Buckland & Coope 1985 for a review of the older literature).

The only objective evidence of evolution based on Quaternary fossil data is the elegant study by Matthews (1977) in which he showed that there had been a reduction in the already vestigial flying wings and slight shortening of the elytra, recognized by statistical analysis, of the beetle *Tachinus apterus* since the Lower Pleistocene in Alaska: namely in over a million generations. This may only be evolution to the status of subspecies. Evolution to the species level can be detected amongst Miocene fossil Coleoptera from the north-west territories of Canada. It is not easy to give a precise age to these fossils but they are of the order of 10 million years old (each year may be read as one generation). Even after the lapse of so much time, the resemblance to modern species is remarkably close (Matthews 1976). There can be little doubt that some of these species are the actual precursors of those of the present day, whose evolution has been through 'fine tuning' rather than abrupt change. Associated with these species, there are others which, though distinctive, would seem to have left no descendants and are probably extinct. However, extinction in small animals is notoriously difficult to establish with certainty. Recently a student of mine discovered one of Matthews' 'extinct' species, a small staphylinid beetle called *Micropeplus hoogendorni*, in Quaternary deposits in the English Midlands that are at least 6 million years more recent than the original locality of its discovery (L.S. Holdridge, personal communication). These Miocene fossil assemblages of coleoptera are different from any living communities, not only in their inclusion of these apparently extinct forms, but also because the descendent species do not occur together anywhere in the world at the present time; they have, today, widely separate geographical ranges. They would thus appear to have changed physiologically as well as undergoing morphological evolution in the last few million years. This is an important observation because it contrasts markedly with Quaternary fossil assemblages which are, of course, very much younger. Almost all of these assemblages have analogues in present-day communities, suggesting that the demonstrable morphological constancy of species in Quaternary time has been accompanied by physiological constancy. In most cases species kept the same company in the past as they do today because their environmental requirements have remained the same for tens or even hundreds of thousands of generations.

So far, studies of Quaternary entomology have been restricted to the north temperate latitudes where the large scale climatic oscillations of the glacial/interglacial cycles were most intense. The response of insect communities to these drastic climatic events has largely been a matter of rearranging their component species by extensive alterations in their geographical distributions. Since these climatic changes were remarkably sudden (Coope & Brophy 1972, Coope & Joachim 1980, Atkinson *et al.* 1986) they were accompanied by local, but not global, mass extinctions that were then followed by invasions and colonizations by suites of species suited to the new conditions. When the climate changed once again to its previous mode, communities that had been locally exterminated became re-established.

CLIMATIC ANALYSIS OF QUATERNARY FOSSIL BEETLE ASSEMBLAGES

A fossil assemblage of coleoptera, namely all specimens obtained from a single stratigraphical horizon at a particular locality, is the closest that we can get to a fossil community, i.e. all those species of beetle that lived together at a particular time and place. However, community structure can rarely be quantified on the basis of fossil assemblages since we hardly ever know the size of the sampling area from which the fossils were derived nor the precise time span of our sample. In Quaternary contexts the time sampled may be as little as a decade but sometimes it may be as much as a century or more. Although some of the specimens in our fossil assemblages have undergone a minor amount of post-mortem transportation, most come from the immediate neighbourhood of the sedimentary trap in which they were buried. Our samples are thus analogous to the light-trapped or vacuum accumulated assemblages of insects studied by present day entomologists. In all such assemblages it is difficult to distinguish between a species that is truly rare in the environment from one that is rare in the assemblage because it was living in abundance at some distance from the collecting site. (This problem is elaborated upon by Collinson & Scott from a palaeobotanical viewpoint in Chapter 18). Thus, although we can talk with confidence about species that are abundant in our assemblages (communities), we find rarity a much more intractable problem. This is illustrated by the difficulty of distinguishing between species that are generally rare but have sporadic outbursts of abundance from those that are persistently present in low numbers. Both species could give the same frequency in our fossil assemblages because each sample includes the accumulated individuals that lived over a number of years. This type of problem could, in theory, be resolved if samples were taken at finer and finer time intervals, but this procedure is difficult in practice unless we find deposits of very high accumulation rate —no such deposit has yet been investigated.

A problem that is peculiar to the interpretation of fossil assemblages stems from the fact that fossils represent the most decomposition resistant parts of organisms (e.g. pollen grains or insect cuticle) and thus may be eroded from an earlier layer and redeposited, leading to assemblages made up of species that lived at different times. Fortunately these mixed assemblages are usually recognizable by their internal ecological or biological inconsistencies. However, there is good reason to believe that Quaternary fossil insects are rapidly decomposed, probably by fungi, as soon as they are exposed at the surface, and are thus rarely redeposited. Thus we find that in the numerous sequences of insect-bearing deposits that we have investigated the fossil assemblages change very rapidly in response to climatic change, with no evidence of any smearing of the contrast that would be expected if there had been redeposition of fossils from earlier horizons. As a further illustration of this point: a magnificent sequence of deposits at Tattershall, Lincolnshire, yielded a rich insect assemblage from the Last Interglacial (120 000 years ago) on the eroded surface of which was laid down a layer with an exclusively cold-adapted suite of beetles. This in its turn was overlain by a temperate assemblage of species dating from about 43 000 years ago (Girling 1980). Although fossils such as hazel nuts were redeposited in this sequence there is no evidence of redeposition of the insect fossils in spite of local erosional events that should have provided ample opportunity for it to occur.

Quaternary fossil beetle assemblages may thus be looked upon as reliable samples of the species present in a particular area during a relatively short period of time. In a real sense they represent the communities of coleoptera and their successive changes provide clues to our understanding of the history of the communities that are the immediate predecessors of those of the present day.

In order to record the changes in insect communities in the geologic past, it is essential to be able to date the fossiliferous deposits accurately. Normal stratigraphic methods provide relative ages only and little information about absolute time. It is fortunate, therefore, that radiocarbon dating techniques can give us an absolute measure of time (on a scale of radiocarbon years which are not the exact equivalent of calender years) which spans the last 45 000 years, a period that includes a number of large-scale climatic fluctuations. A vast array of geological and palaeontological data is now available that establishes the detailed sequence of climatic events during this period (Lowe & Walker 1984). One of these events led to the expansion of continental ice sheets onto the lowlands of central Europe, over most of Canada and the northern United States reaching its maximum extent as recently as 18 000 years ago. As these massive glaciers waned there were several short but intense climatic oscillations, each of which had a drastic effect on the biological communities along the southern margins of the ice. There can be little doubt that there were comparable climatic changes in tropical latitudes, though it is likely that these were less intense than they were in the zones that experienced the full force of the glacial/interglacial cycles. However, the investigation of Quaternary palaeontology in equatorial regions is still in its infancy and almost nothing is known of their fossil insect assemblages at the present time, but equatorial peat beds are known and they may well repay examination for their fossil insects.

Because our knowledge of the Quaternary history of insect communities is most complete for the British Isles and because of the limitations imposed by radiocarbon dating techniques, the discussion which follows will be restricted to changes in fossil beetle assemblages from Britain dating from the last 45 000 years. No attempt is made to list the species upon which the conclusions are based, nor is the stratigraphical context of the sites referred to, but adequate references are provided to give access to this detailed information.

BIOGEOGRAPHICAL COMPOSITION OF COLEOPTERAN ASSEMBLAGES

For the most part, the major changes in fossil beetle assemblages from Britain during the last 45 000 years take the form of changes in species composition that appear to have been induced by marked fluctuations in the regional climate. Thus, arctic assemblages often with exclusively Siberian species, are rapidly replaced by temperate ones that may include species now found only in south-central Europe. Just as rapidly the assemblages may change back again. The most convenient way to summarize these faunal changes is to express them as 'biogeographical spectra', two of which are illustrated here in Figs. 19.1 and 19.2.

Biogeographical spectra are constructed in the following manner. All

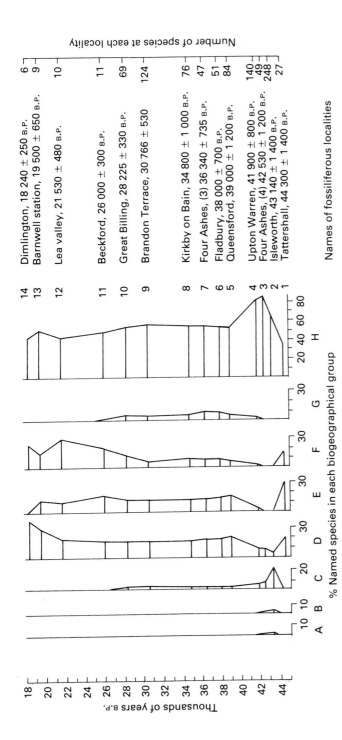

Fig. 19.1. Composite biogeographical spectrum made up from fourteen coleopteran assemblages from sites in England, dating from the latter part of the Devensian (Last) glaciation. For definitions of the biogeographical groups, see p. 429. The relative numbers of species in each group is shown as a bar whose length is proportional to its percentage of the sum of species in all groups. (Sources of data: Tattershall, Girling 1980; Isleworth, Coope & Angus 1975; Four Ashes, Morgan 1973; Upton Warren, Coope, Shotton & Strachan 1961; Queensford, Briggs, Coope & Gilbertson 1975; Fladbury, Coope 1962; Kirby on Bain, Girling 1980; Brandon Terrace, Coope 1968b; Great Billing, Morgan 1969; Beckford, Briggs, Coope & Gilbertson 1975; Lea Valley, Coope & Tallon 1983; Barnwell Station, Coope 1968a; Dimlington, Penny, Coope & Catt 1969).

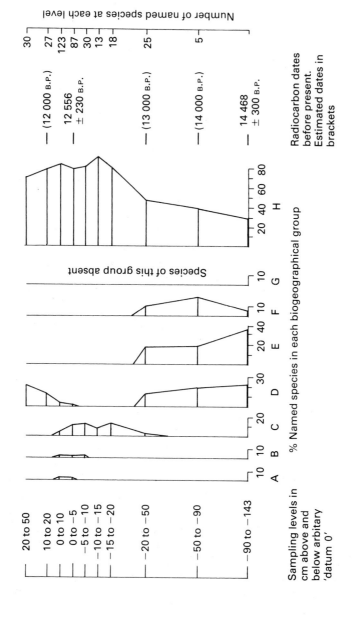

Fig. 19.2 Biogeographical spectrum of a sequence of stratigraphically superposed coleopteran assemblages from the Lateglacial site at Glanllynnau, nr. Criccieth, North Wales. For definitions of biogeographical groups, see p. 429. (Source of data: Coope & Brophy 1972).

recognized species from a particular assemblage can be grouped into one of the following eight exclusive categories whose limits are based entirely on the present-day geographical ranges of the species, i.e. without any reference to the fossil occurrences of the species:

A   southern European species;
B   southern species whose normal ranges just fail to reach Britain;
C   southern species whose normal ranges are south of central Britain;
D   widespread species whose normal ranges are north of central Britain;
E   boreal and montane species whose normal ranges extend down into the upper part of the coniferous forest belt;
F   boreal and montane species whose normal ranges are above the tree line;
G   eastern Asiatic species, some of which also range into North America;
H   cosmopolitan species with wide geographical ranges.

The number of species in each group is then expressed as a percentage of the sum of all named species in the assemblage, and each percentage is drawn up as a bar of proportional length. It should be noted that only the presence of a species is recorded and no account has been taken of their relative abundance. This is because the abundance of a species is often a reflection of the local environmental conditions, and it may be further complicated by the varying distances between the sedimentary traps and the habitats in which it actually lived. Inclusion of relative abundance could have partially masked any regional picture of faunal change with its broad climatic implications.

A view of the historical sequence of faunal changes may then be obtained by arranging the biogeographical percentage bars for each assemblage in a temporal succession with the oldest at the base of the diagram. Time can be either expressed as radiocarbon years or as depth in the sedimentary sequence.

It is convenient here to divide the last 45 000 years into two parts; an early period between 45 000 and 18 000 years ago represented on Fig. 19.1 and a later period between 14 500 and 11 500 years ago represented on Fig. 19.2. The reason for the time gap between these two periods is because, shortly after the ice sheets had begun to melt, there was an episode of severe climate when polar deserts covered much of north-west Europe and the flora and fauna was so meagre that it left almost no palaeontological data for several thousands of years. From about 15 000 years ago to the present day, there is a well-documented record of the insect faunas of the British Isles, Fig. 19.2 represents the faunal changes as the ice-age conditions were in final contest with those of the temperate post-glacial. Curiously enough, we are not yet in a position to describe in detail the faunal history of the post-glacial period (the present interglacial) that started about 10 000 years ago, because the pattern of development of the insect fauna has been complicated by human interference.

Figure 19.1 is a composite biogeographical spectrum constructed from beetle assemblages from fourteen sites in England, all of which have been dated by radiocarbon at between 45000 and 18000 years ago. If we ignore the widespread species of group H, treating them as background noise, a pattern of faunal change becomes immediately apparent in the varying fortunes of the other, more restricted biogeographical groups. This pattern may be summarized as follows. An early phase (Tattershall) was dominated by high boreal or boreo-montane species groups (D,E,F) but the diversity of species at this time was very low, testifying to the severity of the climate. This northern element was suddenly replaced by a rich temperate assemblage of coleoptera (groups A,B,C) with considerable diversity that included 248 named species of beetle and a wealth of unidentified fragments of other orders of insect. This temperate interlude was apparently very short, maybe lasting for as little as a thousand years, and it reached its thermal maximum about 43000 years ago. It was followed by a gradual return to colder conditions with the reappearance of the northern species and a rapid elimination of the most southern element from the fauna. At this time the boreal species were accompanied by a number of Asiatic species (group G) many of which occurred in considerable numbers in Britain. For instance, the commonest large dung beetle in this country for much of the time between 40000 and 25000 years ago, was *Aphodius holdereri*, a species that is now confined to the high plateau of Tibet and adjacent north-western China (Coope 1973). These cold continental assemblages continued unchanged in Britain for the next 15000 years, after which the faunas were once again dominated by boreal and boreo-montane species with the loss of the Asiatic element. As the period of maximum expansion of the ice sheets approached, there was a progressive reduction in the numbers of species in each fossil assemblage, implying a real impoverishment of the insect faunas in the ice-free areas of southern Britain leading to the phase of almost complete polar desert in which the beetle fauna may have been eliminated totally from these islands.

Changes in the beetle fauna of Britain during the terminal stages of the last glaciation have been investigated from a number of sites at which long vertical sequences have been available. The biogeographical spectrum from one of these localities is illustrated in Fig. 19.2; namely an infilled pond left by the retreating ice at Glanllynnau, near Criccieth, North Wales (Coope & Brophy 1972). Here there was sufficient thickness of sediment to provide a sequence of fossil assemblages which could be stacked to show the changes in insect assemblages in the area. In Fig. 19.2 the vertical axis has been drawn proportional to sediment thickness, though, of course, there would be no problem in converting this to a time dimension if enough radiocarbon dates were available. On the right hand side of Fig. 19.2 points fixed by radiocarbon dating are shown and interpolated dates have been introduced assuming a constant sedimentary rate.

The biogeographical spectrum from Glanllynnau again shows sudden changes in the beetle assemblages. There is an early phase dominated by boreal and boreo-montane species. Then at around 13 000 years ago, all the northern species were suddenly exterminated, to be replaced rather gradually by a diverse suite of temperate ones. The rise of the cosmopolitan species (group H) at the same time is also an indicator of increased faunal diversity. In its turn, this temperate assemblage disappeared with equal suddeness shortly before 12 000 years ago. This episode of local extermination involved a wide spectrum of ecological types: terrestrial and aquatic species, carnivores and phytophages (Coope & Brophy 1972). By far the most likely explanation for these sudden faunal changes is that they were in response to abrupt and intense climatic changes, particularly changes in the thermal environment.

It is unfortunate that the Glanllynnau sequence does not extend above 11 000 years ago, but there are numerous other insect assemblages known that continue the picture of climatic deterioration at the top of Fig. 19.2 until it reaches its thermal minimum between 11 000 and 10 000 years ago (Osborne 1972, Coope *et al.* 1979, Coope & Joachim 1980). At this time the beetle fauna of Britain was dominated by a wide variety of Arctic species.

The post-glacial climatic warming must have been very sudden since, by 9 500 years ago, all the high boreal species had completely disappeared from lowland Britain and temperate species of beetle had already reached south-west Scotland (Bishop & Coope 1977). From the presence of relatively southern species in Scotland at this time, the climate must have been as warm as, or even warmer than that of the present day.

QUANTIFICATION OF CLIMATIC CHANGE INFERRED FROM FOSSIL COLEOPTERAN ASSEMBLAGES

Numerous attempts have been made to quantify the climatic changes outlined above (Morgan 1973, Coope 1977). Most successful of these has been the recent application of a computer program which compares the present day geographical ranges of species with the disposition of various climatic variables. This method is reviewed by Atkinson *et al.* 1986. Fig. 19.3 shows the mean temperature values obtained by using this program on a

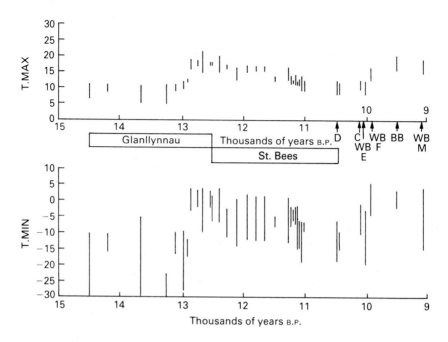

FIG. 19.3 Climatic reconstruction for central Britain during the period 15 000 to 9000 BP (i.e. years before present), estimated by the mutual climatic range method (Atkinson *et al.* 1986). T.MAX refers to the mean temperature of the warmest month of the year (usually July) and T.MIN refers to the mean temperature of the coldest month of the year (usually January). The actual temperature lay within the limits of the bars, though not necessarily at their mid-points. (Sources of data: Glanllynnau, Coope & Brophy 1972: St. Bees, Coope & Joachim 1980; D = Drumurcher, Coope *et al.* 1979; C = Croydon, Peake & Osborne 1971; WB = West Bromwich (E, F & M refer to assemblages from different stratigraphical horizons at the same site), Osborne 1980; BB = Brighouse Bay, Bishop & Coope 1977).

sequence of beetle assemblages from the terminal stages of the last glaciation and the post-glacial period. The fossil assemblages used in this reconstruction are chiefly from Glanllynnau (Coope & Brophy 1972) and from St Bees, Cumbria, (Coope & Joachim 1980) with additional assemblages to complete the coverage to the post-glacial conditions.

It is clear from Fig. 19.3 that the climatic amelioration at 13 000 years ago involved a rise of about 7°C in the mean July temperature and at the same time a rise of about 20°C in mean January temperatures; all within an immeasurably short period of time, probably less than a century. Although the thermal maximum was soon over, with a deterioration evident at 12 200 years ago, it was not till about 11 500 years ago that there was any marked cooling. Lowest temperatures were reached during the thousand years between 11 000 and 10 000 years ago when fully glacial conditions were reimposed for a short time. The climatic recovery at about 10 000 years ago was apparently as sudden as that at 13 000 years ago. Almost immediately the present interglacial became as warm as, or even warmer than it is today in the British Isles.

DISCUSSION

One of the most significant features of the climatic changes during the Late Quaternary is their abruptness and intensity, inferred from the sharp changes in the insect assemblages outlined above. Episodes of wholesale extinction must have been almost synchronous with the events that caused them; namely the sudden rise or fall of the regional temperature. In other words, there was no lag in this aspect of the faunal response. This contrasts with the gradual appearance of colonizing species exploiting the new conditions, whose arrival at a particular location will depend, not only on the suitability of the environment, but also on the routes and rates of spread of the various species from their distant refuges. Thus, colonizing species often show a lag in their response to the climatic stimulus. Some idea of this difference in response may be gained from Fig. 19.2 in which the sudden extinction events stand out sharply, whilst the gradual colonization afterwards takes much longer. This protracted recovery may take many centuries. Herein lies the nub of an interesting palaeoecological problem. After these sudden climatic changes, communities of animals and plants sometimes arise that have no equivalent in species composition to any at the present day. They owe their unusualness to their historical contexts rather than to the peculiarities of the environments in which they lived.

One of these anomalous assemblages dates from about 43 000 years ago (Morgan 1973, Girling 1974, Coope & Angus 1975). At that time an open, treeless grassland covered most of north-west Europe suggestive of the present day tundra (Pennington 1969). But the insect assemblages were entirely temperate and included many species that do not reach as far north as the British Isles at the present day. The nearest equivalent modern habitat would seem to be that of the open ground stripped bare of its topsoil by human activity such as in gravel pits or large-scale civil engineering sites, where many of the insects and plants that characterized this period may be found today. Similar habitats must have dominated almost all of north-west Europe at that time. There is palynological evidence that trees did extend into eastern Europe (West 1977) but the climate seems to have deteriorated severely before the trees could reach Britain. Thus, the flora and fauna of this country does not seem to have had time to progress beyond the pioneer stage of development.

A second group of anomalous assemblages come from the period immediately after the sudden warming at 13 000 years ago. At this time there is evidence that tree birches reached Britain (Pennington 1977) but they were patchy in their occurrence and did not establish true woodlands for almost a thousand years after the climatic amelioration. Lags due to invasion problems are thus inadequate to explain the colonization difficulties, and it is likely that raw soils presented nutritional barriers to the spread of the trees (van Geel, Lange & Wiegers 1984) and that lack of soil humus causing poor moisture retention probably inhibited seedling maturation (Coope & Joachim 1980). The incoming beetles were less influenced by these difficulties. Many of the first arrivals were active flying carnivores (mostly Carabidae) whose food webs can be traced back, via the Collembola, to the algae and lichens that must have been amongst the earliest colonizers of the newly available landscape.

A third group of anomalous communities are found after the sudden warming that ushered in the post-glacial period (i.e. 10 000 years ago). Here again a diverse suite of temperate insects moved into Britain at a time when the only trees on the landscape were birches. It is interesting to note the presence in these early post-glacial woodlands of the carabid beetle *Calosoma inquisitor* which today climbs oak trees to feed on defoliating caterpillars (Lindroth 1985). Apparently in the absence of oak it could make do with birch trees (Osborne 1974).

The gradual build up of the post-glacial forest in Britain was thus not a reflection of gradual climatic warming, but rather a succession of arrivals whose order of precedence was determined by their rates of spread and the varying distances that they had come. After the amelioration of the climate at the beginning of the post-glacial period, the communities of animals and plants took about 3000 years to approach equilibrium with the physical environment.

Given the frequency and intensity of Quaternary climatic changes, these periods when the biota was out of harmony with the physical environment must have been similarly frequent. They show that, in our understanding of fossil communities, and perhaps of living ones also, it is essential to know their historical contexts as well as the environmental conditions in which they actually lived.

This investigation into the response of communities to Quaternary climatic changes shows that communities did not react *en bloc* to these changes, but individual species altered their geographical ranges as each found conditions intolerable or new circumstances and new territories became available (Coope 1979). Of course those groups of species that were ecologically tied to one another by mutual dependence must have moved at the speed of the slowest member. It is interesting to speculate that the immense diversity of phytophagous insects to which oak trees are host today (see Southwood, Chapter 1), may be partially due to the relative slowness with which these trees were able to move in response to climatic warming. Their rate of progress was slow enough for even the most laggard members of the entourage to keep pace.

For the most part, however, the numerous and intense Quaternary climatic changes caused the communities of plants and animals to be repeatedly dismantled and reassembled bit by bit elsewhere. The community might be compared with an itinerant symphony orchestra whose members can either perform as individuals, as interactive small groups or as a full harmonious company. In temperate latitudes, communities must have been structured in such a manner that their component species could respond as individuals, or small groups of individuals, to the Quaternary climatic oscillations. There is thus a sense in which living communities must be adapted to conditions of the recent geologic past as they are to those of the present day. The geological environment requires that insect species be capable of undergoing the frequent forced marches imposed by drastic climatic changes, which may involve them in shifts of hundreds or even thousands of kilometres. Under such conditions, intricate co-evolutionary partnerships would be hard to sustain. Furthermore, too rigid an attachment to a particular latitude, for example by precise dependence on photoperiod, would make a species a likely candidate for extinction when the ice sheets swept down from the north. This would be important when considering the role of seasonality in structuring communities (see Wolda, Chapter 4).

Although we have very little evidence from tropical areas, it might be surmised that the existence there of so many intricate co-evolutionary relationships is indicative that the more subdued climatic events there had less of a disruptive impact on the plant and animal communities than in the temperate latitudes.

It is intriguing that this climatically stressful environment apparently did not lead to frequent speciations and extinctions (other than amongst the terrestrial mammals, which seem to be exceptional in this respect—see Webb, Chapter 20). I have discussed elsewhere (Coope 1978) the possibility that it may have been the instability of the Quaternary climate that was the cause of this evolutionary stasis. By repeatedly compelling species to move from place to place, the gene pools were kept well stirred. Under such ebullient conditions the establishment of geographic isolation for long enough to produce genetic isolation may have been almost unattainable, and evolutionary change brought to a complete standstill. Species that could not track the changing positions of suitable environments no doubt must have become extinct when major climatic oscillations began in the late Tertiary. John Matthews' (1976) evidence of extinct Miocene species bears out this suggestion.

Island populations are, however, in a very different position. There, the option of tracking acceptable environments across the globe is not available and all changes must be endured on the spot. When confronted by climatic change, evolution may then have been the sole alternative to extinction. Island communities thus reflect a different historical context compared with their continental counterparts; rapid speciation and frequent extinctions might be expected under these special circumstances to be the result of Quaternary climatic fluctuations.

CONCLUSIONS

Studies of fossil insects from geologically recent deposits show that the species that are living today have *not* undergone any evolutionary change for hundreds of thousands of generations. Furthermore, the demonstrable morphological constancy seems to have been associated with a similar degree of physiological stability. In many cases fossil assemblages can be looked on as representatives of past communities in that they show what species were present in an area at a particular time. However, numbers of individuals in a fossil assemblage are difficult to translate into terms of real abundance that can be compared with similar figures for present-day communities.

In the last few hundred thousand years the major influence on community structure in the temperate latitudes has been the large-scale climatic changes of glacial/interglacial status. Each of these has had a drastic effect on the insect communities of the times by causing widespread local extinctions and great changes in the geographical ranges of species. Thus individual species adjust to changing conditions by tracking the appropriate environment geographically rather than by evolutionary change. Viewed on a geological timescale, species must be adapted to deal with these major climatic events just as the same species must be equipped to meet the lesser variations in historic (ecological) time. This long-term view of adaptation must likewise colour our understanding of community responses to the same large-scale climatic changes. It is clear that an insect community did not react *en bloc* as an integrated whole, but each of its component species responded by moving at its own rate as the climate underwent rapid and intense changes. At times this led to the development of novel species associations in the scramble for newly available space and thousands of years sometimes elapsed before equilibrium with the physical environment was approached.

REFERENCES

Atkinson, T.C., Briffa, K.R., Coope, G.R., Joachim, M.J. & Perry, D.W. (1986). Climatic calibration of coleopteran data. *Handbook of Holocene Palaeoecology and Palaeohydrology,* (Ed. by B.E. Berglund) 851–8. Wiley, Chichester.

Bishop, W.W. & Coope, G.R. (1977). Stratigraphical and faunal evidence for lateglacial and early Flandrian environments in south-west Scotland. *Studies in the Scottish Lateglacial Environment,* (Ed. by J.M. Gray & J.J. Lowe) pp. 61–88. Pergamon, Oxford.

Briggs, D.J., Coope, G.R. & Gilbertson, D.D. (1975). Late Pleistocene terrace deposits at Beckford, Worcestershire, England. *Geological Journal,* **10,** 1–16.

Briggs, D.J., Coope, G.R. & Gilbertson, D.D. (1985). *The Chronology and Environmental Framework of Early Man in the Upper Thames Valley.* British Archaeological Reports, 137, Oxford.

Buckland, P.C. & Coope, G.R. (1985). *A Bibliography and Literature Review of Quaternary Entomology.* Department of Geography, University of Birmingham, Working Paper Series No. 20, 1–90.

Coope, G.R. (1962). A Pleistocene coleopterous fauna with arctic affinities from Fladbury, Worcestershire. *Quarterly Journal of the Geological Society of London,* **118,** 103–23.

Coope, G.R. (1968a). Coleoptera from the 'Arctic Bed' at Barnwell Station, Cambridge, *Geological Magazine,* **105,** 482–6.

Coope, G.R. (1968b). An insect fauna from Mid-Weichselian deposits at Brandon, Warwickshire. *Philosophical Transactions of the Royal Society of London, B,* **254,** 425–56.

Coope, G.R. (1973). Tibetan species of dung beetle from Late Pleistocene deposits in England. *Nature (London)* **245,** 335–6.

Coope, G.R. (1977). Fossil coleopteran assemblages as sensitive indicators of climatic changes during the Devensian (Last) cold stage. *Philosophical Transactions of the Royal Society of London, B,* **280,** 313–37.

Coope, G.R. (1978). Constancy of insect species versus inconstancy of Quaternary environments. *Diversity of Insect Faunas,* (Ed. by L.A. Mound & N. Waloff) 176–87. (Symposia of the Royal Entomological Society of London, 9) Blackwell Scientific Publications, Oxford.

Coope, G.R. (1979). Late Cenozoic Fossil Coleoptera: Evolution, Biogeography and Ecology. *Annual Review of Ecology and Systematics,* **10,** 247–67.

Coope, G.R. & Angus, R.B. (1975). An ecological study of a temperate interlude in the middle of the last glaciation, based on fossil Coleoptera from Isleworth, Middlesex, *Journal of Animal Ecology,* **44,** 365–91.

Coope, G.R. & Brophy, J.A. (1972). Late Glacial environmental changes indicated by a coleopteran succession from North Wales. *Boreas,* **1,** 97–142.

Coope, G.R., Dickson, J.H., McCutcheon, J.A. & Mitchell, G.F. (1979). The lateglacial and early postglacial deposit at Drumurcher, Co. Monaghan. *Proceedings of the Royal Irish Academy, B,* **79,** 63–85.

Coope, G.R. & Joachim, M.J. (1980). Lateglacial environmental changes interpreted from fossil Coleoptera from St. Bees, Cumbria, NW England. *Studies in the Lateglacial of North West Europe,* (Ed. by J.J. Lowe, J.M. Gray & J.E. Robinson) pp. 55–68. Pergamon, Oxford.

Coope, G.R., Shotton, F.W. & Strachan, I. (1961). A Late Pleistocene flora and fauna from Upton Warren, Worcestershire. *Philosophical Transactions of the Royal Society of London, B,* **244,** 379–421.

Coope, G.R. & Tallon, P. (1983). A Full Glacial insect fauna from the Lea Valley, Enfield, N. London. *Quaternary Newsletter* **40,** 7–10.

Geel, B. van, Lange, L. de & Wiegers, J. (1984). Reconstruction and interpretation of the local vegetational succession of a Lateglacial deposit from Usselo (the Netherlands), based on the analysis of micro and macro fossils. *Acta Botanica Neerlandica,* **33,** 535–46.

Girling, M.A. (1974). Evidence from Lincolnshire of the age and intensity of the mid-Devensian temperate episode. *Nature (London)*, **250**, 270.

Girling, M.A. (1980). *Two Late Pleistocene insect faunas from Lincolnshire.* Unpubl. PhD Thesis, University of Birmingham.

Lindroth, C.H. (1985). The Carabidae (Coleoptera) of Fennoscandia and Denmark. *Fauna Entomologica Scandinavica Vol. 15, part 1.* 1–225, E.H. Brill, Scandinavian Science Press Ltd., Leiden & Copenhagen.

Lowe, J.J. & Walker, M.J.C. (1984). *Reconstructing Quaternary Environments.* 1–399. Longman, London.

Matthews, J.V. Jr (1976). Insect fossils from the Beaufort Formation: geological and biological significance. *Geological Survey of Canada, Papers,* **76-1B**, 217–27.

Matthews, J.V. Jr. (1977). Coleopteran fossils: their potential value for dating and correlation of late Cenozoic sediments. *Canadian Journal of Earth Sciences,* **14**, 2339–47.

Morgan, A. (1969). A Pleistocene fauna and flora from Great Billing, Northamptonshire, England. *Opuscula Entomologica,* **34**, 109–12.

Morgan, A. (1973). Late Pleistocene environmental changes indicated by fossil insect faunas of the English Midlands. *Boreas* **2**, 173–212.

Osborne, P.J. (1972). Insect faunas of Late Devensian and Flandrian age from Church Stretton, Shropshire. *Philosophical Transactions of the Royal Society of London, B,* **263**, 327–67.

Osborne, P.J. (1974). An insect assemblage of Early Flandrian age from Lea Marston, Warwickshire and its bearing on the contemporary climate and ecology. *Quaternary Research,* **4**, 471–86.

Osborne P.J. (1980). The late Devensian/Flandrian transition depicted by serial insect faunas from West Bromwich, Staffordshire, England. *Boreas,* **9**, 139–47.

Peake, D.S. & Osborne, P.J. (1971). The Wandle gravels in the vicinity of Croydon. *Proceedings of the Croydon Natural History and Scientific Society,* **14**, 147–75.

Pennington, W. (1969). *The History of the British Vegetation,* 1–152. Modern Biology Series, The English Universities Press Ltd., London.

Pennington, W. (1977). The Late Devensian flora and vegetation of Britain. *Philosophical Transactions of the Royal Society, B,* **280**, 247–71.

Penny, L.F., Coope, G.R. & Catt, J.A. (1969). Age and insect fauna of the Dimlington Silts, East Yorkshire. *Nature, (London),* **224**, 65–7.

van Geel, B., Lange, L. de & Wiegers, J. (1984). Reconstruction and interpretation of the local vegetational succession of a Lateglacial deposit from Usselo (the Netherlands), based on the analysis of micro and macro fossils. *Acta Botanica Neerlandica,* **33**, 535–46.

West, R.G. (1977). Early and Middle Devensian flora and vegetation. *Philosophical Transactions of the Royal Society of London, B,* **280**, 229–46.

20. COMMUNITY PATTERNS IN EXTINCT TERRESTRIAL VERTEBRATES

S. DAVID WEBB

Florida State Museum, University of Florida, Gainesville, Florida 32611, USA

INTRODUCTION

The earth yields a rich, albeit fragmentary, record of terrestrial vertebrate history. For nearly two centuries palaeontologists have collected fossil vertebrates in a systematic manner, and other earth scientists have added diverse data regarding the ages and environments of the ancient tetrapods. Increasingly, the grand perspective of this 400 million year record compensates for its unevenness. Vertebrate palaeontology sheds light on the history of trophic structure on the continents, on changing patterns of diversity, on the balance between stability and faunal turnover, and on the convergence of communities in time and space.

Why should an ecologist study palaeoecology when the biosphere loses so much of its essence upon death? The answer is that a few generations of scientific study do not span an adequate interval in which to observe the full panoply of major ecological processes. Records of ancient life vastly extend the ecologist's time dimension. As Lewontin (1969, p. 13) observed '... an exact theory of the evolution of communities of organisms ... must explain in some sense the present state of the biosphere, but must also contain statements about the past history of living communities and about their future as well.' A dynamic view of the biosphere requires comparative study of all possible community configurations through time and space.

In this essay I summarize the fossil record of terrestrial vertebrates with special emphasis on faunal dynamics. This is one paleontologist's view special emphasis on faunal dynamics. This is one palaeontologist's view selected from a considerable body of detailed palaeontological and palaeo-ecological studies. It is condensed to about 20 million years of history per page. I dwell more on the Cenozoic mammal fauna of North America

METHODS

The recorded pattern of taxonomic richness gains credibility as the fossil record of tetrapods continues to accumulate and as geochronometric methods become more sophisticated. Taxonomic richness patterns consist of two components: the signal, reflecting evolutionary history; and the noise, introduced by the vagaries of the record. The general problem is to discriminate between the two components and then to eliminate the noise. The use of higher taxonomic categories increases the signal/noise ratio, but also reduces the sensitivity of the analysis to finer details of diversity changes in space and time. Previous compilations of taxonomic richness and stratigraphic ranges of fossil vertebrates have been prepared at the level of order, family and even genus by Simpson (1952), Romer (1966), in co-ordinated contributions by Thomson (1977), Carroll (1977), Bakker (1977) and Gingerich (1977), and these were reviewed and updated by Padian & Clemens (1985).

Figure 20.1 indicates the number of non-marine tetrapod families from successive geologic periods. The records for Amphibia follow Carroll (1977); the Reptilia are modified from Padian & Clemens (1985) by removing the marine groups and also the Synapsida, which I count as Mammalia; and the numbers for the rest of the Mammalia are derived from Savage & Russell (1983). The genus-level treatment of North American Cenozoic herbivores, summarized in Fig. 20.3, is taken from the faunal lists for successive land-mammal ages in Savage & Russell (1983).

Most of the available record of vertebrate palaeoecology consists of death-assemblages of bones and other hard parts in water-laid sediments. Major advances in methods of collecting and interpreting such accumulations have taken place in the last two decades often under the rubric of 'taphonomy'. (See also Collinson & Scott, Chapter 18). Actually taphonomy focuses on the biases inherent in fossil deposition; it serves as a means to a more important end which is palaeoecology. As Behrensmeyer & Hill (1980, p. 222) state'...the ultimate aims of work in taphonomy and paleoecology [are] the the reconstruction of whole paleocommunities and the study of their changes through time'. A number of taphonomic studies on specific kinds of assemblages have done much to indicate how to increase the signal/noise ratio in interpreting the terrestrial vertebrate record. Increasingly sophisticated studies of trace fossils (e.g. Bown & Kraus 1981) and fossil soils (e.g. Retallack 1983) offer keys to recognizing missing parts of ancient terrestrial communities. Other sedimentological and geochemical data also play important roles in reconstructing the environments of ancient biotas. Although it is sometimes convenient to speak of a 'vertebrate community', it is understood that a community, in its purest sense, consists of the whole biota that once lived in a given place. This is recognized in principle, even if, as in many fossil localities, the evidence of that biota consists only of vertebrate hard parts. Floras and other organisms formerly

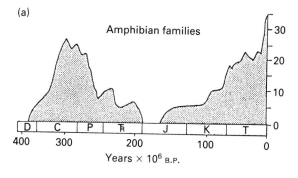

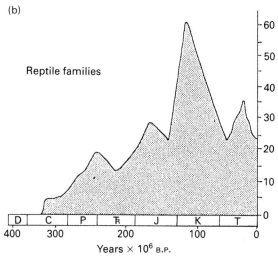

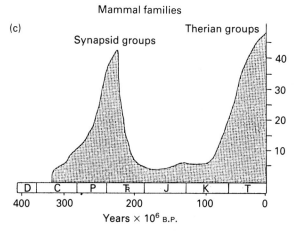

FIG. 20.1. Numbers of families of (a) Amphibia, (b) non-marine Reptilia, and (c) non-marine Mammalia. Note bimodality of the amphibians and the mammals offsetting the acme of the reptiles in the Jurassic and Cretaceous. See 'Methods' for further discussion. See Collinson & Scott, Fig: 18.1 for details of geologic periods.

associated with vertebrate faunas are of critical importance, although often they are preserved under different conditions in different sedimentary regimes.

In this chapter I attempt to characterize some of the large-scale changes in terrestrial community structure that are suggested by the record of fossil vertebrates. The first part briefly outlines the history of tetrapod faunas with emphasis on trophic and phylogenetic organizaton. The second part features some regularities in this history that presumably reflect community patterns in terrestrial ecosystems.

FOUR TETRAPOD DYNASTIES

In its broadest outline the history of tetrapods may be viewed as a progression of four dynasties: (i) the amphibian dynasty, late Devonian through late Carboniferous (about 400–280 million years ago); (ii) the synapsid mammal dynasty, early Permian through early Triassic (about 280–220 million years ago); (iii) the diapsid reptile dynasty, middle Triassic through late Cretaceous (about 220–65 million years ago); and (iv) the therian mammal dynasty, Cenozoic Era (about 65 million years ago to the present). As in any history such subdivisions are somewhat arbitrary and could be multiplied; for example, Bakker (1977) proposed eight tetrapod dynasties in which the synapsid mammal dynasty and the diapsid reptile dynasty each consisted of three subdivisions.

The amphibian dynasty

The earliest tetrapods of the late Devonian and Carboniferous were mainly top predators in a freshwater food-chain. The principal taxa were labyrinthodont and lepospondyl amphibians. They were preserved most abundantly in coal swamp settings, mainly in Euramerica near the palaeo-equator. A total of about 100 genera is recognized during this dynasty including a good representation of the generally large-sized labyrinthodonts but a poor sample of the small lepospondyls (Carroll 1977). As larvae, most taxa presumably were aquatic; as adults many remained aquatic or semi-aquatic.

In coal swamp deposits fully terrestrial tetrapods occur relatively rarely as erratics transported from upland settings (Milner 1980, Olson 1985). Probable terrestrial taxa include some edopoids, some dissorophoids and a few anthracosaurs such as *Seymouria*. Besides their rarity in aquatic settings, these groups are recognized as terrestrial amphibians by such morphological features as long limbs, ossified phalanges, deep skulls and absence of lateral line systems. They are generally regarded as carnivores or insectivores. Possibly the prior evolution of a rich terrestrial arthropod fauna provided the resource that permitted the early tetrapods to become fully terrestrial.

The synapsid mammal dynasty

The earliest reptiles and the earliest mammals appear in the late Carboniferous (Stephanian), often together in the same deposits. They are surely sister groups as evidenced by the shared innovation of a land-laid egg. This is formalized by referring both classes to the Amniota. They played an essential role in establishing the synapsid dynasty, but took until the late part of the early Permian to complete the transition.

The earliest amniotes established themselves in the early tetrapod communities in two ways: first, they supplanted the amphibia that had served as terrestrial insectivores and carnivores; and second, they founded new roles as large, terrestrial herbivores. Reptilian groups such as captorhinids, protorothyrids and procolophonids took up the roles of insectivores and small carnivores, while larger synapsid predators, such as *Dimetrodon*, usurped the top carnivore title from amphibia such as *Eryops*. Meanwhile there appeared new tetrapod herbivores including diadectids, a family of large anapsid reptiles, and edaphosaurs, a family of very large synapsid mammals. During most of its history the second tetrapod dynasty was dominated by large herbivorous synapsids.

Olson (1961, 1966) argues persuasively that the evolutionary success of therapsid mammals in the Permo-Triassic is linked to their success as primary consumers in terrestrial communities. The early history of this dynasty is recorded in the Permian of the USA and the USSR (Olson 1962), and its later history is magnificently preserved in the Permo-Triassic sediments of Africa and South America (Pitrat 1973; Kitching 1978; Bonaparte 1982). Evidently the general progression of tetrapod evolution during this dynasty was shared by all continents (including Antarctica), a consequence of their coalescence as Pangaea.

The opportunity for large, herbivorous tetrapods to dominate terrestrial ecosystems stemmed from the emergence of 'open' habitats during the Permian. The older coal swamp forests had produced a dense canopy which shaded out the ground (Milner 1980), but the more upland Permian floras, rich in conifers and cordaites, fostered a fertile ground layer that was available to tetrapod herbivores (Scott 1980). This shift occurred in about the middle of the Permian sequence in Texas and Oklahoma, when the 'Caseid Chronofauna' supplanted the 'Permo-Carboniferous Chronofauna' and when as Olson (1975, p.376) notes '...somewhat drier conditions developed...' In South Africa, farther from the palaeo-equator than Euramerica, Hiller and Stavrakis (1984, p.1) note that the Karroo Group of the late Permian and early Triassic '...records a general change to warmer climatic conditions following the widespread glaciation of the early to middle Permian...with later deposits being laid down under an increasingly arid regime.' Thus the rise of the synapsid dynasty appears to be broadly linked to the Permian opening of habitats and the success of major new groups of seed plants (Collinson & Scott, Chapter 18).

The number of tetrapod families and their turnover rate are much higher in the synapsid dynasty than in the amphibian dynasty (Fig. 20.1). Bakker (1977) attributed these differences to precocious attainment of endothermy in many early amniote groups, but this seems an unlikely and an unnecessary postulate. I propose two alternative hypotheses that involve ecological rather than physiological factors, and are therefore more practical to test. First, as primary consumers the herbivorous synapsids and anapsids captured a much larger share of total terrestrial energy flow, leading to more complex and evolutionarily more volatile tetrapod assemblages. Second, the shift in vertebrate faunas from mainly aquatic to mainly terrestrial ecosystems led them into generally less stable physical environments, with a consequent increase in taxonomic turnover rates.

The diapsid reptile dynasty

By the late Triassic, diapsid reptiles had replaced synapsid mammals as dominant herbivores and carnivores on all continents, and so they persisted until the end of the Cretaceous. The succession of vertebrate faunas from the Triassic of South America provides the fullest available record of the transition from the second to the third dynasty, although evidence from other continents suggests that this transition was world-wide (Romer 1975, Bonaparte 1982). As Romer (1975, p.469) noted, the faunas of intermediate age '... are not simply transitional in nature but have positive distinctive features such as an abundance of gomphodont cynodonts and rhynchosaurs, together with a variety of thecodonts and a few early dinosaurs.'

Explanations to account for the replacement of synapsid-dominated faunas by diapsid-dominated faunas during the mid-Triassic are varied. One is that increasingly xeric climates prevailed over most continents and favoured reptiles which were physiologically better-equipped to withstand aridity (Robinson 1971). Their physiological advantages include the uric acid nitrogen-excretion pathway, which is far superior in water conservation to the urea-producing pathway of mammals, and generally lower metabolic rates. Possibly a major shift in terrestrial plant formations, generally dominated by gymnosperms, for example from the *Glossopteris* flora to the *Dicroidium* flora, gave archosaurian herbivores a selective advantage over therapsid herbivores (Bonaparte 1982). Finally it should be noted that no competitive scenario is necessary: either of these suggested changes (one in the physical, the other in the biotic environment) might have produced a major extinction of synapsid mammals, thus giving the diapsid dynasty its opportunity (Benton 1983).

The renaissance of dinosaur palaeobiology during the past decade has helped spark studies into the taphonomy and palaeoecology of terrestrial archosaur faunas. These include work by Gradzinski (1970), Dodson (1971), Dodson *et al.* (1980), Lucas (1981) and Russell (1983). A central question concerns the metabolic level (or levels) maintained by various archosaurs, and the answer to this question has important implications for the probable trophic structure of various diapsid-dominated faunas. Although the results are still far from conclusive, they do not tend to support Bakker's early claim that in dinosaur faunas large carnivore biomass was as low in relation to large herbivore biomass as in modern large mammal communities. Russell (1983) found in the rich samples from the Oldman Formation in Alberta that carnosaurs (*Albertosaurus*) accounted for about 6% of the estimated large vertebrate biomass, rather than the 1% expected from the mammalian (endothermic) model. Likewise, Lucas (1981) in comparing a coastal lowland community (the *Parasaurolophus* community) with an inland community (the *Alamosaurus* community) in the late Cretaceous of New Mexico, found a high frequency of tyrannosaurs in the inland setting. Along with screenwashing for associated small vertebrates, palaeoecological studies of Mesozoic continental sediments promise valuable new insights into the community structure of diverse dinosaur faunas (Van Valen & Sloan 1977; Jacobs & Murry 1980; Archibald & Clemens 1984).

The therian mammal dynasty

The Cenozoic record of terrestrial vertebrates, dominated by therian mammals, is more extensively documented than any other part of vertebrate history. Padian & Clemens (1985) found 4000 fossil mammal genera, more than all other tetrapods combined, of which only about 100 genera had pre-Cenozoic records. The best sampled periods from the preceding tetrapod dynasties, namely the Permian and the Cretaceous, each yield about 10% of that number of genera. Recognizable stratigraphic divisions are shorter as one approaches the Recent Epoch, and geographic coverage tends to be wider and denser. The Quaternary Period is especially valuable for its richly refined record of the flux in fossil mammal faunas.

Mesozoic

The origins of the therian dynasty are traceable back into the Mesozoic, and this record may hold the explanation for their sudden success in the Palaeocene. The first abundant teeth of Mesozoic mammals were discovered a century ago in anthills in the late Cretaceous Lance Formation of Wyoming and broadly similar samples, known collectively as the Lance Fauna, occur in adjacent states and provinces. In the last two decades a distinctive new mammal fauna has been discovered. Known as the 'Bug Creek Fauna', it is described by Clemens *et al.* (1979, p.47) as '... incomparably the richest known deposit of Mesozoic mammals of any age, anywhere in the world.' In both late Cretaceous faunas the three major groups of mammals are Allotheria (multituberculates), Metatheria (marsupials), and Eutheria (placentals), but the Lance Fauna has five genera of placentals, while the Bug Creek adds three more placental genera. These additional placentals, namely *Procerberus, Protungulatum,* and *Purgatorius,* were probable new immigrants from Asia and evidently gave rise to many of the Palaeocene orders.

The ecological setting of the Bug Creek Fauna differed from that of the Lance. After analysing the seventy-three non-mammalian vertebrate genera associated with these faunas, Estes & Berberian (1970, p.1) found that while the Lance fauna '... was probably deposited within the general environment of a swamp forest with relatively small watercourses, the Bug Creek Fauna seems to have been laid down in the relatively deeper waters of major rivers issuing from those lowland swamps.' Thus, the additional Bug Creek placentals were probably associated with better-drained habitats set back from the swamps in which the Lance Fauna accumulated. Van Valen & Sloan (1977) further distinguished these two vertebrate faunas and depositional facies, restoring from the Bug Creek Fauna the '*Protungulatum* Community' and from the Lance Fauna the '*Triceratops* Community'. The former was thought to be associated with a cooler temperate forest which had spread from the north, and which eventually displaced the swamp community along with its dinosaurian inhabitants.

Palaeocene

The Palaeocene mammal fauna of North America consisted of a richly varied group of mainly arboreal mammals. Nearly all were small to medium-sized. Four orders carried over from the Cretaceous, the extinct Multi-tuberculata and Condylarthra and the extant Insectivora and Primates, comprise most of the Palaeocene fauna. The rodent-like multituberculates were longer-lived than any other order (from late Jurassic to early Oligocene); Jenkins & Krause (1983) have shown in some detail that *Ptilodus* had elaborate arboreal adaptations resembling those of a tree squirrel. After very limited diversity in the early Palaeocene the primates soon branched into several successful omnivorous and frugivorous groups. By far the most abundant and diverse order were the condylarths which evidently doubled the number of their genera every few million years, so that even within the Puercan (earliest Palaeocene) there were several dozen genera (Van Valen 1978). They probably included in one form or another the ancestry of all subsequent ungulate orders. Most Palaeocene condylarths were broadly comparable to modern hyraxes. Most were evidently arboreal, and their diets ranged from omnivorous to herbivorous. The Mesonychidae, now often placed in their own order, Acreodi, were carnivorous. Another truly carnivorous order of archaic mammals, the Creodonta, appeared in the mid-Palaeocene. The only large herbivores belonged to the three relatively rare orders, Taeniodonta, Pantodonta and Dinocerata, which appeared in North America during the mid and late Palaeocene. The pantodonts were partly amphibious molluscivores; the other two orders consisted of browsing herbivores.

The predominant Palaeocene settings were evidently cypress swamps and multistoried subtropical forests (see Chapter 18), and the mammalian fauna tends to corroborate this picture. For a few million years in the late Palaeocene (specifically the Tiffanian) the climate cooled, as evidenced by an increasing percentage of deciduous trees and a decrease in floral diversity in the Rocky Mountain Region. Likewise Rose (1981, p.386) found in Tiffanian mammal faunas of the Bighorn Basin in Wyoming 'dramatically lower species richness and evenness'.

Eocene

The Eocene is marked by the appearance of the Rodentia, the richest of all mammalian orders, and by the two modern ungulate orders, the Perissodactyla and the Artiodactyla. New families continued to appear almost as rapidly as they had in the Palaeocene. The rich woodland communities that are most regularly sampled are dominated by arboreal rodents and primates. Winkler (1983) showed by bulk screen-washing in the Willwood Formation of the Clarks Fork Basin, Wyoming, that small, probably arboreal mammals are far more numerous than surface sampling could detect, although some subsamples represented a coherent assemblage of larger terrestrial herbivores.

The equable climatic circumstances that prevailed during the Eocene are emphasized by new collections from within the Arctic Circle on Ellesmere Island (West & Dawson 1978). There the predominant species are prosimian primates and an extinct Dermopteran related broadly to living 'flying foxes' of the Old World tropics; and with these mammals are found warm-adapted groups of reptiles as well as a subtropical flora. A related feature is the remarkably high degree of faunal resemblance between the early Eocene biota of western North America and that of western Europe, suggesting the absence of climatic and physical barriers between them.

In the middle Eocene the first indications of seasonal aridity appear, and by the late Eocene there is convincing evidence that woodland savanna had become the predominant biome in North America, displacing the subtropical forests that had prevailed for the first 17 million years of Tertiary history. As the vast Lake Gosiute and other Green River lakes retreated during the middle Eocene they formed seasonal evaporites and were encroached upon by deeply oxidized redbed deposits. About half of the rich Green River flora consists of species with small, compound leaves; the families Leguminosae, Sapindaceae and Anacardiaceae predominate; and the first grass pollen appears. MacGinitie (1969) called the flora 'Orizaban-subtropical' by analogy with the seasonally-arid woodland savanna presently living on the slopes of Mt. Orizaba near the Tropic of Cancer in Mexico.

By the late Eocene the mammalian fauna reflected still further trends toward seasonal aridity and scrubby habitats. The same three orders that appeared in the early Eocene now played a leading role in producing new families of precocious savanna-adapted herbivores. Among the rodents were the eomyids and zapodids with five-crested lophodont dentitions. The diversity of large herbivores increased markedly: among the new groups of perissodactyls were helaletid tapiroids, amynodont rhinocerotids and chalicotheriids; to the bunodont artiodactyls were added pig-like entelodonts, and for the first time a great diversity of selenodont artiodactyl families appeared, including camelids, hypertragulids, leptomerycids and oreodonts. As Cifelli (1981) showed, Perissodactyla diversified earlier in the Eocene, whereas Artiodactyla caught up with them and began to surpass them by the late Eocene when the selenodont radiation took place. The first rabbits appeared in North America in the late Eocene. A majority of these new groups of herbivores can be traced to Asiatic origins.

Oligocene

In North America the early Oligocene is a time of very high faunal turnover. About 60% of early Oligocene genera are new (Savage & Russell 1983). Among new families of rodents are the Heteromyidae, Cricetidae, Geomyidae, Castoridae, Sciuridae, and Cylindrodontidae; and new families of large mammals include the Canidae, Felidae, Mustelidae, Tapiridae, Rhinocerotidae, Anthracotheriidae and Tayassuidae. Most of the new herbivore families had attained truly hypsodont dentitions (Webb 1977).

Climatic trends toward cooler mean annual temperature, greater seasonal temperature range, and lower precipitation level, marked especially by more severe arid seasons, continued from the late Eocene into the Oligocene. Hutchison (1982) recognized the effects of such trends, particularly increasing aridity, in the steeply declining diversity curve for aquatic reptiles in the Rocky Mountain region during the Eocene and first half of the Oligocene. Likewise, the percentage of entire-margined leaves in such rich floras as the Florissant in Colorado decreased dramatically, indicating a drop of some 10° in mean annual temperature (Wolfe 1978). The early Oligocene is recognized in many parts of the world as a time of climatic rigour, when marine biotas were decimated, sea surface temperatures plunged, and eustatic sea levels dropped as much as 200 m.

Miocene

The most striking feature of the Miocene savanna fauna in North America was its diversity of grazing and browsing herbivores. By the late Miocene, when ungulate diversity reached its peak, there were a dozen genera of horses, including three browsers and nine grazers, and nearly as many camel genera which also included browsers and grazers. The ancestries of these native ungulates can be traced back into the Eocene, but their major diversification began in the early Miocene. From then until almost the end of the Miocene, the number of native ungulate genera doubled every 3 to 5 million years. A nearly equal contribution to the Miocene herbivore fauna of North America was made by immigrant groups from Asia including the ancestral stocks of three ruminant families, the Antilocapridae, the Palaeomerycidae, and the Moschidae, two proboscidean families, Mammutidae and Gomphotheriidae, and the rodent taxon that gave rise to the vast Neogene radiation of New World cricetids, many of which soon spread into South America (Webb 1985).

About 20 million years ago in the early Miocene the mid-continental North American biome shifted from woodland savanna to grassland savanna. The richest mid-continental Miocene flora is the Kilgore Flora from Nebraska; from it MacGinitie (1962, p.83) deduced that the plains between streams supported a grassland savanna composed of '... small live oaks, pines, blackberry, and persimmon, with shrubs of Mahonia, currant, hawthorne, sagebrush and relatively abundant species of composites.'

Kurten (1971) calls the Miocene the 'Epoch of Revolutions' and emphasizes the worldwide trend toward glacial conditions with their radical fluctuations and generally cooler and drier climates. The now familiar glacio-eustatic events associated with the Messinian (latest Miocene) stage in Europe, when the Mediterranean dried up, represent an important phase in such climatic deterioration. The sudden decline of the rich savanna ungulate fauna in North America at about the same time suggests that the same cooling and drying trends that produced the savanna fauna, when carried farther, led to its demise.

Plio-Pleistocene

The final step in the secular trend toward deterioration of Cenozoic climates came during the last few million years when glaciers began to form in the Northern Hemisphere. Mid-continental North America saw the final transition from savanna to steppe as the predominant biome. Deserts were established over a considerable expanse of Mexico and the south-western United States, while in the north a unique steppe-tundra biota developed across all of Beringia (the land area now partly beneath the Bering Sea but also including adjacent parts of northern Asia and northern North America). During the Pliocene and Pleistocene, provincial differences from one part of the continent to another became far more profound than in any previous epochs. The ratio of allochthonous (immigrant) to autochthonous (native) new taxa vastly increased during successive stages of the Pliocene and Pleistocene. Because of these complexities it is extremely difficult to summarize faunal trends during the last 5 million years in North America. Nevertheless, one of the principle trends was for smaller herbivores to replace larger herbivores, as extinction episodes cut down the latter (Webb 1969).

PATTERNS IN TETRAPOD HISTORY

This section offers some interpretations of regularities perceived in the history of tetrapod faunas outlined above. The underlying assumption is that palaeontological patterns can be made to reveal large-scale, long-term ecological processes. Such patterns may provide useful lessons in the following subjects: (i) dynasties, chronofaunas, and transitions; (ii) disassembly and reassembly rules; (iii) diversity and faunal equilibria; (iv) continental configurations and area effects; (v) ecomorphs and replicate communities; and (vi) evolutionary relays and replacements.

Dynasties, chronofaunas and transitions

Each of the first three tetrapod dynasties reigned for about 100 million years and was then largely replaced by a fundamentally different dominant taxonomic group. Perhaps the same can be expected of the fourth (our own) dynasty. Within dynasties some associations of vertebrates persisted with

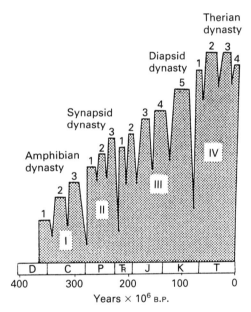

FIG. 20.2. Synopsis of tetrapod history including four successive dynasties (roman numerals) and subsidiary stable faunal intervals or 'chronofaunas' (arabic numerals). Standing generic diversity, which apparently increased in the course of tetrapod history, is suggested subjectively by the height of each column. The arabic numerals represent the following 'chronofaunas' (from left to right): Amphibian dynasty: 1 = rhachitome faunas; 2 = pelycosaur-anapsid-amphibian faunas; 3 = caseid-pareiasaur-deinocephalian faunas; Synapsid dynasty: 1 = gorgonopsian-anomodont faunas; 2 = lystrosaur-cynodont faunas; 3 = dicynodont-gomphodont faunas; Diapsid dynasty: 1 = gomphodont-rhynchosaur faunas; 2 = prosauropod faunas; 3 = sauropod-stegosaur faunas; 4 = iguanodont-deinonychid faunas; 5 = hadrosaur-ceratopsian faunas; Therian dynasty (in North America): 1 = Palaeogene subtropical forest faunas; 2 = late Eocene-Oligocene woodland savanna faunas; 3 = Miocene grassland savanna faunas; and 4 = Plio-Pleistocene steppe faunas.

only minor modifications in taxonomic composition and apparent trophic structure; vertebrate palaeontologists have followed Olson (1952) in calling these long-term associations '*chronofaunas*'. The best-known examples, including various Permian synapsid chronofaunas (Olson 1962) and various Neogene mammal faunas (Webb 1977), endured about 10 million years each. If one assumes that the known record is a heavily abridged version of the complete vertebrate history, then it is possible that such persistent terrestrial communities were even more commonplace than the explicit studies of chronofaunas suggest. As it stands, the record presents a 'herky-jerky' view of vertebrate history (Fig. 20.2). On the one hand it points to the great durability and apparent stability of certain associations; on the other it indicates that the transitions were relatively brief intervals of active reorganization.

The pulse of chronofaunas and transitions may be represented most completely by the Tertiary record of primary consumer genera in North America indicated in Fig. 20.3. The four stable peaks of herbivore genera are probably 'real' plateaux of diversity (not artifacts of the record), as are the intervening valleys of low diversity. A good indication of their reality is that they span more than one stratigraphic interval (land mammal age); for example, the low numbers that occur in the interval that had its mid-point about 56 million years ago are also found in the interval that had its mid-point about 51 million years ago. The subtropical woodland savanna fauna of the Oligocene and the grassland savanna fauna of the Miocene each endured for about 10 million years. These two peaks appear to be more diverse and to last longer than the subtropical forest fauna of the Palaeocene or the steppe fauna of the Plio-Pleistocene. The transitions between chrono-faunas last about 5 million years and are thus of shorter duration than the chronofaunas themselves. As discussed above for the Cenozoic Era, these transitions appear to be correlated with pulses of climatic change. The transitions between dynasties may have similar or even shorter durations. The last interdynastic transition, at the end of the Cretaceous, may have lasted only about 1 million years or less (Archibald & Clemens 1984).

Disassembly and reassembly rules

Each transition between dynasties appears to move from the base of the trophic system upward. Following some abiotic trigger mechanism, the fundamental evolutionary changes that foment revolutions in terrestrial ecosystems are probably innovations in the mode of plant reproduction (see Chapter 18). These in turn may change the selective advantages of com-peting primary consumer strategies, which in turn may alter other pathways

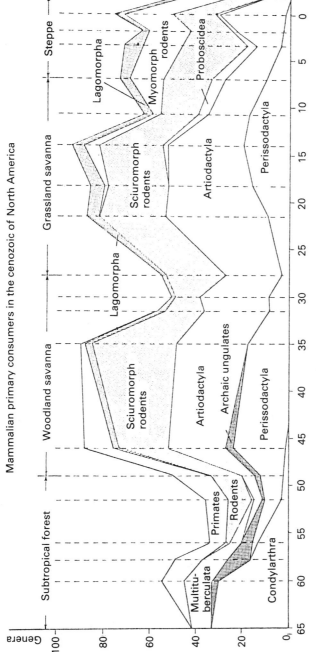

Fig. 20.3. Recorded numbers of primary consumer genera during successive land-mammal ages of North America (after Savage & Russell 1983). Divisions between ages indicated by vertical dashed lines from 65 million years ago to present. Genera are grouped into orders to indicate waxing, waning and possible replacement patterns. Note plateaux of high diversity (about eighty-five genera) and valleys of low diversity (about sixty genera) spanning more than one land-mammal age.

of energy flow at higher trophic levels. Thus, disassembly of a dynasty begins at the base of the food web.

Following their conquest of the land early in the Palaeozoic, tracheophytes experienced two major evolutionary radiations: the Carboniferous–Permian diversification of seed plants; and the mid-to-late Cretaceous radiation of angiosperms (Niklas, Tiffney & Knoll 1985; Collinson & Scott, Chapter 18). Each of these tracheophyte transitions preceded a new tetrapod dynasty by about half a geologic period. I have noted above the possible correlation between the rise of the synapsid dynasty and 'upland' seed plant communities of the Permian. The relationship between the angiosperm revolution that began in the mid-Cretaceous, and the rise of therian dynasty was probably mediated by the insects which the mammals ate. Lillegraven, Kraus & Bown (1979) found it highly probable that the diversification of therian mammals represented a response to new evolutionary opportunities presented by flowering plants and coadapted insects.

Since the early Permian the best sampled parts of successive tetrapod faunas have been the primary consumers. This is evident to vertebrate palaeontologists whether the herbivores are measured by their volume in collections or by their sample sizes in statistical studies. For example, Russell (1983, p.7) estimated that some 100 million years ago the environments now represented by the Oldman Formation in Alberta supported fifty-one duckbilled dinosaurs and twenty-eight horned dinosaurs per square kilometre, in contrast with only nine tyrannosaurs and lower densities of smaller animals. Such results are consistent with the basic principles of community energetics. There are exceptions such as the La Brea Tar Pits in California with their extensive samples of sabercats and dire wolves, but such predator-rich sites are rare and generally obvious. The natural abundance of primary consumers tends to be amplified by the fact that most fossil vertebrate accumulations are biased in favour of larger-bodied animals, and they usually consist mainly of primary consumers.

Despite their abundance, the large herbivores are the most vulnerable to extinctions during times of large-scale environmental change. No tetrapods greater than 10 kg survived the late Cretaceous (Padian & Clemens 1985). Likewise, large herbivores bore the brunt of a string of late Cenozoic extinctions which culminated in the well-known late Pleistocene extinctions (Martin & Klein 1984). As primary consumers they gain efficiency by growing large. That they have repeatedly done so is manifest in the vertebrate fossil record, and has been 'codified' as Cope's 'Law'. Yet paradoxically the large primary consumers recurrently become the most vulnerable set of terrestrial organisms to major changes in vegetational structure. And as they became extinct the conspicuous secondary consumers followed rapidly.

The reassembly rules for a new dynasty differ markedly from the disassembly rules for an old dynasty to judge from the tetrapod record. It is noteworthy that none of the founding taxa of new tetrapod dynasties appears *de novo* in the fossil record. Living inconspicuously among the dinosaur faunas of the Cretaceous were therian mammals that represented the approximate ancestry of most of the orders that radiated in the Palaeocene. Similarly, when the therapsid mammals held sway in the early to mid-Triassic, thecodont archosaurs (the 'pretenders' to the synapsid throne) existed and in some environments were diverse and abundant. The historical evidence suggests that each of these groups already had the potential to seize the dominant roles. If, as noted above, the transitions between dynasties are relatively brief, about 5 million years and in some instances possibly less, it is not surprising that pre-existing groups are rapidly pressed into service. As suggested above, the trigger for a change of dynasties was evidently a major change in vegetational characteristics with consequent disruption in the hegemony of the dominant herbivores. Whether such a biotic revolution was ultimately triggered by an extra-terrestrial event or some more ordinary earthly events is largely irrelevant to discussions of proximate changes within the food web.

Reassembly of a new tetrapod dynasty evidently moves downward from the top of the food web. Thus the heralds of a new dynasty are not the new, dominant herbivores, but rather they are relatively inconspicuous insectivores or small carnivores. During the rise of the therian dynasty in the latest Cretaceous, the first evolutionary developments stemmed from small, insectivorous placental groups, often broadly grouped in the order Insectivora (technically divisible into several distinct orders). The Bug Creek Fauna of the late Cretaceous reveals *Procerberus* as the progenitor of omnivorous to frugivorous primates and *Protungulatum* as the most ancient of the herbivorous to omnivorous condylarths. A few million years later in the early Palaeocene intense diversification among many groups, especially among the medium-sized herbivores, rapidly got under way.

This 'top-down' reassembly pattern can be observed at the emergence of each new tetrapod dynasty. The key progenitors of successive dynasties are respectively rhipidistian crossopterygians, sphenacodont pelycosaurs, the-codont archosaurs, and, as just noted, insectivorous therians. Each of these groups of secondary consumers gives rise to a major new group of large primary consumers. In each case also, the same major taxonomic group spawns the large secondary consumers that prey upon the primary consumers. Thus, the major predators upon large synapsid herbivores are large synapsid carnivores, and the top carnivores in early Tertiary mammal communities are collateral descendants of the same insectivorous ancestors that produced the herbivores.

Diversity and faunal equilibria

There are many indications that tetrapod diversity, even at the generic level, held steady over long intervals. Simpson (1969, p.169), in a study that ranged from the ordinal down to the generic level, stated that 'the total variety of amphibians and reptiles remained essentially level from Permian to Cretaceous...', approximately half of the total history of the tetrapods. In one of the best sampled parts of the record, the flat-topped pattern of mammalian primary-consumer genera in Fig. 20.3 indicates asymptotic growth (or a recurrent faunal equilibrium) throughout the Cenozoic of North America. Because each of these plateaux is sampled from several successive mammal ages they are unlikely to be artifactual. Other studies of diversity in favourable parts of the land mammal record have concluded that rates of origination and extinction have remained in dynamic equilibrium (Webb 1969; Lillegraven 1972; Gingerich 1977). Likewise, the mixed results of Mark & Flessa (1977) can be obviated by combining their large herbivore and small herbivore categories, the two groups having been complementary throughout the late Cenozoic (Simpson 1969; Webb 1969).

The scepticism of Padian & Clemens (1985) about such equilibrium results led them to suggest that apparent rates of originations and extinctions rise and fall together as mere artifacts of faunal sample sizes. This view misses the point that any segment of the fossil record may be regarded as a statistical sample of diversity trends during its time interval. If, for example, land mammals had tended to increase their diversity during most intervals, the records of originations from most samples would then tend to exceed the corresponding records of extinctions. The actual prevalence of a balance between originations and extinctions during much of the tetrapod record, especially in the best-sampled parts of the land mammal succession, strongly suggests that tetrapod diversity was governed by a dynamic equilibrium most of the time.

A somewhat broader question to be asked of the tetrapod record is whether the numbers of taxa have increased or remained steady throughout their entire history. In its broadest terms, the answer must be that the kinds of tetrapods have experienced a net increase from the Carboniferous to the Tertiary. This answer is warranted at the family level by the data summarized in Fig. 20. 1. An early marked increase in tetrapod diversity, recognizable at all taxonomic levels from the genus level upward, came with the emergence of the synapsid dynasty in the early Permian, probably reflecting the development of large primary consumers. Other net growth in diversity came in the Cenozoic. Simpson (1969, p.169) explained the net increase in orders of tetrapods then with two observations: first that '...Tertiary mammals ...spread into new ecological situations...'; and secondly that '...flying vertebrates add another major adaptive type, intricately subdivisible, to the terrestrial faunas'.

The fossil record of tetrapods seems to embody evolutionary progress, if one accepts the major transitions from aquatic to amphibious, to fully terrestrial, and even to aerial modes of life, as a kind of progression from primitive to derived. Furthermore, the chronological sequence of tetrapods introduces successively more homoeostatic physiological mechanisms, for example warm-bloodedness, larger brain sizes and better water retention. And with each new phylogenetic group that appears, some members of the earlier groups also survive alongside it.

How can this last conclusion, that tetrapods progressed and gained in net diversity, be reconciled with the evidence, cited above, that points to the prevalence of faunal equilibria? The answer seems to be that faunal equilibria are profoundly disrupted now and then. The transitions between dynasties (and similarly between chronofaunas) when diversity is low provide magnificent opportunities for new adaptive radiations to burst forth. Faunal equilibria are diversity-dependent, and thus times of low faunal diversity, following major extinction episodes, permit exponential diversification.

The history of tetrapod diversity thus resembles the game of 'musical chairs'. During most of the game the players march in an orderly manner to stately music. But once in a while, unexpectedly, the music stops and everyone is launched into a mad scramble. The principal difference is that in the game the number of available chairs continues to shrink, whereas in evolutionary history the opportunities available to tetrapods seem to have increased.

The equilibrium number of species on recent islands, according to the hypothesis of MacArthur & Wilson (1967), depends in part on the rate of immigration from source pools (such as an adjacent mainland biota) which in turn depends on distance and other factors influencing vagility. In the larger perspective of long-term, large-scale faunal dynamics, immigrations (and vicariance events) ordinarily play a lesser role, probably because their driving function is overtaken by that of autochthonous originations. One of the most useful natural experiments on the effects of immigration on faunal equilibria is the Great American Interchange of the late Pliocene and Pleistocene (Stehli & Webb 1985). By contrasting present South American land mammal diversity during continental suturing with land mammal diversity during the continent's prior isolation, one can show the same equilibrium effects, albeit more impressive, as the more familiar island biogeographic examples.

Other examples of how immigrant genera become established may be recognized in the North American mammal fauna. Immigration waves occur in remarkably uneven episodes which generally coincide with intervals of low diversity (see Fig. 20.3). The highest immigration rate for land mammal genera in the Palaeogene (early Tertiary) was during the Tiffanian (late Paleocene), when diversity dropped and the fauna appeared to undergo reorganization. Another example occurs in the early Miocene, especially in the Hemingfordian valley, when an extremely large cohort of immigrant genera came from the Old World. Conversely, during the next two mammal ages (Barstovian and Clarendonian), the peak of generic diversity for the entire Cenozoic, the immigration rate was at its nadir. Finally, the largest stream of immigrant genera entered North America during the late Pliocene and Pleistocene, an interval of intermediate and fluctuating diversity characterized by diverse and changing environments. Thus there appears to be a broad correlation between immigration episodes and intervals of low diversity in the North American Cenozoic record of land mammals. This pattern suggests that faunal equilibria ordinarily limit the opportunities available to immigrants (Webb 1985).

Continental configuration and area effects

In recent years the large-scale relationship between area and diversity has captured the attention of palaeontologists as a corollary of the global history of plate tectonics. The hypothesis advanced by Schopf (1974) to account for the most devastating mass extinction in the history of marine organisms was the worldwide reduction in shallow shelf area in the late Permian. The fact that terrestrial vertebrate diversity did not experience a late Permian mass extinction, and indeed was rapidly increasing through that interval (as shown in Fig. 20.1 for reptiles and especially mammals), supports Schopf's hypothesis; for as Pangaea coalesced and sea-level dropped, land area and terrestrial diversity would be expected to increase.

In the early Mesozoic, tetrapod faunas, like land plant floras, had begun to separate into northern and southern hemispheric distributions, following the breakup of Pangaea. In the Jurassic, terrestrial vertebrate faunas record fundamental phylogenetic divisions along geographic lines, for example, Estes' (1983) recognition of the dichotomy between iguanians in Gondwanaland and all other lizard groups in Laurasia in the early Jurassic. By the late Cretaceous, the southern continents were well-separated, and there were ecogeographic filters betwen North and South America and between North America and Asia (Lillegraven, Kraus & Bown 1979; Bonaparte 1984).

During the Cenozoic Era continental biotas frequently experienced more physical isolation from one another than during previous times in tetrapod history. Kurten (1969) was the first to suggest that this geographic disjunction might produce greater provincialism in terrestrial biotas and thus account for the greater diversity of tetrapods in the therian mammal dynasty. This influence alone might explain the fact that there are many more orders and genera of birds and mammals than of amphibia or reptiles. On a continent by continent basis, Flessa (1975) confirmed that land area can precisely predict generic, familial and ordinal diversities of continental mammal faunas. From the Permian to the Cenozoic the trend has been toward greater continental dispersion, and this trend partly explains tetrapod diversity increases.

Ecomorphs and replicate communities

The prevalence of faunal equilibria, noted above, implies the probability of highly-structured (interactive) tetrapod communities. Under similar environmental circumstances they might be expected to produce, at different times or in different places, replicate communities and analogous species (ecomorphs). Such analogous systems, recognized among both ancient and living communities, warrant careful study for the special insights they may offer into the structure of communities (see also Chapter 10). Many such examples from the fossil record involve stable marine communities such as coral reefs. Among ancient tetrapod faunas the fossil land mammal assemblages of North America offer a set of possible analogues to faunas living on other continents. For example Webb (1983) compared the grassland savanna fauna of the Miocene in North America with that living in Africa, although that comparison was limited to the ungulate guild. Concordant results were obtained for Oligocene and younger guilds of terrestrial carnivores and scavengers by Stanley, Van Valkenburgh & Steneck (1983).

In the case of the convergent savanna faunas of North America and Africa, comparable niches appear to have been filled in most cases by groups that were phylogenetically distinct at the level of order, family or at least subfamily. For example, the major ungulate radiation in Africa involves Bovidae; that family is not even present in North America during most of the Miocene but the diverse Equidae are the ecological vicars of the Bovidae. In the North American savanna 'the giraffe ecomorph' is *Aepycamelus*, a camelid, and 'the hippo ecomorph' is *Teleoceras,* a short-legged rhinocerotid genus. Discovery by Voorhies & Thomasson (1979) of fossil grass anthoecia on the hyoid bones of well-preserved *Teleoceras* skeletons confirms the postulated grazing fare of that extinct genus, thereby giving direct evidence that its feeding habits converged with those of the hippo.

Furthermore, the savanna ungulate guild of the North American Miocene bears a number of structural resemblances to that living in Africa. Five extensive quarry samples from mid to late Miocene North American sites were compared with four East African park faunas with respect to ungulate body weights, feeding types (grazers, browsers and mixed feeders), and estimated biomass in each feeding category. Estimated biomass distribution attributed about 80% of the ungulate biomass to medium to large-sized grazers. In both fossil North American and modern African faunal samples browsing ungulates represent a relatively small percentage of ungulate biomass (typically about 5%), and it was invested in only two or three species of very large ungulates (giraffe or giraffe-camel weighing 1000 to 3000 kg). And finally, alpha diversity, taken as the maximum number of ungulate taxa sampled in a single North American fossil quarry or in a single African game region, is about sixteen to eighteen species. Thus, the ungulate faunas in widely separated but ecologically similar settings evolved along closely comparable patterns, even though the analogous species had ancestries in different orders that lived on different continents during different epochs.

Relays and replacements

An important pattern often cited by paleontologists involves apparent replacement of one higher taxon (family, order or class) by another. Three basic criteria, often explicitly discussed, but at least implicit in such instances, are that the fossil ranges of the two groups overlap (or closely abut one another) in space and time, that they exhibit an inverse correlation in numbers of individuals, species and/or genera through time and space, and that the two groups occupy more or less similar adaptive zones (or appear to utilize similar resources). To the extent that such cases are well-documented in the fossil record, they provide rather convincing evidence of a structured ecosystem in which an adaptive zone is defined, regardless of which higher taxon fills it.

One can postulate two classes of replacement phenomena. The first involves extensive competition. The other class of replacements does not involve competition: an ecological opportunity simply becomes available to a new group after an older group goes extinct. The cause or causes of that extinction need not be known or specified, although they might include extinction by physical changes in the environment, or by such biological causes as predator pressure, allelopathy or disease. In cases where these two classes can be distinguished, the competitive encounters should be called 'displacements' and the non-competitive replacements, 'relays'.

The best-documented case of replacement in the history of North American land mammals is probably the displacement of rodent-like Multituberculata by Rodentia principally during the early Eocene. Rodents had appeared one age earlier and multituberculates lasted several million years later, but only as rare samples of a few very small species. Van Valen & Sloan (1966) showed that the early Eocene was the principal time of rapidly-changing inverse abundances of rodents and multituberculates. As these authors concluded, it is virtually impossible to know which of several possible adaptive or environmental advantages may have given the crucial selective advantage to rodents.

Three other examples of probable replacements may be cited briefly. First the Eocene displacement of Condylarthra by the modern ungulate orders Artiodactyla and Perissodactyla appears convincing from the complementary pattern of diversity, although no more detailed studies have been presented (see Fig. 20.3). Second in the late Cenozoic of North America, Webb (1969) proposed that the immigration and diversification of grazing rodents (microtines and many cricetids) represented replacement of grazing ungulates (such as Camelidae and Equidae) that had held sway in the Miocene. In this case it would appear from the record that the successive declines in the number of large herbivore genera preceded the successive increases in the number of small herbivore genera, and are thus a chain of relays. And third, the displacement of South American native ungulate stocks by the immigration of North American ungulate groups has been suggested by a number of authors (e.g. Simpson 1969), although disputed by others (see discussion in Cifelli 1985).

It may be significant that each of these suggested replacements of one group by its ecological vicars began during an interval of low diversity and was then fully realized during the next plateau of high diversity as shown in Fig. 20.3. For example, rodents are represented by one probable immigrant genus at one site in the Clarkforkian, but do not displace the multituberculates until the Wasatchian peak of diversity is attained. That transition interval of some 2 or 3 million years evidently favoured the preliminary diversification and spread of revolutionary new immigrant taxa. More generally one may suppose that transitions between tetrapod dynasties and between chronofaunas were probably non-equilibrium punctuated intervals that fostered 'relays' and 'displacements'.

CONCLUSIONS

1   The most favourable parts of the fossil vertebrate record are marked by stable peaks of generic numbers and persistent lineages that seem to represent coadapted sets. During such chronofaunal intervals, which typically last for 10 million years, the direct influence of abiotic environmental factors appears relatively unimportant. On the other hand, abiotic mechanisms are strongly implicated as the causes of the great turnovers between dynasties. The three tetrapod dynasties, prior to our present one, endured for about 100 million years each. Transitions between dynasties are relatively brief events, lasting about 5 million years or less.

2   Disassembly of an old dynasty appears to move upward from the base of the food web. Revolutions in land plants, notably the Carboniferous-Permian rise of seed plants and the early to mid-Cretaceous rise of flowering plants, may have triggered the dynastic changes. The large herbivores were the most susceptible to disruption, and their extinction led to that of the large carnivores. Reassembly, on the other hand, was from the top down. Each new dynasty emerged from insectivores or small carnivores that had existed inconspicuously (in terms of size and abundance) in the prior dynasty. These founders then gave rise to wholly new groups of dominant herbivores, and in each new dynasty the large carnivores branched from the same phylogenetic stem as the large herbivores.

3   The total numbers of orders, families and genera of tetrapods tend to be constant over long periods of time. More detailed analyses also indicate that origination and extinction rates of genera fluctuate together, suggesting that tetrapod faunal diversity is governed by a MacArthur–Wilson equilibrium. One way to reconcile the net gain in numbers of terrestrial vertebrate taxa from the Carboniferous to the Cenozoic with the predominance of faunal equilibria is to postulate that explosive diversification was concentrated in transitions between chronofaunas and dynasties when diversity-dependence was temporarily relaxed. The total array of adaptive types and the total numbers of tetrapod taxa (at all levels) certainly have increased in the course of their history.

4   From the Permian, when the continents were coalesced as Pangaea, to the Cenozoic, when the continents were more widely dispersed, terrestrial vertebrate faunas have experienced a net trend toward increased provinciality. The number of orders of birds and mammals may be greater than that of amphibians and reptiles because their basic radiation took place in a framework of more dispersed continents.

5   Replicate communities appear in different parts of the world, at different times, and consist of different taxa, but nevertheless live in similar environments and have key species that play similar adaptive roles. In the case of convergent savanna ungulate guilds in the Miocene of North America and the Recent of Africa, apparent structural resemblances include the alpha diversity (eighteen species), the distribution of body sizes among browsing, mixed-feeding and grazing sub-guilds, and the proportion of total ungulate biomass in each of these three groups (80% grazer biomass). Furthermore, the most abundant grazer and the largest browser in the fossil ungulate fauna from North America each shows remarkable structural resemblance to its living African ecomorph, the hippo and the giraffe respectively.

6   Replacement patterns, in which one higher taxon appears to occupy the same adaptive zone that another has just departed, also indicate highly-structured ecosystems. Often such replacements or relays begin during times of low diversity. These are also the times of increased intercontinental immigrations. Together these patterns suggest that tetrapod communities were highly-structured, and that many important evolutionary changes were concentrated during relatively brief and relatively unstructured transition intervals.

ACKNOWLEDGMENTS

I thank E.C. Olson, Mike Rosenzweig, Bob Carroll, Doug Jones, Bruce MacFadden and the late George Simpson for helpful comments, and the National Science Foundation (Grant BSR 8314649) for its support of my faunal studies. This is University of Florida contribution to vertebrate paleontology no. 254.

REFERENCES

Archibald, J.D. & Clemens, W.A. (1984). Mammal evolution near the Cretaceous–Tertiary boundary. *Catastrophes and Earth History: the New Uniformitarianism* (Ed. by W.A. Berggren & J.A. Van Couvering), pp. 339–71. Princeton University Press, Princeton, New Jersey.

Bakker, R.T. (1977). Tetrapod mass extinctions — a model of the regulation of speciation rates and immigration by cycles of topographic diversity. *Patterns of Evolution, as Illustrated by the Fossil Record* (Ed. by A. Hallam), pp.439–68. Elsevier, Amsterdam.

Behrensmeyer, A.K. & Hill, A.P. (Eds) **(1980).** *Fossils in the Making: Vertebrate Taphonomy and Paleoecology.* University of Chicago Press, Chicago.

Benton, M.J. (1983). Large-scale replacements in the history of Life. *Nature,* **302,** 16–17.

Bonaparte, J.F. (1982). Faunal replacement in the Triassic of South America. *Journal of Vertebrate Paleontology,* **2,** 362–71.

Bonaparte, J.F. (1984). Nuevas pruebas de la connexion fisica entre Sudamerica y Norteamerica en el Cretacico tardio (Campaniano). *Actas III Congreso Latino-Americano Palaeontologia y Bioestratigrafia, Porto Alegre,* pp.141–9.

Bown, T.M. & Kraus, M.J. (1981). Vertebrate fossil-bearing paleosol units (Willwood Formation, Lower Eocene, Northwest Wyoming, U.S.A.): implications for taphonomy, biostratigraphy, and assemblage analysis. *Palaeogeography, Palaeoclimatology, Palaeoecology,* **34,** 31 – 56.

Carroll, R.L. (1977). Patterns of amphibian evolution: an extended example of the incompleteness of the fossil record. *Patterns of Evolution as Illustrated by the Fossil Record* (Ed. by A. Hallam), pp.405 – 37. Elsevier, Amsterdam.

Clemens, W.A., Lillegraven, J.A., Lindsay, E.H. & Simpson, G.G. (1979). Where, when and what — a survey of known Mesozoic mammal distribution. *Mesozoic Mammals: the First Two-Thirds of Mammalian History* (Ed. by J.A. Lillegraven, Z. Kielan-Jaworowska & W.A. Clemens), pp.7 – 58. University of California Press, Berkeley, CA.

Cifelli, R.L. (1981). Patterns of evolution among the Artiodactyla and Perissodactyla (mammallia). *Evolution,* **35,** 433 – 40.

Cifelli, R.L. (1985). South American ungulate evolution and extinction. *The Great American Biotic Interchange* (Ed. by F.G. Stehli & S.D. Webb), pp.249 – 66. Plenum, New York.

Dodson, P. (1971). Sedimentology and taphonomy of the Oldman Formation (Campanian), Dinosaur Provincial Park, Alberta (Canada). *Palaeogeography, Palaeoclimatology, Palaeoecology,* **10,** 21 – 74.

Dodson, P., Behrensmeyer, A.K., Bakker, R.T. & McIntosh, J.S. (1980). Taphonomy and paleoecology of the dinosaur beds of the Jurassic Morrison Formation. *Paleobiology,* **6,** 208 – 32.

Estes, R. (1983). The fossil record and early distribution of lizards. *Studies in Herpetology and Evolutionary Biology, Essays in Honor of Ernest Edward Williams* (Ed. by A. Rhodin & K. Miyata), Cambridge, Massachusetts.

Estes, R. & Berberian, P. (1970). Paleoecology of a late Cretaceous vertebrate community from Montana. *Breviora, Museum of Comparative Zoology,* **343,** 1 – 35.

Flessa, K.W. (1975). Area, continental drift and mammalian diversity. *Paleobiology,* **1,** 189 – 94.

Gingerich, P.D. (1977). Patterns of evolution in the mammalian fossil record. *Patterns of Evolution as Illustrated by the Fossil Record* (Ed. by A. Hallam), pp. 469 – 500. Elsevier, Amsterdam.

Gradzinski, R. (1970). Sedimentation of dinosaur-bearing Upper Cretaceous deposits of the Nemegt Basin, Gobi Desert. *Palaeontologia Polonica,* **21,** 147 – 229.

Hiller, N. & Stavrakis, N. (1984). Permo-Triassic fluvial systems in the Southeastern Karroo Basin, South Africa. *Palaeogeography, Palaeoclimatology, Palaeoecology,* **45,** 1 – 21.

Hutchison, J.H. (1982). Turtle, crocodilian, and champsosaur diversity changes in the Cenozoic of the North-central region of western United States. *Palaeogeography, Palaeoclimatology, Palaeoecology,* **37,** 149 – 64.

Jacobs, L.L. & Murry, P.A. (1980). The vertebrate community of the Triassic Chinle Formation near St. Johns, Arizona. *Aspects of Vertebrate History: Essays in Honor of Edwin Harris Colbert* (Ed. by L.L. Jacobs), pp. 55 – 72. Museum of Northern Arizona Press, Flagstaff.

Jenkins, F.A., Jr. & Krause, D.W. (1983). Adaptations for climbing in North American multituberculates (mammalia). *Science,* **220,** 712 – 15.

Kitching, J.W. (1978). The stratigrafic distribution and occurrence of South African fossil amphibia in the Beaufort beds. *Palaeontologia Africana,* **21,** 101 – 12.

Kurten, B. (1969). Continental Drift and Evolution. *Scientific American,* **220,** 54 – 64.

Kurten, B. (1971). *The Age of Mammals.* World Naturalist Series, Weidenfeld & Nicolson Press, London.

Lewontin, R.C. (1969). The meaning of stability. *Diversity and Stability in Ecological Systems,* pp. 13 – 24. Brookhaven Symposia in Biology, 22. Brookhaven National Laboratory, Upton, New York.

Lillegraven, J.A. (1972). Ordinal and Familial Diversity of Cenozoic Mammals. *Taxon*, **21**, 261–74.

Lillegraven, J.A., Kraus, M.J. & Bown, T.M. (1979). Palaeogeography of the World of the Mesozoic. *Mesozoic Mammals: the First Two-Thirds of Mammalian History* (Ed. by J.A. Lillegraven, Z. Kielan-Jaworowska & W.A. Clemens), pp. 277–308. University of California Press, Berkeley, California.

Lucas, S.G. (1981). Dinosaur communities of the San Juan Basin: a case for lateral variations in the composition of Late Cretaceous dinosaur communities. *Advances in San Juan Basin Paleontology*, (Ed. by S.G. Lucas, J.K. Rigby, Jr. & B.S. Kues), pp. 337–93. University of New Mexico Press, Albuquerque.

MacArthur, R.H. & Wilson, E.O. (1967). *The Theory of Island Biogeography.* Princeton University Press, Princeton, New Jersey.

MacGinitie, H.D. (1962). The Kilgore Flora: a Late Miocene flora from northern Nebraska. *University of California Publications in Geological Sciences*, **35**, 67–158.

MacGinitie, H.D. (1969). The Eocene Green River flora of northwestern Colorado and northeastern Utah. *University of California Publications in Geological Sciences*, **83**, 1–203.

Mark, G.W. & Flessa, K.W. (1977). A test for evolutionary equilibria: Phanerozoic brachiopods and Cenozoic mammals. *Paleobiology*, **3**, 17–22.

Martin, P.S. & Klein, R.G. (1984). *Quaternary Extinctions: A Prehistoric Revolution.* University of Arizona Press, Tucson.

Milner, A.R. (1980). The tetrapod assemblage from Nyrany, Czechoslovakia. *The Terrestrial Environment and the Origin of Land Vertebrates* (Ed. by A.L. Panchen), pp. 439–96. Systematics Association Special Volume 15, Academic Press, London.

Niklas, K.J., Tiffney, B.H. & Knoll, A.H. (1985). Patterns in vascular land plant diversification: an analysis at the species level. *Phanerozoic Diversity Patterns: Profiles in Macroevolution* (Ed. by J.W. Valentine), pp. 97–128. Princeton University Press, Princeton, New Jersey.

Olson, E.C. (1952). The evolution of a Permian vertebrate chronofauna. *Evolution*, **6**, 181–96.

Olson, E.C. (1961). Food chains and the origin of mammals. *International Colloquium on the Evolution of Lower and Non-specialized Mammals* (Ed. by G. Vandebroek), pt. 1, pp. 97–116. Koninklijke Vlaamse Academie voor Wetenschappen, Letteren en Schone Kunsten van Belgie, Belgium.

Olson, E.C. (1962). Late Permian terrestrial vertebrates, U.S.A. and U.S.S.R. *Transactions of the American Philosophical Society.* **52**, 2–224.

Olson, E.C. (1966). Community evolution and the origin of mammals. *Ecology*, **47**, 291–302.

Olson, E.C. (1975). Permo-Carboniferous Paleoecology and Morphotypic Series. *American Zoologist*, **15**, 371–89.

Olson, E.C. (1985). Vertebrate paleoecology: a current perspective. *Paleogeography, Paleoclimatology, Paleoecology*, **50**, 83–106.

Padian, K. & Clemens, W.A. (1985). Terrestrial vertebrate diversity: episodes and insights. *Phanerozoic Diversity Patterns: Profiles in Macroevolution* (Ed. by J.W. Valentine), pp. 41–96. Princeton University Press, Princeton, New Jersey.

Pitrat, C.W. (1973). Vertebrates and the Permo-Triassic extinctions. *Palaeogeography, Palaeoclimatology, Palaeoecology*, **14**, 249–64.

Retallack, G.J. (1983). A Paleopedological approach to the interpretation of terrestrial sedimentary rocks: the mid-Tertiary fossil soils of Badlands National Park, South Dakota. *Geological Society of America Bulletin*, **94**, 823–40.

Robinson, P.L. (1971). A problem of faunal replacement on Permo-Triassic continents. *Palaeontology*, **14**, 131–53.

Romer, A.S. (1966). *Vertebrate Paleontology*, (3rd edn.). University of Chicago Press, Chicago, Illinois.

Romer, A.S. (1975). Intercontinental correlations of Triassic Gondwana vertebrate faunas. *Gondwana Geology*, pp. 469–73. International Union of Geological Sciences, Commission on Stratigraphy, Sub-committee on Gondwana Stratigraphy and Palaeontology, Gondwanaland Symposium Proceedings, Paper 3.

Rose, K.D. (1981). Composition and species diversity in Paleocene and Eocene mammal assemblages: an empirical study. *Journal of Vertebrate Paleontology*, **1**, 367–88.

Russell, D.A. (1983). A Canadian Dinosaur Park. *Terra, the Natural History Museum of Los Angeles County*, **21**, 3–9.

Savage, D.E. & Russell, D.E. (1983). *Mammalian Paleofaunas of the World*. Addison-Wesley Publishing Company, Reading, Massachusetts.

Schopf, T.J.M. (1974). Permo-Triassic extinctions: relation to sea-floor spreading. *Journal of Geology*, **82**, 129–43.

Scott, A.C. (1980). The ecology of some Upper Paleozoic floras. *The Terrestrial Environment and the Origin of Land Vertebrates* (Ed. by A.L. Panchen), pp. 87–115. Systematics Association Special Volume 15, Academic Press, New York.

Simpson, G.G. (1952). Periodicity in Vertebrate Evolution. *Journal of Paleontology*, **26**, 359–70.

Simpson, G.G. (1969). The first three billion years of community evolution. *Diversity and Stability in Ecological Systems*, pp. 162–77. Brookhaven Symposia in Biology, 22. Brookhaven National Laboratory, Upton, New York.

Stanley, S.M, Van Valkenburgh, B. & Steneck, R.S. (1983). Coevolution and the fossil record. *Coevolution* (Ed. by D.J. Futuyma & M. Slatkin), pp. 328–49. Sinauer Associates, Inc., Sunderland, Massachusetts.

Stehli, F.G. & Webb, S.D. (Eds) (1985). *The Great American Biotic Interchange*. Topics in Geobiology, Vol.4, Plenum, New York.

Thomson, K.S. (1977). The pattern of diversification among fishes. *Pattern of Evolution, as Illustrated by the Fossil Record* (Ed. by A. Hallam) pp. 377–404. Elsevier, Amsterdam.

Van Valen, L. (1978). The beginning of the age of mammals. *Evolutionary Theory*, **4**, 45–80.

Van Valen, L. & Sloan, R.E. (1966). The extinction of the multituberculates. *Systematic Zoology*, **15**, 261–78.

Van Valen, L. & Sloan, R.E. (1977). Ecology and extinction of the dinosaurs. *Evolutionary Theory*, **2**, 37–64.

Voorhies, M.R. & Thomasson, J.R. (1979). Fossil grass anthoecia within Miocene rhinoceros skeletons: diet in an extinct species. *Science*, **78**, 331–3.

Webb, S.D. (1969). Extinction-origination Equilibria in late Cenozoic Land Mammals of North America. *Evolution*, **23**, 688–702.

Webb, S.D. (1977). A history of savanna vertebrates in the New World. Part I: North America. *Annual Review of Ecology and Systematics*, **8**, 355–80.

Webb, S.D. (1983). The rise and fall of the late Miocene ungulate fauna in North America. *Coevolution* (Ed. by M.H. Nitecki), pp. 267–306. University of Chicago Press, Chicago.

Webb, S.D. (1985). Main pathways of mammalian diversification in North America. *The Great American Biotic Interchange* (Ed. by F.G. Stehli & S.D. Webb), pp. 200–17. Plenum, New York.

West, R.M. & Dawson, M.R. (1978). Vertebrate Paleontology and the Cenozoic History of the North Atlantic Region. *Polarforschung*, **48**, 103–19.

Winkler, D.A. (1983). Paleoecology of an early Eocene mammalian fauna from paleosols in the Clarks Fork Basin, Northwestern Wyoming (U.S.A.). *Palaeogeography, Palaeoclimatology, Palaeoecology*, **43**, 261–98.

Wolfe, J.A. (1978). A paleobotanical interpretation of Tertiary climates in the Northern Hemisphere. *American Scientist*, **66**, 694–703.

IV
NEW PERSPECTIVES

21. COMMUNITY ORGANIZATION FROM THE POINT OF VIEW OF HABITAT SELECTORS

MICHAEL L. ROSENZWEIG

Department of Ecology and Evolutionary Biology, University of Arizona, Tucson, Arizona 85721, USA

INTRODUCTION

Ecologists have long recognized that habitat selection is and ought to be influenced by population densities (Svardson 1949; Morisita 1950; Mac-Arthur & Pianka 1966; Fretwell & Lucas 1970; Schoener 1975; Lawlor & Maynard Smith 1976; Diamond 1978; Inouye 1978; Rosenzweig 1979a; Werner & Hall 1979; Cody 1981; Brew 1982; O'Connor, Chapter 8). Fretwell's work (see also Fretwell 1972) provided a theoretical basis for understanding this influence within a species. I have been trying to do the same interspecifically (Rosenzweig 1985).

The thesis of this paper is that developing the capacity to predict habitat selection as a function of population densities leads to a useful framework for the rules of community organization. I shall argue for this thesis by trying to use it. Examples of theories that predict behaviours in terms of variable population densities will be described, and three field tests will be reviewed. One of these will point to the need for expansion of theory, and two ways of accomplishing this will be explored. The suggestions emerging from this exercise will be combined into a framework of organizational rules. Finally, rules that other authors have suggested for various communities will be laid upon the framework.

Many of the contributions to this symposium, including this one, address the question of community organization by investigating interrelationships among a very small number of species. Often this number is two. Is it possible to learn anything useful about large systems of interactions by focusing on one?

Population dynamics provides an analogy and perhaps a clue. Interactions of two and three species exhibit some of the kinds of dynamics seen in large systems. Moreover, Takens' theorem (1981) proves that one can make accurate inferences about the dynamics of entire systems just by studying (carefully) the dynamics of any single species that belongs to it (Schaffer 1985).

However, arguing by analogy is not good science, so I would rather leave the question unanswered. Looking at two species and their possible relationships can be done in the hope, but not the certainty, that they will teach us something useful.

For me, the best indication of usefulness is to be able to apply the lessons to the real world. Do two-species subsystems in nature behave like isolated two-species systems in theory? Do the principles garnered from isolated small systems help us to organize, to understand and to explain what we see in larger systems? I think the answer to both questions is often yes, and I shall try to review the evidence which has convinced me in the present case. It remains for future investigation to determine what, if any, rules govern the extension of principles drawn from small systems to large ecosystems.

ISOLEG THEORIES

Isoleg theories are a type of optimal foraging theory (Rosenzweig 1985). They are maps of optimal habitat selection on a state space. (Each axis of the state space is the density of one of the species being studied.) For example, suppose there are two species, A and B, and only two habitat types, I and II. An isoleg theory would predict the four proportions $P_{A,I}$; $P_{A,II}$; $P_{B,I}$; $P_{B,II}$ at each state space point, (N_A, N_B), where $P_{J,K}$ is the proportion of patches of type K that are accepted by an individual of species J, and where N_J is the density of species J.

Isoleg theories are so named from the Greek *iso* (equal) and *lego* (choose), because their qualitative features may best be summarized by drawing lines on a state space at each point of which the optimal behaviour of one of the species is a constant.

The most important isolegs have an additional property. Many theories of optimal foraging predict sudden shifts of optimal behaviour if variables which control that behaviour are gradually altered. For example, a species may change suddenly from complete rejection to complete acceptance of a habitat type if a competitor's density grows past a certain threshold. Such behaviours are called disjunct behaviours (Brown & Rosenzweig 1986). The additional property of some isolegs is that at every point on them, a specific sudden shift is predicted. Those are the isolegs which I will discuss in this chapter.

In addition to the usual sorts of assumptions which every biological theory must specify, isoleg theories specify the rules that foragers must follow in assessing their habitats. Some sets of foragers rank habitat types similarly; this is the shared preference rule. The distinct preference rule states that forager species have very different habitat preference.

The biological basis for a shared preference is a quantitative niche axis. The niche axis may be productivity or some other measure of habitat-richness. It may also be its inverse, like a gradient of predator pressure. In either case, the resources of the gradient differ only in amount, not kind. (A predator is a sort of negative resource, like a pollutant.) All foragers—at least when alone—should, therefore, prefer the richest end of the axis and rank habitats solely according to their richness.

The biological basis for a distinct preference is a qualitative niche axis. Resources in the habitats of such an axis do differ in kind. The best-known case of such an axis is a food size axis (although food size axes are not generally correlated with habitats). Qualitative axes may be continuous—as in food size—but they need not be, as some of the examples in the discussion will make clear.

An isoleg theory of shared preferences with interference competition

Assume that all patches (samples) of the same habitat type are functionally homogeneous in space. That is, the patches may vary temporally, but at any given instant, all samples of a habitat type are identical, or, at least, cannot be told apart when encountered by foragers. There are various biological situations that can produce such a circumstance (Brown & Rosenzweig 1986). For example, individuals may be able to live very long portions of their lives in a single patch; then, any spatial heterogeneity within a patch type would provoke resettlement into the more profitable samples until all patches were spatially homogeneous and no one could profit from moving (Fretwell's (1972) Ideal Free Distribution). Another situation that produces this circumstance is the forager that can identify only the habitat type of a patch, but not its relative value compared to other patches of its type. Many nectarivores may be like that: they can identify a species of flower, but probably cannot tell how much nectar is in its nectaries except by actually removing it. Such a forager must treat all samples of a habitat type as if they were the same.

Now, we assume again the two species A and B and the two habitats I and II. We add the assumptions that habitats I and II are ranked the same by species A and B: both prefer I. By preference, I mean something quite density-specific: a species prefers a habitat, I, if and only if, as its density and that of its competitor(s) approach zero, the limit of its fitness in I is greater than it is in II. Of course, biologists generally replace fitness with the surrogate variable, 'net rate of energy accumulation,' but we should never lose sight of the ultimate variable we claim to be studying.

In this model, A and B do not ignore each other. We shall assume that A interferes with B (by aggressive behaviour, chemistry or however). Thus we may speak of A as the dominant, and B as the subordinate species.

To make this case interesting, there has to be a trade-off. Otherwise (i.e. if A could dominate B *and* exploit both I and II more effectively than B), it would be hard to imagine B's survival. So, we shall assume that the subordinate is the better exploiter.

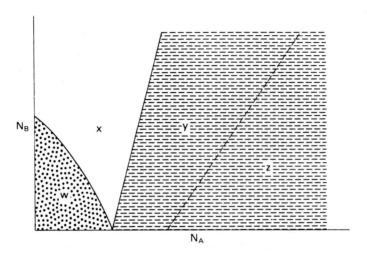

FIG. 21.1. The isoleg model of shared preferences when all patches of a single habitat type are homogeneous in space and there is interference competition. The density of the dominant species is N_A. The density of the subordinate species is N_B. The lower-case letters code the behavioural regions predicted by the theory. w: A and B both choose the better habitat. x: B chooses both habitats; A only the better. y: B chooses the poorer; A the better habitat. z: A chooses both; B the poorer habitat. (After Rosenzweig 1985.)

Figure 21.1 is a map of isolegs based on this set of assumptions (Rosenzweig 1979a, 1985; Pimm, Rosenzweig & Mitchell 1985). Let us now read this map as an example. All its isolegs reflect sudden shifts in disjunct behaviours. In the triangular region of Fig. 21.1 (region w), both A and B use all the I patches they encounter, and reject all of the II patches. The diagonal line (it need not be straight) which borders region w is an isoleg. But it identifies a shift in the behaviour of only one of the species: B, the subordinate. A point (N_A^*, N_B^*) is on this isoleg if and only if any increment, however small, in either N_A^* or N_B^*, results in B's accepting all the secondary patches (rather than none of them) (region x).

A second B-isoleg intersects the A axis at the same point as the first, but this isoleg has a very steep positive slope. A point is on this isoleg if and only if any increment in A or decrement in B (moving the system to region y) results in B's rejecting all the primary patches.

Finally, there will be an isoleg for the dominant species if it can use the secondary habitat at all profitably. This line will also have a steep positive slope. A point is on it if any increment in A or decrement in B (moving the system to region z) causes A to accept all secondary patches.

Figure 21.1. makes a set of qualitative predictions. It shows that behaviour will exhibit sudden shifts. It allows for only two dominant and three subordinate behaviours. These behaviours should occur in only four combinations instead of the six imaginable. Moreover, the four allowable combinations should occupy distinct regions of the state space (w,x,y,z) with a certain geometric relationship to each other, and with borders matching the previously described isolegs. No doubt, each of the qualitative predictions is too probable to be significant by itself. As a set, however, they are far less likely.

Distinct preferences with exploitative competition

The assumptions of this model differ in one important detail: the species rank the habitats inversely (distinct preferences). A prefers I; B prefers II. Another different assumption made by this model is that the species compete only exploitatively. This situation leads to a different isoleg map (Fig. 22.2; Rosenzweig 1981).

Again, all isolegs map sudden behavioural shifts. However, this time there are only two, and they both have positive slope. Other differences in

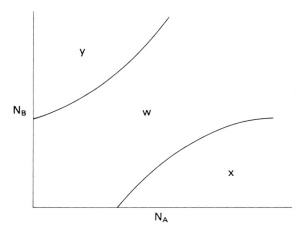

FIG. 21.2. The isoleg model of distinct preferences when all patches of a single habitat type are homogeneous in space. The densities of the species are N_A and N_B. w: A chooses its better habitat (I) and B chooses its better habitat (II). x: A chooses both; B its better habitat. y: B chooses both; A its better habitat.

predictions are manifest. In region w near the origin, the species again restrict themselves to their preferred habitats, but this time those habitat types will be distinct, preventing interspecific competition. This model predicts only two behaviours per species: selection of the preferred habitat only, and opportunism. So there are four imaginable combinations of behaviour. Brown & Rosenzweig (1986) have shown that the isolegs of this model cannot cross. Hence, of the four imaginable combinations, only three should exist. Finally, it is possible to analyse the population dynamics of this model and show that at equilibrium, the species do not compete (Pimm & Rosenzweig 1981; Brown & Rosenzweig 1986). That is why this model was the original bearer of the epithet 'the ghost of competition past'.

FIELD TESTS

The hummingbird isolegs

Pimm, Rosenzweig & Mitchell (1985) performed the first test of isoleg theory on a system of three species of hummingbirds in south-eastern Arizona. Of the three, the bluethroated hummingbird, *Lampornis clemenciae*, is strongly territorial and behaviourally dominates the other two (Lyon, Crandall & McKone 1977; Pimm 1978). The other two differ greatly in size: the smaller one is the blackchinned hummingbird, *Archilochus alexandri*; the larger one, almost as large as the dominant species, is Rivoli's hummingbird, *Eugenes fulgens*. These two do not interact in our experimental situation, and therefore we separated our analyses into two pairs of interactions: bluethroated with Rivoli's and bluethroated with blackchinned.

Pimm, Rosenzweig & Mitchell (1985) set up experimental sites with either 0.35 M or 1.2 M sucrose solution. By mist-netting and holding birds in an aviary, they varied the birds' densities and explored their behaviour at a wide variety of points in the state space.

Observers recorded the durations that each species actually had their beaks in the feeders. They also recorded the duration of territorial occupation by bluethroated. These data produce the variable TBTH, total bluethroated time, which is the sum of bluethroated feeding and territorial activity. Other variables with more obvious definitions were also generated.

The results of this work have been presented in two stages. The first (Pimm, Rosenzweig & Mitchell 1985) demonstrated several things. First, the behaviour of neither blackchinned nor Rivoli's is influenced by the density of the other, supporting our decision to study the three species as two separate pairs. Second, it showed strong effects of bluethroated activity density (TBTH) on the behaviours of all three species. Neither bluethroated behaviour nor that of Rivoli's was influenced by anything else we measured, but blackchinned behaviour was also affected by its own activity density.

The data agree with the shared preference isoleg model in two other particulars. One is that the dominant species had feeding behaviours which included only strong restriction to the richer habitat type, and more or less evenhanded use of both. But both the two subordinate species also exhibited restriction to the poorer patch type. This phenomenon is well known from systems of interference competition (e.g. Bovbjerg 1970; Murray 1981), so our results merely help to confirm that the hummingbird system is indeed so structured. However, the second point of agreement is much more interesting because, as we shall now see, it bears upon the validity of the isoleg model which we claim underlies such an interaction.

The interference isoleg model (Fig. 21.1) predicts that the behaviour of the subordinate will be controlled in different ways in different regions of the state space. Near the origin, adding subordinates should increase their use of the poorer patch type. But at higher densities of dominants, adding subordinates should decrease their use of the poorer patch or not affect it much at all. This predicted change in control of the subordinate behaviour should and did show up as a significant interaction term (of appropriate sign) in the mutliple linear regression of blackchinned and bluethroated densities on blackchinned behaviour. (The data do not support this conclusion for Rivoli's, probably because it was never common enough for us to detect any intraspecific effects at all.) To my knowledge, this alteration of control as a function of the system's position in the state space was never previously detected, probably because no previous model had predicted it and therefore no one had looked for it.

Other predictions of the interference isolegs are weakly supported by the first stage of analysis. The evidence in favour of them becomes more readily appreciated, however, in a second stage of data reduction (Rosenzweig 1986).

One should like to know whether the hummingbird behaviours are disjunct as predicted, and whether these disjunct behaviours are located in the predicted regions of the state space. Because the behaviours may be expressed as proportions, i.e. the proportions of feeding time a species spends feeding from the richer patches, Kendall & Moran's (1963) formula for the distribution of a finite number of points on a unit line was extended and used. The extended formula specifies the probability that if n points are distributed randomly on the line, there will be k or more gaps (a gap is an interval with no point) of size x or larger. If one finds that the gaps are too large to be accounted for by chance, then the behaviours most probably are clustered into groups as theory predicts.

All three species showed significant clustering at the 5% level of probability. The dominant species exhibits two clusters as predicted: one of very high proportional use of rich patches, the second of rather evenhanded use. Rivoli's, a subordinate, should have had three clusters and did: high use of rich; high use of poor, and evenhanded use. Blackchinned, also a subordinate, should have had these same three classes and did, but it also had a fourth at about 33% use of the richer patch. This fourth class is unpredicted, but it turned out to be a likely result of our field methods.

Behaviours can now be classified as group 1, 2, 3 or 4 for each species. The convention followed is that 1 indicates greatest use of rich patches, 2 next greatest, etc. Of course, bluethroated uses only 1 and 2 (it had only two clusters), and only blackchinned uses all four. The cluster number was then entered on the state space at the point which represented the activity densities which obtained on the day it was observed. We can now examine the results for evidence of isolegs.

Beginning with the simplest graph, that of bluethroated (Fig. 21.3), we see clearly the division of the space into two regions as predicted. There was no overlap. The isoleg is the line which divides the space.

The second graph is that of blackchinned (Fig. 21.4). Here we see only three regions, as predicted, instead of the four which might have been

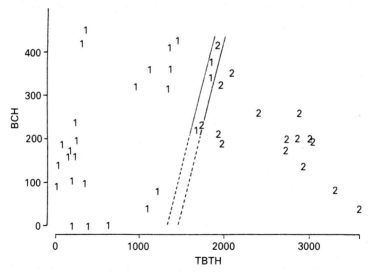

FIG. 21.3. Bluethroated hummingbird behaviour plotted over the activity densities of black-chinned hummingbirds (BCH) and bluethroated hummingbirds (TBTH). The numbers code the category of bluethroated behaviour (see text). The lines estimate the bluethroated isoleg. (From Rosenzweig 1986; © 1986 by Springer-Verlag.)

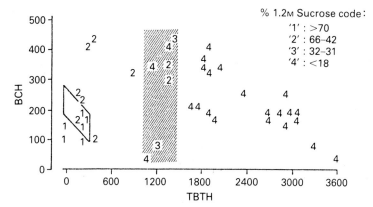

FIG. 21.4. Blackchinned hummingbird behaviour plotted over the activity densities of black-chinned hummingbirds (BCH) and bluethroated hummingbirds (TBTH). The numbers code the category of blackchinned behaviour (see text). The parallelogram and rectangle estimate the blackchinned isolegs. (From Rosenzweig 1986; © 1986 by Springer-Verlag.)

present, given that there are four clusters. (Cluster number three does not occupy its own separate region; instead it is subsumed in the boundary area between 2's and 4's). Thus we can draw only the two isolegs of Fig. 21.1: one separates preference for rich patches from evenhandedness, the other, evenhandedness from preference for the poor patches. The combination of

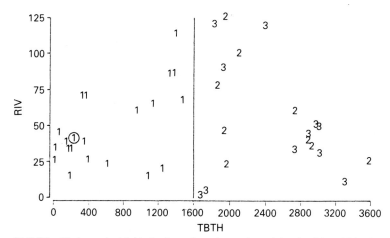

FIG. 21.5. Rivoli's hummingbird behaviour plotted over the activity densities of Rivoli's hummingbirds (RIV) and bluethroated hummingbirds (TBTH). The numbers code the category of Rivoli's behaviour (see text). The line is the estimated Rivoli's isoleg. Note that no isoleg can be drawn to separate behaviour type 2 from behaviour type 3. (From Rosenzweig 1986; © 1986 by Springer-Verlag.)

the results of bluethroated (Fig. 21.3) and blackchinned (Fig. 21.4) yields an isoleg diagram enough like Fig. 21.1 to make a new figure redundant.

A theory which cannot be falsified is untestable. Hence it is interesting that only one of the three *Eugenes* clusters occupies a separate region of state space (Fig. 21.5). The other two share a region for as yet unexplained reasons.

A Bombus Isoleg

Inouye (1978) studied the foraging of two species of bumblebee, *Bombus flavifrons* and *B. appositus,* in the high Rocky Mountains of Colorado. During the warmest month of the summer (*c.* 14 July to 14 August), their most important sources of nectar are monkshood (*Aconitum columbianum*) and a tall species of larkspur (*Delphinium barbeyi*).

Usually, *B. flavifrons* concentrates on the *Aconitum* whereas *B. appositus* concentrates on the larkspur (see also Pyke 1982). Inouye was able to produce niche shifts, however, by reducing the densities of one or the other species of *Bombus*. The bee allowed to remain at full density increased its use of the other's flower.

Because interspecific aggressiveness had not been observed between these species, they are good candidates for a test of the distinct preference model. Accordingly, individual foragers are being timed as they visit standardized arrays of cut inflorescences in 0.1 M sucrose solution. Details of the method will appear when the entire set of experiments is completed.

An isoleg of *B. appositus* has already been located. As predicted, it has a significant positive slope. This is known not from a careful classification of behaviours such as was done with the hummingbirds (there is not yet enough data for that), but from an ordinary multiple linear regression whose results demonstrate that the activity densities of both species affect *B. appositus* behaviour (intraspecific interaction: $P < 0.013$; interspecific: $P < 0.037$), and do so in the predicted directions.

GERBIL ISOLEGS—CENTRIFUGAL ORGANIZATION

Rosenzweig & Abramsky (1986) studied the isolegs of two species of psammophilic gerbil in the Negev Desert, Israel. Because the two species are nocturnal, their behaviour must be studied indirectly. We used a method which relies upon the distribution of their populations to reveal their selectivity. If most individuals are found in a restricted set of habitats, we conclude they are being selective; if, instead, they are randomly distributed among habitats we conclude that they are being opportunistic.

Multiple regression showed that distributions of both species were affected significantly by the population densities of both species. The gerbil system, in short, must have isolegs. (Remembering the case of behaviours 2 and 3 in Rivoli's hummingbird, we note that this conclusion must never be taken for granted.) But there is a problem.

The signs of the intraspecific gerbil effects are each in the right direction: the more populous a gerbil species, the less selective it is about its habitat. But the signs of the interspecific effects are both negative. This was an unpredicted phenomenon. For distinct preferences, they should both have been positive. For interference and shared preferences, one should have been negative, one positive. We concluded that at least one other isoleg model must be developed.

Natural history of the gerbils suggests a model which we have termed centrifugal community organization or centrifugal preferences (Rosenzweig & Abramsky 1986). It adds a third habitat type to the set of assumptions. Then it asumes that both species prefer the same one of the three, but have distinct secondary preferences. This accords with our observations of the gerbils, because both species are found in abundance where both perennial cover and open space for annuals are common. But if open space is limited, then *G. pyramidum* is rare. And if cover is sparse (i.e. less than 15 or 20%), then *G. allenbyi* is rare.

The core habitat in a centrifugally-organized set of species is the one preferred by all. But this is irrelevant to their coexistence. The business of coexistence is accomplished in the secondary habitats. Here preferences are distinct and the classical trade-off principle is at work.

One way to understand centrifugal organization is to imagine the environment as a set of stresses (not just physiological stresses but adversities of all sorts). The stresses are qualitatively unlike each other. Though all species should be able to thrive in an environment free of all types of stress, each may need to specialize on the type of stress it can most efficaciously withstand. The stress on gerbils in an open habitat may be exposure to predators. In a habitat covered by perennials, it may be shortage of the annuals on which gerbils depend for seeds to eat.

Another case of such organization may be the consumers of conifer seeds (Smith & Balda 1979). All require the seeds for successful reproduction, but each appears to have a different manner of surviving during the frequent periods when seeds are absent.

Finally, Shimada & Fujii (1985) have uncovered a case in the laboratory. Two species of parasitoid wasp both preferred late fourth instar larvae of the azuki bean weevil. But they had distinct secondary preferences to which they expanded when parasitoid populations were high.

SPATIAL HETEROGENEITY WITHIN PATCH TYPES

Previous isoleg theories have assumed that there is no detectable spatial heterogeneity among samples of a single patch type. But Brown (1986) has been investigating the effects of allowing just such heterogeneity. In such circumstances, individuals forage in a patch if and only if the patch yields profits above a minimum rate; they continue foraging in it until its resources are reduced to the point where the minimum rate is reached; then they leave. The critical resource level is called the 'giving-up-density.' The sample patch must have time to recuperate from giving-up-density before a forager will again accept it. Note that giving-up-density varies inversely with forager density. Brown's theory is a more general form of the classic marginal value theory of Charnov (1976).

Brown & Rosenzweig (1986) have added the assumption of intratypical spatial heterogeneity to the distinct preference model (Fig. 21.6). The results are markedly different from the previous distinct-preference model (Fig. 21.2).

The isolegs of a distinct preference model with detectable spatial heterogeneity begin as straight lines perpendicular to the density axis of the

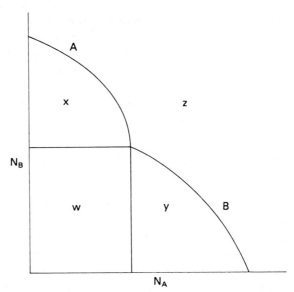

FIG. 21.6. Distinct preference isolegs in an environment in which foragers can recognize spatial variability within a patch type. Compare with Fig. 21.2 where patches of a given habitat type do not differ from each other in space. The lower case letters code the behavioural regions as in Fig. 21.2 with the addition of region z: both species choose both habitats. (After Brown & Rosenzweig 1986).

species they represent. They maintain this direction until they intersect, after which they have negative slopes. Whereas the isolegs cannot cross in the previous case of distinct preference (Fig. 21.2), they *must* in this one. This generates a new region (z) of the graph; in it, both species use both patch types.

Models with detectable intratypical spatial heterogeneity generate two sorts of isoleg. The isolegs of Fig. 21.6 are much the same sort as previous isolegs; they mark sudden shifts in disjunct behaviours. But because of the variability among patches of a single type, not all a species' secondary patches are accepted when it gets across these isolegs. The richer samples are accepted and the poorer ones rejected. This leads to partial-preference isolegs. Brown & Rosenzweig (1986) have investigated them, but they are absent from Fig. 21.6 and will not be discussed here because they are not yet known to be germane to the questions of community organization.

DISCUSSION

Organization as a set of rules

Traditionally, community organization is described as a set of patterns. Many of them relate some morphological property of consumers to a similar measure of their resources or habitat. Pearson & Mury (1979), for example, show that tiger beetle mandible size correlates positively with tiger beetle food size, and Fitter, (Chapter 6) shows how plant roots are specialized on the size of soil interstices).

Many such patterns are now known and make considerable sense. Yet I believe that it is a mistake to expect such regular patterns in every community. Often the rules of community organization will demand not fixed patterns, but fluid relationships. A major lesson of isoleg analysis is that it is possible to analyse these relationships and thus decipher the rules of organization of the community.

The theory of limiting similarity may be the premier case of searching for pattern where none is expected (see Abrams 1983 for review). This theory displays the utilization functions of each species on one or more niche axes, and predicts the overlap of the functions. But such parameters should not be static; they should vary depending upon conditions. We know from such studies as isoleg analysis that resource and habitat utilization functions should depend upon variables like population sizes. So, to assume that these functions keep a fixed pattern is to assume that densities and other conditions are always at equilibrium. Werner & Platt (1976), studying six species of *Solidago*, provide an excellent example: utilization overlaps differ markedly between old fields and prairies despite the fact that the rules by which the guild is organized seem to remain the same. Perhaps it is not too speculative to guess that fixed pattern is to be expected only if morphological features are subjected to rules; both physiological and behavioural features may be too plastic for such an outcome except in the most stable environments.

If the rule's the thing, what sort(s) ought we to look for? Isoleg theory suggests that there are at least four types. Some sets of species subdivide niche axes qualitatively, some quantitatively. Some types of habitat appear homogeneous to species, whereas the relative richness of patches of other types can be recognized by foragers. Because the pair of alternatives (functional homogeneity/detectable heterogeneity and quantitative/qualitative) are not known to depend upon each other, they may combine in four possible ways. Let us now devote a paragraph of recapitulation to each alternative.

A quantitative niche axis is an array of habitats whose resources are the same and are obtained with the same foraging techniques. The habitats, from the vantage point of the foragers, vary only in the *amounts* of resources they offer. Thus, all forager species which subdivide the axis rank the habitats in the same order, suggesting the term 'shared preferences.'

Qualitative niche axes offer habitats requiring specific adaptations. To forage better in one means to forage more poorly in the others. Forager species which manage to coexist in a community should therefore rank habitats differently, whence the term 'distinct preferences'.

Note that it is the outlook of the forager that counts in this abstraction. The same axis may afford one set of species a quantitative axis, another a qualitative one. For example, a set of flowering plant species arrayed along a productivity gradient may treat it quantitatively, but its flowers may provide a qualitative axis for some set of nectarivores. Levins (1979) has strongly emphasized the need for ecologists to realize that environmental variation cannot be assessed independently of the organisms which use the environment.

The forager's point of view is again paramount in distinguishing functionally homogeneous from detectably heterogeneous habitat types. A functionally homogeneous habitat is one whose patches cannot usefully be discriminated from each other by foragers. Possibly, the patches are all in fact the same, and there is no heterogenity to recognize. This would be the case, for example, in an ideal free distribution (foragers reside in each patch in whatever density is required for that patch to yield the average rate of return for all patches) (Fretwell 1972). But patches may also be quite different from each other. For instance, termite nests will not each yield the same reward to the consumer who does the work of invading them. But the consumer must do the work before it finds out the yield. Perhaps it can tell this species of termite from that, and this substrate from that. But once it knows these things, once it determines its habitat type, it can only assume the particular sample under attack will yield the average for the habitat.

Detectable heterogeneity is the opposite. The forager can estimate both average yield from a habitat type and how a particular patch compares to the average. Note that fine-grained vs. coarse-grained is not the same distinction as homogeneous vs. heterogeneous. The fine-grained species treats different habitat types as the same; the species unable to detect heterogeneity discriminates among habitat types, but not among different patches of a single habitat type.

The ability to distinguish different samples of the same habitat type has profound consequences. Levins (1979) goes so far as to point out that heterogeneity to which organisms can adapt actually provides a niche axis. In his words, consumers can 'feed on the variance.' Species which can detect heterogeneity live in an environment of greater effective complexity and should be able to coexist at higher diversities (*ceteris paribus*); foragers cannot partition habitats that cannot be distinguished.

Even if species do not use a difference in ability to detect heterogeneity so as to coexist, we must know whether they can detect it to understand their community organization. For instance, species which detect heterogeneity but subdivide a qualitative niche axis should never be haunted by the ghost of competition past, that is, their population dynamics should always reflect their competitive relationship (Brown & Rosenzweig 1986); species in a functionally homogeneous environment which are arrayed on a qualitative axis should evolve to use entirely non-overlapping habitat types and not exhibit competitive dynamics (Pimm & Rosenzweig 1981; Brown & Rosenzweig 1986). Also, detection of heterogeneity can affect the direction of niche shifts: because it changes the slopes of isolegs from positive to negative (compare Figs. 21.2 and 21.6), it predicts that the reduction of a competitor's density may actually decrease niche breadth (Rosenzweig & Abramsky 1986).

Relationship to other schemes of community organization

Can the suggestions emanating from isoleg analysis be related to other suggested plans for abstracting the rules of community organization?

Plant–animal interactions

The terms qualitative and quantitative have been used before in the context of community ecology. They were introduced to describe the constitutive defenses of plants against insect herbivores (Rhoades & Cates 1976). Some plants defend themselves with digestibility-reducing substances such as tannins; the more concentrated they are in a plant tissue, the less desirable it is to a herbivore. These are the quantitative defenses and they set up a niche axis which is essentially the same as the shared preference axes of isoleg theory. Other plants use substances which are toxic even in small quantities. Toxins should evoke specialized herbivores adapted to detoxify or even to utilize them. But each toxin is expected to require a qualitatively different mechanism of neutralization. Hence an array of toxins is a distinct-preference niche axis from the herbivore's point of view. Feeny (1976), simultaneously offering the same ideas, termed the defenses 'apparent' (quantitative) and 'inapparent' (qualitative). The ideas have proved useful in studies of plant defenses (e.g. Fox 1981).

Significantly, Rhoades (1985), although he reaffirms the categories of qualitative and quantitative defences, finds it necessary to erect another pair: stealth and opportunism. These he attaches to the herbivores. Stealthy herbivores have evolved to minimize the effectiveness of food plant defences. Their populations are relatively small and fairly steady. Opportunists strike when plant defences are down. Their populations are highly variable. Rhoades suggests that these strategies are rather independent of qualitative and quantitative defences.

Rhoades' second pair of categories corresponds nicely to the heterogeneous/homogeneous dichotomy of isoleg theory. Opportunists recognize and capitalize on the times and places which are unusually rewarding. Stealthy herbivores take what they can get, and thus get the average.

The principal difference between opportunists and species which recognize heterogeneity appears to be that opportunists concentrate their activities in time, recognizers in space. But this is no real difference. Brown (1986) has shown the great theoretical similarity of space and time recognition. Moreover, his data suggest both are used by desert rodents in Arizona (some of which have highly seasonal activity patterns).

Plant communities in time and space

Connell & Slatyer (1977) propose that the mechanisms of succession should fall into three categories: facilitation, tolerance and inhibition. In facilitation, each sere prepares the environment of its successor. Species are adapted only to the conditions of the sere they belong to. Thus, this mechanism is a rule for subdividing habitats qualitatively.

Similarly, tolerance is interpretable as a quantitative rule. All species can tolerate early successional conditions. These early conditions are characterized by high resource levels, so that, in the absence of competitors, all species should grow best in early environments. But as time passes, resources are depleted and more and more species fall away, unable to tolerate the lower resource levels.

Finally, inhibition is also a qualitative rule. Its crux is that early species have higher vagility and make better colonists, whereas later species are more tolerant of physical extremes or natural enemies. The underlying axis on which the trade-off might be based is unknown.

Another important system for abstracting plant rules is Grime's (1973, 1977). Grime arrays plant strategies along two axes. The stress axis is a productivity axis: high stress is defined to be abiotic conditions leading to poor average production. The disturbance axis reflects the temporally patchy presence of negative biotic and abiotic influences. Low disturbance means such influences are rare and minor.

If disturbance is low, plants adapted to high stress (stress resistors) differ from those adapted to low stress (competitors) along a clear quantitative axis, productivity. This axis duplicates in space the tolerance axis of Connell & Slatyer in time. Grime (1977) himself points out that his distinctions are really founded upon continuous variation in the environment.

There is nothing that precludes qualitative differences among adaptations of plant species based upon qualitatively different stresses. This would parallel the situation called centrifugal community organization in animals (Rosenzweig & Abramsky 1986) and is quite close to the idea of competition for enemy-free space (Holt 1977, 1984; Jeffries & Lawton 1984, 1985). Plant 'competitors' can also be adapted to qualitatively different opportunities, but this would parallel the rules of distinct preference isolegs.

It remains to consider ruderals. These are plants adapted to productive, but intermittent habitats. It seems to me that they differ from competitors in exactly the same way that recognizers and non-recognizers of heterogeneity differ. Recognizers, like ruderals, take advantage of habitat heterogeneity by concentrating their activities in habitats when they are paying high rewards. They are consuming the variance. Non-recognizers treat each habitat type as homogeneous from sample to sample in space.

Ruderals may also be organized along qualitative axes. For example, winter and summer annuals each respond to one of the two rainfall modes of the Sonoran Desert (Shreve & Wiggins 1951). If I may be permitted an animal case, dung beetles (e.g. Hanski 1980b) and carrion-breeding flies (e.g. Kneidel 1984) responding to their two very different temporal opportunities, are another example of qualitatively different ruderals.

In sum, the scheme of Grime is readily joined to the isoleg scheme. Both have a quantitative axis and a heterogeneity axis, and the qualitative option, although missing from the former, is easily appended.

Fundamental niche overlap patterns

Although we must be cautious about expecting pattern among realized niches, it may be found more easily among fundamental niches. Colwell & Fuentes (1975) have suggested that fundamental niche relationships among species can be reduced to three types. The first, coextensive niches, has no ecological cause but depends on the need for species self-recognition during mating (Zwolfer 1974; Rosenzweig 1979b; Colwell 1986). The second, reciprocal overlap, is clearly a case of qualitative rules at work; each species thrives in only its own limited part of niche space. The third, included niches, has the earmarks of quantitative rules; all species can live and thrive at one end of niche space, but fewer and fewer do well as we pass to the other end. Often it can even be shown that the several species have the same optimum habitat (e.g. Grace & Wetzel 1981).

Exceptions

Not all properties of communities have such ready parallels with the isoleg scheme. For example, Cody (1973) has shown that Chilean bird communities have high beta diversity, whereas Californian bird communities of similar overall diversity in a similar climate have high alpha diversity. Another prominent case is the food web analyses of Pimm & Lawton (see Pimm 1984). Properties such as the prevalence of herbivory are related to the relative strength of population interaction and appear to belong to a set of rules not covered in this chapter.

The most telling exceptions are communities which are not organized. The sociological adjective 'anomic,' meaning disorganized (from the Greek for lawless) describes them well. Anomic communities are so fleeting that rules of organization do not have time to do their work. Shorrocks & Rosewell, (Chapter 2) and Atkinson & Shorrocks (1984) have explored such a situation in *Drosophila* species. Carrion flies may also be anomic (e.g. Hanski & Kuusela 1977; Kneidel 1984), although dung beetles —inhabiting what appears to be a similarly ephemeral habitat —do seem to be organized at least partly (Hanski 1980a,b,c; Holter 1982; Yasuda 1984 and unpublished; Doube, Chapter 12). Again, plants suggest their own examples: Shmida & Ellner (1984) present a detailed plant model founded upon the assumption that there are pockets of ephemeral habitat in deserts.

This is not meant to provoke dismay. So what if ecologists give up on a general field theory? We have, all of us, always suspected such a thing would never come to pass, and most of us, I believe, relish the notion of our science's intractability. We are all not-so-secret worshippers of diversity, and so a diversity of explanations and patterns suits our collective personality. But we must not carry a good thing too far. After all, we are scientists. So, reducing the bewildering array of ecological phenomena into a small number of abstractions should also give us considerable aesthetic pleasure.

ACKNOWLEDGMENTS

Thanks to Joel Brown, D. Inouye, S. Louda, C. Smith, M. Shimada, T. Vincent for many ideas and suggestions. The editors deserve much credit for suggesting alterations in the manuscript. Both the US National Science Foundation (DEB-7910506 and BSR-8103487) and the US–Israel Binational Science Foundation supported the research.

REFERENCES

Abrams, P. (1983). The theory of limiting similarity. *Annual Review of Ecology and Systematics,* **14,** 359–76.

Atkinson, W.D. & Shorrocks, B. (1984). Aggregation of larval Diptera over discrete and ephemeral breeding sites: the implications for coexistence. *American Naturalist,* **124,** 336–51.

Bovbjerg, R.V. (1970). Ecological isolation and competitive exclusion in two crayfish (*Orconectes virilis* and *Orconectes immunis*). *Ecology,* **51,** 225–36.

Brew, J.S. (1982). Niche shift and the minimisation of competition. *Theoretical Population Biology,* **22,** 367–81.

Brown, Joel S. (1986). Coexistence on a resource whose abundance varies: a test with desert rodents. Ph.D. thesis, University of Arizona.

Brown, Joel S. & Rosenzweig, M.L. (1986). Habitat selection in slowly regenerating environments. *Journal of Theoretical Biology,* **123,** 151–71.

Charnov, E.L. (1976). Optimal foraging, the marginal value theorem. *Theoretical Population Biology.* **9,** 129–36.

Cody, M.L. (1973). Parallel evolution and bird niches. *Mediterranean Type Ecosystems: Origin and Structure* (Ed. by F. DiCastri & H.A. Mooney), pp. 307–38. Ecological Studies Series No. 7. Springer-Verlag, New York.

Cody, M.L. (1981). Habitat selection in birds: vegetation structure, competition, and productivity. *Bioscience,* **31,** 107–13.

Colwell, R.K. (1986). Community biology and sexual selection: lessons from hummingbird flower mites. *Community Ecology* (Ed. by J. Diamond & T. Case) pp. 406–24. Harper & Row, New York.

Colwell, R.K. & Fuentes, E.R. (1975). Experimental studies of the niche. *Annual Review of Ecology & Systematics,* **6,** 281–310.

Connell, J.H. & Slatyer, R.O. (1977). Mechanisms of succession in natural communities and their role in community stability and organization. *American Naturalist,* **111,** 1119–44.

Diamond, J.M. (1978). Niche shifts and the rediscovery of interspecific competition. *American Scientist,* **66,** 322–31.

Feeny, P.P. (1976). Plant apparency and chemical defense. *Recent Advances in Phytochemistry,* **10,** 1–40.

Fox, L.R. (1981). Defense and dynamics in plant–herbivore systems. *American Zoologist,* **21,** 853–64.

Fretwell, S.D. (1972). *Populations in a Seasonal Environment,* Princeton University Press, Princeton, New Jersey.

Fretwell, S.D. & Lucas, H.L., Jr. (1970). On territorial behavior and other factors influencing habitat distribution in birds. I. Theoretical development. *Acta Biotheoretica,* **19,** 16–36.

Grace, J.B. & Wetzel, R.G. (1981). Habitat partitioning and competitive displacement in cattails (*Typha*): experimental field studies. *American Naturalist,* **118,** 463–74.

Grime, J.P. (1973). Control of species density in herbaceous vegetation. *Journal of Environmental Management.* **1,** 151–67.

Grime, J.P. (1977). Evidence for the existence of three primary strategies in plants and its relevance to ecological and evolutionary theory. *American Naturalist,* **111,** 1169–94.

Hanski, I. (1980a). Patterns of beetle succession in droppings. *Annales Zoologica Fennici,* **17,** 17–25.

Hanski, I. (1980b). Spatial variation in the timing of the seasonal occurrence in coprophagous beetles. *Oikos,* **34,** 311–21.

Hanski, I. (1980c). The community of coprophagous beetles (Coleoptera, Scarabaeidae and Hydrophilidae) in northern Europe. *Annales Entomologica Fennici,* **46,** 57–73.

Hanski, I. & Kuusela, S. (1977). An experiment on competition and diversity in the carrion fly community. *Annales Entomologica Fennici,* **45,** 108–15.

Holt, R.D. (1977). Predation, apparent competition and the structure of prey communities. *Theoretical Population Biology,* **12,** 197–229.

Holt, R.D. (1984). Spatial heterogeneity, indirect interactions, and the coexistence of prey species. *American Naturalist,* **124,** 377–406.

Holter, P. (1982). Resource utilization and local coexistence in a guild of scarabaeid dung beetles (*Aphodius* spp.). *Oikos,* **39,** 213–27.

Inouye, D.W. (1978). Resource partitioning in bumblebees: experimental studies of foraging behaviour. *Ecology,* **59,** 672–78.

Jeffries, M.J. & Lawton, J.H. (1984). Enemy-free space and the structure of ecological communities. *Biological Journal of the Linnaean Society,* **23,** 269–86.

Jeffries, M.J. & Lawton, J.H. (1985). Predator–prey ratios in communities of freshwater invertebrates: the role of enemy-free space. *Freshwater Biology,* **15,** 105–12.

Kendall, M.G. & Moran, P.A.P. (1963). *Geometrical Probability.* Charles Griffin, High Wycombe, Buckinghamshire.

Kneidel, K.A. (1984). Competition and disturbance in communities of carrion-breeding Diptera. *Journal of Animal Ecology,* **53,** 849–65.

Lawlor, L. & Maynard Smith, J. (1976). The coevolution and stability of competing species. *American Naturalist,* **110,** 79–99.

Levins, R. (1979). Coexistence in a variable environment. *American Naturalist,* **114,** 765–83.

Lyon, D.L., Crandall, J. & McKone, M. (1977). A test of the adaptiveness of interspecific territoriality in the blue-throated hummingbird. *Auk,* **92,** 448–54.

MacArthur, R.H. & Pianka, E.R. (1966). On the optimal use of a patchy environment. *American Naturalist,* **100,** 603–9.

Morisita, M. (1950). Dispersal and population density of a water-strider, *Gerris lacustris* L (in Japanese). *Contributions in Physiological Ecology.* Kyoto Univ., no. 65.

Murray, B. (1981). The origin of adaptive interspecific territorialism. *Biological Reviews,* **56,** 1–22.

Pearson, D.L. & Mury, E.J. (1979). Character divergence among tiger beetles (Coleoptera: Cicindelidae). *Ecology,* **60,** 557–66.

Pimm, S.L. (1978). An experimental approach to the effects of predictability on community structure. *American Zoologist,* **18,** 797–808.

Pimm, S.L. (1984). Food webs, food chains and return times. *Ecological communities: Conceptual Issues and the Evidence.* (Ed. by D.R. Strong, D.S. Simberloff, L.G. Abele & A.B. Thistle) pp. 397–412. Princeton University Press, Princeton, New Jersey.

Pimm, S.L. & Rosenzweig, M.L. (1981). Competitors and habitat use. *Oikos,* **37,** 1–6.

Pimm, S.L., Rosenzweig & Mitchell, W. (1985). Competition and food selection: field tests of a theory. *Ecology,* **66,** 798–807.

Pyke, G.H. (1982). Local geographic distributions of bumblebees near Crested Butte, Colorado: competition and community structure. *Ecology,* **63,** 555–73.

Rhoades, D.F. (1985). Offensive–defensive interactions between herbivores and plants: their relevance in herbivore population dynamics and ecological theory. *American Naturalist,* **125,** 205–38.

Rhoades, D.F. & Cates, R.G. (1976). Toward a general theory of plant antiherbivore chemistry. *Recent Advances in Phytochemistry,* **10,** 168–213.

Rosenzweig, M.L. (1979a). Optimal habitat selection in two-species competitive systems. *Fortschritte der Zoologie,* **25,** 283–93.

Rosenzweig, M.L. (1979b). Three probable evolutionary causes for habitat selection. *Contemporary Quantitative Ecology and Ecometrics. Vol. 12. Statistical Ecology.* (Ed. by G.P. Patil, & M.L. Rosenzweig), pp.49–60. International Co-operative Publishing House, Fairland, Maryland.

Rosenzweig, M.L. (1981). A theory of habitat selection. *Ecology,* **62,** 327–335.

Rosenzweig, M.L. (1985). Some theoretical aspects of habitat selection. *Habitat Selection in Birds.* (Ed. by M.L. Cody) pp. 517–40. Academic Press, New York.

Rosenzweig, M.L. (1986). Hummingbird isolegs in an experimental system. *Behavioural Ecology and Sociobiology,* **19,** 313–22.

Rosenzweig, M.L. & Abramsky, Z. (1986). Centrifugal community organization. *Oikos,* **46,** 339–48.

Schaffer, W.M. (1985). Order and chaos in ecological systems. *Ecology,* **66,** 93–106.

Schoener, T.W. (1975). Presence and absence of habitat shift in some widespread lizard species. *Ecological Monographs,* **45,** 233–58.

Shimada, M. & Fujii, K. (1985). Niche modification and stability of competitive systems. I. Niche modification process. *Researches on Population Ecology,* **27,** 185–201.

Shmida, A. & Ellner, S. (1984). Coexistence of plant species with similar niches. *Vegetatio,* **58,** 29–55.

Shreve, F. & Wiggins, I.L. (1951). *Vegetation and Flora of the Sonora Desert.* Carnegie Institution of Washington Publication No. 591, Vol. 1. Washington, D.C.

Smith, C.C. & Balda, R.P. (1979). Competition among insects, birds and mammals for conifer seeds. *American Zoologist,* **19,** 1065–83.

Svardson, G. (1949). Competition and habitat selection in birds. *Oikos,* **1,** 157–74.

Takens, F. (1981). Detecting strange attractors in turbulence. *Lecture Notes in Mathematics* (Ed. by D.A. Rand & L.-S Young) pp. 366–81. Springer-Verlag, New York.

Werner, E.E. & Hall, D.J. (1979). Foraging efficiency and habitat switching in competing sunfishes. *Ecology,* **60,** 256–64.

Werner, P.A. & Platt, W.J. (1976). Ecological relationships of co-occurring goldenrods (*Solidago*: Compositae). *American Naturalist*, **110**, 959–71.

Yasuda, H. (1984). Seasonal changes in the number and species of scarabaeid dung beetles in the middle part of Japan. *Japanese Journal of Applied Entomology and Zoology*, **28**, 217–22. (Japanese with English summary).

Zwolfer, H. (1974). Das Treffpunkt-prinzip als Kommunikationstrategie und Isolations-mechanismus bei Bohrfliegen (Diptera: Trypetidae). *Entomologica Germanica*, **1**, 11–20.

22. SUPPLY SIDE ECOLOGY: THE ROLE OF PHYSICAL TRANSPORT PROCESSES

JONATHAN ROUGHGARDEN[1], STEPHEN D. GAINES[2]
AND STEPHEN W. PACALA[3]

[1]*Department of Biological Sciences, Stanford University,
Stanford, California 94305, USA*
[2]*Hopkins Marine Station, Stanford University,
Pacific Grove, California 93950, USA and*
[3]*Ecology Section, Biological Sciences Group, University of Connecticut,
Storrs, Connecticut 06268, USA*

In both population and community ecology it is increasingly clear that data taken within a study site have limited power to explain what happens in the site. At some scale most ecological systems are open systems, and the control exerted by physical transport processes on population and community dynamics matches the effect of local processes, such as predation and competition among the residents of the site. Here, we briefly review our studies with intertidal barnacles from central California and with insectivorous lizards from Caribbean islands to illustrate this generalization. Offshore circulation processes control both the population and community ecology of the rocky intertidal zone, and plate tectonic motion controls the community ecology of Caribbean islands. This controlling role of transport processes implies that open local models and hierarchical regional models should increasingly be formulated and analysed, and that field studies should invest as much effort in characterizing the transport processes affecting a system as in doing experiments to detect the local biological interactions within the system.

BARNACLES OF THE ROCKY INTERTIDAL ZONE

Barnacles from the rocky intertidal zone illustrate a population's dynamics with open subsystems. A typical study site comprises a small area of rock, from say 0.1 to 10 m^2. Barnacle larvae spend weeks in the water column and are transported far beyond the site where they were released. Hence, the population at the study site is an open system, and this is why oceanic processes play an important role.

Small scale

Consider an area of rock in the intertidal zone. At high tide larvae from the water column land on the rock. Then they find unoccupied spots where they metamorphose into sessile adults. The population dynamics on the substrate are determined by the kinetics of larval settlement, and by the growth and mortality of the already settled organisms. Growth consumes space making it unavailable for recruitment, while mortality converts occupied space back into vacant space. Roughgarden, Iwasa & Baxter (1985) present a simple model to predict the outcome of these three processes. It predicts a qualitatively different picture of population dynamics depending on the settlement rate. Recall from standard demography that a population's age distribution approaches a stable age distribution from any initial condition. This classic result depends on the mixing of reproductive output from all cohorts. For example, if a population is started as a single cohort (all individuals of the same age), then a baby boom occurs as the cohort passes through its reproductive age. When these offspring later pass through the reproductive age themselves, another boom is produced, but not as pronounced as the first because the ages of those producing the second are not quite the same. Gradually, the successive booms decay away leaving the stable age distribution. In contrast, the dynamics of sessile organisms on a rock may not approach a stable age distribution because a cohort may preempt the available space, thereby preventing its mixing with subsequent cohorts. Whether this happens depends on the settlement rate and on the shape of the 'cohort area function', a curve describing the area covered by a cohort as a function of the cohort's age. Specifically, if *both* the settlement rate is high *and* a cohort's area expands before eventually shrinking, then a cohort preempts space on the surface. This causes the population dynamics to be strongly oscillatory, and a stable age distribution is approached slowly if at all.

Figure 22.1 (top) presents the cohort area function for the intertidal barnacle, *Balanus glandula* from the rocky intertidal zone of central California (Gaines & Roughgarden 1985). The cohort area function clearly increases before eventually decreasing. Therefore, an oscillatory component is expected to linger at high-settlement sites, and a relatively smooth approach to a stable age distribution is expected at low-settlement sites. In Fig. 22.1 (bottom), the serial correlation function for the amount of vacant space in the site is presented for two sites, one of which (PR) invariably has a higher settlement rate than the other (KLM). A statistically significant oscillatory component with a period of 30 weeks is evident at the high settlement site (PR), and the oscillatory component is absent at the low settlement site (KLM).

The settlement rate also affects the community ecology of the rocky intertidal zone. A frequent observation at high settlement sites in temperate latitudes is that the species can be ranked in a dominance hierarchy with respect to interference competition (Paine 1984). For example, mussels (M) overgrow any barnacle, and of the barnacles, the balanid (B) overgrows the chthamalid (C), leading to a hierarchy represented as M > B > C. If the settlement rate is sufficiently high for all these species, the system tends to a

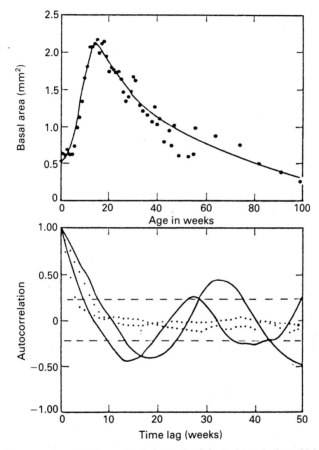

FIG. 22.1. Demography of the barnacle, *Balanus glandula*, in the rocky intertidal zone at Hopkins Marine Station in central California. (Top) Cohort area function. The basal area occupied by a cohort of barnacles is plotted as a function of the age of the cohort in weeks. The cohort area is expressed as the expected area occupied by an individual that settles in the site. (Bottom) Autocorrelation function for amount of vacant space through time. Solid lines refer to two quadrats at a high settlement site (PR) and dotted lines to two quadrats at a low settlement site (KLM). The horizontal dashed lines are the 95% confidence limits for the null hypothesis of white noise fluctuation in amount of vacant space. The autocorrelation functions for PR show a statistically significant oscillatory component.

monoculture consisting of M. If so, disturbance, a generic term for mechanisms of mortality that generate vacant space, prevents the system from culminating in a monoculture, and preserves a mixed community instead (Paine 1966; Dayton 1971). This role of moderate disturbance in promoting species diversity clearly depends on the settlement rate being sufficiently high, because interference competition only occurs in high densities where each organism has extensive physical contact with others. Thus disturbance from wave action, or from a predator such as the starfish, *Pisaster ochraceus*, that preferentially preys upon the competitively dominant prey, has a qualitatively different effect depending on the settlement rates into the local habitat.

Similarly, the classic zonation described between balanid and chthamalid barnacles resulting from interference competition (Connell 1961a,b; Wethey 1983) depends on high settlement rates. Indeed, in central California, where settlement rates are generally low, the zones of both genera usually overlap, and vacant space is present as well, so little interference competition develops (Gaines & Roughgarden 1985).

Medium scale

Because the settlement rate is so important in controlling what happens at local sites within the intertidal zone, the question moves to what determines the settlement rate. Why do some sites regularly have a higher settlement rate than others? Part of the answer lies in the relation between the substrate's chemical and physical propertics and the larvae's preferences, and ability to express those preferences, under field conditions (Crisp 1974, 1976; Scheltema 1974; Lewis 1977). For a particular species of barnacle larvae and type of substrate, however, the settlement rate is also an increasing function of the concentration of larvae in the water. Sites located in places where more larvae are encountered enjoy a high settlement rate for this reason alone. Indeed, the sites at Hopkins Marine Station with a high settlement rate (PR) are situated seaward from those with a low settlement rate (KLM) (Gaines, Brown & Roughgarden 1985). The upstream rocks cast a 'settlement shadow' on the downstream rocks because larvae tend to settle out on the first substrate they encounter. There is nothing inherently unsuitable about the substrate at KLM that leads to low settlement there. In fact, survivorship is actually higher at KLM than at PR because predation is lower—fewer predators are attracted to areas with a low abundance of barnacles.

A further question is what affects the quantity of larvae reaching the margin of the intertidal habitat to begin with. In central California, intertidal habitat is often fringed by kelp forests that harbour fish and invertebrates that prey upon zooplankton. The barnacle larvae pass through these forests twice; when entering the water column as nauplii, and on their return to shore as cyprids. Bray (1981) has shown that fish predation, primarily from the blacksmith, *Chromis punctipinnis*, in kelp forests of southern California, causes a gradient in zooplankton concentration between the incurrent and excurrent sides of the forest. This observation has been confirmed with the kelp forests adjacent to the intertidal habitat at Hopkins Marine Station (Gaines & Roughgarden 1987). The fish causing the predation are juvenile rockfish, *Sebastes* spp. During the summer, all components of the zooplankton are affected, including barnacle larvae. The cyprids are seventy times more abundant outside the kelp community than inside the kelps adjacent to the rocky intertidal zone. The juvenile rockfish abandon the kelp forest late in the fall, and the pattern breaks down, indicating that predation from fish, rather than from invertebrates, is the primary cause of the pattern.

The year-to-year variation in the settlement rate of barnacles seems to be explained by considering three factors: the amount and direction of surface water transport, the size of the kelp forests, and the size of the rockfish stocks (Table 22.1). Barnacle settlement during the years 1982–85 was

TABLE 22.1. Factors affecting barnacle recruitment

	1982	1983	1984	1985
Ekman Currents[1]	46.8 (8.2)	−75.3 (7.7)	52.3 (3.3)	72.1 (6.0)
Kelp Area[2]	36000	8500	33000	46500
Rockfish Stock[3]	Average[4]	0.7 (0.2)	7.2 (2.0)	236.2 (24.6)
Barnacle Recruitment[5]	1.1 (0.1)	3.7 (0.3)	1.4 (0.1)	0.05 (0.01)

[1] Mean (S.E.) expressed as $m^3 s^{-1}$ 100 m^{-1} coastline, + is offshore, − is onshore transport, for January through June, unpublished data from A. Bakun, National Oceanic and Atmospheric administration, used with permission.

[2] Expressed as m^2 at time of peak abundance, rounded to nearest 500 m^2, calculated from aerial photographs supplied by B. Van Wagenen of Ecoscan Resource Data.

[3] Mean (S.E.) expressed as number/min diver observation, unpublished data from E. Hobson, National Oceanic and Atmospheric Administration, used with permission.

[4] Abundance comparable to 1984, personal communication, R. Lee, California Fish and Game, and as shown in bird diets, personal communication, D. Ainley, Point Reys Bird Observatory.

[5] Mean (S.E.) expressed as number cm^{-2} week^{-1} averaged through April to July.

successively moderate, high, moderate, and low. During the same period, the Eckman surface transport driven by wind and Coriolis forces was successively offshore, onshore, offshore, and strongly offshore. The kelp forests were successively extensive, reduced (from storm damage), extensive, and extensive. Finally, the size of the juvenile rockfish stock in the kelp forest was successively average, low, average, and high. Indeed, these factors may be somewhat redundant. The combination of the Eckman transport with the size of the kelp habitat may predict the rockfish stock, and this, in turn, may predict the barnacle recruitment. In any case, these medium-scale processes control the quantity of barnacle larvae that arrive at the margin of the intertidal habitat, and therefore control the overall settlement rates realized throughout the intertidal habitat itself.

Large scale

The species population for a barnacle can be modelled as a metapopulation subdivided into local populations that each has space-limited recruitment. Each local population sends its reproductive output into a common larval 'pool' that, in turn, resupplies the local populations with recruits. For example, with H local populations indexed by y, and one larval pool, the adult dynamics are

$$\partial n_y(x_a, t)/\partial t + \partial n_y(x_a, t)/\partial x_a = -\mu_y(x_a)n_y(x_a, t) \tag{1a}$$

$$n_y(0, t) = F_y(t)\int_0^\infty c_y(x_l)L(x_l, t)dx_1 \tag{1b}$$

$$A_y \equiv F_y(t) + \int_0^\infty a_y(x_a)n_y(x_a, t)dx_a, \qquad (y = 1\ldots H) \tag{1c}$$

and the larval dynamics are

$$\partial L(x_l, t)/\partial t + \partial L(x_l, t)/\partial x_l = -v(x_l)L(x_l, t) - \sum_{y=1}^{H} F_y(t)c_y(x_l)L(x_l, t) \tag{1d}$$

$$L(0, t) = \sum_{y=1}^{H} \int_0^\infty m_y(x_a)n_y(x_a, t)dx_a. \tag{1e}$$

A simplifying assumption is that larval age, x_l, and adult age, x_a, are independent. A larva's age is measured from the time of its release into the water column; when it settles, the clock is restarted, so an adult's age refers to the time since settlement. Equations (1a–c) are the dynamics of the adults in the local populations. The $n_y(x_a,t)$ is the number of organisms in the local population y whose age is between $x_a + dx_a$; $\mu_y(x_a)$ is the instantaneous mortality rate of organisms of age x_a in local population y; $F_y(t)$ is the vacant space, and A_y the total area of substrate in local population y; and $a_y(x_a)$ is the basal area of a barnacle of age x_a in local population y. Equation (1a) describes the ageing and mortality of barnacles after settlement, equation (1b) describes recruitment to vacant space, and equation (1c) stipulates that the total area within each local population is conserved. The integral in (1b) indicates that the settlement rate to vacant space in the yth local population is proportional to the quantity of larvae in the water column, where $c_y(x_1)$ is the constant of proportionality for larvae of age x_1 relative to the substrate at y. Equation (1d–e) are the dynamics in the larval pool. Equation (1d) describes the ageing and loss of larvae, with mortality occurring at the instantaneous rate of $\nu(x_l)$, and the remaining loss representing the larvae that settle out into the local populations. Equation (1e) shows that larvae are added to the pool by summing the reproductive output from all the local populations, where $m_y(x_a)$ is the rate of larval production by adults of age x_a in local population y. The single-species version of this model is introduced and analysed in Roughgarden & Iwasa (1986), the extension to include genetic variation for the ecological parameters in Iwasa & Roughgarden (1985), and the extension to two and more competing species in Iwasa & Roughgarden (1986).

This model predict that coexistence requires habitat specialization in a sense roughly comparable to niche specialization in classical competition models. The details, however, are new and complex. Let $b_{s,y}$ represent, for species s, the expected larval production from an individual that has settled at site y times the probability of landing there from the water column, and divided by the larval mortality rate in the water column (see Iwasa & Roughgarden 1986 for details). This theory predicts that coexistence requires each species to have at least one local site at which it is better than all the rest (i.e. has the highest $b_{s,y}$). For example, consider two species with about the same degree of larval mortality while in the water column. If both species have about the same range along the coast, then their larvae should settle at different depths (e.g. one intertidal, the other subtidal). Alternatively, if both settle at about the same depths, their co-occurrence at a local site should represent a section of the coast where the species ranges overlap, but the centres of the species distributions should not closely coincide.

This idea possibly applies to coexistence among balanid and chthamalid barnacles in Great Britain. As mentioned earlier, zonation between balanid and chthamalid barnacles, as on the west coast of Scotland (Connell 1961a,b), results from interference competition under conditions of high settlement. This high settlement may be caused by relatively confined circulation in the narrow waters between Scotland and Ireland. Both species settle at about the same depths, and the co-occurrence of both species should therefore represent a region of overlap between the ranges of species with different centres of distribution. In fact, *Chthamalus* is absent from the cold waters of Scandinavia and the east coast of Britain where *Semibalanus* dominates, and *Semibalanus* is less common in Wales, Cornwall and the south coast of Britain where *Chthamalus* dominates (Southward 1976; Crisp, Southward & Southward 1981). The regional picture is that *Semibalanus* is a northern form, while *Chthamalus* is a southern form extending from the Mediterranean to the south and west of Britain (see Southward & Crisp 1954). The west coast of Scotland has two special features: the ranges of both genera overlap there with approximately equal abundance, and local settlement rates are high. The account of intertidal ecology obtained from the west coast of Scotland thus represents the rocky intertidal zone in a special area of Europe.

The geographic distribution of oceanic circulation patterns perhaps explains the fundamentally different pictures of rocky intertidal ecology reported from different coasts around the world. Parrish, Nelson & Bakun (1981) have discovered a conspicuous latitudinal gradient in the extent of Ekman onshore transport along the western coast of North America. Northern latitudes have a markedly higher onshore transport than more southern locations. Central California, in particular, faces primarily strong offshore transport most of the year. The distribution of kelps also follows a latitudinal gradient, with the greatest kelp biomass occurring in southern and central California (Foster & Schiel 1985). The latitudinal gradients in both factors should contribute to producing a latitudinal gradient in settlement rates, and hence, the coast of Oregon and Washington should have higher settlement rates than in northern and central California. The settlement rate at Boiler Bay, Oregon (T. Farrell, personal communication) is 200 times as large as that at Hopkins during 1985, motivating speculation that more intensive sampling along the Pacific Coast would reveal a continuous latitudinal gradient in settlement rate. Such a latitudinal gradient could explain why rocky intertidal studies from Washington and Oregon emphasize post-settlement processes as controlling community structure, and rely on concepts like the intermediate disturbance principle and the keystone predator (Dayton 1971; Menge & Sutherland 1976; Paine 1984) that are seen only in high settlement conditions. In contrast, our studies from central California find settlement itself to be the rate-limiting step in community dynamics, and the controls of community structure there act via their effect on the settlement rate. Similarly, the observations of Underwood, Denley & Moran (1983) from the rocky coast of New South Wales in Australia seem to represent a settlement-limited community.

A theoretical task for the future is to explore specific submodels for the coupling among the local populations founded on hydrodynamic transport processes. For example, the subpopulations can be arrayed along a hypothetical coast. The larval pool offshore can be taken as a spatially distributed population of larvae, and the coupling among the local populations of adults mediated by diffusion with the superposition of directional water transport, in a manner reminiscent of the model offered by Jackson & Strathmann (1981). In this way one can make a tentative first step toward integrating the population biology of benthic organisms with the descriptions of offshore physical oceanographic processes now appearing for shelf seas around the British Isles (Pingree & Griffiths 1980) and for the California Current System (Hickey 1979; Huyer 1983; Ikeda & Emery 1984; Mooers & Robinson 1984; Koblinsky, Simpson & Dickey 1984).

ANOLIS LIZARDS OF EASTERN CARIBBEAN ISLANDS

Unlike barnacle populations, a lizard population in a typical study site (say 0.1 to 10 ha) is effectively a closed system. But the community at the site may be open, thus providing the *entrée* for physical transport processes to assume a major role. In marine systems the water column is a habitat itself; feeding and dispersal co-mingle as larvae spend a major fraction of their life-span in the water column. In contrast, nothing remains airborne for very long. This simple fact governs the level (population or community) and scale at which marine and terrestrial ecological systems are open, and the degree to which geographically separated systems are functionally isolated from each other.

Small scale

In the Lesser Antilles, anoles replace ground-feeding insectivorous birds. Their abundance typically varies between 0.25 to 4 lizards per m^2 depending on the quantity of insects in the habitat, and on the season. Avian predation is low and, unusually for a lizard, anoles themselves are virtually top predators. Island anoles have exceptional population-dynamic stability based on census records taken at least once a year on St. Maarten from 1977 through the present. The abundance varies by a factor of two during the year, as juveniles recruit to the population during the winter; however, year-to-year variation in abundance compared at the same point in the seasonal cycle (summer) is less than 10% (Table 30.1 in Roughgarden 1986). An anole on St. Maarten can grow to maturity in one year, and the maximum life span is perhaps 5 years. Food limits the abundance of anoles. Anole abundance correlates with insect abundance (Roughgarden, Heckel & Fuentes 1983); food supplementation causes higher growth rates (Licht 1974; Stamps 1977); experimentally increasing the abundance of anoles lowers lizard growth rates (Pacala & Roughgarden 1985); and experimentally removing anoles increases the number of insects on the forest floor by a factor of two and of spiders in the woods by a factor of 20 to 30 (Pacala & Roughgarden 1984).

Most of the islands in the Lesser Antilles have only one species of anole, some have two, and none has three or more in natural habitat, although 'enclaves' of introduced anoles near houses or beaches may boost an island's total species count above two. All the islands with two species of *Anolis* have species that differ greatly in body size, except for St. Maarten. Comparative community studies suggested that strong interspecific competition was occurring today on St. Maarten, and experimental studies during the last 5 years with introductions to offshore cays, and with exclosures placed in natural habitat, have confirmed that suggestion (Roughgarden, Pacala & Rummel 1984; Pacala & Roughgarden 1985; Rummel & Roughgarden 1985a). Interspecific competition between anoles depends on similarity in body size; the closer the body sizes, the stronger the competition. Body size correlates with prey size, and a difference in body size indicates a partitioning of the insect food supply with respect to prey size. Further partitioning correlated with body size results from differences in foraging location. Larger lizards perch above smaller lizards, presumably to obtain a better vantage point from which sight to their larger, preferred prey. This leads to larger lizards foraging further from the base of trees than small lizards, thereby amplifying the degree of partitioning revealed by prey size differences alone.

The competition experiments show that the distance in morphological niche space over which competition can be detected is short. Competition is barely detectable between species whose adults differ, on the average, by 30 mm in length. Provided their carrying capacities are roughly equal, two species of *Anolis* can probably coexist if they differ by only 10 mm in length (the 'limiting similarity'). Since the competitive distance is short, the Lesser Antillean anole faunas are probably undersaturated. That is, perhaps three species, for example, a 50 mm species, an 80 mm species, and a 110 mm species could coexist on most of the islands. If so, the guilds are still open, and what is in them is determined to a large extent by physical transport processes.

Medium scale

The scale between the local study sites discussed above and the entire Caribbean theatre is the community that is geographically distributed across an island. If the community in an area of, say, 1 km² or less, is weakly coupled to the communities in adjacent areas of about the same size, then the island-wide community can be pieced together from information taken at these adjacent areas, as though assembling a quilt (Karlin & McGregor 1972).

This approach underlies a hypothesis for the species distributions of St. Maarten (Roughgarden, Pacala & Rummel 1984). Strong competition occurs there today between *A. gingivinus* and *A. pogus*. *A. gingivinus* is about 10 mm larger than *A. pogus*. Also, *A. gingivinus* occurs throughout the island while *A. pogus* occurs only in the central hills. We also know (i) that *A. gingivinus* can tolerate a higher body temperature than *A. pogus* before showing signs of heat stress, even though both species have the same body temperatures on the average and perch in the same microclimate, and (ii) from introduction experiments, that *A. pogus* can survive and reproduce in the hotter habitat near sea-level where it does not occur. If we then hypothesize that hotter temperatures, while far from lethal, nonetheless reduce the daily activity for *A. pogus* relative to *A. gingivinus* we expect that the *ratio* of the carrying capacities, K_p/K_g, is lowest at the coast in xeric habitat, and increases continuously into the relatively mesic and cool hills. Since *A. pogus* does coexist in the hills, the competition coefficient (α) of *A. gingivinus* there evidently satisfies $K_p/K_g > \alpha_{p,g}$, whereas at sea-level, where *A. pogus* is competitively excluded by *A. gingivinus*, the reverse inequality must hold, $K_p/K_g < \alpha_{p,g}$. We hypothesize therefore, that the edge of the range for *A. pogus* is at the elevation where the carrying capacity ratio roughly equals the competition coefficient. This approach would not be valid if we had to treat the island-wide community as though it were assembled from strongly coupled subsystems, as is necessary in marine systems.

Also at this intermediate scale, the two-species anole guilds were discovered to have two distinct patterns of species replacement along intra-island environmental gradients (Roughgarden, Heckel & Fuentes 1983). In the northern islands (with Bimaculatus Group anoles), both species vary in parallel, with the smaller species being more abundant than the larger in all habitats. Both are relatively scarce in dry habitat and abundant in middle-

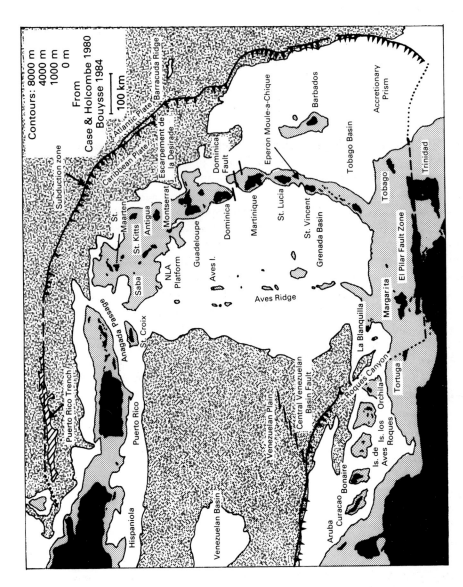

Fig. 22.2. Eastern Caribbean region. Solid areas are above sea level today.

elevation woods and forest. But in the southern islands (with Roquet Group anoles), the smaller species is more abundant than the larger at sea-level, and *vice versa* in middle-elevation woods and forest. Thus, in the south the species replace each other along gradients, and do not vary in parallel. This finding was hypothesized to represent the evolution of habitat specialization in the south, where the upland habitat is more extensive than in the north, as suggested by the theory of density-dependent selection along a cline. This issue needs to have alternative hypotheses developed, however, and to receive further study.

Large scale

Most geologists and biologists believe the Lesser Antilles to be relatively recent oceanic volcanic islands. The extensive present-day volcanic activity on the Lesser Antilles befits their placement at the leading edge of the eastward-moving Caribbean Plate (see Fig. 22.2), and almost all of the exposed rock on the islands is Eocene or later. All this is consistent with viewing the Lesser Antilles as a homogeneous group of rather young geologic structures. As such, each species on them must have colonized independently by some mechanism of over-water dispersal and the fauna should consist of taxa with good colonizing abilities.

Accordingly, the first hypothesis for the assembly of anole guilds (Williams 1972) supposed that an island was consecutively colonized by species with the same body size followed by evolutionary divergence in body size (ecological character displacement; Brown & Wilson 1956). Moreover, the Lesser Antilles were hypothesized to represent early stages in the buildup of complex communities, as further invasions would presumably generate a diverse fauna.

The character displacement hypothesis is not, however, consistent with phylogenetic data, data on the ecology of invasions, or the fossil record. St. Maarten, in particular, does not have any species that has recently colonized from an adjacent area. This is required by the hypothesis that two similar-sized species represent the stage before character displacement has evolved. Moreover, both accidental and experimental introductions (Wingate 1965; Roughgarden, Pacala & Rummel 1984) show that an invading species cannot 'take' when introduced into a habitat already occupied by a resident the same size as the invader. Finally, fossil evidence from the Antigua Bank (Etheridge 1964; Steadman, Pregill & Olson 1984) and Anguilla Bank (J. Roughgarden & S. Pacala, unpublished data) both show large lizards becoming smaller since the Pleistocene, not medium-sized lizards becoming larger. Thus, similarity in body size is a derived condition resulting from, not an initial condition preceding, co-evolution.

An alternative hypothesis, but still in the faunal-buildup *genre,* was based on a model from competition theory (Roughgarden, Heckel & Fuentes 1983; Rummel & Roughgarden 1985b). The model consists of alternating episodes of invasion and coevolution. A co-evolutionarily stable fauna is invaded, and then the augmented fauna comes to co-evolutionary equilibrium itself. These episodes are repeated until the fauna cannot be further invaded, or until a cycle results.

The invasion episode was modelled with the Lotka–Volterra competition equations. The initial condition for an invasion is a resident fauna at population-dynamic and co-evolutionary equilibrium, and a rare invader. No evolution is assumed to occur during the invasion. Different choices of invaders lead to branches in a graph representing the various pathways of faunal assembly. After the invader has established, the body sizes in the augmented fauna co-evolve such that each species climbs its own 'adaptive surface' (Wright 1931), that is, evolves in body size in a direction leading to an increase in mean fitness. After each generation the adaptive surface for each species is recomputed to take account of all evolutionary changes. Eventually, the adaptive surfaces stop changing and each species is near a peak of its adaptive surface, and so a co-evolutionary equilibrium is achieved. Then the fauna is tested again to find all the points where another invasion is likely. The model does not take account of a possible change in niche widths; it only pertains to the mean body size, and presumes that this is relatively invariant.

Perhaps the most spectacular result from this faunal-buildup model is that, with asymmetrical competition, species may become extinct during co-evolution. Thus, in theory, co-evolution does not necessarily stabilize a community by spacing out the niches of the residents; instead, stability following co-evolution may simply reflect the sparseness of the community that remains after the extinctions have occurred.

A special case of the faunal-buildup model suggested the hypothesis that a cyclic process, analogous to the taxon cycle of Wilson (1961), was happening in the northern Lesser Antilles. An island with one species is supposed to be colonized by a second species that is larger than the original resident. This second species then evolves a smaller body size, converging upon the original resident, and eventually causing its extinction. At this point, another invader could, in principle, trigger another cycle.

The taxon cycle hypothesis is consistent with the phylogenetic relations among eastern Caribbean anoles, the ecology of colonization, and the changes in body size since the Pleistocene revealed by the fossil record. The current information does differ in some significant details from the original hypothesis (namely by removing Marie Galante from the cycle). Nonetheless, as Fig. 22.3. shows, all the available evidence points to the following sequence: (i) a community in the northern Lesser Antilles with solitary-sized species is invaded by a larger species from the vicinity of Guadeloupe, (ii) the invader subsequently converges to the solitary size itself, (iii) as a result, the range of the original resident contracts, and it eventually becomes extinct. Different islands in this region show various pieces of the sequence, and different islands start the sequence at different points depending on the

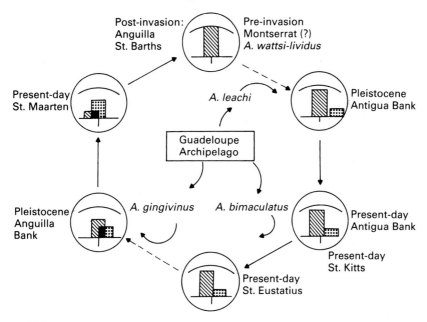

FIG. 22.3. Taxon-cycle on the NLA platform. Circles depict stages in the cycle. Within a circle, the horizontal axis represents body size, and the vertical axis represents relative abundance. The overlying curve represents the resource spectrum. An island with a solitary-sized species, like Montserrat, is invaded by a larger species from the Guadeloupe Archipelago. On the Antigua Bank fossil remains demonstrate that the larger anole, *A. leachi* has become smaller since the Pleistocene, although it is still quite large. On the St. Kitts Bank, the larger anole, *A. bimaculatus* is smaller on St. Eustatius than on the rest of the bank, which is interpreted as a derived condition. On the Anguilla bank fossil remains indicate an invasion of the larger species, *A. gingivinus,* about 4000 years B.P. It was larger than its present size, but smaller than the larger anoles of the other two banks. Today, the smaller species of the Anguilla bank, *A. pogus,* is found only on St. Maarten where its range has constricted to the hills in the centre of the island, and *A. pogus* has become extinct during historic times on Anguilla itself. An extinct species of the lizard genus *Leiocephalus* has been found on two of the three island banks, and the extinction appears to have occurrred about 6000 years B.P. on the Anguilla bank; however, no wholesale extinction of the herpetofauna has occurred since the Pleistocene. Human habitat destruction may have contributed to extinctions during historic times, and endemic species of the lizard genera, *Ameiva* and *Iguana,* and the snake, *Alsophis,* are endangered today by the introduced monogoose, *Herpestes,* on many islands (J. Roughgarden & S. Pacala, unpublished data).

size of their invader. A very large invader begins at step (i) as on the Antigua Bank, and an invader of only moderate size to begin with starts at step (ii) as on the Anguilla Bank.

The largest scale

The issue now arises of whether the conceptual framework of models pertaining to the colonization of oceanic islands by over-water dispersal is itself applicable to much of the Lesser Antilles. Are faunas really 'assembled' from propagules of over-water colonists whose resource requirements squeeze in between the resource use of the residents? Or, are faunas 'inherited' with the habitat itself; are the faunas chips of a parental fauna accompanying the fragmentation of a geologic entity? To answer this fundamental issue, we now sketch a new hypothesis for the geologic origin of the eastern Caribbean region (Roughgarden 1987). It will emerge that the taxon-cycle sequence just discussed is appropriate to a special section of the northern Lesser Antilles to be called the NLA platform. Over-water dispersal of anoles is feasible, and appears to have occurred, among the islands within this limited area. Otherwise, the composition of anole guilds is primarily determined by the plate tectonic processes that underlie the origin and fragmentation of geologic entities.

Anoles can serve as 'living strata' to aid in reconstructing the geologic origin of the Lesser Antilles. For a biological taxon to mark a geologic entity, it must readily colonize any newly formed entity, it must rebuff further colonizations by the members of adjacent geological entities for otherwise the labels would mix, and it must not be prone to extinction, especially from the demographic stochasticity associated with a small population size. Anoles are well known as hardy survivors on tropical cays; the competition between anoles retards cross invasion between adjacent entities, and prevents cross invasion altogether between species of the same body size; and the huge population size on a Lesser Antillean island (estimated at 10^8 for a 400 km^2 island) precludes chance extinction and increases the likelihood of contributing a propagule to newly opened habitat. Moreover, anoles have been in the Caribbean theatre for a long time. A complete fossil *Anolis* in amber dated at 20–23 million years ago has been found in the Dominican Republic (Rieppel 1980). It is a juvenile lizard indistinguishable from the green anoles now living there, *A. chlorocyanus* and *A. aliniger*. Finally, the lineage (clade) consisting of the present-day Iguanidae, Agamidae, and Chamaeleonidae, and including *Anolis*, is an old line, existing back into the middle Jurassic (175 million years ago) (Estes 1983). In contrast, old rock strata in the Caribbean are scarce, as Jurassic and

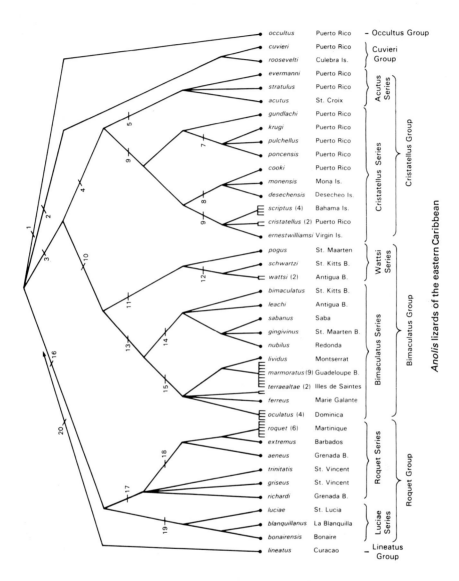

FIG. 22.4. Major lineages of *Anolis* lizards in the eastern Caribbean. The classified taxa are the populations on present-day island banks. Populations showing geographical variation that resulted in the naming of subspecies (geographical races) are indicated as horizontal lines with vertical tick marks to denote each subspecific name that has been defined. (From Roughgarden 1987.)

Cretaceous materials have been lost to subduction or covered by more recent magmatic activity. The systematics of these lizards may therefore provide sorely needed clues about the origin of the Caribbean to supplement geologic data.

Figure 22.4 is a phylogenetic tree for eastern Caribbean anoles with the species arranged, so far as possible, in clockwise order beginning with Puerto Rico and ending with Curacao in the Netherlands Antilles (Roughgarden 1987). The tree expresses a consensus of systematic data from squamation, karyotypes, the electrophoretic assay of proteins from over 20 loci, and the immunological analysis of albumins, as based on the research of H. Dessauer, G. Gorman, H. Heatwole, R. Henderson, P. Hummelinck, J. Lazell, A. Schwartz, D. Shochat, R. Thomas, G. Underwood, and E. Williams during the last four decades.

The northern Lesser Antilles are populated by the Bimaculatus Group, a sister lineage to the Cristatellus Group of Puerto Rico. The Bimaculatus Group itself contains the Wattsi Series, all of which are small, brown lizards that perch within a few feet of the ground, and the relatively arboreal Bimaculatus Series which are large or medium-sized usually green or grey-green lizards with white, blue or red accent markings. On the northern island banks with two species (Anguilla, Antigua, and St. Kitts) the smaller species is always a member of the Wattsi Series, and the larger always a member of the Bimaculatus Series.

The southern Lesser Antilles are populated by the Roquet Group, a lineage found only on islands in a triangular region of the south-eastern Caribbean extending from Martinique in the Lesser Antilles to Bonaire in the Netherlands Antilles. The Roquet Group itself contains the Roquet Series, all of which are in the Lesser Antilles and the Luciae Series that, curiously, is found on Bonaire and La Blanquilla in the Nether lands–Venezuelan Antilles and on St. Lucia, which today lies squarely between two islands with Roquet Series anoles. Curacao, adjacent to Bonaire, has an anole related to those in Jamaica and the Cayman islands; it is not at all closely related to the species on Bonaire.

Finally, the species on the central islands of the Lesser Antilles appear quite old in having geographical variation in body size, body colour, squamation, and secondary sexual characters such as head and tail crests. Early systematists recognized this geographical variation by naming twelve 'subspecies' within the Guadeloupe archipelago, four in Dominica, and six in Martinique. The species of other islands are internally homogeneous, and tend to vary geographically only in background hue.

The phylogenetic tree contributes to refuting the hypothesis that anoles colonize the Lesser Antilles by successive over-water dispersal from Puerto

Rico in the North, and from Venezuela in the south. By this hypothesis, islands close to the source regions should have more anole species than those far from the source regions (the classic distance effect of island biogeography) *and* the more recent invasions from the source region should still be closely-related to taxa in the source region. Instead, in both the north and south, single-species islands lie between the putative source areas and the two-species islands. Also, the anoles in the north are more closely related to the species on Guadeloupe at the centre of the arc than to anything in Puerto Rico, and in the south anoles are more closely related to the species on Martinique in the centre of the arc than to anything in Venezuela.

The tiny cays on the Puerto Rico Bank (British Virgin Islands) provide the signature of habitat fragmentation (vicariance). The species diversity of reptiles and amphibians can be represented as nested subsets relative to island area. The smallest cays (< 1 ha) have *A. cristatellus*, slightly larger cays also have a gecko, *Sphaerodactylus macrolepis*, still larger cays also have *A. stratulus*, and so forth (Lazell 1983). During the Pleistocene 15000 years ago, the Puerto Rico Bank extending to Anegada was above water. As the glaciers melted, slight hilltops became the tiny cays of the British Virgin Islands today. The lizards on these cays presumably did not have to disperse there, as they were there to begin with. The fauna on the cays consists of species that have the property of not becoming extinct when the size of their habitat shrinks. They have, so to speak, a good 'bottle-necking' ability, and colonizing ability is probably irrelevant to their presence on the cays of the Puerto Rico Bank. Precisely, the Puerto Rican taxa from the smallest cays are the sister taxa to the radiations in the northern Lesser Antilles. This implies that the fauna in the northern Lesser Antilles is a differentiated fragment of the Puerto Rican fauna.

A relatively obscure geologic fact is especially important to reconstructing the geologic origin of the northern Lesser Antilles. La Desirade, on the Guadeloupe archipelago, has Jurassic rock (Fink 1972). After repeated study, no doubt remains as to its magmatic origin, similar in appearance, chemical composition, and mineralogy to the basal units of the Greater Antilles, and especially Puerto Rico. (Bouysse, Schmidt-Effing & Westercamp 1983). Although a tiny sliver of land on an aerial map (≈ 30 km^2),La Desirade is the tip of an 'iceberg' marking the edge of a cliff that drops 5000 m to the ocean floor in less than 10 km horizontal distance, and is the site of a distinct local positive gravity anomaly (Bourguer anomaly). Tomblin (1975, p. 471) wrote '...in view of the long time interval betwen these and the next oldest rocks in the Lesser Antilles and the large horizontal movements which undoubtedly took place in the region during this interval ...it is possible that the crust now exposed in Desirade represents a small block which was much closer to the Virgin Islands in Late Jurassic time and which subsequently separated from this area and moved relatively eastward.' More recently Bouysse (1979, 1984), Speed (1985), and Donnelly (1985) have all referred to the origin of the northern Lesser Antilles as conjoined in some way with Puerto Rico.

The phylogenetic tree of *Anolis* strongly supports this type of geologic hypothesis; the Bimaculatus Group of the northern Lesser Antilles is sister to the Cristatellus Group of Puerto Rico. The bifurcation in the phylogenetic tree where these lineages split would seem to coincide with the movement of a small block representing proto-Guadeloupe away from the Puerto Rican area. This block would carry on it the ancestor of *A. cristatellus*, just as the small cays of the Puerto Rico bank today retain *A. cristatellus* itself. Moreover, the entire range of the Bimaculatus Group coincides exactly with the geologic entity bounded by the 1000 m depth contour encompassing the area from the Anegada passage to the passage between Dominica and Martinique.

We term the structure that contains the present-day islands of the Anguilla Bank down through Dominica as the 'NLA platform'. The taxon cycle involving *Anolis* seems to be occurring on the Anguilla, Antigua, and St. Kitts Banks toward the northern part of this platform as a result of dispersal from the Guadeloupe archipelago, which is the oldest portion of the platform.

Continuing south, the phylogenetic tree of *Anolis* shows a major dichotomy between the Cristatellus–Bimaculatus lineage of Puerto Rico and the NLA platform versus the Roquet Group of the south-eastern triangle of the Caribbean region. The top corner of this triangle possibly marks a remnant of the enigmatic plate boundary between the North and South American plates. This remnant may intersect the Lesser Antillean arc at about 15°N, between Dominica and Martinique, where Tomblin (1975), Vierbuchen (1979), and Dorel (1981) have noted the deepest seismicity of the Lesser Antilles; the seismicity occurs there with an east–west trend, and Stein *et al.* (1982) show that strike slip faulting occurs there at depth. This fault is labelled as the 'Dominica Fault' in Fig. 22.2. A line extending from this fault to some point between Bonaire and Curacao appears to coincide with a major difference in the crustal structure of the ocean floor in the Venezuelan Basin. Biju-Duval *et al.* (1978) discovered three parallel faults trending east-north-east that they termed the 'central Venezuelan Basin fault zone'. Diebold *et al.* (1981) confirmed this discovery and characterized the very different crustal structure on opposite sides of this fault zone. Moreover, this south-eastern corner is a magnetically quiet zone, in contrast to the rest of the Venezuelan Basin where north-east–south-west trending magnetic anomalies parallel the central Venezuelan Basin Fault (Ghosh, Hall & Casey 1984). Although differing somewhat from the hypothesis offered here, Bouysse (1984) has suggested that the northern Lesser Antilles has developed separately from the southern Lesser Antilles, and Speed (1985) has alluded to a similar possibility. The data on *Anolis* strongly support this kind of hypothesis. Presumably, this south-eastern triangle, that harbours the Roquet Group of *Anolis,* is a piece of the South American plate that has become sutured to the present-day eastward-moving Caribbean plate.

Within the south-eastern triangle there is further heterogeneity. The coast of Venezuela, including the Netherlands–Venezuelan Antilles, appears to represent the remnants of an island arc that collided with the coast of South America in the late Cretaceous (Maresch 1974). Curacao and Bonaire have different basements, formed at a mid-ocean spreading ridge and at a subduction zone respectively (Beets & Mac Gillavry 1977; Beets *et al.* 1984), and the anoles on these adjacent islands are not at all closely related. In the Lesser Antilles, the chemical composition of the rocks on St. Lucia differ markedly from those of its neighbours (Fig. 2 in Tomblin 1975) just as St. Lucia's anole is not closely related to those of its neighbours.

Bucher (1952) suggested that the motion of the Caribbean plate between the North and South American plates is analogous to the motion of a glacier in a valley. The analogy explains the opposite symmetry of pull-apart basins with the strike–slip motion at faults in the north and south boundary zones of the Caribbean plate (Mann & Burke 1984). We suggest extending the analogy by viewing the southern Lesser Antilles as the 'terminal moraine' at the southern half of the leading edge of the Caribbean plate. As this part of the Caribbean plate has moved eastward, it may have accumulated some islands in its path. The magmatic activity at the subduction zone adds an enormous amount of material tending to mask nuclei that may have both heterogeneous origins and times of entry into the island arc moraine. Specifically, St. Lucia (near Eperon Moule-a-Chique) may have a different geologic origin from its present-day neighbours of Martinique and St. Vincent, and instead, may have originated with Bonaire and La Blanquilla, both of which are now lodged on the Venezuelan coast.

New discoveries point to an older age for the basement of the Lesser Antilles than previously believed. In 1975, the oldest material known in the Lesser Antilles, apart from the then controversial La Desirade find, was Eocene ($\approx$50 million years ago), with large deposits on the NLA platform, and on the Grenada Bank. These facts contributed to the acceptance of a Tertiary origin for the Lesser Antilles. Now, Upper Cretaceous ($\approx$ 70 million years ago) basement material has been obtained from both slopes of the Anegada Passage (Speed, Gerhard & McKee 1979; Bouysse, Andreieff & Westercamp 1980; Bouysse *et al.* 1985), and on Union Island in the Grenadines (Westercamp *et al.* 1985). Thus, the basement of the Lesser Antilles is certainly older than the Tertiary, and its origin therefore reflects the origin of the Caribbean region itself. These older ages and the geologic heterogeneity discussed above can now be interpolated into plate tectonic reconstructions of the geologic origin of the Caribbean that have assumed a relatively recent origin for the structures of the eastern Caribbean (e.g. Pindell & Dewey 1982).

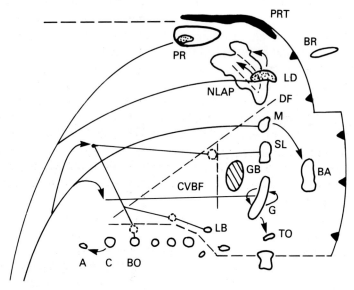

FIG. 22.5. Schematic for proposed geologic origin of the eastern Caribbean. Abbreviations are: PR, Puerto Rico; PRT, Puerto Rico Trench; BR, Barracuda Ridge; LD, La Desirade; NLAP, Northern Lesser Antilles Platform; DF, Dominica Fault; M, Martinique; SL, St. Lucia; BA, Barbados; GB, Grenada Basin; G, Grenada-St. Vincent; TO, Tobago; CVBF, Central Venezuelan Basin Fault; LB, La Blanquilla; BO, Bonaire; C, Curacao; A, Aruba. Lines ending in an arrow indicate *Anolis* populations. Lines without arrows indicate the path traced by *Anolis* populations as they occupied basal elements of geologic structures that today are in the eastern Caribbean.

Figure 22.5 presents a reconstruction of the geologic origin of the eastern Caribbean that takes into account the points raised above. The two halves of the Lesser Antilles are hypothesized to develop separately. In the north, Puerto Rico and La Desirade split during the Late Cretaceous. As they drift apart, the NLA platform, containing the present-day Anguilla, Antigua, and St. Kitts banks, forms as an accretionary prism. Sometime in the late Tertiary, first the *A. wattsi* lineage, and then the *A. bimaculatus* lineage colonize the exposed areas on the platform by overwater dispersal. A taxon cycle as outlined in Fig. 22.3 appears to be occurring here. In the south matters are certainly more complex. An ancient island arc consisting of basal elements of Bonaire, La Blanquilla, and St. Lucia is hypothesized to lie in the path of material that contributes to Martinique and Grenada-St. Vincent. As this material moves east, it squashes some islands into the Venezuelan coast, and abducts St. Lucia. Then back-arc spreading brings the southern Lesser Antilles in line with the islands in the north, and leaves the Aves Ridge as a residue. The now fully-consolidated Caribbean plate carries the Lesser Antilles eastward as a single unit. Obviously this hypothesis needs to be 'chewed over', but something along these lines seems necessitated by the biologic and geologic heterogeneity now known in the Lesser Antilles, and by the age of the material itself.

These studies suggest that the composition of *Anolis* guilds in the Eastern Caribbean results from the combination of competition, habitat bottle-necking, and plate tectonics. Ecological competition prevents cross-colonization by species of the same or very similar body size, but otherwise the strength of competition between anoles is slight. When a geologic fragment splits from a larger unit a very particular set of species is retained on the fragment; those with good 'bottle-necking' ability. This special extraction then develops in isolation and never represents an early stage in the buildup of a complex fauna.

The role of plate tectonics in determining the composition of communities would seem to motivate new theoretical questions in community ecology. The 'disassembly of communities' is mysterious. Why does habitat bottle-necking produce a species diversity pattern involving nested subsets? Can habitat bottle-necking serve as a model for mass extinction? Suppose communities on separate fragments collide. When will the two communities fuse to form one system, and what will that system look like? Or conversely, when will the communities remain as a regional mosaic, with a faunal break marking a geologic suture zone?

CONCLUSION

The new findings reviewed here show the ecology of barnacles from the marine intertidal zone, and lizards from Caribbean Islands, to be controlled by physical transport processes. The investigation of such processes falls in the domains of oceanography, geology, and perhaps also meteorology. Ecological studies, except those on very long-lived organisms, tend to regard the earth sciences at arm's length; earth-science processes merely set the backdrop for ecological processes, but do not play a leading role. But earth-science processes actually control both the population and community ecology of the intertidal zone and the community ecology of Caribbean islands to an extent that matches the effect of any biological mechanisms that operate within these ecological systems. The reason is that these ecological systems are, at some level and scale, open systems. And it is the supply of juveniles to populations in a section of rocky intertidal habitat, and of propagules to communities on Caribbean islands, that is the rate-limiting step.

This importance of large-scale physical transport processes implies that ecology will never be completely experimental. During the last decade, small-scale species interactions have been increasingly investigated with field experiments in both marine and terrestrial environments. The larger-scale processes are not similarly amenable to experiments, because the enthusiasm of the Welsh for rerouting the Gulf Stream, and of Californians for provoking the San Andreas Fault, is likely to be restrained. Nonetheless, the union of ecological approaches, such as natural experiments, field perturbation experiments, theoretical models, and systematics, with methods for large-scale phenomena from earth sciences appears sufficient to solve the questions ecologists are responsible for answering. And finally, it seems to us there is more uniformity of method in ecology than the tenor of some of its internal disputes suggests. By using all the available methods, and no less, ecology today is successfully explaining the abundance, distribution, and phenotype of organisms.

ACKNOWLEDGMENTS

We thank Michael Novacek and Eugene Gaffney of the American Museum, and Robert May and Andrew Dobson of Princeton University, for hospitality while the research on the geological origin of the Caribbean was begun, and Joel Cracraft, David Larue, and James Case for invaluable discussion on historical analysis and on the geology of the Caribbean. We also thank the US Department of Energy and the National Science Foundation for supporting research on barnacle population dynamics and on the community ecology of Caribbean lizards respectively. Finally, J.R. gratefully acknowledges support from the Guggenheim Foundation while this article was written.

REFERENCES

Beets, D.J. & Mac Gillavry, H.J. (1977). Outline of the Cretaceous and Early Tertiary history of Curacao, Bonaire, and Aruba. *Guide to Geological Excursions on Curacao, Bonaire, and Aruba*, pp. 1–6, STINAPA Documentation Series No. 2, Caribbean Marine Biological

Institute, Curacao. (Also published as GUA Papers of Geology, Series. 1, No. 10; Stichting GUA, Geologisch Institut, Amsterdam).

Beets, D.J., Maresch, W.V., Klaver, G.T., Mottana, A., Bocchio, R., Beunk, F.F. & Monen, H.P. (1984). Magmatic rock series and high-pressure metamorphism as constraints on the tectonic history of the southern Caribbean. *Geological Society of America Memoir,* **162,** 95 – 130.

Biju-Duval, B., Mascle, A., Montadert, L. & Wanneson, J., (1978). Seismic investigations in the Columbia, Venezuela and Grenada basins, and on the Barbados Ridge for future IPOD drilling. *Geologie en Mijnbouw,* **57,** 105 – 16.

Bouysse, P. (1979). Caracteres morphostructuraux et evolution geodynamique de l'arc insulaire des Petites Antilles. *Bureau de Recherches Géologiques et Minières Bulletin Section IV, 3/4,* 185 – 210.

Bouysse, P. (1984). The Lesser Antilles island arc: structure and geodynamic evolution. *Initial Reports, Deep Sea Drilling Project* (United States Government Printing Office) 78A, 83 – 103.

Bouysse, P., Andreieff, P. & Westercamp, D. (1980). Evolution of the Lesser Antilles island arc, new data from the submarine geology. *Transactions of the 9th Caribbean Geological Conference* (Santo Domingo, August 1980) pp. 75 – 88.

Bouysse, P., Andreieff, P., Richard, M., Baubron, J., Mascle, A., Maury, R. & Westercamp, D. (1985). Geologie de la Ride d'Aves et des pentes sous-marines du nort des Petites Antilles et esquisse bathymetrique a 1/1.000.000 de l'Est-Caraibe. *Documents du Bureau de Recherches Géologiques et Minières,* 93.

Bouysse, P., Schmidt-Effing, R. & Westercamp, D. (1983). La Desirade island (Lesser Antilles) revisited: lower Cretaceous radiolarian cherts and arguments against an ophiolitic origin for the basal complex. *Geology,* **11,** 244 – 7.

Bray, R.N. (1981). Influence of water currents and zooplankton densities on daily foraging movements of blacksmith, *Chromis punctipinnis,* a planktivorous reef fish. *Fishery Bulletin,* **78,** 829 – 41.

Brown, W. L. Jr. & Wilson, E.O. (1956). Character displacement. *Systematic Zoology,* **7,** 49 – 64.

Bucher, W.H (1952). Geological structure and orogenic history of Venezuela. *Geological Society of America Memoir,* **49.**

Connell, J.H. (1961a). The influence of interspecific competition and other factors on the distribution of the barnacle *Chthamalus stellatus. Ecology,* **42,** 710 – 13.

Connell, J.H. (1961b). Effects of competition, predation by *Thais lapillus,* and other factors on natural populations of the barnacle *Balanus balanoides. Ecological Monographs,* **31,** 61 – 104.

Crisp, D.J. (1974). Factors influencing the settlement of marine invertebrate larvae. *Chemoreception in Marine Organisms* (Ed. by P.T. Grant & A.N. Mackie) pp. 177 – 265. Academic Press, London.

Crisp, D.J. (1976). Settlement responses in marine organisms. *Adaptation to Environment: Essays on the Physiology of Marine Animals* (Ed. by R.C. Newell), pp. 83 – 124, Butterworths, London.

Crisp, D.J., Southward, A.J. & Southward, E.C. (1981). On the distribution of the intertidal barnacles *Chthamalus stellatus, Chthamalus montagui* and *Euraphia depressa. Journal of the Marine Biological Association of the United Kingdom,* **61,** 359 – 80.

Dayton, P.K. (1971). Competition, disturbance, and community organization: the provision and subsequent utilization of space in a rocky intertidal community. *Ecological Monographs,* **41,** 351 – 89.

Diebold, J.B., Stoffa, P.L., Buhl, P. & Truchan, M. (1981). Venezuela Basin Crustal Structure. *Journal of Geophysical Research,* **86,** 7901 – 23.

Donnelly, T.W. (1985). Mesozoic and Cenozoic plate evolution of the Caribbean region. *The Great American Biotic Interchange* (Ed. by F.G. Stehli & S.D. Webb), pp. 89–121, Plenum, New York.

Dorel, J. (1981). Seismicity and seismic gaps in the Lesser Antilles arc and earthquake hazards in Guadeloupe. *Geophysical Journal of the Royal Astronomical Society*, **67**, 679–96.

Estes, R. (1983). The fossil record and early distribution of lizards. *Advances in Herpetology and Evolutionary Biology* (Ed. by A.G.J. Rhodin & K. Miyata), pp. 365–98, Museum of Comparative Zoology, Harvard University, Cambridge, Massachusetts.

Etheridge, R. (1964). Late Pleistocene lizards from Barbuda, British West Indies. *Bulletin Florida State Museum*, **9**, 43–75.

Fink, L. K. Jr. (1972). Bathymetric and geologic studies of the Guadeloupe region, Lesser Antilles island arc. *Marine Geology*, **12**, 267–88.

Foster, M.S. & Schiel, D.R. (1985). *The Ecology of Giant Kelp Forests in California: A Community Profile.* Fish and Wildlife Service, Biological Report 85.

Gaines, S.D., Brown, S., & Roughgarden, J. (1985). Spatial variation in larval concentrations as a cause of spatial variation in settlement for the barnacle, *Balanus glandula*. *Oecologia*, **67**, 267–72.

Gaines, S.D. & Roughgarden, J. (1985). Larval settlement rate: A leading determinant of structure in an ecological community of the marine intertidal zone. *Proceedings of the National Academy of Sciences* (USA) **82**, 3707–11.

Gaines, S.D. & Roughgarden, J. (1987). Fish in offshore kelp forests affect recruitment to intertidal barnacle populations. *Science*, **235**, 479–81.

Ghosh, N., Hall, S.A., & Casey, J.F. (1984). Seafloor spreading magnetic anomalies in the Venezuelan Basin. *Geological Society of America Memoir* **162**, 65–80.

Hickey, B.M. (1979). The California Current System—hypotheses and facts. *Progress in Oceanography*, **8**, 191–279.

Huyer, A. (1983). Coastal upwelling in the California Current System. *Progress in Oceanography*, **12**, 259–84.

Ikeda, M. & Emery, W.J. (1984). Satellite observations and modeling of meanders in the California Current System off Oregon and Northern California. *Journal of Physical Oceanography*, **14**, 1434–50.

Iwasa, Y. & Roughgarden, J. (1985). Evolution in a metapopulation with space-limited subpopulations. *IMA Journal of Mathematics Applied in Medicine and Biology*, **2**, 93–107.

Iwasa, Y. & Roughgarden, J. (1986). Interspecific competition among metapopulations with space-limited subpopulations. *Theoretical Population Biology.* **30**(2), 194–214.

Jackson, G.A. & Strathmann, R.R. (1981). Larval mortality from offshire mixing as a link between precompetent and competent periods of development. *American Naturalist*, **118**, 16–26.

Karlin, S. & McGregor, J. (1972). Polymorphisms for genetic and ecological systems with weak coupling. *Theoretical Population Biology*, **3**, 210–38.

Koblinsky, C.J., Simpson, J.J. & Dickey, T.D. (1984). An offshore eddy in the California Current System. Part II: Surface manifestation. *Progress in Oceanography.* **13**, 51–69.

Lazell, J.D. Jr. (1983). Biogeography of the herpetofauna of the British Virgin Islands, with description of a new anole (Sauria: Iguanidae). *Advances in Herpetology and Evolutionary Biology* (Ed. by A.G.J. Rodin & K. Miyata), pp. 99–117, Museum of Comparative Zoology, Harvard University, Cambridge, Massachusetts.

Lewis, C.A. (1977). A review of substratum selection in free-living and symbiotic ciripeds. *Settlement and Metamorphosis of Marine Invertebrate Larvae* (Ed. by F.S. Chia & M. Rice), pp. 207–17. Elsevier, Amsterdam.

Licht, P. (1974). Response of *Anolis* lizards to food supplementation in nature. *Copeia* 1974, 215–21.

Mann, P. & Burke, K. (1984). Neotectonics of the Caribbean. *Reviews of Geophysics and Space Physics,* **22,** 309–62.

Maresch, W.V. (1974). Plate tectonics origin of the Caribbean mountain system of Northern South America: Discussion and proposal. *Geological Society of America Bulletin,* **85,** 669–82.

Menge, B.A. & Sutherland, J.P. (1976). Species diversity gradients: synthesis of the roles of predation, competition, and temporal heterogeneity. *American Naturalist,* **110,** 351–69.

Mooers, C.N.K. & Robinson, A.R. (1984). Turbulent jets and eddies in the California Current and inferred cross-shore transports. *Science,* **223,** 51–3.

Pacala, S. & Roughgarden, J. (1984). Control of arthropod abundance by *Anolis* lizards on St. Eustatius (Neth. Antilles). *Oecologia,* **64,** 160–2.

Pacala, S. & Roughgarden, J. (1985). Population experiments with the *Anolis* lizards of St. Maarten and St. Eustatius. *Ecology,* **66,** 129–41.

Paine, R.T. (1966). Food web complexity and species diversity. *American Naturalist,* **100,** 65–75.

Paine, R.T. (1984). Ecological determinism in the competition for space. *Ecology,* **65,** 1339–48.

Parrish, R.H., Nelson, C.S. & Bakun, A. (1981). Transport mechanisms and reproductive success of fishes in the California current. *Biological Oceanography,* **1,** 175–203.

Pindell, J. & Dewey, J.F. (1982). Permo-Triassic reconstruction of western Pangea and the evolution of the Gulf of Mexico/Caribbean region. *Tectonics,* **1,** 179–211.

Pingree, R.D. & Griffiths, D.K. (1980). Currents driven by a steady uniform wind stress on the shelf seas around the British Isles. *Oceanologica ACTA,* **3,** 227–36.

Rieppel, O. (1980). Green anole in Dominican amber. *Nature,* **286,** 486–7.

Roughgarden, J. (1986). A comparison of food-limited and space-limited animal competition communities. *Community Ecology* (Ed. by J. Diamond & T. Case), pp. 492–516. Harper & Row, New York.

Roughgarden, J. (1987). *The Anoles of the Eastern Caribbean: Competition, Coevolution and Plate Tectonics.* Cambridge University Press. (In preparation.)

Roughgarden, J. & Iwasa, Y. (1986). Dynamics of a metapopulation with space-limited subpopulations. *Theoretical Population Biology,* **29**(2), 235–61.

Roughgarden, J., Heckel, D. & Fuentes, E.R. (1983). Coevolutionary theory and the biogeography and community structure of *Anolis. Lizard Ecology, Studies of a Model Organism* (Ed. by R.B. Huey, E.R. Pianka & T.W. Schoener), pp. 371–410. Harvard University Press, Cambridge, Massachusetts.

Roughgarden, J., Iwasa, Y. & Baxter, C. (1985). Demographic theory for an open marine population with space-limited recruitment. *Ecology,* **66,** 54–67.

Roughgarden, J., Pacala, S. & Rummel, J. (1984). Strong present-day competition between the *Anolis* lizard populations of St. Maarten (Neth. Antilles). *Evolutionary Ecology* (Ed. by B. Shorrocks), pp. 203–20, Blackwell Scientific Publications, Oxford.

Rummel, J. & Roughgarden, J. (1985a). Effects of reduced perch-height separation on competition between two *Anolis* lizards. *Ecology,* **66,** 430–44.

Rummel, J. & Roughgarden, J. (1985b). A theory of faunal buildup for competition communities. *Evolution,* **39,** 1009–33.

Scheltema, R.S. (1974). Biological interactions determining larval settlement of marine invertebrates. Thalassia Jugoslavia, **10,** 263–96.

Southward, A.J. & Crisp, D.J. (1954). Recent changes in the distribution of the intertidal barnacles *Chthamalus stellatus* Poli and *Balanus balanoides* (L.) in the British Isles. *Journal of Animal Ecology,* **23,** 163–77.

Southward, A.J. (1976). On the taxonomic status and distribution of *Chthamalus stellatus* (Cirripedia) in the North-east Atlantic region: with a key to the common intertidal barnacles of Britain. *Journal of the Marine Biological Association of the United Kingdom,* **56,** 1007 – 28.

Speed, R.C. (1985). Cenozoic collision of the Lesser Antilles arc and continental South America and the origin of the El Pilar Fault. *Tectonics,* **4,** 41 – 69.

Speed, R.C., Gerhard, L.C. & McKee, E.H. (1979). Ages of deposition, deformation, and intrusion of Cretaceous rocks, eastern St. Croix, Virgin Islands. *Geological Society of America Bulletin I,* **90,** 629 – 32.

Stamps, J.A. (1977). Rainfall, moisture, and dry season growth rates in *Anolis aeneus. Copeia* 1977, 415 – 19.

Steadman, D.W., Pregill, G.K. & Olson, S.L. (1984). Fossil vertebrates from Antigua, Lesser Antilles: evidence for late Holocene human-caused extinctions in the West Indies. *Proceedings of the National Academy of Sciences* (USA) **81,** 4448 – 51.

Stein, S., Engeln, J., Wiens, D., Fujita, K. & Speed, R. (1982). Subduction seismicity and tectonics in the Lesser Antilles arc. *Journal of Geophysical Research,* **87** (B10), 8642 – 64.

Tomblin, J.F. (1975). The Lesser Antilles and the Aves Ridge. *The Ocean Basins and Margins. Vol. 3.* The Gulf of Mexico and the Caribbean (Ed. by A.E.M. Nairn & F.G. Stehli) pp. 467 – 500. Plenum, New York.

Underwood, A.J., Denley, E.J. & Moran, M.J. (1983). Experimental analyses of the structure and dynamics of mid-shore rocky intertidal communities in New South Wales. *Oecologia,* **56,** 202 – 19.

Vierbuchen, R.C. Jr. (1979). *The tectonics of north-eastern Venezuela and the southeastern Caribbean Sea.* Ph. D. thesis. 193 pp. Princeton University, Princeton, New Jersey.

Westercamp, D., Andreieff, P., Bouysse, P., Mascle, A. & Baubron, J. (1985). Geologie de l'archipel des Grenadines (Petites Antilles meridionales). *Documents du Bureau de Recherches Géologiques et Minières,* **92.**

Wethey, D. (1983). Geographic limits and local zonation: The barnacles *Semibalanus* (*Balanus*) and *Chthamalus* in New England. *Biological Bulletin* **165,** 330 – 41.

Williams, E.E. (1972). The origin of faunas. Evolution of lizard congeners in a complex island fauna: a trial analysis. *Evolutionary Biology,* **6,** 47 – 88.

Wilson, E.O. (1961). The nature of the taxon cycle in the Melanesian ant fauna. *American Naturalist,* **95,** 169 – 93.

Wingate, D. (1965). Terrestrial herpetofauna of Bermuda. *Herpetologica,* **21,** 202 – 18.

Wright, S. (1931). Evolution in Mendelian populations. *Genetics,* **16,** 97 – 159.

23. THE ANALYSIS OF COMMUNITY ORGANIZATION: THE INFLUENCE OF EQUILIBRIUM, SCALE AND TERMINOLOGY

PAUL S. GILLER[1] AND JOHN H. R. GEE[2]

[1]*Department of Zoology, University College, Cork, Eire and*
[2]*Department of Zoology, University College of Wales, Aberystwyth,*
SY23 3DA

INTRODUCTION

Conceptual models can play central roles in areas of science that are rich in descriptive detail but poor in generally accepted theory; one such area is community ecology. At the very least the models can form frameworks for the organization of information. In their fullest and most useful form they can depict the casual processes behind patterns in the structure of communities. Stripped of the distracting complexity and embellishment that characterizes reality and gives pleasure to the naturalist in every ecologist, these conceptual models can provide the bases for predictions that are testable by experiment.

It can be argued that the single most important development in community ecology has been the emergence of a structural model of communities based on the Hutchinsonian concept of niche. This model is the basis of a large body of theory which regards the community as the outward and visible sign of a notional space filled to saturation with niches, each overlapping its neighbours to a limited extent (see Giller 1984). A saturated community represents a state of equilibrium governed largely by competitive interactions between species. Until quite recently, this beguilingly attractive model was accepted with little dissent, and still occupies an unchallenged position in most ecology textbooks. But in the latter part of the seventies and the early eighties it became clear that some of the testable predictions find little support in certain communities. In others, support for the predictions depends on the spatial and temporal scales of the analysis. Furthermore, some predictions fail to discriminate clearly between the Hutchinsonian model and models founded on different principles. In general, equilibrium models also tend to obscure important relationships between intrinsic and extrinsic processes in communities (Brown 1988).

519

With the universality of the Hutchinsonian model under attack, a variety of alternative views appeared in the literature. Each identified different processes which were influential in shaping the structure of one community or another. Many of these alternatives are reviewed in the recent symposium volumes on community ecology (Price *et al.* 1984; Strong *et al.* 1984; Diamond & Case 1986). Their diversity leads to the worrying prospect that communities might each be unique in structure, moulded individually by one or more of a wide range of factors. The hope for a predictive science of community ecology lies with the possibility that communities might instead be rationally arranged on a restricted number of axes. Ideally, each axis would be directly or indirectly related to measurable characteristics of the organisms or their environment. Schoener (1986) has made a start on such a system of classification.

The preceding chapters have reviewed the essential features of the patterns and processes of communities from a very wide range of habitats. Their constituent organisms come from almost every branch of the taxonomic tree. Our purpose here is not to add to Schoener's pluralistic system (we do not have enough detail on individual communities to attempt that task), but to highlight three interconnected areas of interest which arise from the proceedings of this symposium. These relate to the ultimate goal of developing an agreed core of theory in community ecology. Two are recurring conceptual themes, the other is a matter of terminology. We emphasize that it is not our intention that the exclusion of any aspect from this discussion should cast a shadow on its present or future importance in our understanding of communities.

EQUILIBRIUM OR NON—EQUILIBRIUM?

At least some of the differences in organization between the communities discussed in earlier chapters can be accounted for in terms of their position along a continuum between what we shall call equilibrial and non-equilibrial states. These differences are related to suites of ecological characteristics which intergrade as the continuum is traversed (Fig. 23.1). Wiens (1984a) has proposed a comparable gradient (although our definition of the end-points differs) and similar ideas have been discussed elsewhere (e.g. Chesson & Case 1986).

Unfortunately 'equilibrium' is one of several ecological terms that have no precise consensus definition (see below for others). To some ecologists an equilibrium community is one that has the property of remaining unchanged in its species abundances as long as environmental conditions are constant. This equilibrium point might be unstable in the face of perturbation, or it

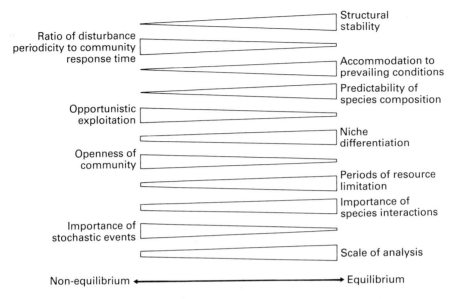

FIG. 23.1. The relative importance of various characteristics of communities along a continuum from an equilibrium to a non-equilibrium state.

might have stability of a neighbourhood or global kind (*sensu* Maynard Smith 1974). A non-equilibrium community would then be one in which species abundances were continually changing and the structure of the community was entirely unpredictable, despite the constancy of environmental conditions. This purist view of equilibrium/non-equilibrium has most relevance in theoretical models where environmental conditions can be held absolutely constant. In reality conditions fluctuate, predictably and unpredictably, with a variance that is often related to geographic location. In fluctuating conditions, an equilibrium community might be expected to change predictably as it converges on a new equilibrium point, but a non-equilibrium community would again behave in an entirely unpredictable fashion.

There is little evidence of truly non-equilibrial communities, as defined above, although the tropical trees described by Hubbell & Foster (1986) may come close. However, even this system does have stable or predictable structure despite its state of flux at the level of individual species abundances. On the other hand we believe that there are many communities that would change, as a result of the ensuing interactions between species, if the environmental conditions prevailing at any instant were held constant. Given sufficient time, these communities would eventually reach an equilibrium point, although that point might conceivably have zero abundances for some or, exceptionally, all species. Therefore, when we use the term non-equilibrial it is not in the pure, theoretical sense outlined above.

Equilibrial communities

For our purposes we define an equilibrium in terms of constancy or predictability of community structure. An equilibrial community is one that is accommodated to the prevailing environmental conditions. In an environment that remains constant over periods measured in multiples of the generation time of the characteristic taxa, the essential structure of the community will remain stable. This will be reflected in a constancy of the relative abundances of species, guilds and functional groups, and in the strength of the functional links between them. To borrow from Southwood's definition of the community (Chapter 1), the morphology of the spider's web of interactions between the species is unchanging. The common feature of equilibrial communities is that their structure is essentially determined by these interactions (which also provide a degree of stabilizing feedback), and stochastic extrinsic events play a relatively minor role (see Fig. 23.1). Our definition of equilibrium differs from the concept of 'steadiness' used by Wiens (1984a) because in our view steadiness in either population abundances or species composition is contingent on environmental constancy. In a changing environment an equilibrial community will not be 'steady', but will track changing conditions with a minimum time lag. Nevertheless it will maintain its overall structure and be organized primarily by biotic interactions.

In an equilibrial community the species may show the classic patterns of niche partitioning that follow from the Hutchinsonian model, or their use of resources may be determined by more subtle processes and rules (e.g. Chapters 8 and 21). It is not necessary that the community be saturated with species or that the dominant interaction be interspecific competition. Communities of insects on bracken (Lawton 1984) would be equilibrial by the criterion of constancy of structure and composition but there are unaccountable differences in species-richness between some bracken communities. There is also little evidence of interspecific competition, at least among the herbivores.

Non-equilibrial communities

Non-equilibrial communities are defined as those that have a varying and largely unpredictable structure. They are not accommodated to prevailing environmental conditions but are responding to conditions at some time in the past. There is therefore an expectation of change in the existence or thickness of parts of Southwood's community web (Chapter 1), even if the environment remains constant. Whilst changing, these non-equilibrial communities are likely to be approaching an equilibrium state from 'above', by the thinning or loss of parts of the web, or from 'below', by the addition or strengthening of parts. Figure 23.1 lists the general characteristics of these communities. Biotic interactions are likely to be transitory, and the evidence of their presence will be masked by direct and delayed responses to environmental factors. Stochastic extrinsic factors are likely to dominate the organization of such communities and the responses of some species may be quite independent of the rest of the community. Nevertheless non-equilibrial communities are not random assemblages, the equivalent of the non-interactive null model, lacking any recognizable structure. Instead they will have a trophic structure (in terms of relative abundance of specific functional groups) and, although species composition may vary, a set of core species (Hanski 1982) that remain dominant over space and time (e.g. herbivorous insects, Chapter 1; fungi, Chapter 11; dung and carrion beetles, Chapter 12; freshwater benthic invertebrates, Chapter 16). They may also show predictable community responses to given sets of environmental conditions (e.g. freshwater plankton, Chapter 14) or over environmental gradients (e.g. freshwater benthic invertebrates, Chapter 16) even though there are likely to be only loose patterns in resource partitioning and niche spacing.

The structure of non-equilibrial communities is also likely to reflect the influence of periods in the past during which resources were restricted and biotic interactions intense (ecological 'crunches', Schoener 1982). Such periods may be important in establishing species distribution patterns, community composition and ecological adaptations. Their appearance can be stochastic in space and time, as seen during periods of food limitation in zooplankton and phytoplankton when competition becomes important (Chapter 15). They can also occur seasonally (Chapter 4). Larger scale ecological and evolutionary crunches of climatic, geologic and extra-terrestrial origin have profound effects on species pools from which both equilibrial and non-equilibrial communities are drawn (Chapters 5 and 20).

Disturbance and community response

The distinction that we have drawn between the ends of the equilibrium/ non-equilibrium continuum has centred on the persistence of the ecological structure of the community. At a different level, the distinction is probably best made on the ratio of the periodicity of community disturbance to the community response time. The key to explaining much of the pattern and process of communities may be the relationship of the time between ecological disturbances to the time taken for the community to approach an equilibrium following disturbance. Community response time is closely approximated by the r values of the core species, in turn reflecting generation time, fecundity and longevity, but it also includes the influence of colonization by propagules from elsewhere (Chapter 22). The notion of the ratio between disturbance periodicity and community response time is akin to Southwood's (1977) durational stability and has been used by Sutherland (1981) in connection with community stability. A related idea is behind Huston's (1977) hypothesis explaining community diversity in terms of the relationship between the frequency of density independent reduction of populations and the rate of competitive displacement.

In the context of Schoener's (1986) scheme for classification of communities, the periodicity of disturbance appears most closely related with 'primitive' axes E5 (long-term climatic variation), E1 (severity of physical factors) and O5 (homeostatic ability). Community response time is most closely related to generation time (O3), but recruitment (O2) and mobility (O4) are also important.

Communities such as the phytoplankton of temperate lakes (Chapter 14) and the continental shelf (Chapter 15) lie near the non-equilibrium end of the continuum; in both of these the interval between disturbances appears to be less than the community response time. Whilst both communities show responses to temporal and spatial patterns in environmental conditions on many scales, biotic interactions appear to be of minor importance in community organization for most of the time (with the possible exception of grazing in freshwater plankton). Similarly, the benthic communities of small streams that are subject to frequent spates (Chapter 16) tend toward the non-equilibrium end of the spectrum. Frequent disturbance also appears to be a key factor in the organization of some communities of green plants (Chapter 5) and some fungi (Chapter 11). The communities in which disturbance plays an important role tend to be either spatially limited or they occupy structurally-simple, three-dimensional habitats, and the prediction of species composition is likely to be poor. On the other hand, interactions between species are more important in systems where disturbance occurs less frequently in relation to community response time, such as in desert rodents (Chapter 9), tropical primates (Chapter 10), some fungi (Chapter 11), aquatic microbes (Chapter 13), and phytoplankton in some tropical lacustrine systems (Chapter 14). Benthic communities from acid streams enjoying a stable, low discharge, change little in species composition with time and show food webs with peculiarly high connectance values (Chapter 16). These have the characteristics of communities that belong toward the equilibrium end of the continuum (Fig. 23.1).

The influence of spatial and temporal variation in resources

Resources may be defined as environmental factors that are directly used by an organism and may potentially influence individual fitness (Wiens 1984b). There are many correlational, observational and experimental data sets that suggest that resources are important factors structuring natural communities (Tilman 1982). The literature abounds with evidence that the greater the diversity of resources offered by a habitat, the greater the species-richness that it can support. Spatial heterogeneity in a habitat can enhance species richness by simply providing a larger number of resource types that are partitionable amongst potentially competing species. Alternatively, a heterogeneous habitat may provide a range of ratios of supply rates of a small number of resources for which species compete in the manner suggested by Tilman (1982). Relationships between the spatial hetero geneity of habitats and their species-richness have been shown in plants (Chapters 5 and 6), herbivorous insects (Chapters 1 and 7), farmland birds (Chapter 8), desert rodents (Chapter 9) and marine plankton (Chapter 15).

When there is large-scale (in relation to organism size and foraging range) spatial variation in resources, or in abiotic environmental factors, there is potential for spatial partitioning of resources between species. Often this spatial variation takes the form of a stable habitat gradient which produces a predictable pattern in community organization, at least in terms of trophic structure and function. This is clearly shown in all the aquatic habitats discussed in earlier chapters. In rivers (Chapter 16) the spatial variation takes the form of longitudinal pattern in the nature of energy inputs; in the marine benthos (Chapter 17) community structure responds to horizontal gradients in organic enrichment; in marine sediments there are microscale vertical gradients in chemical resources for microbes (Chapter 13); and in the open sea organization of planktonic communities reflects vertical gradients of light and nutrients associated with the thermocline (Chapter 15). Similarly, community responses to resource gradients are described for many terrestrial communities including plants (Chapters 5 and 6), desert rodents (Chapter 9) and lizards (Chapter 22).

The spatial distribution of resources can also influence community structure in another way, independent of the number of resource types present. Given a particular relationship between periodicity of disturbance and community response, the spatial distribution of resources will influence the likelihood that a community will be structured by biotic interactions. If resources are patchy in space, competition between species need not result in competitive elimination in the way that might be expected if Lotka–Volterra dynamics applied in a uniform environment (Levin 1974; Slatkin 1974). Such a reduction in the influence of competition can result in higher regional diversities, although the consequences of resource patchiness for the community depend on the relationship between the rates of population processes occurring within patches and those occurring between patches (Hanski 1983). This dependence on the relative rates of processes is reminiscent of the effect of the balance between frequency of disturbance and rate of response on the equilibrial status of communities.

Unpredictable variation in resource availability with time is a form of disturbance in the sense discussed above. However, predictable changes with time can be accommodated by organisms with appropriate life histories. For instance, diurnal changes in resources can be tracked by species with sufficiently short generation times, such as bacteria and protozoa (Chapter 13). Predictable variation in resource availability is also an aspect of seasonality, and a factor to which longer-lived populations and communities can adapt (Chapter 4). Seasonal change is a dominant feature in the organization of planktonic communities (Chapters 13, 14 and 15). These predictable temporal changes provide an opportunity for partitioning resources in time, so that there is some point in the seasonal cycle at which every species can persist and grow in population size in the face of competition and predation. An important condition is that each species should have some means of bridging the gaps between favourable periods; this might be by use of alternative resources (primates, Chapter 10), by migration to a refuge (birds, Chapter 8), by possessing a persistent resting stage in its life history (invertebrates in temporary pools, Chapter 16), or simply by having a time scale of competitive elimination from the community that is longer than the gap. The time course of resource availability (Fig. 23.2) may affect the likelihood that a resource becomes the object of intense competition within or between species (Price 1984). In general, competition is likely to be intense when the rate of population response (mainly dependent on intrinsic growth rate) is fast relative to the rate of increase of resource availability. Naturally, it will also be intense when demand exceeds supply as a resource declines in availability. Conversely, competition is likely to be of less importance when populations cannot keep up with the rate of increase in resource availability. The resource systems typical of the communities discussed in the preceding chapters are arranged according to their temporal characteristics in Table 23.1.

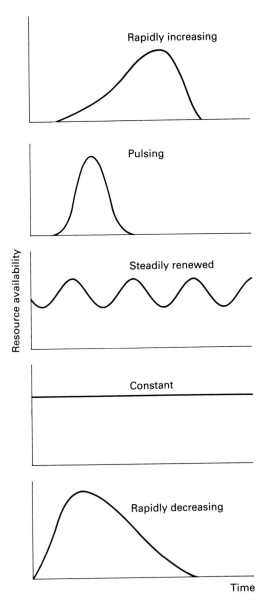

FIG. 23.2. Classification of resource types according to their time course of availability (redrawn from Price 1984). Rapidly increasing — resources that increase over much of the active season of consumers, then decline rapidly. Pulsing — resources that increase and decline rapidly. Steadily renewed — resources produced over prolonged periods and not readily overexploited. Constant resources — often physical, uninfluenced by exploitation or seasonal variation. Rapidly decreasing — resources produced over a short period each season, with a subsequent prolonged decline. For examples of these resource types see Table 23.1.

TABLE 23.1. The resources and exploiting organisms described preceding chapters, categorized after Price (1984)

Resource availability	Resource types	Exploiters
Rapidly increasing	Temperate deciduous foliage	Herbivores
	Flowers	Generalist pollinators
	Temperate insects	Insectivores
Pulsing (ephemeral)	Phytoplankton	Zooplankton
	Herbaceous plant parts	Specialized herbivores
	Fruits	Frugivores (e.g. primates)
	Fruits (fallen)	Detritivores — insects, fungi
	Temperate flowers	Specialist pollinators
	Plant parasites/leaf-miners	Parasitoids
	Freshwater macrophytes	Epiphytic insects
	Dung and carrion	Detritivorous and scavenging invertebrates
	Phytotelmata and temporary pools	Freshwater invertebrates and vertebrates
Steadily renewed	Gut contents	Internal parasites
	Marine phytoplankton	Zooplankton
	Tropical flowers	Pollinators
	Tropical foliage	Folivores
	Nectar	Nectivorous insects and birds
	Chemical species in marine sediments	Bacteria
	Terrestrial plant nutrients	Herbaceous plants and trees
	Organic marine sediments	Infaunal invertebrates
Constant	Space	Plants, intertidal organisms, territorial animals
	Nesting sites	Farmland birds
	Blood	External parasites
Rapidly decreasing	Seeds in deserts	Granivorous insects, mammals, and birds
	Insect life stages	Specialized parasitoids
	Dung and carrion	Detritivores and scavengers
	Epilimnetic nutrients	Phytoplankton

Temporal changes in resources, or in the physical environment, that take place on a temporal scale that is much longer than generation times, may be sufficiently severe to require that species either shape up (evolve) or ship out (migrate). The alternative is global extinction. During the Pleistocene, long-term variations of climate at particular sites accompanied the expansion and regression of the polar ice sheets. Often the effect on communities was a geographic reshuffling of their membership, as species responded to the new conditions at different rates (e.g. insects, Chapter 19). On still longer time scales, the origination of resources has had profound effects. The evolution of branching megaphyllous leaves and the arborescent habit through the Devonian and early Carbonaceous lead to an increase in terrestrial food resources and possibly to the evolution of new and diverse forms of animals (Chapter 18). The development of extensive grasslands also had pronounced effects on evolution of tetrapod communities (Chapter 20).

Independent aggregation of the individuals on resources that are both spatially patchy and ephemeral can create probability refuges which will delay the point at which, as the community fills with species and individuals, competition acts as an organizing force (Chapter 2). Ephemeral resource patches such as the fruiting bodies of fungi and decaying fruit (Chapter 2), carrion and dung (Chapter 12), and some phytotelmata (Chapter 16) may be similar in this regard. Rain pools that habitually form in the same place, but are only temporarily suitable for growth and reproduction, differ in that their resources may be pre-empted by species that leave behind resistant propagules or diapausing stages. The probability refuges that would otherwise permit the persistence of the less competitive species are thus rendered unavailable.

SCALE

The spatial and temporal scales of community investigations determine the range of patterns and processes that may be detected, and therefore the level of understanding and explanation that can be achieved. They also influence our perception of the nature of the community. It seems probable that the point occupied by the community on the equilibrium/non-equilibrium continuum depends on the scale of space or time on which it is being viewed (Fig. 23.3). At some spatial or temporal scale a community which originally appeared best placed towards one end of the continuum may seem better placed at the other. Community ecologists generally have been insensitive to this effect of scale of resolution (Wiens 1984a, 1986), and some vociferous disagreements have arisen as a result of the choice of different scales of investigation (e.g. coral reef fish communities, Sale & Dybdahl 1975; Anderson *et al.* 1981). We have already seen that spatial and temporal environmental variability has major effects on community patterns and organization; these effects are scale dependent. Events may seem random on one scale, but new phenomena and patterns may appear if the system is examined on different scales. This will apply not only to spatial and temporal scales, but also to taxonomic scale (*i.e.* taxonomic breadth of the study).

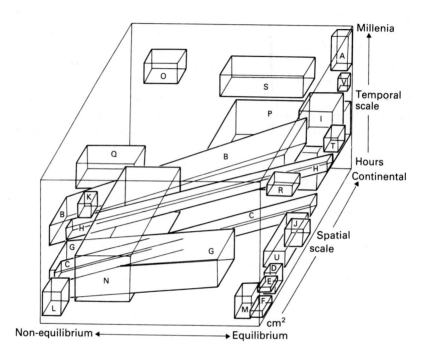

Non-equilibrium ←⎯⎯⎯⎯⎯⎯⎯⎯⎯⎯→ Equilibrium

FIG. 23.3. A pictorial representation of the equilibrium/non-equilibrium continuum indicating the relative positions of various communities discussed in earlier chapters, viewed at different spatial and temporal scales. Communities have been placed on the axes on the basis of their attributes relative to the equilibrium/non-equilibrium extremes (Fig. 23.1). Note that the position occupied by many communities depends on the scale of analysis. Key to Communities: A = tetrapod genera, high altitude coniferous forest, and taxodiaceous coniferous swamps; B = tropical rain forest trees; C = coral reef fish; D = fungi of rotting wood; E = marine planktonic bacteria and protozoa; F = marine sediment bacteria; G = terrestrial herbivorous insects; H = birds in habitats lacking vertical structure; I = Caribbean lizards; J = plankton in ice-covered and tropical lakes; K = temperate marine and freshwater plankton; L = communities in patchy, ephemeral resource habitats; M = Communities in perennial and patchy resource habitats; N = North American intertidal invertebrates; O = Late Quaternary terrestrial Coleoptera; P = herbaceous plants; Q = freshwater invertebrates of unstable, stony, lotic habitats; R = freshwater invertebrates of stable, acid streams; S = subtidal marine benthos; T = birds in vertically-structured habitats; U = desert rodents; V = marine plankton of the North and South Central Pacific Gyres.

Small scale v. large scale

Given that scale of investigation has an important effect on procedures, observations, results and conclusions, how is the choice of scale to be made? Does it depend on the question under investigation? If so, which questions are most likely to lead to a better understanding of community organization?

In general, reducing the scale of analysis makes a study more tractable, facilitates replication in experimental design, and often ensures 'prompt and clean' results (Dayton & Tegner 1984). However, many ecological phenomena operate over large spatial and temporal scales and may only be glimpsed incompletely at small scales of investigation. It is becoming increasingly clear, for instance, that observations made within one study site may have limited power to explain the structure of the community at that site (Chapter 22). Temporal activity patterns often make it necessary to investigate a system over a lengthy period to explore niche differentiation (e.g. Chapters 6 and 12), and it is essential to encompass many generations of the dominant organisms to take account of genetic change. On the other hand, studies at a large scale tend to average out patchiness in space and time that may have an important functional role. Resolution and detail are also poor at the larger scales (see Chapters 18 and 19 for further discussion). As an example, 'local' (small scale) and 'regional' (large scale) studies of communities in water-filled tree holes reveal different aspects of community structure. The former emphasize stochastic processes which shape communities in individual holes, the latter the deterministic processes controlling the regional fauna from which the local communities are drawn (Kitching 1987). Thus controlling processes tend to interact over a wide range of spatial scales, as is also clearly shown in desert rodents (Chapter 9).

Openness of communities

Community composition at one scale is often influenced by processes that operate at a different scale—at some scale most ecological systems are open. The extent of this openness varies between systems and is influenced as much by the nature of the surrounding habitat as by the nature of extrinsic physical forces. This openness is evident on both spatial and temporal axes. Supply-side ecology (Chapter 22; Lewin 1986) suggests that the control of community dynamics that is exerted by physical transport processes operating on grand scales of time and space often matches the effects of local, short-term processes such as competition and predation. Migrations in response to adverse seasonal conditions lead to dismantling and reassembly of communities through the year. Similar changes, but on a much longer time scale, occurred during the Pleistocene when species moved at different rates and with different time lags in response to the climatic changes that accompanied the waxing and waning of the polar ice (Chapter 19).

Identification of species interactions

When two species that are potential competitors first meet, behaviourally mediated avoidance is a likely outcome of competition between them. At this stage competition would be detectable by removal experiments, with the expectation that niche expansion by the remaining species would follow the removal. With time, it is probable (but not inevitable) that phenotypic plasticity would be restricted by genetic adaptation to a reduced habitat range, and that the mutual avoidance would become less dependent on behaviour. Eventually a point might be reached at which the remaining species could not expand its range after the removal of a competitor, and evidence of the interaction would only be accessible by summoning the 'ghost of competition past' (Rosenzweig 1979; Connell 1980). Equally, a substantial fraction of structure in some communities may be an inheritance from a previous ecological crunch. Studies that are conducted on extensive scales of time or space, such as those of Diamond (1986), Knoll (1986) and Webb (Chapter 20), offer a new perspective on this historic role of species interactions in community organization. Southwood (Chapter 1) and Claridge (Chapter 7) show that increasing the scale of investigation can reveal subtle competitive linkages, some of which are between species active at different times and are mediated through a common host.

The idea of differing scales of study can also be applied to taxonomic range. From a practical point of view, it is much easier to study a small taxonomic group, such as a restricted feeding guild, than one that spans a wider range of taxa. The problem is that the taxonomic scale of a study also influences the ability to detect species interactions. On the one hand, no amount of evidence that competition affects the densities of small groups of selected species necessarily says anything about the likelihood that competition causes large-scale patterns at the community level (Moulton & Pimm 1986). On the other, painting with a broader taxonomic brush may include distantly related species that are unlikely to interact at all and which 'dilute' the patterns created by interactions (Chapter 1). This would make it difficult to distinguish between a community in which interactions had an influence on community structure and a null model of the same community. Nevertheless, reports of important interactions spanning major taxonomic boundaries are accumulating (e.g. Carpenter 1979; Hurlbert, Loayza & Moreno 1986; Schluter 1986) and the community ecologist adopts a narrow taxonomic view at his peril.

Interaction of scales

Differing scales of spatial and temporal heterogeneity may separate sub-communities within certain areas (Fenchel, Chapter 13). It is unlikely that species populations that differ considerably in size and generation time will show strong direct interactions, such as predator–prey relationships or competition for common resources. Thus physical space may be thought of as harbouring a heirarchy of sub-communities, each representing a few trophic levels and relating to characteristic scales of time and space, albeit overlapping to some extent.

The regional species pools from which local communities are drawn are often seen to reflect the action of processes operating on evolutionary scales of time and geographic scales of space (Chapters 5, 9, 12 and 22). However, the composition and maintenance of the pool also depends on the cumulative effects of processes operating on restricted scales of time and space. Patterns in the fossil record represent a large space–time integration and provide, in effect, a time-averaged measure of gamma diversity (Chapter 1, p.11).

Scale gradients

Most changes in ecological systems are gradual rather than abrupt, so it is sensible to consider the effects of scale in communities in terms of scale gradients. Overviewing systems on this basis leads to an appreciation of changing patterns and processes with changing scale.

Spatial scales

The spatial axis largely depends on whether the organisms under consideration are active or sedentary. Table 23.2 indicates six distinct points along the gradient and the relevant chapters that explore these spatial scales in the proceedings. On the largest scales, properties of local patches and conditions at different sites are averaged out and lost, but one can examine larger trends involving species' climatic limits, responses to different sets of predators and competitiors, evolutionary potential and community convergence and non-convergence. Different measures of diversity are also evident as one increases the spatial scale of analysis (see Chapter 1). The widely documented species–area relationship may be also seen as an expression of changing scale of analysis. Several examples are provided in Chapters 7 and 16.

TABLE 23.2. Scales of spatial analysis as exemplified by studies of communities in this symposium (after Wiens *et al.* 1986). Chapter references are in brackets

Sedentary organisms	Scale	Determining factors	Scale	Motile organisms
Herbaceous vegetation (6) Marine bacteria (13) Stream invertebrates (16)	Micro- $0.1–10\ m^2$	Species and mobility or size	Single organism's living area	Birds (8, 21)
Vegetation (5) Intertidal and infaunal invertebrates (17, 22) Endophytic insects (7)	Meso- 0.1–10 Ha	Species, population density, patch size Species, dispersal, resource distribution	Local patch occupied by many individuals Habitat including many patches	Decomposers (2, 11, 12, 13) Epiphytic insects (1, 16) Decomposers (2, 11, 12, 13) Plankton (14,15) Birds (8, 21) Rodents (9) Insects(7)
Vegetation (5) Intertidal and infaunal invertebrates (3, 17, 22) Endophytic insects (7)	Macro- $10–100\ Km^2$	Species, dispersal, habitat distribution	Region large enough to close system from emigration/immigration	Primates (10) Rodents (9) Decomposers (11, 12) Aquatic benthic invertebrates (16, 17) Insects (7)
Regional vegetation (5) Intertidal invertebrates (22)	Mega- $10^3–10^5\ Km^2$	Barriers to dispersal, physiological adaptation	Biogeographic or intracontinental	Rodents (9) Lizards & barnacles (22) Primates (20)
Intercontinental biome vegetation (5, 10)	Global $>10^5\ Km^2$	Evolutionary history, barriers to dispersal	Intercontinental	Primates (10) Lizards (22)

Temporal scales

Contemporary communities represent only one slice of a time continuum, and the extension of the temporal frame of analysis is important in the development and testing of robust models of community organization. At one extreme, changing conditions over long periods affect the composition of species pools by influencing the probabilities of colonization, speciation and extinction. On a shorter time frame, the probabilities of migration, birth and death of individuals and hence the maintenance of species populations influence the composition of local assemblages (Brown, Chapter 9). Extending the scale of analysis over geological and evolutionary time is an important departure from traditional community ecology. Examination of the fossil record not only provides evidence of the pattern of community change over long periods, but also provides rarer insights into patterns on shorter time scales. It can indicate which of many processes controlling community structure in ecological time are important in the development of communities over geologic time (Chapter 18). Comparative studies of contemporary communities can also feed back to help interpret the structure of fossil communities, for example in marine infaunal invertebrate communities (Chapter 17).

In the preceding chapters we have seen variation in generation time from 20 minutes to hundreds of years, and variation in size from $0.5\,\mu m$ to $100\,m$, across a taxonomic range from bacteria to trees. Similarly variation in living area of single organisms can range from mm^2 for bacteria to km^2 for active mammals, birds and insects. Clearly species of different size and life history will respond in different ways to environmental variations over space and time.

TERMINOLOGY

Community ecology may be unique amongst the branches of science in lacking a consensus definition of the entity with which it is principally concerned. A random sample of definitions of a community would be likely to show an inverse relationship between specificity and popularity. Many authors have expressed concern that communities rarely exist as naturally definable units. In reality it is the study of the community level of organization that is important. Whether a natural unit or a level of organization, the definition of community suffers the problem of diversity of scales discussed in another context above. The community has been given such a variety of meanings and used to describe so many different levels of species associations, that it borders on being meaningless. At least some

TABLE 23.3. Range of community definitions used in the ecological literature, listed in order of least restrictive to most restrictive within each group

Definitions	Examples
Locational	
All living organisms in habitat or prescribed area	Pond, intertidal zone, oak wood
All organisms in stratum of habitat	Soil, forest canopy
Any assemblage of populations living in prescribed area	Woodland birds
Groups of species living closely enough for the potential of interaction	Rainpool insects
Trophic	
Species in all trophic levels	Plankton
All species in adjacent trophic levels	Plants + herbivores Herbivores + predators
All species on single trophic level	Herbivores
All species with similar resource requirements	Desert granivores Nectivores
Species using similar resources in similar ways	Granivorous rodents Hummingbirds
Taxonomic	
All species in prescribed area	
Large scale taxonomic assemblages	Fungi, fish, birds, insects
All species in more restricted taxon	Primates, *Drosophila*
Species of one taxon with similar resource requirements	Frugivorous primates
Life-form	
All species of particular life-form	Rainforest trees, seaweeds

of the differences of opinion in community ecology may be consequences of this failure to adopt consistent and unequivocal definitions to cover the community level of analysis. The failure is perhaps less apparent among plant ecologists than animal ecologists (to use a distinction which should rightly be regarded as outmoded), but the differences between the ways in which these two groups have used the same terms have also lead to a good deal of confusion in the past (Krebs 1978).

The differences in definitions relate to different emphases on organizational as against locational aspects, on spatial and taxonomic scale, and on the diversity of resource utilization (Table 23.3). Some ecologists believe that the community is, and should remain, a broad term, used to designate natural assemblages of different sizes (Odum 1971) that have intrinsic interest to community ecologists (Macarthur 1972). Strong *et al.* (1984) define ecological communities as groups of species living closely enough for the potential of interaction, without defining taxonomic boundaries or interaction type. Roughgarden & Diamond (1986) cast the net even further, describing the community as a set of species defined in various ways, whose relative values depend on the questions being asked. However, the practical function of a scientific term is to represent *unambiguously* an agreed set of properties (Macfadyen 1963). Given that the apparent properties of a community depend on scale, there is a need for clear distinctions between scales of analysis in the use of terms associated with communities.

In our view the basis of a set of terms is provided by Southwood's definition of the community level of organization (Chapter 1) and is validated independently by Fenchel (Chapter 13). In this definition the important properties are the wide taxonomic and functional affinities in the set of populations to be included, the spatial and temporal limits, and the combined presence of horizontal (i.e. competitive), vertical (predatory) and, possibly, diagonal (mutualistic) interactions amongst these species. *Assemblage* seems an appropriate term for a set of organisms whose pattern of organization is unknown.

To be sure, taxonomic groupings (insects, birds, herbaceous plants, etc.) will be evident within most communities as defined above. Nevertheless they should be seen as parts of the whole community (sometimes termed component or sub-communities) rather than as communities in their own right. Designations of taxonomically restricted communities (e.g. primate communities, herbaceous plant communities) are acceptable as indications of the level of analysis, but the entities referred to are portions of a whole community and show different properties, or perhaps subsets of the properties, of the community of which they are part. The taxonomic labelling of these sub-communities should be sufficient to indicate that this is so. Equally, a set of species that is functionally defined will have properties which differ from those of the community from which it is drawn.

Taxonomic and functional sub-communities themselves consist of groups of species that utilize the same types of resources in similar ways. These groups are usefully referred to as *guilds* (Root 1967), e.g. granivorous desert rodent guilds, herbivorous stem-boring insect guilds. Where study is restricted to a smaller taxonomic unit, the term *taxon guild* has been used (Schoener 1986). A heirarchy of useful terms in community ecology is given in Table 23.4. Schoener (1986) has also coined the term 'similia-community' for a grouping of co-occurring species that are similar in 'crucial organismic and environmental traits'. Although these traits have yet to be described, it appears that a similia-community represents a subset of a taxon guild. The usefulness of this awkward term has yet to be established.

The observable patterns of structure, composition and dynamics, and the dominant processes involved in their formation, will differ markedly between these levels of analysis. Failure to appreciate these differences will follow from the lack of a set of consistent definitions in community ecology and will undoubtedly continue to fuel contentious debates. An understanding of communities may develop by studying the patterns and processes among guilds and gradually building upwards toward a more holistic appreciation of communities (the bottom-up approach). Alternatively, ecologists can examine and compare patterns at the whole community level and work down through the various components to reach a deeper understanding of the underlying processes (the top-down route). However, little headway toward generalizations concerning community organization will be made by comparisons between different levels of the community heirarchy.

CONCLUSIONS

Equilibrium and scale emerge as recurring themes in the proceedings of this symposium. Natural communities differ in their closeness to a presumed equilibrium state, and in the degree to which biotic interactions can be regarded as responsible for their organization. Differences of scale in community investigations affect the information that can be gleaned and lead to different interpretations and conclusions for the same community (e.g. Wiens, Rotenberry & Van Horne 1987). As we have seen, one of the effects of scale in community ecology is to change the apparent equilibrial status of the system under investigation.

Two important points follow from the acceptance of the crucial influence of scale. First, the scale of investigation, relative to organism size, activity area and generation time, should be made abundantly clear in any community study. Second, no single scale can yield complete understanding of community organization. It is dangerous to generalize from small scale findings, and there are limitations to developing an understanding of proximate causes affecting relationships when viewed over large scales (Wiens 1986). Clearly we should explore each community over a wide spectrum of scales in order to obtain a satisfactory explanation of species composition, dynamics and diversity. It will often be impossible for any one ecologist to work at a wide variety of scales. Instead individuals will tend to

TABLE 23.4. Hierarchy of levels of organization in community ecology

Level	Patterns	Definition
Community	Trophic structure/food webs Competitive interactions Taxonomic diversity	Group of organisms generally of wide taxonomic affinities occurring together, many of which will interact within a framework of horizontal and vertical linkages
Component or sub-community	Habitat preference Resource partitioning Horizontal but few or no vertical interactions	Taxonomically or functionally restricted group of species occurring together, some of which will interact, mainly with horizontal linkages
Guild	Species size patterns Horizontal linkages Qualitative or quantitative resource preferences	Group of two or more co-occurring species populations utilizing the same type of resources in similar ways and interacting with horizontal linkages
Taxon guild	Species size patterns Horizontal linkages Qualitative or quantitative resource preferences	Species populations in one taxonomic subclass belonging to the same guild

work on small components of communities, or on small spatial and/or small temporal scales. The hope will be that these restricted investigations can be pieced together to reveal something about entire systems over long periods of time. As a small contribution to preventing the misunderstandings that are wont to arise when the same system is analysed at different scales, we have made a plea for consistent terminology. In doing so we echo the sentiments of May (1984) and Wiens (1984a).

To return briefly to conceptual models in community ecology, DeAngelis & Waterhouse (1987), in a review of mathematical models, arrive at conclusions similar to those discussed above. They describe a continuum between equilibrial and non-equilibrial systems, but one which differs in that the placement of a system at the non-equilibrial end can result from biotic instability as well as from stochastic effects. This biotic instability is the consequence of strong internal feedbacks and expresses itself in the form of limit cycles, chaos or competitive exclusion in mathematical models, and instability in simple laboratory ecosystems. Communities characterized by biotic instability or stochastic domination may yet persist in models through the influence of disturbance in the first case and though a variety of compensatory mechanisms in the second. But a state of equilibrium can be attained at the landscape scale even when biotic instability and stochastic domination are unrestrained at the level of the local patch. This presents a pleasing parallel with our empirically-based analysis of the effect of spatial scale on the perception of the equilibrium status of a community.

ACKNOWLEDGMENTS

We thank Mike Begon and Alan Hildrew for their perceptive comments on an earlier draft.

REFERENCES

Anderson, G., Ehrlich, A., Ehrlich, P., Roughgarden, J., Russell, B. & Talbot, F. (1981). The community structure of coral reef fishes. *American Naturalist* 117, 476–95.

Brown, J.H. (1988). Species diversity. *Analytical Biogeography* (Ed. by A.A. Myers & P.S. Giller). Chapman & Hall, London, in press.

Carpenter, F.L. (1979). Competition between hummingbirds and insects for nectar. *American Zoologist* 19, 1105–14.

Chesson, P.L. & Case, T.J. (1986). Overview: Nonequilibrium community theories: chance variability, history and coexistence. *Community Ecology* (Ed. by J.M. Diamond & T.J. Case), pp. 229–39. Harper & Row, New York.

Connell, J.H. (1980). Diversity and the coevolution of competitors, or the ghost of evolution past. *Oikos* 35, 131–8.

Dayton, P.i. & Tegner, M.J. (1984). The importance of scale in community ecology: a kelp forest example with terrestrial analogs. *Ecological Communities: Conceptual Issues and the Evidence* (Ed. by D.R. Strong, D. Simberloff, L.G Abele & A.B. Thistle), pp. 457–81. Princeton University Press, Princeton, New Jersey.

DeAngelis, D.L. & Waterhouse, J.C. (1987). Equilibrium and nonequilibrium concepts in ecological models. *Ecological Monographs* 57, 1–21.

Diamond, J.M. (1986). Evolution of ecological segregation in the New Guinea montane avifauna. *Community Ecology* (Ed. by J.M. Diamond & T.J. Case), pp. 98–125. Harper & Row, New York.

Diamond, J.M. & Case, T.J. (Eds) (1986). *Community Ecology.* Harper & Row, New York.

Giller, P.S. (1984). *Community Structure and the Niche.* Chapman & Hall, London.

Hanski, I. (1982). Dynamics of regional distribution: the core and satellite species hypothesis. *Oikos,* 38, 210–21.

Hanski, I. (1983). Coexistence of competitors in patchy environment. *Ecology* 64, 493–500.

Hubbell, S.P. & Foster, R.B. (1986).Biology, chance, and history and the structure of tropical rain forest tree communities. *Community Ecology* (Ed. by J.M. Diamond & T.J. Case), pp. 314–29. Harper & Row, New York.

Hurlbert, S.H., Loayza, W. & Moreno, T. (1986). Fish-flamingo-plankton interactions in the Peruvian Andes. *Limnology and Oceanography* 31, 457–68.

Huston, M. (1977). A general hypothesis of species diversity. *American Naturalist* 113, 81–101.

Kitching, R.L. (1987). Spatial and temporal variation in food webs in water-filled treeholes. *Oikos* 48, 280–8.

Knoll, A.H. (1986). Patterns of change in plant communities through geological times. *Community Ecology* (Ed. by J.M. Diamond & T.J. Case), pp. 126–41. Harper & Row, New York.

Krebs, C.J. (1978). *Ecology: the Experimental Analysis of Distribution and Abundance.* Harper & Row, New York.

Lawton, J.H. (1984). Non-competitive populations, non-convergent communities and vacant niches: the herbivores on bracken. *Ecological Communities: Conceptual Issues and the Evidence* (Ed. by D.R. Strong, D. Simberloff, L.G. Abele & A.B. Thistle), pp. 67–100. Princeton University Press, Princeton, New Jersey.

Levin, S.A. (1974). Dispersion and population interactions. *American Naturalist* **108**, 207–28.

Lewin, R. (1986). Supply side ecology. *Science* **234**, 25–7.

Macarthur, R.H. (1972). *Geographical Ecology.* Harper & Row, New York.

Macfadyen, A. (1963). *Animal Ecology, Aims and Methods.* Pitman, London.

May, R.M. (1984). An overview: real and apparent patterns in community structure. *Ecological Communities: Conceptual Issues and the Evidence* (Ed. by D.R. Strong, D. Simberloff, L.G. Abele & A.B. Thistle), pp. 3–16. Princton University Press, Princeton, New Jersey.

Maynard Smith, J. (1974). *Models in Ecology.* Cambridge University Press.

Moulton, M.P. & Pimm, S.L. (1986). The extent of competition in shaping an introduced avifauna. *Community Ecology* (Ed. by J.M. Diamond & T.J. Case), pp. 80–97. Harper & Row, New York.

Odum, E.P. (1971). *Fundamentals of Ecology.* W.B. Saunders, Philadelphia.

Price, P.W. (1984). Alternative paradigms in community ecology. *A New Ecology: Novel Approaches to Interactive Systems* (Ed. by P.W. Price, C.N. Slobodchikoff & W.S. Gaud), pp. 353–83. Wiley, New York.

Price, P.W., Slobodchikoff, C.N. & Gaud, W.S. (Eds) (1984). *A New Ecology: Novel Approaches to Interactive Systems.* Wiley, New York.

Root, R.B. (1967). The niche exploitation pattern of the blue-gray gnatcatcher. *Ecological Monographs* **7**, 317–50.

Rosenzweig, M.L. (1979). Optimal habitat selection in two-species competitive systems. *Fortschritte der Zoologie* **25**, 283–93.

Roughgarden, J. & Diamond, J.M. (1986). Overview: the role of species interactions in community ecology. *Community Ecology* (Ed. by J.M. Diamond & T.J. Case), pp. 333–43. Harper & Row, New York.

Sale, P.F. & Dybdahl, R. (1975). Determinants of community structure for coral reef fishes in an experimental habitat. *Ecology* **56**, 1343–55.

Schluter, D. (1986). Character displacement between distantly related taxa? Finches and bees in the Galapagos. *American Naturalist* **127**, 95–102.

Schoener, T.W. (1982). The controversy over inter-specific competition. *American Scientist* **70**, 586–95.

Schoener, T.W. (1986). Overview: kinds of ecological communities—ecology becomes pluralistic. *Community Ecology* (Ed. by J.M. Diamond & T.J. Case), pp. 467–79. Harper & Row, New York.

Slatkin, M. (1974). Competition and regional coexistence. *Ecology* **55**, 128–34.

Southwood, T.R.E. (1977). Habitat, the templet for ecological strategies? *Journal of Animal Ecology,* **46**, 337–65.

Strong, D.R., Simberloff, D., Abele, L.G. & Thistle, A.B. (Eds) (1984). *Ecological Communities: Conceptual Issues and the Evidence.* Princeton University Press, Princeton, New Jersey.

Sutherland, J.P. (1981). The fouling community at Beaufort, North Carolina: a study in stability. *The American Naturalist,* **118**, 499–519.

Tilman, D. (1982). *Resource Competition and Community Structure.* Princeton University Press, Princeton, New Jersey.

Wiens, J.A. (1984a). On understanding a non-equilibrium world: myth and reality in community patterns and processes. *Ecological Communities: Conceptual Issues and the Evidence* (Ed. by D.R. Strong, D. Simberloff, L.G. Abele & A.B. Thistle), pp. 439–57. Princeton University Press, Princeton, New Jersey.

Wiens, J.A. (1984b). Resource systems, populations and communities. *A New Ecology: Novel Approaches to Interactive Systems* (Ed. by P.W. Price, C.N. Slobodchikoff & W.S. Gaud), pp. 397–436. Wiley, New York.

Wiens, J.A. (1986). Spatial and temporal variation in studies of shrubsteppe birds. *Community Ecology* (Ed. by J.M. Diamond & T.J. Case), pp. 154–72. Harper & Row, New York.

Wiens, J.A., Addicott, J.F., Case T.J. & Diamond J.M. (1986). Overview: the importance of spatial scale in ecological investigations. *Community Ecology*. (Ed. By J.M. Diamond & T.J. Case), pp. 145–53.

Wiens, J.A., Rotenberry, J.T. & Van Horne, B. (1987). Habitat occupancy patterns of North American shrubsteppe birds: the effects of spatial scale. *Oikos* **48**, 132–47.

AUTHOR INDEX

Page numbers shown in *italics* refer to the list of references

SCOTTISH AGRICULTURAL COLLEGE

AUCHINCRUIVE

LIBRARY

SUBJECT INDEX

Abra sp., food and location, 385
Acer spp., 105
  speciation, 111
Aconitum columbianum, nectar source, 478
Adès distribution, 39
Aepycamelus, 459
African guenons, *see Cercopithecus* spp.
African leaf monkeys, *see Colobus* spp.
African talapoin, *see Micropithecus talapoin*
Aggregation,
  and community, 529
  of populations, 350
Agropyron, root penetration, and shoot
    elongation, 125
Agrostis capillaris
  interaction with *Sanguisorba*, 132, 134
  productivity, 132
  root competition, 125
Alamosaurus community, 445
Albertosaurus, 445
Aleurotrachelus sp., 5–6
Algae, effect of herbivores, 353–4
  see also Plankton, bacteria; Plankton,
    freshwater; Plankton, marine
'Allocation rules', *see* Community
Allochthonous
  organic matter
    freshwater, 357
    marine, 375
  taxa, 450
Allotheria, 446
Alouatta spp., 212
Ammonium, diel changes, 333–4
Amniota, 443
Ampharete spp., food and location,
    385
Amphibia, fossil families, 402, 440–3
Amphibian dynasty, 442–3
Amphiura sp., food and location, 385
Anabaena sp., properties, 300
Angiosperms
  fossil history, 400, 406–11
  pollen records, early, 407
  see habit, 405
Anolis spp.
  community organization
    large scale, 503–13
    medium scale, 501–3
    small scale, 499–501
  competition, 500–1

guild composition, 513
role of physical transport processes,
    499–513
systematics, 507
Anoxic environment, microbial
    communities, 285–8
Antarctic ocean, productivity, 378–9
Anthracosaurs, 442
Anthracotheriidae, 449
Antilles, Lesser, NLA platform, 506
Antilocapridae, 449
Ants, resource partitioning, 198
Aonidiella spp., 152
Aphanocapsa spp., properties, 310
Aphanothece spp., properties, 310
Aphinizomenon spp., properties, 310
Aphodius holdereri, 430
Aphytis spp., 152–3
Aquatic assemblages
  freshwater benthic, 347–71
  freshwater plankton, 297–318
  marine benthic, 373–97
  marine plankton, 327–45
  microbial communities, 281–93
  vascular plants, 358–9
Arachnothera sp., 74
Arborescence, evolution of, 402, 405
Archidochus alexandri, habitat selection,
    474–8
Archisaurs, 445, 455
Arctic
  and Alpine species, 104
  coniferous forests, 409
  Eocene mammals, 448
  ocean productivity, 378
Arrhenatherum elatius
  phosphorus uptake, 133
  productivity, 132
Arthropods, ancient, 404
  spore dispersal, 404
Ascomorpha spp., generation time,
    315
Asian leaf monkeys, *see Presbytis* spp.
Assemblage, definitions, 4, 9, 163, 537
Asterionella spp.
  competitive ability, 302–3
  light absorption, 299
  properties, 310
Autochthonous
  production, marine, 375